HEAT AND MASS TRANSFER IN ROTATING MACHINERY

PROCEEDINGS OF THE INTERNATIONAL CENTRE FOR HEAT AND MASS TRANSFER

ZORAN P. ZARIĆ, EDITOR
C. L. TIEN, SENIOR CONSULTING EDITOR

HEAT AND MASS TRANSFER IN ROTATING MACHINERY

Edited by

Darryl E. Metzger
Department of Mechanical and Aerospace Engineering
Arizona State University, Tempe

and

Naim H. Afgan
International Centre for Heat and Mass Transfer
Belgrade, Yugoslavia

● **HEMISPHERE PUBLISHING CORPORATION**

Washington   New York   London

DISTRIBUTION OUTSIDE NORTH AMERICA

SPRINGER-VERLAG

Berlin   Heidelberg   New York   Tokyo

HEAT AND MASS TRANSFER IN ROTATING MACHINERY
Copyright © 1984 by Hemisphere Publishing Corporation. All rights reserved.
Printed in the United States of America. Except as permitted under the
United States Copyright Act of 1976, no part of this publication may be
reproduced or distributed in any form or by any means, or stored in a data
base or retrieval system, without the prior written permission of the publisher.

1 2 3 4 5 6 7 8 9   B C B C   8 9 8 7 6 5 4 3

Library of Congress Cataloging in Publication Data

Main entry under title:

Heat and mass transfer in rotating machinery.

(Proceedings of the International Centre for Heat
and Mass Transfer; 16)
"Papers from the XIV symposium of the International
Centre for Heat and Mass Transfer, held September 1982 in
Dubrovnik, Yugoslavia."—Pref.
Bibliography: p.
Includes index.
1. Heat—Transmission—Congresses. 2. Mass transfer—
Congresses. 3. Turbomachines—Congresses. I. Metzger,
Darryl E., date. III. Afgan, Naim. III. Inter-
national Center for Heat and Mass transfer.
TJ260.H385 1984   621.402'2   83-12595
ISBN 0-89116-294-1 Hemisphere Publishing Corporation
ISSN 0272-880X

DISTRIBUTION OUTSIDE NORTH AMERICA:
ISBN 3-540-12976-6 Springer-Verlag Berlin

Contents

EXPERIMENTAL TECHNIQUES

GAS TURBINES

STEAM TURBINES

Preface

This volume contains papers from the XIV Symposium of the International Centre for Heat and Mass Transfer, held September 1982 in Dubrovnik, Yugoslavia.

The symposium was organized to focus attention on heat and mass transfer processes especially associated with rotating machinery components. The understanding of such processes plays an increasingly significant role in the continued development of many different types of rotating machinery.

The objective of the Dubrovnik meeting was to bring together researchers and practitioners in a forum for exchange of information both on research topics and on design problems and strategies. Contributed papers were organized into sessions on generic research areas and on specific types of machines. The same general format has been followed in arranging this volume, although in some cases papers span more than a single category and placement is therefore somewhat arbitrary.

The editors would like to acknowledge the contribution of the following organizing committee members and session chairmen: G. Bois, Ecole Centrale de Lyon, France: M. E. Elovic, General Electric Company, USA; B. Gal'Or, The Technion, Israel; M. Hirata, University of Tokyo, Japan; M. Majcen, University of Zagreb, Yugoslavia; P. J. Marto, U.S. Naval Postgraduate School, USA; R. E. Mayle, Rensselaer Polytechnic Institute, USA; W. D. Morris, University of Hull, UK; J. M. Owen, University of Sussex, UK; D. B. Spalding, Imperial College of Science and Technology, UK; S. L. K. Wittig, University of Karlsruhe, FRG; C. H. Wu, Engineering Thermophysics Research Institute, People's Republic of China.

Darryl E. Metzger
Chairman, Symposium Committee

Naim H. Afgan
Scientific Secretary, ICHMT

ROTATING TUBES
AND CHANNELS

Secondary Flows and Enhanced Heat Transfer in Rotating Pipes and Ducts

Y. MORI
Department of Physical Engineering
Tokyo Institute of Technology
Meguro-ku, O-okayama, Tokyo, Japan

W. NAKAYAMA
Mechanical Engineering Research Laboratory
Hitachi, Ltd.
Kandatsu, Tsuchiura, Ibaraki, Japan

ABSTRACT

The objective of this paper is to give a general review of secondary flows and enhanced heat transfer in rotating pipes and ducts. The secondary flows are caused by body forces such as Coriolis force, that are caused by density variation in a centrifugal field or the resultant of that by density in a centrifugal field and centrifugal force due to curvature of a duct. Heat transfer in rotating ducts is enhanced by the secondary flow and its performance varies with the shape of the duct cross-section and the intensity and orientation of the body force.

Based on these understandings, theoretical and experimental works on featuring secondary flows and heat transfer performances in rotating and revolving pipes and ducts are summarized. As theoretical ones, the two analytical methods and numerical works are explained and their results are compared with experiments. Data for helium and two phase flows are shown as getting attractive, but more work in these fields is required in the future.

NOMENCLATURE

a	:	radius of pipe or characteristic length of pipe (m)
C_p	:	isobaric specific heat (J/kg K)
$\vec{F}$	:	body force (Pa/m)
f	:	nondimensional body force, or friction factor
G	:	temperature distribution in the cross-section
g	:	nondimensional temperature $= G/\tau a$
K	:	dynamical parameter
k	:	thermal conductivity (W/(m·K))
Nu	:	Nusselt number
P	:	pressure (Pa)
p	:	nondimensional pressure $= (a^2/\nu^2)(P/\rho)$
Pr	:	Prandtl number
R	:	radius of curvature (m)
Ra	:	Rayleigh number
Re	:	Reynolds number
r	:	radial coordinate (m)
T	:	temperature (K)
Tw	:	wall temperature (K)
$\vec{V}$	:	velocity vector (m/s)
U,V,W	:	components of $\vec{V}$ in X, Y and Z directions (m/s)

```
u,v,w : nondimensional velocity  = (U,V,W)×(a/ν)
X     : axial coordinate (m)
Y     : coordinate in the body force direction (m)
Z     : coordinate in the cross-section (m)
x,y,z : nondimensional coordinate  = (X,Y,Z)×(1/a)
β     : expansion coefficient (1/K)
δ     : boundary layer thickness (m)
η     : nondimensional radial coordinate  = r/a
μ     : viscosity (Pa s)
ν     : kinematic viscosity  = μ/ρ (m²/s)
ρ     : density (kg/m³)
τ     : temperature gradient in the axial direction (K/m)
ψ     : angular coordinate in the cross-section (rad.)
Ω     : angular velocity (1/s)
ω     : nondimensional angular velocity  = 2a²Ω/ν
κ     : thermal diffusivity (m²/s)
```

Superscript

```
—
      : mean value in Z direction
'
      : perturbed component by secondary flow
```

Suffix

```
δ     : outer edge of boundary layer
1     : main flow
```

Other Symbols

```
< >   : mean value in Y·Z section
```

1. INTRODUCTION

   Heat transfer in rotating ducts and pipes has become a subject of great importance for engineers in various industries. Research works on coolant flow and heat transfers in turbine blades have been intensified with the aim of raising the operating temperature of gas turbines. As the electric generator in power stations has increased in capacity, cooling of the rotating field windings has called for important design consideration. Recently, the research and development of a superconducting generator is being carried out in several countries, for which the flow of liquid helium in the cryogenic rotor is an important research subject. Besides those advanced technologies, conventional rotating machines also pose cooling problems as they are housed in noiseproof or dustproof containers.

   Those diverse industrial needs have motivated the present authors to conduct theoretical and experimental studies in the past years. This paper presents a summary of our works, together with an updated review of the works made by other investigators.

   In the coolant passages of rotating systems, the fluid is subject to a centrifugal force or a Coriolis force. Those forces, where they act in the direction of fluid passage, either accelerate or retard the flow. Where the forces act in the transverse direction to the flow, they cause secondary flows, which increase the heat transfer coefficient on the passage wall and the resistance to a coolant flow. The secomdary flows are the most distinguishing phenomena in heat transfers in rotating pipes and ducts, and are our primary concern

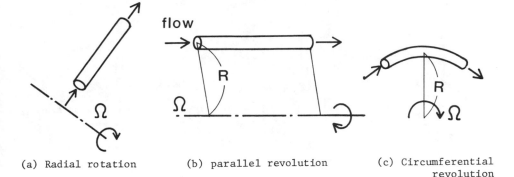

(a) Radial rotation   (b) parallel revolution   (c) Circumferential
                                                                  revolution

Fig. 1   Principal orientation of rotating and revolving duct flows

	radial rotating duct	parallel revolving duct	circumferential revolving duct
source of body force	Coriolis force acting on a primary velocity	density variation in a centrifugal field	density variation in a centrifugal field + centrifugal force due to curvature of duct
dynamical parameter	$Re \cdot \omega = (W_m d/\nu) \cdot (2a^2\Omega/\nu)$	$Ra = R\Omega^2 \beta \Delta Ta^3/\kappa\nu$	$Ra = R\Omega^2\beta\Delta Ta^3/\kappa\nu$ $Re\sqrt{a/R} = (\frac{W_m d}{\nu})\sqrt{\frac{a}{R}}$

Table 1   Body forces and dynamical parameters for single-phase flow

in the present paper, because they pose challenging problems to researchers and the accurate estimate of their effects on heat transfer performances is important for cooling system designs.

Fig. 1 shows three principal orientations of flow passage against the axis of rotation taking a circular pipe as a representative passage. We call those passages (a) a radial rotating duct, (b) a parallel revolving duct, and (c) a circumferential revolving duct. A coolant passage in real rotating machines is equivalent to any of those ducts, or it is formed by joining them in a series of bends. The body forces of primary importance which cause secondary flows are listed in Table 1. For single phase flows, the dynamical parameters listed in Table 1 signify the intensity of secondary flows. They and conventional parameters of Reynolds number, Prandtl number and Graetz number make up a set of parameters needed to define flow and temperature fields.

As for the cross-sectional geometry of the coolant passage, the configurations of practical importance are those illustrated in Fig. 2, (a) circular, (b) square, and (c) narrow rectangular. In general, the body force F acts in any direction depending on the posture of a cross-section with respect to the axis of rotation. However, the practically important directions of forces are those of Fig. 2.

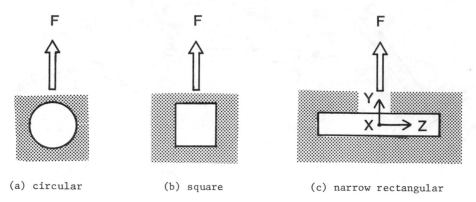

(a) circular   (b) square   (c) narrow rectangular

Fig. 2   Cross-section of flow passage and action of body force

In the following section 2, the fundamental cases of fully developed flows will be discussed. That is followed by a summary of recent numerical studies. In the section 3, the experimental works will be reviewed. The section 4 is devoted to a review of recent works on two-phase cooling schemes.

2. SINGLE-PHASE FLOW AND HEAT TRANSFER — THEORY

Equation of momentum and energy for the flow under the influence of a body force $\vec{F}$ are shown by equations (1) and (2);

$$\rho(\vec{V} \cdot \nabla)\vec{V} \;=\; -\nabla P + \mu\nabla^2\vec{V} + \vec{F} \tag{1}$$

$$\rho\, c_p\, (\vec{V} \cdot \nabla)T \;=\; k\nabla^2 T \tag{2}$$

respectively, where $\vec{V}$ is the velocity vector, T the temperature, P the pressure, and other notations are listed in NOMENCLATURE. The assumption of constant physical properties is adopted for the sake of simplicity to explain the features of secondary flow in rotating ducts.

Equation (1) is resolved into the equation for a primary flow (velocity component u) and those for secondary flow (v, w). In the cases of a radial rotating duct and a circumferential revolving duct, the momentum equation is solved taking the body force into account except the effects of physical proper-ty variation due to temperature distributions. On the other hand for a parallel revolving duct, the energy and momentum equations are coupled through the body force term. In order to solve a set of those coupled equations, one must resort to either one of the following analytical methods.
(1)   Perturbation method.
(2)   Boundary layer method.
(3)   Energy balance and Entropy balance methods.
(4)   Direct numerical integration by either finite difference methods or finite element methods.

The perturbation method suits the analysis of flows in circular pipes ac-companied by weak secondary flows. Fig. 3(a) shows the secondary flow pattern, the axial velocity and temperature distributions displaced only slightly from the axially symmetric distributions. When the secondary flow is caused with a considerable intensity, the velocity and temperature distributions are distorted remarkably from symmetric ones, as shown in Fig. 3(b). In such cases, the boundary layer modeling proves to be a powerful analytical tool, and this has been used extensively by the present authors for laminar as well as turbulent flows [1, 2, 3, 4]. For the cross-sectional geometries of square and rectangle, the perturbation method is difficult to apply because the expansion in terms of the perturbation parameters becomes complex. The boundary layer method requires a rather complex modeling. The method of more integral nature based on the balance of energy and entropy was proposed to handle the problems of square and rectangular ducts [5]. The above analytical methods have merits in yielding concise and general correlations among friction factors, heat transfer coefficients and the dynamical parameters. The recent advances in numerical analysis described later on have given a well founded proof to the boundary layer modeling, and also provide one with detailed features of flow and temperature fields. Numerical analyses have successfully correlated the results obtained by perturbation methods for small dynamical parameter and boundary layer method for large parameter.

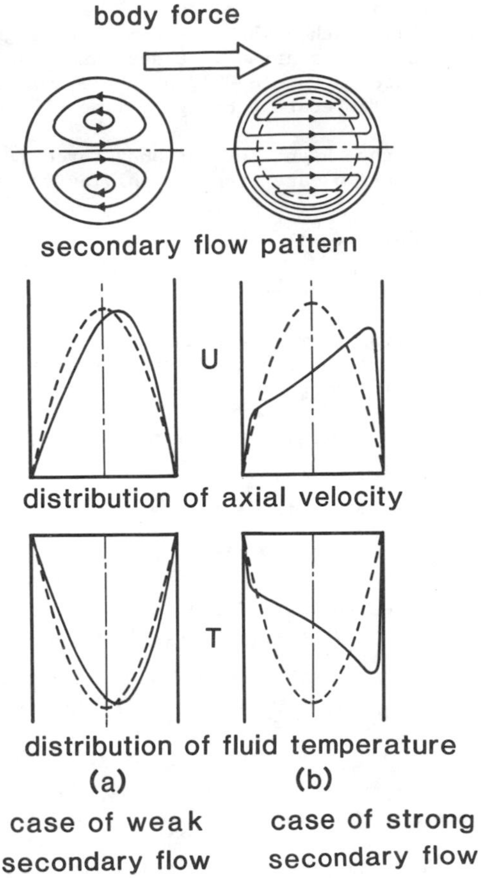

In what follows, the dynamical parameter will be denoted by ε or K for the sake of conciseness. It should be noted, however, although equations can be formulated in such general terms, quantative results differ for different cases. This is also pointed out experimentally by Trefethen, as quoted in [6]. Recently, progresses for numerical analysis have been extending the applicable region of numerical results to large dynamical parameter for which so far the boundary layer method has been only the way of analysis of secondary flows. However, it should be noted that the boundary layer method can produce general relations for secondary flows when numerical analysis is not easily applicable.

body force

secondary flow pattern

distribution of axial velocity

distribution of fluid temperature

(a)   (b)

case of weak   case of strong
secondary flow   secondary flow

Fig. 3   Secondary flow pattern and
distributions of axial
velocity and temperature

2.1. Fully-Developed Flows with Weak Secondary Flows

(1) Perturbation method. For laminar and fully developed flows, the non-dimensionalization of equations (1) and (2) leads to expansion of the velocity components and the temperature in terms of a dynamical parameter listed in Table 1, denoted here by ε to signify the smallness of the parameter. In terms of the characteristic length a, the kinematic viscosity ν, the non-dimensional

velocity is defined as $\vec{v}(u,v,w) \equiv a\vec{V}/\nu$. The fully developed condition of temperature is reached when the temperature gradient in the axial direction (X-direction) becomes constant. Under this condition the non-dimensional temperature θ
is defined as $\theta \equiv (T_W - T)/\tau a$, where τ is the axial temperature gradient and T_W
the wall temperature changing linearly in the axial direction.

In the absence of the body force ($\vec{F} = 0$), the axial velocity u_0 and the
temperature θ_0 are found by solving the following equations:

$$\nabla^2 u_0 = \left(\frac{\partial P_0}{\partial x}\right) = \text{constant} \tag{3}$$

$$\nabla^2 \theta_0 = \Pr u_0 \frac{\partial \theta_0}{\partial X} \tag{4}$$

Having known the solutions u_0 and θ_0, one finds perturbations from the
basic solutions by substituting the following expansion series into equations
(1) and (2).

$$\begin{aligned} u &= u_0 + \varepsilon u_1 + \varepsilon^2 u_2 + \cdots \\ v &= \varepsilon v_1 + \varepsilon^2 v_2 + \cdots \\ w &= \varepsilon w_1 + \varepsilon^2 w_2 + \cdots \end{aligned} \tag{5}$$

$$\theta = \theta_0 + \varepsilon \theta_1 + \varepsilon^2 \theta_2 \cdots \tag{6}$$

The succesive approximate equations are obtained by equating terms having the
equal power of ε. They are linear and solvable with the boundary condition on
the pipe wall:

$$\begin{aligned} u_{n-1} &= v_n = w_n = 0 \quad (n = 1, 2, \cdots) \\ \theta_{n-1} &= 0 \end{aligned} \tag{7}$$

As one proceeds to higher approximations, the computation becomes increasingly
tedious. The practical limit is up to about fourth order approximation.

The friction coefficinet f and the Nusselt number Nu, when expressed in the
ratio to those for a stationary pipe (f_0, Nu_0) are given as :

$$\frac{f}{f_0} = 1 + C_2 \varepsilon^2 + C_4 \varepsilon^4 + \cdots \tag{8}$$

$$\frac{Nu}{Nu_0} = 1 + C_2' \varepsilon^2 + C_4' \varepsilon^4 + \cdots \tag{9}$$

In equations (8) and (9), C_2, C_4, C_2', C_4' are constants determined from solutions for u_2, v_2, w_2, θ_2, $\cdots$. The solutions for the equations of odd power
give no contributions to f and Nu as their effects vanish when taken mean values
over the whole cross-section.

The perturbation method was used by Barua [7] and Morris [8] for the radial
rotating pipe and the parallel revolving pipe, respectively. The Morris's
result (for $\Pr = 0.7$ and negligible effect of gravity) is shown below as an
illustration.

$$\frac{Nu}{Nu_0} = 1 + 1.76 \times 10^{-8} (RaRe)^2 - 1.16 \times 10^{-16} (RaRe)^4 + \cdots \tag{10}$$

(2) <u>Energy balance and entropy balance method</u>. In a narrow rectangular
duct (Fig. 2(c)), the onset of secondary flows is prompted by the growth of

naturally present disturbance. The stability problem analogous to that in a gravity field [9] needs to be solved prior to the analysis of velocity and temperature fields. The linear stability analysis for the case of a parallel revolving duct yields the critical rotational Rayleigh number of 1708 [10]. The integral balance method described in this section is useful to this kind of analysis of the case when the dynamical parameter exceeds the critical value by a small magnitude marking the onset of secondary flow.

We begin with general formulation of equations taking the coordinate system shown in Fig. 2(c); the critical value for the circular pipe (a) is zero and that for the square duct (b) is taken up by the similar way with the rectangular case. The X-axis is taken in line with the duct axis, the Y-axis in the direction of the body force, and the Z-axis being perpendicular to the body force. The first step is the decomposition of the velocity, pressure, temperature and body force into the components averaged over Z and the perturbation components as shown below:

$$\vec{V} = \overline{\vec{V}}(Y) + \vec{V}'(Y,Z) \qquad\qquad \vec{F} = \overline{\vec{F}}(Y) + \vec{F}'(Y,Z)$$

$$P = \overline{P}(X,Y) + P'(Y,Z) \qquad\qquad T = \overline{T}(X,Y) + T'(Y,Z) \tag{11}$$

The averaged component marked with an overbar will hereafter be called 'the primary component'. For fully-developed flows, the primary velocity component $\overline{\vec{V}}$ and the body force component $\overline{\vec{F}}$ are functions of Y alone, whereas $\overline{P}$ and $\overline{T}$ are linear functions of X.

Substituting equations (11) into equations (1) and (2), we obtain for perturbation components;

$$\rho_0\{(\vec{V}'\cdot\nabla)(\overline{\vec{V}}+\vec{V}') - \frac{\partial}{\partial Y}(\overline{U'V'},\overline{V'^2},\ \overline{V'W'})\} = -\nabla P' + (\vec{F}-\overline{\vec{F}}) + \mu\nabla^2\vec{V}' \tag{12}$$

$$\rho_0 c_p\{(\vec{V}'\cdot\nabla)(\overline{T}+T') - \frac{\partial}{\partial Y}(\overline{V'T'})\} = k\nabla^2 T' \tag{13}$$

for primary components:

$$\rho_0 \frac{\partial}{\partial Y}(\overline{U'V'},\ \overline{V'^2},\ \overline{V'W'}) = -\nabla\overline{P} + \overline{\vec{F}} + \mu\nabla^2\overline{\vec{V}} \tag{14}$$

$$\rho_0 c_p\{(\overline{\vec{V}}\cdot\nabla)\overline{T} + \frac{\partial}{\partial Y}(\overline{V'T'})\} = k\nabla^2\overline{T} \tag{15}$$

Equations (12) and (13) are further reduced to the space-averaged equations to calculate the amplitude of the secondary flow by the following process. The X-component of equation (12) is multiplied by U' (the perturbation component of the velocity in the axial direction) and integrated over the Y-Z domain. After eliminating the terms less than triple products of perturbation component, we obtain;

$$-\rho_0 < U'V' \frac{\partial\overline{U}}{\partial Y} > + \mu < U'\vec{i}(\nabla^2\vec{V}') > = 0 \tag{16}$$

where V' is the perturbation component of the velocity in the Y-direction, $\vec{i}$ the unit vector in the X-direction. The symbol < > means the average value in the Y-Z domain. The terms in equation (16) have explicit physical meanings, the first term being the kinetic energy transported from the primary flow $\overline{U}$ through the Reynolds stress U'V', and the second term representing the viscous dissipation.

In a similar manner, the kinetic energy balance is derived from the Y and Z-components of equation (12);

$$<(V'\vec{j} + W'\vec{k})(\vec{F} - \overline{\vec{F}})> + \mu<(V'\vec{j} + W'\vec{k})(\nabla^2\vec{V}')> = 0 \qquad (17)$$

where $\vec{j}$ and $\vec{k}$ are the unit vector in the Y and Z-directions, respectively. The terms in the first bracket of equation (17) represent the kinetic energy introduced to the perturbation components by the body force, and the second bracket expresses the viscous dissipation.
The energy equation (13) is multiplied by T' and averaged over the Y-Z domain and we have:

$$-\rho_0 c_p < U'T'\frac{\partial \overline{T}}{\partial X} + T'\frac{\partial \overline{V'T'}}{\partial Y} > + k < T'\cdot\nabla^2 T' > = 0 \qquad (18)$$

The terms in the first bracket in equation (18) represent the entropy production by the interaction of the perturbation components and the primary temperature distribution. The second bracket is the entropy production by heat conduction.

The perturbation components in equations (16), (17) and (18) are approximated by the following expansion series of equation (19).

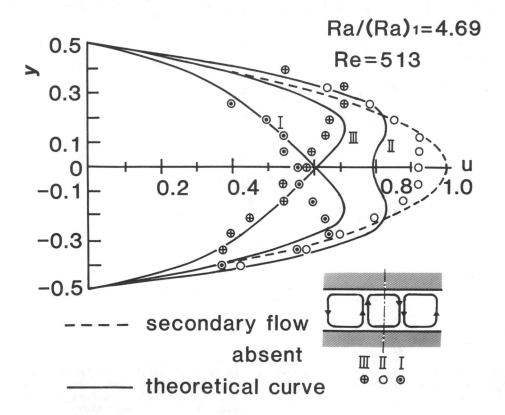

Fig. 4   Velocity distributions of a flow in a parallel channel heated
         from below

$$U' = a_1 U_1'(Y) \cos \alpha_1 Z + a_2 U_2'(Y) \cos \alpha_2 Z + \cdots$$
$$V' = b_1 V_1'(Y) \cos \alpha_1 Z + b_2 V_2'(Y) \cos \alpha_2 Z + \cdots$$
$$W' = b_1 W_1'(Y) \sin \alpha_1 Z + b_2 W_2'(Y) \sin \alpha_2 Z + \cdots \qquad (19)$$
$$T' = c_1 T_1'(Y) \cos \alpha_1 Z + c_2 T_2'(Y) \cos \alpha_2 Z + \cdots$$

where U_n', V_n', W_n' and T_n', $(n=1, 2, \cdots)$ are profile functions to be assumed in order to perform the integrations in the Y-Z domain. A good approximation may be made by the solutions of the linear stability analysis. When equations (19) are substituted in equations (14) and (15), the primary components can be found by solving those equations with unknown constants a_n, b_n, c_n being left for further determination. Those formal solutions of primary components and the expansion series (19) are substituted in equations (16), (17) and (18). After performing integration, we obtain a set of algebraic equations relating the amplitude constants a_n, b_n, c_n to the wave numbers $\alpha_n(L/2\pi)$, $(n=1, 2, \cdots, L=\text{the}$ width of the duct) and the dynamical parameter. The number of simultaneous equations is 3n if the perturbation components are approximated by up to the n-th order harmonics.

The integral balance method applied to the flow in a narrow rectangular duct yields an analytical solution which can be compared with experimental data. Data on rotating or revolving duct flows data have been rarely reported to demonstrate effects of secondary flow by measuring velocity profiles. Therefore, an analogy to the parallel revolving duct is to be recognized by substituting the gravity acceleration to the centrifugal acceleration. The flow in a horizontal flat channel heated from below is taken up. Fig. 4 shows a comparison between the analytical results explained in this section and the experimental data for the flat channels under gravity where Ra is the dynamical parameter.

The distributions of axial velocity component were measured at three locations, a center of the secondary flow cell and the boundaries of the cell. The broken line shows the distribution in the absence of the body force. The computed curves were obtained by taking into account the first components of a secondary flow. A good agreement between theory and experiment can be seen.

2.2. Fully-Developed Flows with Strong Secondary Flows

(1) Boundary layer method. Boundary layer modeling is based on the concept that, when the dynamical parameter is far greater than the critical value, interaction between the secondary and axial flows plays a dominant role in the balance of momentum in most of the flow region. Fig. 3(b) shows an axial velocity distribution profile which is markedly distorted from the symmetrical distribution as a result of strong secondary flow effects. The model assumes a boundary layer on the circumference of the pipe wall, only where molecular or turbulent diffusion is important. The secondary flow is approximated by a uniform flow in the inertia-dominated core region and return currents in the boundary layer. The flow in the core is in the direction of the body force. The temperature distribution also bears a notable feature produced by a strong secondary flow.

Taking the coordinate system shown in Fig. 5, and non-dimensionalizing the velocity, the space coordinates and the pressure by ν/a, a, and $(\nu^2/a^2)\rho$, respectively, the non-dimensional quantities are introduced. The momentum equation for the core region is written as follows when the values there are indicated by the subscript $_1$;

$$\frac{\partial}{\eta \partial \eta}(\eta u_1 v_1) + \frac{\partial}{\eta \partial \psi}(u_1 w_1) = C \qquad (20)$$

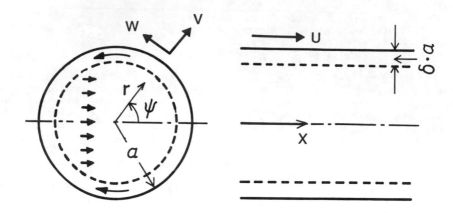

Fig. 5   System of Coordinates

where C is a constant and equal to $-\partial p/\partial x$.

The energy equation is written in terms of the temperature g_1, non-dimensionalized by the constant temperature gradient in the axial direction τ, and the radius of a pipe a as:

$$\frac{\partial}{\eta \partial \eta} (\eta v_1 g_1) + \frac{\partial}{\eta \partial \psi} (w_1 g_1) = u_1 \tag{21}$$

Equations (20) and (21) are satisfied by:

$$u_1 = A + \frac{C}{D} \eta \cos\psi$$

$$v_1 = D \cos\psi \tag{22}$$

$$w_1 = - D \sin\psi$$

$$g_1 = A' + \frac{C}{2D^2} \eta^2 + \frac{A}{D} \eta \cos\psi \tag{23}$$

The six unknown quantities are A, C, D, A', the velocity boundary layer thickness δ and the ratio of the velocity and temperature boundary layer thickness ζ. They are determined from the following relations:
(i)   Overall mass balance relating A, D, δ to Re (Re $\equiv 2aUm/\nu$)
(ii)   Overall force balance relating A, δ to C
(iii)   Overall energy balance relating A', δ, ζ to Re and Pr
(iv)   Continuity condition of secondary flow relating D to δ
(v)   Integral balance of momentum for the circumferential flow in the boundary layer
Variation of δ in the circumferential direction is small enough and from the last relation cited above, the following relation is reduced by using the relation in the core region that the pressure gradient in the circumferential direction is balanced with the body force and by putting $\xi = 1 - \eta$;

$$\tau_{\psi m} = [\int_0^\delta (f_{1\delta} - f) \sin \psi d\xi]_m \tag{24}$$

where f is the nondimensional body force, $\tau_{\psi m}$ is the circumferential shear stress averaged over $\psi = 0 \backsim \pi$ being determined by the circumferential velocity

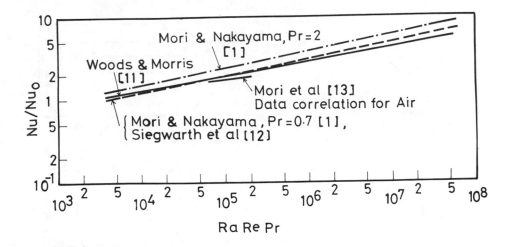

Fig. 6   Comparison of the numerical solution with the analytical solutions

in the boundary layer. The first term in the integral of equation (24) signifies the body force f_1, and the second term f represents the body force in the
boundary layer, which is less than that in the core region due to large variations of velocity and temperature. For the cases shown in Table 1, $f_{1\delta} - f$ is
proportional to $(2a^2\Omega/\nu)$ for Coriolis force and $(R\Omega^2\beta\tau a^4/\nu^2)$ for density variation in a centrifugal force where τ is the temperature gradient
in the axial direction.
        (vi)   Integral balance of energy in the boundary layer

    The boundary layer method allows the expansion of velocity and temperature
in terms of the dynamical parameter, here denoted by K. For example, the intensity of secondary flow D and the boundary layer thickness δ are expanded as:

$$D = D_1K^{1/2} + D_2 + D_3K^{-1/2} + \cdots$$
$$\delta = \delta_1K^{-1/2} + \delta_2K^{-1} + \cdots \qquad\qquad (25)$$

    In case of a parallel revolving duct, $K = (RaRe)^{2/5}$ [1]. The solutions give
the Nusselt number in the retio to Nu_0 of a stationary laminar pipe flow. For
a Prandtl number of 0.7:

$$\frac{Nu}{Nu_0} = 0.182(Ra\ Re\ Pr)^{1/5} \qquad\qquad (26)$$

Fig. 6 shows the comparison between equation (26) and the results of the numerical analysis [11] for parallel revolving pipes. The numerical analysis produced
the Nusselt number ratios, correlated by the solid line for a wide range of Pr
(>1), which do not show a dependence upon Pr as predicted by [1]. Also
shown are the results of the integral method employed by Siegwarth et al for
large Prandtl number fluids [12], and the experimental correlation reported by
Mori et al [13] for air flow in a horizon pipe. The coorelation proposed by
Siegwarth et al [12] almost overlaps the curve of equation (26). As for the
effect of Prandtl number on Nusselt number, we will discuss this subject in the
later section of numerical analysis.

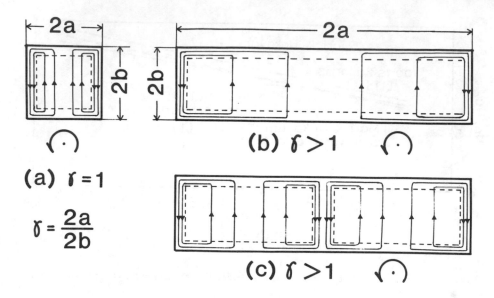

Fig. 7 Parallel revolving ducts

The boundary layer method was also employed by Ito and Nanbu [14] for the analysis of flow in the radial rotating pipe.

(2) <u>Energy balance and entropy balance method.</u> For flows in square or rectangular ducts, the analytical procedure explained above becomes cumbersome. A number of higher harmonic components with respect to ψ are required to describe the secondary flow recirculating in the rectangular cross-section. The integral balance method was used in an analysis of parallel revolving rectangular ducts [15] as shown in Fig. 7, but the analysis is laborious. As shown in Fig. 7(b), when Ra is between the 1st (minimum) critical value and the 2nd critical one, the secondary flow has one pair of vortex, while when Ra exceeds the second critical value, the secondary flow consists of two pairs as shown in (c). The integral balance method also assumes a core region and a boundary layer along the wall. The velocity and the temperature in the core region are given by;

$$\vec{V}_1 = (C_2 + \frac{C}{C_1} Y, C_2, 0) \tag{27}$$

$$T_1 = C_3 + \frac{C_2}{C_1} Y + \frac{\alpha}{2C_1^2} Y^2 \tag{28}$$

where C_1, C_2, C_3 are unknown constants and $C = -\partial p/\partial x$.

Multiplying the axial velocity component U on the equation of momentum in the X-direction, and integrating over the boundary layer region, we obtain:

$$\int\{-\frac{1}{2} \frac{\partial(U \cdot UV)}{\partial Y} + CU + \mu UV^2 U\} dZdY = 0 \tag{29}$$

The first and second terms on the left-hand side of equation (29) represent the rate of kinetic energy transport through the inertial interaction UV and the pressure gradient in the axial direction. The third term represents the viscous dissipation.

Multiplying the equations of momentum in the Y-and Z-directions by V and W, respectively, summing up the two equations, and integrating the resultant equation over the boundary layer region, we obtain:

$$\int \{ - \frac{1}{2}\rho(\frac{\partial(VW^2)}{\partial Z} + \frac{\partial(VW^2)}{\partial Y}) - \frac{\partial(PV)}{\partial Y} + FV + \mu(V\nabla^2 V + W\nabla^2 W)\}dYdZ = 0 \tag{30}$$

Each term on the left-hand side of equation (30) bears a clear physical meaning. The terms in the first bracket are the kinetic energy required to accelerate and retard the secondary flow along the wall. The second term is the transport of energy from the core region through the pressure work. The third expresses the work done by the secondary flow against the body force, and the fourth term viscous dissipation. An assumption of a secondary flow pattern symmetrical with respect to the Z-axis eliminates the terms in the first bracket. Equation (30) is simplified to the one expressing the balance of energy flow from the core region to the boundary layer by way of secondary flow.

Multiplying the energy equation by T' and integrating over the boundary

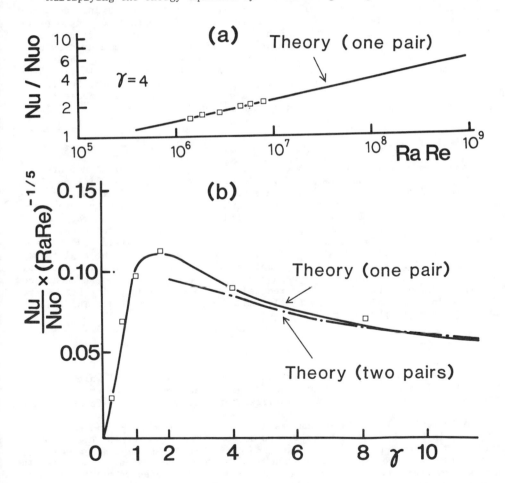

Fig. 8 Enhancement of heat transfer in parallel revolving rectangular ducts

layer region, we obtain:

$$\int \{ -\frac{1}{2} \frac{\partial(VT' \cdot T')}{\partial Y} + UT'\tau + kT'\nabla^2 T' \} \, dYdZ = 0 \tag{31}$$

The first term on the left-hand side of equation (31) represents the entropy production by convective heat transfer of secondary flows. The second term is the entropy production by axial flows. The third term is the production by heat conduction.

Integrations in equations (29), (30) and (31) can be performed by dividing the boundary layer region into sub-regions as illustrated in Fig. 7 and assuming velocity and temperature profiles in the boundary layer. The unknown constants C_1, C_2 and C_3, and the boundary layer thickness δ are determined from equations of continuity and force balance in the axial direction, which are explained in the preceeding section.

The experimental and analytical results [15] by the boundary layer method about the Nusselt number ratio (Nu/Nu_0) for air in parallel revolving rectangular ducts are shown in Fig. 8(a) and the dependence of the ratio on the aspect ratio in Fig. 8(b). Experiments to measure local heat transfer coefficients are not easy, therefore, local mass transfer coefficients were measured by use of naphthalene and from local evaporated amount. Nusselt number was obtained from the correction concerning Prandtl number and Schmidt number. In Fig. 8(b) theoretical predictions for one and two pairs of vortices are shown, but the difference is not so remarkable, and for larger aspect ratios, the Nusselt number ratio for two pairs is a little higher than that for one pair.

2.3. Numerical Analysis

Woods and Morris [11] performed numerical integration of axial momentum, vorticity and energy equations for fully developed flows in a parallel revolving cylindrical pipe by the finite difference method. Their numerical results are correlated over a broad range of Ra Re Pr as shown in Fig. 6. The authors noted that the proposed correlation shown by a solid curve covers a broad range of Prandtl numbers and coincides with the asymptotic correlation obtained by Siegwarth et al [12] for large Pr fluids. The asymptotic analysis was performed neglecting the inertia terms from the axial momentum equation. The negligence of the convective terms leads to a parabolic distribution of the axial velocity component, a situation where the secondary flow is weak. This can be understood by inspecting the definition of the reduced secondary vorticity $\psi' = Pr\psi$, where $\psi \sim 0$ (the intensity of secondary flow), and equation (12) in [11]. The inertia terms expressed in the form;

$$\frac{1}{Pr} \frac{1}{r} \frac{\partial(\psi', u)}{\partial(r, \theta)}$$

can be neglected when ψ' is of order unity, i.e., ψ is of $O(1/Pr)$ which becomes small as Pr increases. The enhancement of heat transfer for high Pr fluids is particularly large even where the dynamical parameter of the body force is small.

The numerical analysis encounters a difficulty when both Pr and Re are large, as implied by the plot for Pr = 60 or 40 in Fig. 3 of [11]. The convective terms become appreciable, and a large change of fluid temperature near the wall requires a fine finite difference resolution of the region. Therefore, the present reviewers hold the view that further study is needed to establish the Nusselt number correlation for high Pr fluids and larger Re flows.

As for the effect of Coriolis forces acting on the secondary velocities, the analysis of Woods and Morris [11] revealed that its effect on the Nusselt number is insignificant in a broad range of the tube radius — to — rotation radius ratios.

Miyazaki [16] reported the results of numerical analysis of flows in a circumferential revolving duct. For the fluid having the Prandtl number of 0.7, the increases in the friction factor f and the Nusselt number Nu are presented in terms of their ratios to those of a stationary straight duct, f_S and Nu_S, respectively. The following summary for a square duct would give a rough estimate on the effects of the relevant dynamical parameters. The ratio f/f_S increases from 1.1 to around 1.5 as ω increases from 5 to 50, Ra from 10^3 to 10^4, Re $\sqrt{a/R}$ from 10 to 1000. The ratio Nu/Nu_S exhibits a similar dependence on Ra and Re $\sqrt{a/R}$, however, it is insensitive to the change of ω.

The reports of developing flows in the entrance region are by now scarce, besides the analyses of flows in impeller passages where heat transfer is not the primary concern. To see a detailed feature of flows and heat transfer in the entrance region, the numerical technique of Patankar and Spalding [17] would be useful. For the case of large body forces, the methods proposed by Pratap and Spalding [18] or Pollard and Thyagaraja [19] can be applied. However, for practical applications, the problem remains of how one can assume with a reasonable approximation the flow at the inlet.

The distinguishing traits of numerical analysis are summarized as follows. Firstly, it can correlate the analytical solutions for the fully developed flow with weak secondary flow and for one with strong secondary flow. Secondly, only the numerical analysis can make clear the flow and heat transfer performance in the entrance region, even though the calculation is not easy and asks ingenious computational techniques. Thirdly, it is well expected that future advanced numerical technology with advanced computers will give us solutions to almost all problems concerning heat transfer in rotating or revolving ducts for larger dynamical parameter and large Reynolds number. However, it is worth noting that when the dynamical parameter is much smaller or larger, the analytical methods described above can often lead us to the solution without much trouble, and that in most of heat transfer problems associated with rotating or revolving ducts equipped in machineries such as electric generators, the effect of heat transfer in the entrance region is not so serious and heat transfer of fully developed flows is very important.

3. SINGLE-PHASE FLOWS AND HEAT TRANSFER — EXPERIMENT

The literature on heat transfer experiments published prior to 1974 are summarized in a review article by Bald and Hands [20]. Their subjects are concerned with the development of superconducting generators, with a majority of references to the previous experimental works conducted using air as heat transferring fluid. Among the cited references, the paper by Mori, Fukada and Nakayama [4] contains the information about the effect of secondary flows in both laminar and turbulent flows. Fig. 9 shows the ratios of the Nusselt number in a radial rotating pipe to that of a stationary pipe. The enhancement of heat transfer is remarkable in laminar flows, whereas it is modest in turbulent flows. Recently, Morris and Dias reported the turbulent flow data obtained with a square duct [21].

Experimental data on water in a parallel revolving pipe, Fig. 1(b), are scarce. The experimental Nusselt numbers found by Morris as cited in [22] fall above the solution of numerical analysis for laminar flows by about 35 percent.

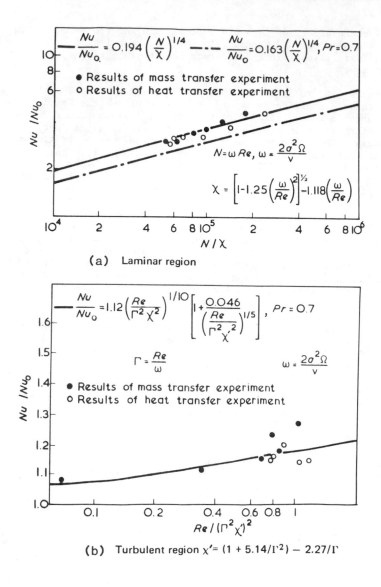

(a)   Laminar region

(b)   Turbulent region $\chi' = (1 + 5.14/\Gamma^2) - 2.27/\Gamma$

Fig. 9   Nusselt number ratio of fully developed flow in radial rotating pipe

The flow in the experiment was supposed to be laminar. Recently, Nakayama, Fuzioka and Watanabe [23] reported the data of water in turbulent regime. The work was motivated by the development of a water-cooled ac generator rotor. The laboratory experiment and the measurements on the prototype machine produced the data shown in Fig. 10. The solid curve represents the correlation [3] given by;

$$\frac{Nu}{Re^{0.8}Pr^{0.4}} = 0.033 \left(\frac{Re}{\Gamma^{2.5}}\right)^{1/30} \left[1 + \frac{0.014}{(Re/\Gamma^{2.5})^{1/6}}\right] \tag{32}$$

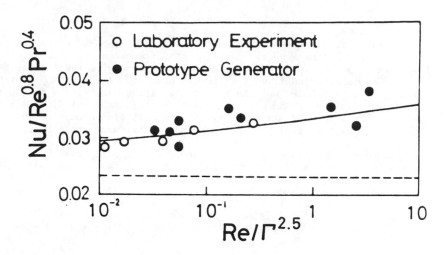

Fig. 10   Nusselt number for turbulent flow in the straight pipe
revolving parallel of the axis of rotation

where $\Gamma = \text{Re}^{22/13}(\text{Gr Pr}^{0.6})^{-12/13}$.

It is seen that equation (32) correlates well the experimental data.   The
broken line is for the $\text{Nu/Re}^{0.8}\text{Pr}^{0.4}$ in a stationary pipe, i.e. 0.023.   It was
noted that the local Nusselt number attains a fully developed value after the

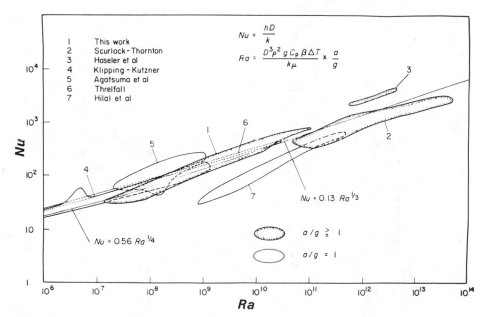

Fig. 11   Summary of helium data of free convection experiment

distance from the inlet exceeds about twenty times the pipe diameter.

The questions about the entrance effect, i.e. how long the entrance region is, and how the local heat transfer coefficient varies along the length, are very difficult to be answered, but particularly important for the laminar flow experiment. The information is very scarce at present. A recent paper by Morris and Woods [24] pointed out the importance of the parameter ω in a parallel revolving pipe. The underlying notion is that the adjustment of flow from the inlet to the fully developed state is governed by Coriolis force more than the body force caused by the density difference in a centrifugal force field.

For superconducting generator applications, it has been pointed out that free convective heat transfer in liquid helium yields very high Nusselt numbers to the order of above 10^3. This is due to a large coefficient of volumetric expansion of liquid helium (more than 0.1 K^{-1} at $T > 2.5$ K) and large centrifugal accelerations in generators (6000 ∿ 7000 times the gravitational one). For a 60 HZ machine, the centrifugal force at 0.40 m radius would be 5800 times the earth's gravitational field [25]. The Nusselt number data are correlated well by a formula;

$$Nu = C\, Ra^n$$

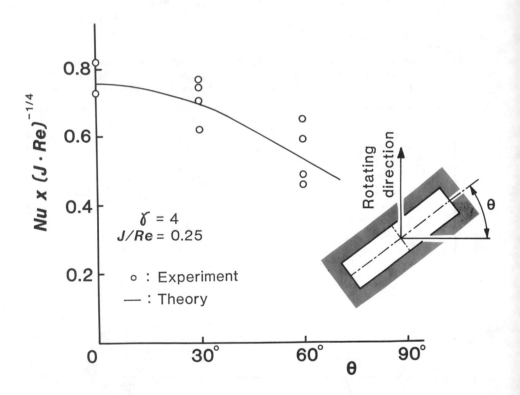

Fig. 12 Variation of Nusselt number by the oblique angle between the
         rotation and the duct wall
         (J is equivalent of ω, with a replaced by (2 X gap width))

where n = 0.25 ∿ 0.3 and the acceleration term in Rayleigh number being replaced by the centrifugal acceleration [26, 27].   Fig. 11 shows the data reported in the literature cluster around the proposed correlations [26].

For parallel or radial rotating square or rectangular ducts, the case when the body force is oblique to the duct axis or duct wall has been rarely studied. The case of radial rotating rectangular duct when the Coriolis force is oblique to the duct wall is found in connection with through-hole cooling of ceramic gas turbine blades. Fig. 12 shows the experimental results [28] about the Nusselt number ratio against the oblique angle θ.

4. TWO-PHASE FLOW HEAT TRANSFER

Two-phase flow heat transfer in rotation systems has been the subject of recent research works [29, 30] as it is a promising cooling technique for high temperature gas turbines. Water-cooled gas turbine blades with through-holes to let the mixture of water and air flow radially have many mechanical and structural advantages. Two-phase flow in a rotating radial passage may have a stratified structure with liquid film pressed on a part of the wall by Coriolis force [29] or form a mist flow when wall temperatures are high. Mori, Hijikata and Yasunaga [30] simulated heat transfer in a rotating passage by that in a

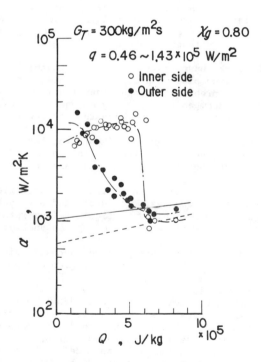

Fig. 13   Heat transfer coefficient of hot coiled tubules cooled by air-water mist flow
(α = heat transfer coefficient, Q = heat load per unit mass of water)

spiral tube. Fig. 13 shows the heat transfer coefficients on the inner wall of the spiral tube marked by open circles and those on the outer wall by solid circles. The solid line represents the correlation for a single-phase flow in a spiral tube [31], and the broken line for a single-phase flow in a straight tube. The data indicate that a large variation of local heat transfer coefficient exists in two-phase flows until the heat flux is raised to the point of complete dry-out of liquid phase. The similar performance could be seen in through-hole mist-cooling of gas turbine blades.

Two-phase flow heat transfer is also important for the thermal design of superconducting generators, where nucleate boiling is expected on the coolant passages not far from the rotation axis. Ogata and Nakayama [26] and Scurlock and Thornton [27] reported insensitive nature of nucleate boiling of helium to centrifugal forces acting toward the heat transfer surface. Ogata and Nakayama [26], however, pointed out that the maximum and minimum heat fluxes are affected by revolution of the surface.

5. CONCLUDING REMARKS

Theoretical and experimental works on flows and heat transfer in rotating pipes and ducts are reviewed and summarized as follows.

(1) The two analytical methods have proved to be powerful tools to obtain correlations between f, Nu and the dynamical parameters pertinent to forced convection in rotating systems; the boundary layer method for flows in circular pipes, and the integral method based on energy balance and entropy balance for flows in rectangular ducts.

(2) Recent numerical works have provided detailed feature of flow and temperature fields in rotating pipes and ducts. An integration scheme has to be devised to solve the problems of high Plandtl number fluids and very high dynamic parameters accompanied by strong secondary flows. A developing flow in the entrance region is also open to future works.

(3) Experimental data on forced convective heat transfer of air in the laminar and turbulent regimes, and those of water in the turbulent regime have verified the usefulness of the correlations based on the boundary layer modeling.
Free convection of liquid helium in centrifugal force fields are correlated by conventional $Nu \propto Ra^n$ formulas. It is pointed out that very high heat transfer coefficients are obtained in liquid helium rotating systems attracting interest of people working on superconducting machines.

(4) Problems of two-phase heat transfer are worth challenging because of their great importance in advanced energy technologies of the future.

Finally the authors would like to emphasize that the present paper is not a mere survey of the previous works. From late 1960 to early 1970, we made fundamental and systematic researches in this field with secondary flows including those in bend pipe and duct flows. Since then many papers have been reported based on scientific interests and requirements of application. We believe that more researches and applications will be made in the future. This is written on the basis of the authors' understanding mentioned above and current research programs in this area of industrial importance.

REFERENCES

1. Mori, Y. and Nakayama, W. 1967. Forced convective heat transfer in a straight pipe rotating around a parallel axis (1st report, Laminar region). International J. Heat and Mass Transfer, Vol. 10, pp. 1179-1194.

2. Mori, Y. and Nakayama, W. 1968. Convective heat transfer in rotating radial circular pipes (1st report, Laminar region). International J. Heat and Mass Transfer, Vol. 11, pp. 1027-1040.

3. Nakayama, W. 1968. Forced convective heat transfer in a straight pipe rotating around a parallel axis (2nd report, Turbulent region). International J. Heat and Mass Transfer, Vol. 11, pp. 1185-1201.

4. Mori, Y., Fukada, T. and Nakayama, W. 1971. Convective heat transfer in a rotating radial circular pipe (2nd report). International J. Heat and Mass Transfer, Vol. 14, pp. 1807-1824.

5. Mori, Y. and Uchida, Y. 1968. Study on forced convective heat transfer in parallel plates under the influence of Coriolis force. Trans. of JSME, Vol. 34, No. 264, pp. 1445-1454 (in Japanese).

6. Kays, W.M. and Perkins, H.C. 1973. Forced convection, Internal flow in ducts. Handbook of Heat Transfer, Section 7, pp. 176-178, McGraw-Hill.

7. Barua, S.N. 1954-1955. Secondary flow in a rotating straight pipe. Proc. Roy. Soc. A, Vol. 227, pp. 133-139.

8. Morris, W.D. 1965. Laminar convection in a heated vertical tube rotating about a parallel axis. J. Fluid Mech., Vol. 21, No. 3, pp. 453-464.

9. Nakayama, W., Hwang, G.J. and Cheng, K.C. 1970. Thermal instability in plane poiseuille flow. Trans. ASME J. of Heat Transfer, Vol. 92, pp. 61-68.

10. Mori, Y. and Uchida, Y. 1965. Study on forced convective heat transfer in horizontal flat plate channels. Trans. JSME, Vol. 31, No. 230, pp. 1511-1520 (in Japanese).

11. Woods, J.L. and Morris, W.D. 1980. A study of heat transfer in a rotating cylindrical tube. Trans. ASME J. Heat Transfer, Vol. 102, pp. 612-616.

12. Siegwarth, D.P., Mikesell, R.D., Readal, T.C. and Hanratty, T.J. 1969. The effect of secondary flow on the temperature field and primary flow in a heated horizontal tube. International J. Heat and Mass Transfer, Vol. 12, pp. 1535-1552.

13. Mori, Y., Futagami, K., Tokuda, S. and Nakamura, M. 1966. Forced convective heat transfer in uniformly heated horizontal tubes. International J. Heat and Mass Transfer, Vol. 9, pp. 453-463.

14. Ito, H. and Nanbu, K. 1971. Flow in rotating straight pipes of circular cross-section. Trans. ASME J. of Basic Engineering, Vol. 93, pp. 383-394.

15. Mori, Y., Fukada, T. and Yanatori, M. 1973. Forced convective heat transfer in a straight pipe rotating around a parallel axis (2nd report). Trans. Jap. Soc. Mech. Engr., Series 2, Vol. 39, No. 324, pp. 2484-2485 (in Japanese).

16. Miyazaki, H. 1973. Combined free-and forced-convective heat transfer and fluid flow in rotating curved rectangular tubes. Trans. ASME J. Heat Transfer, Vol. 95, pp. 64-71.

17. Patankar, S.V. and Spalding, D.B. 1972. A calculation procedure for heat, mass and momentum transfer in three-dimensional parabolic flows. International J. Heat and Mass Transfer, Vol. 15, pp. 1787-1806.

18. Pratap, V.S. and Spalding, D.B. 1976. Fluid flow and heat transfer in three-dimensional duct flows. International J. Heat and Mass Transfer, Vol. 19, pp. 1183-1188.

19. Pollard, A. and Thyagaraja, A. 1979. A new method for handling flow problems with body forces. Computar Methods in Applied Mechanics and Engineering, Vol. 19, pp. 107-116.

20. Bald, W.B. and Hands, B.A. 1974. Cryogenic heat transfer research at Oxford. Cryogenics, Vol. 14, pp. 179-197.

21. Morris, W.D. and Dias, F.M. 1980. Turbulent heat transfer in a revolving square-sectional tube. J. Mech. Eng. Sci., Vol. 22, No. 2, pp. 95-101.

22. Woods, J.L. and Morris, W.D. 1974. An investigation of laminar flow in the rotor windings of directly-cooled electrical machines. J. Mechanical Engineering Science, Vol. 16, No. 6, pp. 408-417.

23. Nakayama, W., Fuzioka, K. and Watanabe, S. 1975. Flow and heat transfer in the water-cooled rotor winding of a turbine generator. IEEE Transactions on Power Apparatus and Systems, Vol. PAS-97, No. 1, pp. 225-231.

24. Morris, W.D. and Woods, J.L. 1978. Heat transfer in the entrance region of tubes that rotate about a parallel axis. J. Mechanical Engineering Science, Vol. 20, No. 6, pp. 319-325.

25. Jones, M.C. and Arp, V.D. 1978, Review of hydrodynamics and heat transfer for large helium cooling systems. Cryogenics, Vol. 18, pp. 483-590.

26. Ogata, H. and Nakayama, W. 1977. Heat transfer to subcritical and supercritical helium in centrifugal acceleration fields, 1. Free convection regime and boiling regime. Cryogenics, Vol. 17, pp. 461-470.

27. Scurlock, R.G. and Thornton, G.K. 1977. Pool heat transfer to liquid and supercritical helium in high centrifugal acceleration fields. International J. Heat and Mass Transfer, Vol. 20, pp. 31-40.

28. Mori, Y. and Hijikata, K. Heat transfer of radial rotating rectangular ducts slanted to rotating direction, (unpublished).

29. Dakin, J.T. and So, R.M.C. 1978. The dynamics of thin liquid films in rotating tubes: Approximate analysis. Trans. ASME J. of Fluids Engineering, Vol. 100, pp. 187-193.

30. Mori, Y., Hijikata, K. and Yasunaga, T. 1981. Study on mist cooling of internal surface of high temperature fine tubes. Trans. JSME, Vol. 419, pp. 1332-1339.

31. Mori, Y. and Nakayama, W. 1967. Study on forced convective heat transfer in curved pipes (2nd report, Turbulent region). International J. Heat and Mass Transfer, Vol. 10, pp. 37-59.

An Experimental Study of Turbulent Heat Transfer in a Tube which Rotates about an Orthogonal Axis

W. DAVIS MORRIS
Department of Engineering Design and Manufacture
The University of Hull
North Humberside HU6 7RX, England

TEOMAN AYHAN
Department of Mechanical Engineering
Trabzon University
Trabzon, Turkey

ABSTRACT

In this paper a programme of work is described which studies the way in which rotation affects local and mean heat transfer in a circular section tube constrained to rotate about an axis orthogonal to its central axis. The results include conditions of radial outward and inward flow. It is demonstrated that centripetal buoyancy adversely affects heat transfer relative to the stationary tube condition when the net flow is radially outwards. The converse is demonstrated for radial inward flow. Coriolis acceleration is shown to improve heat transfer in the case of radially outward flow with the converse for conditions of radial inward flow. The implication of the results when extrapolated to real engine conditions is discussed.

1. INTRODUCTION

The thermodynamic attractions of operating gas turbines at progressively higher levels of turbine entry temperature has resulted in the incorporation of elaborate air cooling passages inside the rotor blades in order to avoid temperature-induced degradation of metal properties. When the blade is not rotating heat transfer from the surface of these internal passages to the coolant is controlled by a predominantly forced convection mechanism although the cross sectional shapes involved may be non-circular. The important question which follows, in the design context, concerns the extent to which this forced convection mechanism is modified by rotation and does the rotation enhance or impair the cooling capability.

Circular-sectioned holes located in the span-wise direction of the rotor blades may be considered as a fundamental cooling geometry in which case the hole rotates about an axis perpendicular to its axis of symmetry. This rotating geometry, referred to as orthogonal-mode rotation, is shown in figure 1 and theoretical and experimental studies of flow and heat transfer have been undertaken by Barua (1955), Benton and Boyer (1968), Mori and Nakayama (1968), Ito and Nanbu (1971), Mori et al (1971), Lokai and Limanski (1975), Metzger and Stan (1977), Skiadaressis and Spalding (1977), Vidyanidhi (1977), Zysina-Molozhen et al (1977) and Morris and Ayhan (1979). These works have collectively given good insight into the physical manifestations of rotation on flow resistance and heat transfer.

All theoretical studies have demonstrated that Coriolis-induced secondary flow can create additional mixing resulting in an increased resistance to flow

25

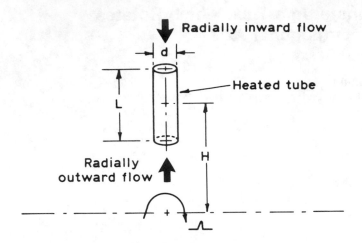

FIGURE 1   ORTHOGONAL-MODE ROTATION

with an attendent enhancement in heat transfer.   Limited experimental data
has tended to confirm these theoretical trends in some instances.   However all
theoretical treatments, to date, have been concerned with constant property
fluids so that the possibility of a buoyant interaction between centripetal
components of acceleration and a temperature-dependent fluid density have been
omitted from the physical modelling.

The validity of ignoring rotational buoyancy has been questioned previously by the present authors, see Morris and Ayhan (1979), when experimental
data was presented for circular-sectioned tubes.   The programme of experiments
was deliberately designed so that the influence of conventional forced convection and Coriolis forces were closely controlled whilst centripetal buoyancy
was systematically varied.   The following dimensional argument aids interpretation of the original work of Morris and Ayhan (1979) and the new data which
forms the main contribution of the present paper.

An examination of the basic conservation equations of mass, momentum and
energy may be used, see Morris (1982), to demonstrate that

$$Nu_m = \phi(Re, Pr, S, Ra, L/d, H/d) \tag{1}$$

where

$$Nu_m = \frac{hd}{k} \qquad\qquad \text{(mean Nusselt number)}$$

$$Re = \frac{w_m d\rho}{\mu} \qquad\qquad \text{(pipe flow Reynolds number)}$$

$$Pr = \frac{\mu c_p}{k} \qquad\qquad \text{(fluid Prandtl number)} \tag{2}$$

$$S = \frac{\Omega d}{w_m} \qquad\qquad \text{(a form of the Rossby number)}$$

$$Ra = \frac{\Omega^2 H \cdot \beta \rho^2 c_p d^3 \Delta T_w}{\mu k} \qquad \text{(rotational Rayleigh number)}$$

and other symbols are as defined in the nomenclature. In equation (1) the Rossby number has its origin in the Coriolis force term in the momentum conservation equations and may be thought of as a measure of the relative strength of the Coriolis to inertial forces acting on the fluid. Further, the rotational Rayleigh number arises from allowing the density of the fluid to vary with temperature in the centripetal-type forces acting on the fluid and may be thought of as a ratio of centripetal buoyancy to viscous forces.

The original experiments of Morris and Ayhan (1979) were conducted with air as the test fluid and, for the range of operating conditions studied, there was no significant variation in the fluid Prandtl number. Consequently for a specified geometric configuration (that is fixed L/d and H/d ratios) it might be expected that

$$Nu_m = \phi(Re, S, Ra) \qquad (3)$$

Using electrically heated tubes, with precise control of the air flow rate and rotational speed, Morris and Ayhan (1979) were able to systematically examine the influence of rotational Rayleigh number by varying the heating rates whilst the Reynolds number and Rossby number remained essentially constant. For L/d = 20.62 and H/d = 63.1, figure 2 shows the typical results which were obtained with a radially outward flow of cooling air. Note in this case that buoyancy produces a significant progressive reduction in mean Nusselt number as the Rayleigh number is increased. Close examination of all their data, obtained over the operating conditions specified in Table 1, enabled these authors to propose the following tentative correlating equation for turbulent flow.

$$Nu_m = 0.022 \, Re^{0.8} \left[\frac{Ra}{Re^2}\right]^{-0.186} S^{0.33} \qquad (4)$$

Although the expected improvement in heat transfer resulting for the influence of Coriolis-induced secondary flow in the tube was confirmed (see the positive exponent in the Rossby number term of equation (4)) this could be offset and reversed at relatively high levels of rotational Rayleigh number. Consequently design calculations based on theoretical and/or experimental data which does not make adequate allowance for buoyancy can seriously overpredict heat transfer for radially outward flow. In this respect figure 2 shows the expected results based on the proposal of Lokai and Limanski (1975) and the basic stationary tube prediction based on the proposal of Kreith (1965).

Because the effect of centripetal buoyancy is to cause the warmer and relatively less dense fluid particles in the near-wall region to migrate towards the rotational axis then this would tend to impair heat transfer if the flow is radially outwards. Alternatively with a radially inward coolant flow rotational buoyancy would be expected to work in conjunction with the forced convection mechanism normally operative in the tube producing an overall improvement in heat transfer. In order to check the validity of this and to confirm the general importance of buoyancy in the turbine blade cooling application the present authors have repeated the entire programme of experiments reported by Morris and Ayhan (1979) but with a radially inward coolant flow. Details of this experimental programme and the salient results obtained now follow.

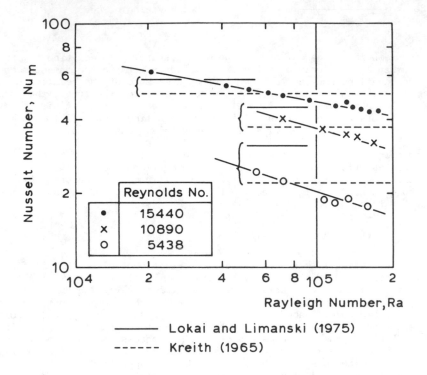

—————— Lokai and Limanski (1975)

------ Kreith (1965)

FIGURE 2   INFLUENCE OF ROTATIONAL BUOYANCY ON HEAT TRANSFER WITH
            RADIALLY OUTWARD FLOW.

Length/diameter ratios (L/d)	Test Section A: 20.62 Test Section B: 10.00
Eccentricity ratio (H/d)	Test Section A: 63.1 Test Section B: 32.8
Reynolds number	5×10^3 – 1.5×10^4
Rossby number	1×10^{-2} – 2.3×10^{-1}
Rayleigh number	2×10^4 – 9×10^6

TABLE 1   RANGE OF VARIABLES COVERED BY MORRIS AND AYHAN (1979) FOR
           RADIALLY OUTWARD FLOW.

2.   APPARATUS AND EXPERIMENTAL PROGRAMME

The apparatus used for the investigation consisted of a rotor arm to which could be attached electrically heated circular-sectioned tubes so that they rotated in the orthogonal-mode depicted in figure 1.   The tubes were made from stainless steel and built into a thermally insulated casing which was bolted to the rotor arm so that a prescribed mid-span eccentricity was effected.   Two test sections were studied and these were instrumented with thermocouples to permit surface temperature measurement and also the coolant inlet and exit temperatures.   These test sections were arranged to give precisely the same geometric features as those used for the radially outward flow experiments and designated as A and B in Table 1.   Air could be sucked radially inwards through the test sections by means of a fan and rotary seal assembly.

Power to the test sections was obtained from a Variac transformer and thermo-couple signals were monitored with a PDP 11/34-controlled data acquisition and processing system.   A full description of the apparatus and the method of data processing to derive variations in local and mean Nusselt number has been pres- ented in the original paper of Morris and Ayhan (1979) and are omitted here in the interest of brevity.   The overall range of experimental variables covered was the same as that given in Table 1 for the case of radially outward flow and figure 3 shows the schematics of the test apparatus.

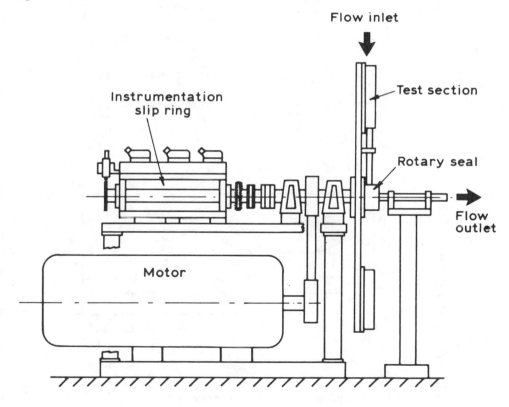

FIGURE 3   SCHEMATICS OF THE APPARATUS.

3. RESULTS AND DISCUSSION

An essential foundation stone in the interpretation of experimentally
determined data with rotating ducts is the behaviour at zero rotational speed.
This is true for heated and unheated flows. Figure 4a illustrates the typical
variation of tube wall and fluid bulk temperature which resulted from the zero
speed experiments and figure 4b a typical variation of the local Nusselt number,
Nu_z, which resulted from the data processing technique (full details of the

method of data processing are given in the authors' original paper, Morris and
Ayhan (1979)). The following features are important. Although individual

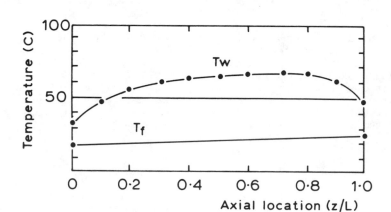

FIGURE 4a   TYPICAL WALL AND FLUID TEMPERATURE VARIATIONS FOR TEST
              SECTION A. (NOMINAL REYNOLDS NUMBER = 5500. NOMINAL
              HEAT FLUX = 2.28 kW/m²) (L/d = 20.62, H/d = 63.1).

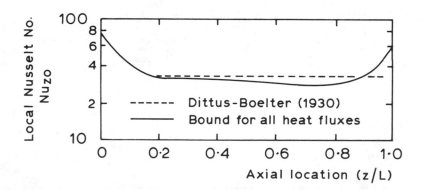

FIGURE 4b   TYPICAL VARIATION OF LOCAL NUSSELT NUMBER AT ZERO
              ROTATIONAL SPEED FOR TEST SECTION A. (NOMINAL
              REYNOLDS NUMBER = 10575) (L/d = 20.62, H/d = 63.1).

tests at specified heat flux levels produce individual wall and bulk fluid temperature signatures for specified through flow Reynolds number values the principles of non-dimensionalisation suggest that data should collapse onto a unique curve if presented in the form of a local Nusselt number, defined using local values of heat flux and wall/bulk fluid temperature difference. Note that the variation of fluid Prandtl number was not large over the range of experiments undertaken. This is clearly evident in the typifying case shown in figure 4b. In this figure the dotted line shows the value for the terminal or fully developed Nusselt number implied by the Dittus-Boelter (1930) equation. Agreement was found to be acceptable in the best approach to fully developed flow available.

When tests were undertaken with radially inward flow and rotation the tendency for local Nusselt number data to collapse into tight bands for all heat flux levels at a specified through flow Reynolds number was not detected. Under these circumstances there was a progressive increase in local Nusselt number as the heat flux levels were increased at a fixed level of through flow Reynolds number and rotational speed. This is shown in figure 5. Because fixed values of through flow Reynolds numbers and rotational speed imply also fixed values of the Rossby number, for a given geometric configuration, then the increase in local Nusselt number is really in response to increases in the rotational Rayleigh number. This tends to confirm the comments mentioned earlier that buoyancy, already found by the authors to reduce heat transfer with radially outward flow, would improve heat transfer in the case of radially outward flow.

This is shown, in terms of mean heat transfer, in figure 6. The mean Nusselt number has been defined using the difference in measured mean wall temperature and the mean bulk temperature of the fluid between the inlet and exit stations and this motivating temperature difference for heat transfer is also used in the definition of the Rayleigh number.

The overall influence of the combined effects of Coriolis forces and centripetal buoyancy with radially inward flow are shown in figure 7 for all data collected in the present series of tests. The results are expressed in the terms of the mean Nusselt number plotted against the quotient of the mean Rayleigh number and the square of the through flow Reynolds number and the individual lines shown represent fixed values of the Rossby number. This representation was used by Morris and Ayhan (1979) for radially outward flow. For comparative purposes the stationary tube predictions resulting from the use of the correlation proposed by Kreith (1965) is shown. For each series of tests there was a definite tendency for the mean heat transfer to increase with increases in the rotational Rayleigh number. A surprising and, as yet not fully explicable, trend concerns the influence of the Coriolis forces quantified via the Rossby number. Whereas all theoretical treatments which exclude buoyancy, have suggested improved heat transfer with increases in Rossby number this is apparently reversed in the case of radially inward flow. Note that, in figure 7, the mean Nusselt number tends to reduce at a fixed level of Ra/Re^2 as the Rossby number increases. This was found in all the tests conducted with both test sections and suggests a much more complex interaction between property variations and the influence of rotation than has hitherto been envisaged.

Although the data shown in figure 7 was obtained with two geometric configurations a simple exponent-type correlation is postulated as a tentative guide for design purposes. Thus it is proposed that the following equation may be used, within the range of variables shown in Table 1, for radially inward flow.

$$Nu_m = 0.036 \, Re^{0.8} \left[\frac{Ra}{Re^2}\right]^{0.112} S^{-0.083} \tag{5}$$

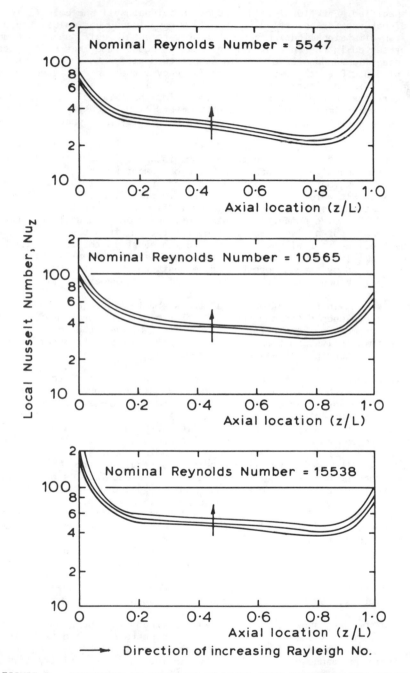

FIGURE 5   INFLUENCE OF ROTATION ON LOCAL HEAT TRANSFER WITH RADIALLY
INWARD FLOW.   (L/d = 20.62,   H/d = 63.1, Ω = 1000 rev/min).

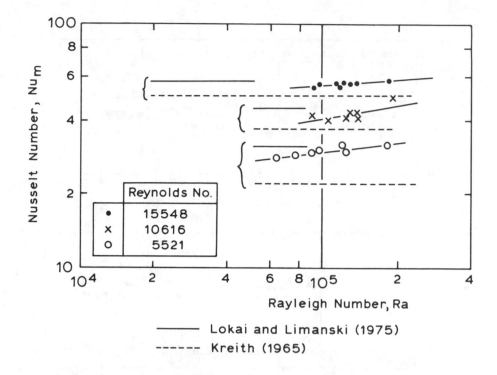

FIGURE 6   INFLUENCE OF ROTATIONAL BUOYANCY ON MEAN HEAT TRANSFER
            WITH RADIALLY INWARD FLOW.   (L/d = 20.62, H/d = 63.1,
            Ω = 1000 rev/min).

Over the range of test conditions investigated the stationary tube correlations of Kreith (1965) for mean turbulent heat transfer are fairly well approximated by

$$Nu_0 = 0.022 \ R^{0.8} \tag{6}$$

Using equation (6) the empirical correlations given by equations (4) and (5) for
radially outward and inward flow respectively may be re-expressed as

$$\frac{Nu_m}{Nu_0} = \left[\frac{Ra}{Re^2}\right]^{-0.186} S^{0.33} = M \qquad \text{(radially outward flow)} \tag{7}$$

and

$$\frac{Nu_m}{Nu_0} = 1.636 \left[\frac{Ra}{Re^2}\right]^{0.112} S^{-0.083} = N \quad \text{(radially inward flow)} \tag{8}$$

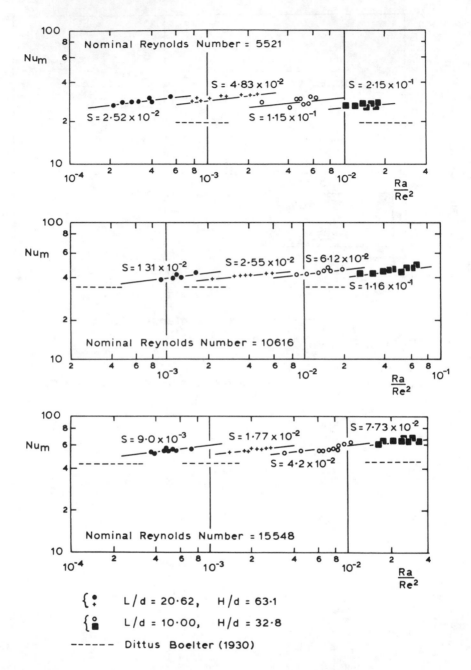

FIGURE 7   INFLUENCE OF CORIOLIS ACCELERATION AND ROTATIONAL BUOYANCY
            ON MEAN HEAT TRANSFER WITH RADIALLY INWARD FLOW.

In conclusion the following points are made concerning the effect of rotation on a notionally turbulent flow with orthogonal-mode rotation and radially inward flow. Rotational buoyancy tends to increase local and mean Nusselt numbers in relation to the stationary tube situation. This is, it is believed, due to the fact that the buoyancy effect is working in conjunction with the forced convection mechanism. For radially inward flow, increases in the Rossby number tend to reduce the heat transfer at a fixed measure of rotational buoyancy. This is a result which has not been fully explained and more detailed studies of the flow field inside the tube are necessary to unravel this feature.

From the design viewpoint great care is needed in the estimation of heat transfer in the turbine rotor blade application and particularly in the use of theoretical or empirical data which does not account for buoyancy. This is clearly shown in Table 2 where, for the purpose of illustration, the implication of using a number of recommended alternative procedures for estimating mean heat transfer are shown.

Engine Type	Industrial gas turbine	Aircraft fan-type gas turbine
GEOMETRIC DETAILS:		
Coolant passage length (mm)	50.0	90.0
Coolant passage diameter (mm)	5.8	1.7
Midspan eccentricity (mm)	305.0	386.0
Rotational speed (rev/min)	10 000	10 000
COOLANT FLOW DETAILS:		
Mean pressure (bar)	6.0	10.0
Mean temperature (K)	603	702
Mean axial velocity (m/s)	114.0	308.0
Mean blade temperature (K)	333	1073
NON-DIMENSIONAL PARAMETERS		
Length/diameter ratio	8.6	52.9
Eccentricity parameter	52.6	227.1
Flow Reynolds number	7.47×10^4	7.61×10^4
Rotational Reynolds number	3.98×10^3	4.40×10^2
Rossby number	5.33×10^{-2}	5.78×10^{-3}
Rotational Rayleigh number	3.15×10^8	1.63×10^7
Rayleigh/Reynolds number^2	5.65×10^{-2}	2.81×10^{-3}
PERCENTAGE CHANGE IN RELATIVE MEAN NUSSELT NUMBER		
Mori, Fukada and Nakayama (1971) (Radially outward and inward flow)	+44.6	+22.0
Lokai and Limanski (1975) (Radially outward and inward flow)	+24.7%	5.2%
Zysina-Molozhan et al (1977) (Radially outward and inward flow)	0%	0%
Morris and Ayhan (1979) (Radially outward flow)	-35.0%	-46%
Morris and Ayhan (present paper) (Radially inward flow)	+60%	+36%

TABLE 2   DESIGN IMPLICATIONS OF EXTRAPOLATING TRENDS REPORTED TO REAL ENGINE OPERATING CONDITIONS.

ACKNOWLEDGEMENTS

The authors wish to express their appreciation to Rolls Royce Limited for financial support for the investigation and to Professor F J Bayley, Director, SERC Centre for Thermo-Fluid Mechanics Research, University of Sussex, for his encouragement during the investigation and for making available the general facilities of the Centre.

NOMENCLATURE

English Symbols

C_p	constant pressure specific heat
d	tube diameter
h	mean heat transfer coefficient
H	eccentricity
k	thermal conductivity
L	tube length
M	function
Nu_m	mean Nusselt number for rotating tube
Nu_o	mean Nusselt number for stationary tube
Nu_z	local Nusselt number for rotating tube
Nu_{zo}	local Nusselt number for stationary tube
N	function
Pr	Prandtl number
Ra	Rayleigh number
Re	Reynolds number
S	Rossby number
T_f	fluid bulk temperature
T_w	wall temperature
w_m	mean axial velocity along tube

Greek Symbols

β	density
ΔT_w	difference in mean wall-fluid bulk temperature
μ	viscosity
ρ	density
ϕ	function
Ω	angular velocity

REFERENCES

1.   Barua S N (1955). Secondary flow in a rotating straight pipe. Proc
        Roy Soc A, 227, 133.

2.   Benton G S and Boyer D (1968). Flow through a rapidly rotating conduit
        of arbitrary cross-section. J Fluid Mech 26, part k, 69.

3.   Dittus F W and Boelter L M K (1930). Univ Calif Publs Engng. 2, 443.

4.   Ito H and Nanbu K (1971). Flow in rotating straight pipes of circular
        cross section. ASME Trans J Basic Eng. 93, (3), 383.

5.   Kreith F (1965). Principles of Heat Transfer. International Textbook
        Company.

6.   Lokai V I and Limanski A S (1975). Influence of rotation on heat and
        mass transfer in radial cooling channels of turbine blades. Izvestiya
        VUZ, Aviatsionnaya Tekhika, 18, No 3, 69.

7.   Metzger D E and Stan R L (1977). Entry region heat transfer in
        rotating radial tubes. AIAA 15th Aerospace Sciences Meeting, Los
        Angeles. Paper No 77 - 189.

8.   Mori Y, Fukada T and Nakayama W (1971). Convective heat transfer in a
        rotating radial circular pipe (2nd report). Int J Heat Mass Transfer,
        14, 1807.

9.   Morris W D (1982). Heat Transfer and Fluid Flow in Rotating Coolant
        Channels. John Wiley & Sons Ltd., Chichester, England. (Research
        Studies Press).

10.   Morris W D and Ayhan T (1979). Observations on the influence of
        rotation on heat transfer in the coolant channels of gas turbine rotor
        blades. Proc Inst Mech Eng., 193, No 21, 303.

11.   Skiadaressis D and Spalding D B (1977). Heat transfer in a pipe
        rotating around a perpendicular axis. ASME Paper No 77-WA/HT-39.

12.   Vidyanidhi V et al (1977). An analysis of steady fully developed heat
        transfer in a rotating straight pipe. Trans ASME Jour Heat Transfer,
        Feb. 148.

13.   Zysina-Molozhen L M et al (1977). Experimental investigation of heat
        transfer in a radially rotating pipe. HGEEE High Temp. 14, 988.

An Experimental Study of Flow and Heat Transfer in a Duct Rotating about a Parallel Axis

PETER L. STEPHENSON
Central Electricity Research Laboratories
Kelvin Avenue
Leatherhead, Surrey, England

ABSTRACT

Measurements have been obtained of turbulent flow and heat transfer in a circular duct rotating about a parallel axis, for conditions found in the cooling systems of the rotors of large turbogenerators. The results are of the variation of local Nusselt number along the test section, and the overall pressure loss coefficient.

NOMENCLATURE

c_p	specific heat at constant pressure
C_t	overall pressure loss coefficient (equation 1)
d	test section diameter
Gr_r	rotational Grashof number ($H\Omega^2 \beta \tau d^4/\nu$)
H	radius of rotation
h	heat transfer coefficient
J	rotational Reynolds number ($\Omega d^2/\nu$)
k	thermal conductivity
Nu	Nusselt number (hd/k)
Pr	Prandtl number ($c_p\mu/k$)
Re	Reynolds number (Wd/ν)
Sw	swirl number ($\Omega H/W$)
W	axial velocity
ΔP	overall pressure drop
β	bulk expansivity
ε	eccentricity parameter (d/H)
μ	dynamic viscosity
ν	kinematic viscosity
ρ	density
τ	axial air temperature gradient
Ω	rotational speed (radians per second)

Subscripts

f	fully developed
i	inlet (i.e. first quarter of test section)
r	rotating
s	stationary

39

1.   INTRODUCTION

The rotor windings of large turbogenerators are normally cooled by passing
hydrogen through hollow copper conductors. As part of a basic study of such
cooling systems, a rig has been built to study flow and heat transfer in a duct
rotating about a parallel axis. The rig uses air, rather than hydrogen, but is
designed to operate at values of the governing nondimensional parameters that
are similar to those occurring in actual generators. Results have been obtained
with the first test section, which was circular in cross-section. The
measurements at the higher flowrates were more relevant to actual generators
and were reported in ref [1]; their implications for the design of rotor cooling
systems were also discussed. It is the aim of the present paper to report the
complete range of data obtained, so as to give a more comprehensive view of the
effects of rotation.

2.   PREVIOUS WORK

Experimental studies of heat transfer for turbulent flow in a circular duct
rotating about a parallel axis are reported in refs [2]-[8]; mass transfer
analogue measurements are also reported in [4] and [8]. These studies show that
the heat transfer is increased by rotation, but to a lesser extent than in
laminar flow. Several studies have also suggested that the effect of rotation
is more pronounced in developing than in fully-developed flow. An approximate
analytical expression for high rotation rates has been reported by Nakayama [9],
and a numerical solution by Majumdar et al. [10]. General studies of the effects
of rotation are given in refs [11] and [12]. Measurements of pressure drop in
isothermal flow for a parallel rotating duct have been reported by Morris [13].
Pressure drop measurements for related geometries are given in refs [14] and
[15].

3.   BRIEF DESCRIPTION OF RIG

The rig consists of a hollow rotor that contains a test section up to 1.5 m
long at a radius of rotation of 0.48 m. This is rotated by an electric motor
at speeds of up to 2200 rpm. Air is supplied via a flowmeter and a coupling to
one end of the shaft, passes along the hollow shaft and via a radial passage to
the test section itself. The air is exhausted axially at the end of the test
section (Fig. 1). Temperatures are measured by thermocouples and the signals
are transmitted to a separate instrument pod. Here they are amplified, multi-
plexed and transmitted via instrument sliprings. The rig has various safety
checks built into its control systems and is operated manually. A PDP 11/40
computer is used to log all the readings and for subsequent off-line data
analysis. The present test section consists of a thick walled aluminium tube,
internal radius 9.53 mm, external radius 14.29 mm. A number of thermocouples
are fitted to the test section and their leads are brought out via grooves in
its outer surface. Round the outside of the test section are wrapped two layers
of insulation tape, an electrical heater, two further layers of insulation and
then a layer of resin insulation out to a radius of 31.06 mm. Three thermocouples are fitted to the outside of this insulation and are used to derive the
radial heat loss. The whole test section then fits into an annulus containing
further thermal insulation. The inlet air temperature is measured by a single
thermocouple. Four thermocouples plus two mixing vanes are provided at the test
section outlet, and the air outlet temperature is taken to be the average of the
two measured values at the second (downstream) mixing vane. The only pressure
measurement made at present is the overall pressure drop between a point
upstream of the air coupling and ambient conditions.

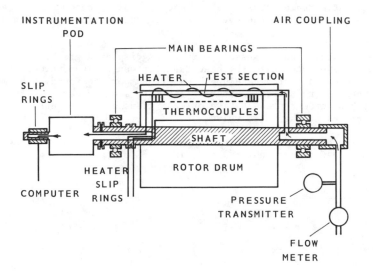

FIG. 1   GENERAL ARRANGEMENT OF RIG

4.   DATA ANALYSIS

Local Nusselt numbers were calculated using a method similar to that of refs [6], [7], and [16]. A polynominal expression (usually tenth order) was fitted to the measured wall temperatures, and readings more than about 1 C from this curve were rejected. This rejection of certain readings was necessary as some thermocouples worked reliably only at lower rotational speeds. From this polynominal expression, the axial variation of air temperature and heat input into the air were calculated. To do this, the axial heat loss was calculated from the tube dimensions, and the radial loss from preliminary experiments with zero flowrate. The local values of Nusselt number were then calculated. The measured air outlet temperature was not used in this procedure. Instead, it was used to provide a check on the overall heat balance, errors in which were always less than 4%.

The axial variation of Nusselt number was displayed graphically to enable the portion of the duct where the heat transfer was fully-developed to be estimated. Hence the fully-developed Nusselt number was found. The mean Nusselt number for the first quarter of the test section was also found. This value was taken to be representative of heat transfer in the entry region, and will be described as the inlet value.

The pressure rise due to rotation was allowed for in calculating the overall pressure loss coefficient, which was defined as follows:

$$Ct = \frac{2\Delta P + \rho H^2 \Omega^2}{\rho W^2} \tag{1}$$

5.   RESULTS AND DISCUSSION

5.1 Flow and Heat Transfer Without Rotation

Tests with zero rotation were made at seven values of Reynolds number in the range 5500 to 33000. The measured values of fully-developed Nusselt number were compared with values from the following correlation [17];

$$Nu = 0.023 \ Re^{0.8} \ Pr^{0.4} \tag{2}$$

The two sets of values differed by 7% for the highest flowrate, and 22% for the lowest. The difference increased as the flowrate dropped, and this was probably because the heat removed by air flowing through the test section became a smaller fraction of the total heat input as the flowrate decreased. This level of agreement was regarded as satisfactory. Typical axial variations of wall temperature are included in Fig. (2); they are qualitatively similar to those of refs [6], [7], and [16]. The overall pressure loss coefficient was found to decrease slowly with increasing Reynolds number; however, as it applied specifically to the present rig, no values were available for comparison. The behaviour of the rig when stationary was thus regarded as satisfactory.

5.2 Flow and Heat Transfer with Rotation

A series of tests was conducted and covered all combinations of the following nominal parameter values:

    (i)   Air flowrates of 1.7, 3.3, 6.7 and 10.0 g/s.

  (ii)   Rotational speeds of 280, 550, 1000, 1600, 1900, 2050 and 2200 rpm.

 (iii)   Zero heat input, and heat input adjusted to give maximum test section wall temperatures of 90 and 130 C.

The tests with rotation and heat transfer therefore covered values of axial Reynolds number, Re, between 5500 and 33000, values of rotational Reynolds number, J, between 600 and 4000, values of swirl number, Sw, between 0.4 and 19, and values of rotational Grashof number between 1700 and 740000. The eccentricity parameter was 0.0397 for all tests, and the test section was 77.2 diameters long.

Typical measured wall temperatures are shown in Fig. (2); as the heat input was adjusted to give the same maximum wall temperature, the main effect of rotation shown here is the increase in air temperature at inlet to the test section. This is because the air is fed via the hollow shaft and so through one of the main bearings. The axial variation of Nusselt number for the highest flowrate (derived from the measurements of Fig. (2)) is shown in Fig. (3). Corresponding results for the lowest flowrate are given in Fig. (4). The variation of fully-developed Nusselt number with J is given in Fig. (5) for the complete set of tests. Corresponding results for the inlet Nusselt number are given in Fig. (6). The percentage increase in Nu with rotation is higher at lower flows. For example, for Re = 5500, the maximum increases in fully-developed and inlet values are 39% and 114% respectively. The corresponding values for Re = 33000 are only 7% and 28%.

For all tests, the values of inlet Nu for the two maximum wall temperatures differ from each other by 14% or less. The values of fully developed Nu for the two maximum wall temperatures generally differ from each other by under 15%. The only exceptions are for the lowest flowrate and speeds of 1900 rpm and higher, and for a flowrate of 3.3 g/s and full speed. For these tests, the two values differ by between 20 and 40%. It is therefore concluded that rotational buoyancy effects are not generally significant, but may become important in fully-developed flow at high rotational speeds and low flowrates. This general lack of dependence on test section temperature plus the large changes occurring near inlet indicate that, for most tests, the major rotational effect is that of inlet swirl combined with Coriolis acceleration.

The absolute increases in inlet Nu are similar for all J values. For this reason, and because J represents the Coriolis acceleration effects, the results have been correlated as follows

$$Nu_{i,r} - Nu_{i,s} = 0.94 \ J^{0.43} \tag{3}$$

A nonlinear least squares fit was used to derive equation (3). To allow for small differences in Re between tests at nominally the same flowrate, the values of inlet Nu for a stationary duct were derived from equation (4).

$$Nu_{i,s} = 0.031 \ Re^{0.79} \tag{4}$$

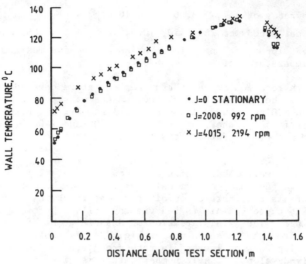

FIG.2 MEASURED WALL TEMPERATURE VARIATION FOR
NOMINAL REYNOLDS NUMBER OF 3.3 x 10⁴

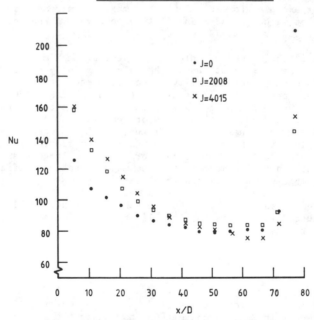

FIG.3 AXIAL VARIATION OF LOCAL NUSSELT NUMBER
RESULTS FOR NOMINAL Re = 3.3 x 10⁴
AND MAXIMUM WALL TEMPERATURE OF 130°C

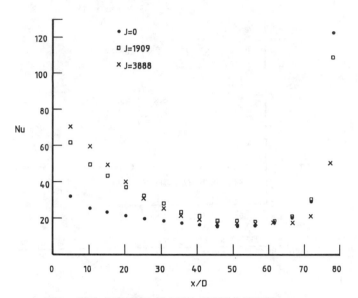

FIG.4   AXIAL VARIATION OF LOCAL NUSSELT NUMBER
RESULTS FOR NOMINAL Re $=5.5 \times 10^3$
AND MAXIMUM WALL TEMPERATURE OF 130^0C

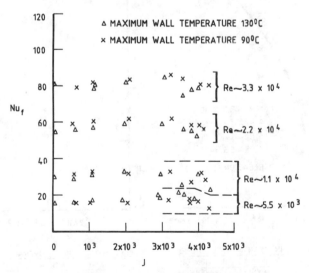

FIG.5   VARIATION OF FULLY DEVELOPED NUSSELT
NUMBER WITH ROTATIONAL REYNOLDS NUMBER

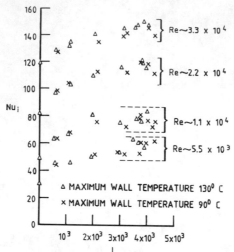

FIG.6   VARIATION OF MEAN NUSSELT NUMBER FOR FIRST QUARTER OF TEST SECTION
WITH ROTATIONAL REYNOLDS NUMBER

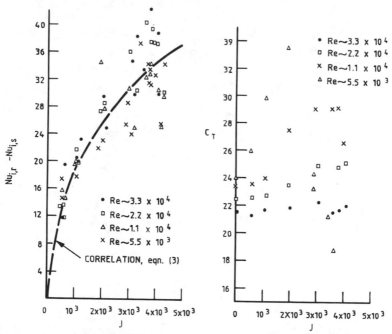

FIG.7   INCREASE IN MEAN NUSSELT NUMBER
FOR INLET REGION AS FUNCTION OF
ROTATIONAL REYNOLDS NUMBER

FIG.8   VARIATION OF OVERALL PRESSURE DROP
WITH ROTATIONAL REYNOLDS NUMBER

Equation (4), in turn, was derived from a nonlinear least squares fit of the measurements for a stationary tube. The maximum difference between measured Nu values and those from equation (4) was 5%. The maximum difference between the measured values for the increase in Nu with rotation and those from equation (3) was 39%. One reason for this comparatively large difference is that the increases in inlet Nu were often small compared with their absolute values, and so were difficult to measure accurately. It can be seen from Fig. (7) that equation (3) represents the general trend of the results.

Measurements of the overall pressure loss coefficient, Ct, are shown in Fig. (8). The results are for a wall temperature of 130°C; similar results were generally obtained for a wall temperature of 90°C, and for an unheated test section. The Ct values for low flows and high rotational rates are difficult to measure accurately because of the need to correct for the pressure rise due to rotation (equation 1); for these cases this pressure rise is comparable with the frictional pressure drop. However, the trends shown in Fig. (8) are likely to be correct.

5.3   Comparison with Earlier Work

The present results are in qualitative agreement with those of Morris and Woods [7]. For example, the effects of rotation are greater near inlet and at lower flows, and in general do not vary systematically with heat input. Morris and Woods reported a mean Nusselt number for the whole test section except the portion controlled by end effects, and correlated it as follows:

$$Nu = 0.0089 \ Re^{0.78} \ J^{0.25} \tag{5}$$

Solely for the purpose of comparing with them, the present results were correlated in the same manner, as follows:

$$Nu_{f,r} = 0.0071 \ Re^{0.88} \ J^{0.023} \tag{6}$$

and

$$Nu_{i,r} = 0.19 \ Re^{0.56} \ J^{0.11} \tag{7}$$

The maximum fractional differences between measured Nu values and those from the above correlations were 38% for equation (6) and 15% for equation (7). The present measurements thus show a much smaller variation of Nu with J than do those of reference [7]. One possible reason for this is the difference in inlet conditions between the two sets of measurements. Another is that the earlier data were for lower values of J (up to 1200) than the present work (J values up to 4000), and Nu appears to vary more rapidly with J for lower J values.

Sakamoto and Fukui [4] obtained heat transfer measurements for fully-developed flow in a rotating parallel duct. The effect of rotation for Re above 10000 was so small that they studied in detail only the results for lower Re. This is in qualitative agreement with the small increases in fully-developed Nu found in the present tests.

Nakayama [9] derived an approximate analytical expression for the fully-developed value of Nu, by assuming a strong secondary flow caused by rotational buoyancy. As the present results show that rotational buoyancy is not, in general, important, it is not surprising that the present measurements give a much smaller increase in Nu with rotation than was predicted by Nakayama.

As the overall pressure loss coefficient applies only to a rig identical with the present one, only qualitative comparison can be made with results of other investigators. The behaviour at higher flows shows the same trends as, for example [6] and [11]. Morris [13] reported friction factors for isothermal flow in a parallel rotating duct, and showed that they were at times lower than for a stationary duct. He suggested that rotation may increase the Re value for transition to turbulent flow, and a similar effect may occur in the present rig for low flows and high rotational effects.

6.   CONCLUSIONS

Heat transfer measurements have been obtained for a duct rotating about a parallel axis and for flow parameters relevant to the cooling of large generator rotors. They show that rotation improves heat transfer, and that the increases are greatest in the developing flow region and for lower flows. At Re = 5500, the maximum percentage increases in Nusselt number were 39% for the fully-developed region and 114% for the inlet region. The corresponding values at Re = 33000 were only 7% and 28%. The overall pressure loss coefficient generally increases with increasing rotation. However, at high rotational speeds and low flowrates it decreases with increasing rotation. Measurements of heat transfer and pressure drop do not, in general, show a significant variation with maximum test section temperature. The only exceptions are for low flowrates and high rotational speeds. This implies that rotational buoyancy is not generally important, so that the dominant effect is of inlet swirl induced by rotation, combined with Coriolis accelerations.

7.   ACKNOWLEDGEMENTS

The help and assistance of Mr F.H. Gawman in conducting the experiments described here is gratefully acknowledged. The work was carried out at the Central Electricity Research Laboratories and is published by permission of the Central Electricity Generating Board.

8.   REFERENCES

1.   Stephenson, P.L. 1982. An experimental study of flow and heat transfer in generator rotor cooling ducts, to be presented at Int. Conf. on Electrical Machines - Design and Applications, IEE, London.

2.   Humphreys, J.F., Morris, W.D. and Barrow, H. 1967. Convection heat transfer in the entry region of a tube which revolves about an axis parallel to itself, Int. J. Heat Mass Transfer, 10, 333.

3.   Le Feuvre, R.F. 1968. Heat transfer in rotor cooling ducts, Thermodynamics and Fluid Mechanics Convention, I. Mech. E., London.

4.   Sakamoto, M. and Fukui, S. 1970. Convective heat transfer of a rotating tube revolving about an axis parallel to itself, Fourth Int. Heat Transfer Conf., Vol. III, Paper FC7.2, Paris-Versailles.

5.   Woods, J.L. 1976. Heat transfer and flow resistance in a rotating duct system. D.Phil. thesis, Sussex University.

6. Morris, W.D. and Dias, F.M. 1976. Experimental observations on the thermal performance of a rotating coolant circuit with reference to the design of electrical machine rotors, Proc. I. Mech. E., 190, 561.

7. Morris, W.D. and Woods, J.L. 1978. Heat transfer in the entrance region of tubes that rotate about a parallel axis, J. Mech. Sci., 20, 319.

8. Mori, Y. 1973. Forced convection heat transfer in a straight rotating pipe in a centrifugal field III, Central Electricity Generating Board Translation CE7440.

9. Nakayama, W. 1968. Forced convection heat transfer in a straight pipe rotating about a parallel axis, Int. J. Heat Mass Transfer, 11, 1185.

10. Majumdar, A.K., Morris, W.D., Skiadaressis, D. and Spalding, D.B. 1976. Heat transfer in rotating ducts, Imperial College Heat Transfer Section report HTS/76/5.

11. Petukhov, B.S. and Polyakov, A.F. 1977. Heat transfer and resistance in rotating pipes (survey), Power Eng., 15, 104.

12. Morris, W.D. 1977. Flow and heat transfer in rotating coolant channels, AGARD Conf. preprint No. 229, 50th meeting of AGARD Propulsion and Energetics Panel, Ankara, Turkey.

13. Morris, W.D. 1981. A pressure transmission system for flow resistance measurements in a rotating tube, J. Phys. E: Sci. Instrum., 14, 208.

14. Kvitovskii, Yu.V. 1971. Hydraulic loss to a liquid flowing through a straight passage rotating about a parallel axis, Central Electricity Generating Board Translation CE7078.

15. Kvitovskii, Yu.V., Gembchinski, Kh. and Gladov, F.L. 1971. Loss of energy due to a bend, an expansion and a contraction of a rotating flow passage, Central Electricity Generating Board Translation CE7079.

16. Lau, S.C., Sparrow, E.M. and Ramsey, J.W. 1981. Effect of plenum length and diameter on turbulent heat transfer in a downstream tube and on plenum related pressure losses, J. Heat Transfer, 103, 415.

17. Knudsen, J.F. and Katz, D.L. 1958. Fluid Dynamics and Heat Transfer, McGraw-Hill, New York.

Pressure Loss Measurements in Circular Ducts which Rotate about a Parallel Axis

ANDREW R. JOHNSON and W. DAVID MORRIS
Department of Engineering Design and Manufacture
The University of Hull
Cottingham Road
Hull, North Humberside HU6 7RX, England

ABSTRACT

Experimental data is presented to illustrate the manner in which flow resistance in circular-sectioned ducts is influenced by duct rotation for the specific case where rotation is about an axis parallel to the central axis of the tube.

1. INTRODUCTION

This paper presents the results of an experimental investigation into the pressure loss experienced by a fluid flowing through a straight circular-sectioned tube which is constrained to rotate about an axis parallel to its central axis, as shown in figure 1. The study forms part of a long term investigation into the influence of rotation on flow geometries which are designed to facilitate the internal cooling of rotating components, see for example Morris (1965, 1969, 1970, 1977, 1981, 1982), Morris and Ayhan (1979), Morris and Dias (1976, 1980, 1981), Morris and Woods (1978), Woods and Morris (1974, 1981). Notable examples of internally cooled rotating components include the rotor windings of large turbine-driven generators and the rotor blades of gas turbines.

In order to illustrate the physical manifestations of rotation it is convenient to refer the motion inside the tube to a reference frame (x,y,z) which rotates with the tube as shown in figure 1. This reference frame is, by definition, non-inertial and correction terms including any translational acceleration of the selected origin together with centripetal and Coriolis components of acceleration must be included in the customary momentum conservation equations to allow for this fact.

To permit the interpretation of experimental data presented later in the paper the effect of these correction terms to the momentum equations will now be discussed using the case of laminar constant property flow by way of exemplification. The general observations which result will also be applicable to turbulent flow.

2. THEORETICAL OBSERVATIONS

The flow system illustrated in figure 1 is geometrically described via the tube diameter, d, the tube length, L, and the eccentricity, H, of the centre line of the tube relative to the axis of rotation. The tube rotates

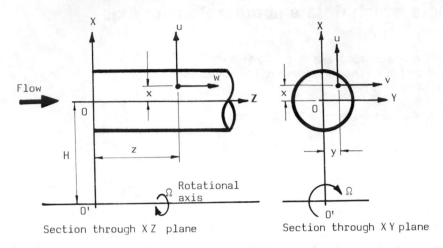

FIGURE 1   FLOW GEOMETRY AND CARTESIAN COORDINATE SYSTEM.

about the axis of rotation with angular velocity, Ω, and the flow through the
tube is taken to be steady and laminar with invariant properties.   In order to
simplify the physical discussion the motion is referred to the Cartesian reference frame shown in figure 1 so that, in general terms, the vectorial representation of the momentum conservation principle (see Morris (1982)) can be expressed as

$$\frac{D\underline{v}}{Dt} \;+\; 2(\underline{\omega} \wedge \underline{v}) \;+\; (\underline{\omega} \wedge \underline{\omega} \wedge \underline{r}) \;+\; \underline{f_o} \;=\; -\;\frac{1}{\rho}\,\nabla p \;+\; \nu\nabla^2\underline{v} \tag{1}$$

where $\underline{r}$ and $\underline{v}$ refer to the position vector and velocity vector of a particle
in the flow measured relative to the rotating reference frame, $\underline{\omega}$ is the angular
velocity vector of the reference frame and f_o represents the translational
acceleration of the selected origin if applicable. Also ρ is the density of
the fluid, ν its kinematic viscosity, the pressure is p and the operator
$\frac{D}{Dt}$ is the usual total derivative.

When equation (1) is applied to the system under consideration we obtain
the following three component equations where u, v and w are the velocity components in the x, y and z directions respectively.

$$\frac{Du}{Dt} \;-\; 2\Omega v \;=\; -\;\frac{1}{\rho}\,\frac{\partial p}{\partial x} \;+\; \Omega^2(H+x) \;+\; \nu\nabla^2 u$$

$$\frac{Dv}{Dt} \;+\; 2\Omega u \;=\; -\;\frac{1}{\rho}\,\frac{\partial p}{\partial y} \;+\; \Omega^2 y \;+\; \nu\nabla^2 v \tag{2}$$

$$\frac{Dw}{Dt} \;=\; -\;\frac{1}{\rho}\,\frac{\partial p}{\partial z} \;+\; \nu\nabla^2 w$$

The non-dimensional parameters which control the flow and pressure fields
and consequently the pressure loss characteristics may be determined by re-

expressing equation set (2) using the transformations

$$
\left.
\begin{array}{l}
X = \dfrac{x}{d} \quad , \quad Y = \dfrac{y}{d} \quad , \quad Z = \dfrac{z}{d} \\[2mm]
U = \dfrac{ud}{\nu} \quad , \quad V = \dfrac{vd}{\nu} \quad , \quad W = \dfrac{wd}{\nu} \\[2mm]
-\phi = \dfrac{p}{\rho w_m{}^2}
\end{array}
\right\} \tag{3}
$$

where w_m is the mean axial velocity through the tube. When equation set (3) is inserted into equation set (2) we get

$$
\left.
\begin{array}{l}
\dfrac{DU}{Dt} - 2J.V = - Re^2 \dfrac{\partial \phi}{\partial X} + J^2(\varepsilon + X) + \nabla^2 U \\[3mm]
\dfrac{DV}{Dt} + 2J.U = - Re^2 \dfrac{\partial \phi}{\partial Y} + J^2.Y + \nabla^2 V \\[3mm]
\dfrac{DW}{Dt} = - Re^2 \dfrac{\partial \phi}{\partial Z} + \nabla^2 W
\end{array}
\right\} \tag{4}
$$

where

$$
\left.
\begin{array}{lll}
Re = \dfrac{w_m d}{\nu} & \text{(Reynolds number)} \\[3mm]
J = \dfrac{\Omega d^2}{\nu} & \text{(Rotation Reynolds number)} \\[3mm]
\varepsilon = \dfrac{H}{d} & \text{(Eccentricity parameter)}
\end{array}
\right\} \tag{5}
$$

Equation set (5) suggests that the flow field will be dependent on the usual pipe flow Reynolds number, the so-called rotational Reynolds number and the eccentricity of the tube relative to the rotational axis. Note that the rotational Reynolds number may be interpreted as a measure of the relative strength of Coriolis forces to viscous forces. Additionally in order to integrate the overall pressure loss along a specified length, L, of tube it is expected that an additional geometric parameter will be necessary involving the length/diameter ratio of the tube.

Equation (1) may also be used to illustrate another important observation for the present rotating geometry. By taking the curl of the momentum conservation equation it is possible to derive the so-called vorticity equation which represents rotations of the flow about the coordinate axes. When this is done for equation (1) it is interesting to note (see Morris (1982) for details) that the centripetal-type terms in the equation vanish identically as a source term for the generation of vorticity relative to the tube. This means that centripetal terms are made manifest in a purely hydrostatic manner similar to the earth's gravitational field. This is not so for the Coriolis acceleration component. If the vorticity vector, $\underline{\xi}$, is defined as $\underline{\xi} = $ curl $\underline{v}$, then on taking the curl of equation (1) we get after some algebraic manipulation invoking the continuity equation that

$$
\dfrac{D\underline{\xi}}{Dt} = (\underline{\xi} \cdot \nabla)\underline{v} + \nu\nabla^2\underline{\xi} + 2(\underline{\omega} \cdot \nabla)\underline{v} \tag{6}
$$

With the exception of the term involving the angular velocity vector and having its origin in the Coriolis acceleration component equation (6) is the usual vorticity equation for an inertial reference frame. Consequently provided $2(\underline{\omega} \cdot \nabla)\underline{v}$ is not zero then Coriolis acceleration can generate vorticity relative to the tube which, for example, indicates that a swirling flow situation can be generated in the axial direction. For the present rotating geometry the Coriolis-induced vorticity source may be expanded by noting that $\underline{\omega} = \Omega \underline{k}$ and $\underline{v} = u\underline{i} + v\underline{j} + w\underline{k}$ where i, j and k are unit vectors in the x, y and z directions. Thus

$$2(\underline{\omega} \cdot \nabla)\underline{v} = 2\Omega\left\{\frac{\partial u}{\partial z} \cdot \underline{i} + \frac{\partial v}{\partial z} \cdot \underline{j} + \frac{\partial w}{\partial z} \cdot \underline{k}\right\} \tag{7}$$

This is an interesting result since it implies that vorticity generation due to Coriolis forces will vanish identically provided the flow is fully developed or in other words when axial gradients of velocity vanish. In turn this suggests that, at distances well away from entrance effects, the pressure loss characteristics described via the Blasius friction factor, C_f, will tend to be unaffected by rotation. Conversely in the immediate entrance region of the duct significant Coriolis-induced increases in frictional pressure loss may be expected. It is against the foregoing physical discussion that the experimental programme described below was undertaken.

3.   EXPERIMENTAL APPARATUS

A rotor system which supported the main flow geometry to be studied was available to the authors as the result of earlier investigations into the effect of rotation on heat transfer in the cooling holes of turbo-generator rotor windings. Figure 2 shows details of the constructional features of the actual test sections which were attached to the rotor and Table 1 gives details of the geometrical parameters over which tests were conducted. Each of the four test sections studied (designated A, B, C and D in Table 1) were nominally 600.0 mm in length and five equispaced pressure tappings were located along this nominal length as indicated in figure 2. Air was used as the test fluid and further details of the rotor system have been adequately given previously by Morris and Woods (1974) and Morris and Dias (1976).

Pressure signals were transmitted from the rotor via a five channel rotary seal unit which was attached to the main rotor via a flexible drive. Rotary sealing was achieved using a system of permanent magnets and a magnetic liquid. Full details of this hydrostatic technique for transmission of rotating pressure signals to stationary measuring equipment has been documented by Morris (1981).

Pressure measurements were made with Furness micromanometers or 'U' tube mercury manometers as appropriate to the level of signal measured. Flow measurements were made with a range of Rotameter flowmeters with care taken to ensure reproducibility by checking individual tests with different flowmeter operating ranges. Similar cross checking of pressure measurements was made with individual pressure drops. The air temperature was monitored using chromel/alumel thermocouples located in the rotating inlet and exit plenum chambers with the signals transmitted via a silver/silver graphite slip ring assembly.

The full range of test geometries, flow rates and rotational speeds over which experiments were undertaken is given in Table 1 together with the range of non-dimensional parameters deemed to control the problem.

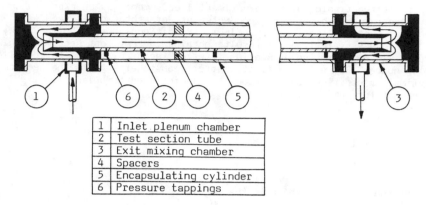

		Inlet plenum chamber
1	Inlet plenum chamber
2	Test section tube
3	Exit mixing chamber
4	Spacers
5	Encapsulating cylinder
6	Pressure tappings

FIGURE 2   SCHEMATIC REPRESENTATION OF THE TEST SECTION.

	TEST SECTION			
	A	B	C	D
Length (mm)	600.0	600.0	600.0	600.0
Bore Diameter (mm)	5.30	6.85	10.24	14.17
Eccentricity (mm)	457.2	457.2	457.2	457.2
1/4 – Span length/diameter ratio	28.3	21.9	14.6	10.6
1/2 – Span length/diameter ratio	56.6	43.8	29.3	21.2
3/4 – Span length/diameter ratio	84.9	65.7	43.9	31.8
Full-Span length/diameter ratio	113.2	87.6	58.6	42.3
Eccentricity Ratio	86.3	66.7	44.6	32.3
Rotational speed range (rev/min)	0–800	0–800	0–800	0–800
Reynolds number range	550 – 82,000	425 – 86,000	300 – 56,500	410 – 41,000
Rotational Reynolds number range	65 – 170	100 – 275	240 – 620	450 – 1200

TABLE 1   RANGE OF EXPERIMENTAL OPERATING CONDITIONS INVESTIGATED.

4.   RESULTS AND DISCUSSIONS

Correct functioning of the experimental equipment was checked via an
initial programme of experiments conducted at zero rotational speed with each
of the four test sections.   For laminar, transitional and turbulent flow,
figure 3 shows a typical set of results obtained.   For test section B this
figure shows the variation of Blasius friction factor C_{fo}, with through flow
Reynolds number, Re, over the final quarter span together with the accepted
theoretical or empirical correlations for laminar and turbulent flow.   In
this respect for laminar flow

$$C_{fo} = 64/Re \tag{8}$$

and for turbulent flow

$$C_{fo} = 0.316/Re^{\frac{1}{4}} \tag{9}$$

represent the full lines shown in the figure.

For Reynolds number values less than 2000 the zero speed data was in good
agreement with the well known theoretical result given by equation (8).   Similarly for Reynolds number values greater than 5000 good agreement with the
empirical turbulent equation (9) was evident.   Transition from laminar to
turbulent flow is generally accepted to occur in the range $2000 \leqslant Re \leqslant 5000$ and
the data shown in figure 3 tends to support this with the well-known 'dip' in
friction factor being clearly in evidence in the transitional zone.

It has been pointed out in section 2 that the influence of Coriolis-induced vorticity relative to the tube is likely to be most pronounced in the
entrance regions of tubes rotating in the so-called parallel mode.   As the
velocity field approaches a fully developed state therefore it is expected
that rotation should have no effect on the flow other than the superposition
of a hydrostatic pressure load arising from centripetal-type terms.   Consequently the friction factor - Reynolds number relationship might be expected
to become less sensitive to rotation as these conditions are approached.   This
is examined in figure 4 where, for test sections A and B, the variation of
friction factor, C_{fR}, with Reynolds number is shown for the final quarter span
of the test sections for a range of rotational speeds.   Note that test sections A and B involve the smallest two values of tube bore diameter and, since
the length of all test sections was standardised at 600.0 mm, represent the
best approach to fully developed flow available from the experiments.   In the
case of test section A a settling length of approximately 100 effective
diameters preceeded the measured friction factor with a corresponding value of
approximately 75 in the case of test section B.

For test sections A and B, data points at each of the rotational speeds
tended to collapse into a tight band which was in fairly close agreement with
equations (8) and (9) for laminar and turbulent flow respectively.   It is important and interesting to note the behavioural pattern in the customary transitional range of Reynolds numbers in figure 4.   The pronounced 'dip' in the
friction factor - Reynolds number relationship, indicative of transition,
dissappeared with a more gradual fairing of the two lines indicated by equations (8) and (9).   Indeed the data shown in figure 4 tended to follow the
turbulent-like line at a lower value of Reynolds number than that with zero
rotational speed.   This was a trend also noted with the other test sections
studied and is in contradiction to a tentative proposal made by Morris (1981),
based on limited data obtained during proving trials for the pressure transmission system used, that rotation might suppress transition.

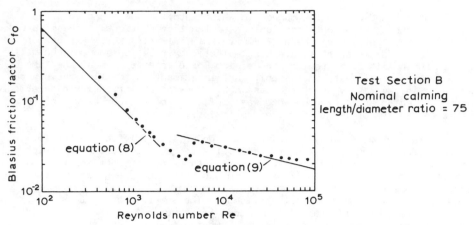

FIGURE 3   TYPICAL FULLY DEVELOPED RESULTS AT ZERO ROTATIONAL SPEED.

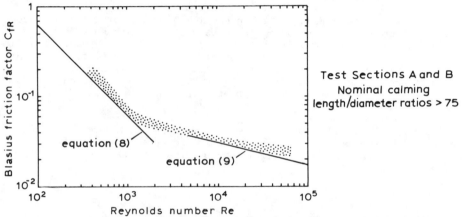

FIGURE 4   BEST APPROACH TO ROTATING FULLY DEVELOPED FRICTION FACTORS AT ALL SPEEDS TESTED.

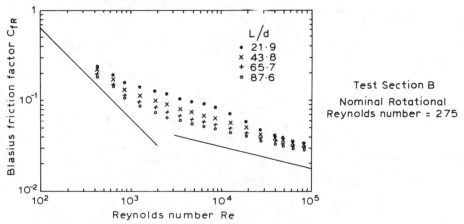

FIGURE 5   EFFECT OF LENGTH/DIAMETER RATIO ON ROTATING FRICTION FACTORS.

Figures 5 and 6 show the manner in which rotation was found to affect friction factors in the entrance regions of the test sections. These curves, obtained with a rotational speed of 800 rev/min, clearly show how tubes of shorter length/diameter ratio are most affected by rotation and consequently confirm the qualitative argument concerning Coriolis effects mentioned above. These figures demonstrate that significant impediment to the flow resistance can occur in relatively short tubes. Again conventional transition behaviour was not detected and a pronounced upward inflexion of the curves was noted at the lowest values of length/diameter ratio level.

Figure 7 shows the way in which rotation affects friction factor for the shortest length/diameter ratio studied over a range of rotational Reynolds numbers. Pronounced increases in flow resistance are evident and the upward distortion of each individual line of constant rotation speed, is extremely marked. As mentioned earlier this was a typical, if not fully explicable, trend found in the shortest test sections. The suppression of this effect in tubes having a greater total distance downstream but not dominated by fully developed conditions is shown in figure 8 where data for test section A is presented (ie a length/diameter ratio of 28.3).

The physical argument presented in section 2 demonstrated that centripetal terms do not create secondary flow and hence would not have an effect on flow resistance. This also suggests that the eccentricity of the tube, since it occurs in the centripetal terms, will not affect the flow resistance. This is confirmed in figure 9. In this figure friction factor data for test sections B, C and D are presented for the ¼-span, ½-span and full-span conditions respectively, which are at a length/diameter ratio within ±2.5% of a nominal value of 43.3.

For the two typical Reynolds number values shown in figure 9 the data appears to be relatively insensitive to eccentricity as suggested from the qualitative discussion of the controlling conservation equations.

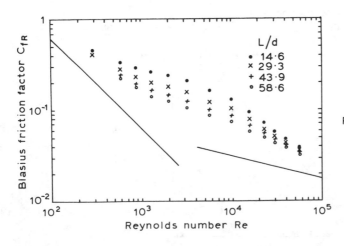

FIGURE 6   EFFECT OF LENGTH/DIAMETER RATIO ON ROTATING FRICTION FACTORS.

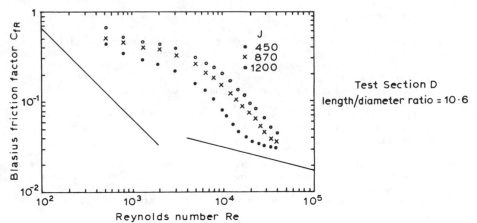

FIGURE 7   EFFECT OF ROTATIONAL REYNOLDS NUMBER ON FRICTION FACTOR IN
            RELATIVELY SHORT TUBES.

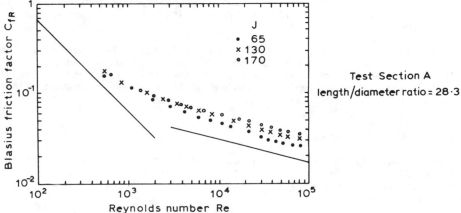

FIGURE 8   EFFECT OF ROTATIONAL REYNOLDS NUMBER ON FRICTION FACTOR FOR
            INTERMEDIATE LENGTH/DIAMETER RATIOS.

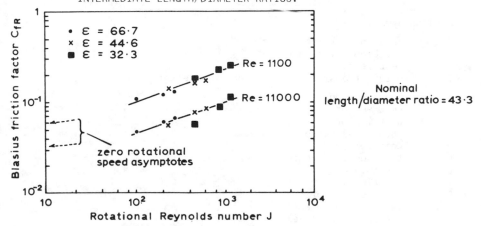

FIGURE 9   EFFECT OF ECCENTRICITY ON ROTATING FRICTION FACTORS.

5.   CONCLUDING REMARKS

This paper is intended to demonstrate that parallel-mode rotation of circular-sectioned ducts can, under certain conditions, dramatically increase flow resistance.   Although the work at this stage does not permit fully quantitative recommendations to be made (work to this end is currently in hand), the following qualitative conclusions may be drawn:

(1) For fully developed flow it is apparent that rotation does not have a significant effect on friction factor in the conventional laminar and turbulent flow regions.

(2) However the usual transitional behaviour pattern for fully developed flow does appear to be influenced by rotation in that the reduction in friction factor noted at zero rotation speed does not occur.   Rather a more gradual change from a laminar-like to turbulent-like flow occurs.

(3) For constant property flow centripetal acceleration does not, in its own right, appear to be important as a source for the creation of secondary flows and hence has no noticeable influence on friction factor.   In this respect its influence is entirely hydrostatic.   This is experimentally confirmed.

(4) Coriolis acceleration has been shown theoretically to be important in the entrance region as a source for the creation of secondary flows and this has also been confirmed experimentally in that test sections having relatively small length/diameter ratios suffer the greatest increase to flow resistance.

ACKNOWLEDGEMENTS

The authors wish to express their appreciation to the Central Electricity Generating Board for financial support for this work and also to Mr B Dibnah and Mr A Hird for their help in the construction of the apparatus and the acquisition of data.

NOMENCLATURE

English Symbols

C_{f_0}	zero speed friction factor
C_{fR}	rotating friction factor
d	test section diameter
$\underline{f_0}$	origin translational acceleration vector
H	test section eccentricity
$\underline{i}$	unit vector
$\underline{j}$	unit vector
J	rotational Reynolds number
$\underline{k}$	unit vector
L	test section length
p	pressure
$\underline{r}$	position vector
$\overline{Re}$	Reynolds number
u	velocity component
U	non-dimensional velocity component
v	velocity component

      $\underline{v}$ velocity vector
      V   non-dimensional velocity component
      w   velocity component
      w_m mean axial velocity
      W   non-dimensional velocity component
      x   coordinate
      X   non-dimensional coordinate
      y   coordinate
      Y   non-dimensional coordinate
      z   coordinate
      Z   non-dimensional coordinate

Greek Symbols

      ρ density
      ν kinematic viscosity
      ϕ non-dimensional pressure
      $\underline{\omega}$ angular velocity vector
      Ω angular velocity
      ϵ eccentricity ratio

REFERENCES

1.   Morris W D (1965a) The influence of rotation on flow in a tube rotating about a parallel axis with uniform angular velocity. Jour Roy Aero Soc 69, 201.

2.   Morris W D (1965b) Laminar convection in a heated vertical tube rotating about a parallel axis. Jour Fluid Mech 21, Part 3, 453.

3.   Morris W D (1969) An experimental investigation of laminar heat transfer in a uniformly heated tube rotating about a parallel axis. Min Tech ARC CP No 1055.

4.   Morris W D (1970) Some observations on the heat transfer characteristics of a rotating mixed convection thermosyphon. Min Tech ARC CP No 1115.

5.   Morris W D (1977) Flow and heat transfer in rotating coolant channels. High Temperature Problems in Gas Turbine Engines, Agard CPP 229, Ankara.

6.   Morris W D (1981) A pressure transmission system for flow resistance measurements in a rotating tube. J Phys Sci Instrum 14, 208.

7.   Morris W D (1982) Heat Transfer and Fluid Flow in Rotating Coolant Channels. John Wiley & Sons Ltd., Chichester, England. (Research Studies Press).

8.   Morris W D and Ayhan T (1979) Observations on the influence of rotation on heat transfer in the coolant channels of gas turbine rotor blades. Proc Inst Mech Eng 193, No 21, 303.

9.   Morris W D and Dias F M (1976) Experimental observations on the thermal performance of a rotating coolant circuit with reference to the design of electrical machine rotors. Proc Inst Mech Eng 190, 46/76, 561.

10.   Morris W D and Dias F M (1980) Turbulent heat transfer in a revolving
       square-sectioned tube.   Jour Mech Eng Sci. 22 (2), 95.

11.   Morris W D and Dias F M (1981) Laminar heat transfer in square-sectioned
       ducts which rotate in the parallel-mode. Power Ind Res. 1.

12.   Morris W D and Woods J L (1978) Heat transfer in the entrance region of
       tubes that rotate about a parallel axis. Jour Mech Eng Sci. 20 (6), 319.

13.   Woods J L and Morris W D (1974) An investigation of laminar flow in the
       rotor windings of directly-cooled electrical machines. Jour Mech Eng
       Sci. 16,   408.

14.   Woods J L and Morris W D (1980) A study of heat transfer in a rotating
       cylindrical tube.   Trans ASME Jour Heat Trans 102, 4, 612.

Turbulent Flow and Heat Transfer in Rotating Channels and Tubes

V. Y. MITYAKOV, R. R. PETROPAVLOVSKY, V. V. RIS,
E. M. SMIRNOV, and S. A. SMIRNOV
Leningrad Polytechnical Institute
Leningrad, USSR

ABSTRACT

This document is a reduction of the author's experimental results on turbulent flow characteristics and heat transfer in rotating channels whose axes are parallel to the plane of rotation. Substantial dissimilarities of longitudinal velocity field profile and pulsational characteristics are caused by effects of stabilization and destabilization and secondary flow production. Local heat transfer coefficients vary over the perimeter of the tube section connecting detected flow peculiarities. It is shown that the increase in rotational intensity caused an increase in the relative dissimilarity of the local heat transfer coefficients and increased their mean value.

1. INTRODUCTION

In isothermic and slightly non-isothermic rotating streams the influence of massive forces upon the velocity field is known to be conditioned by the action of the non-conservative Coriolis force alone [1].

This paper deals with the analysis of experimental data, known or obtained by the authors, on turbulent flow and heat transfer characteristics in channels and tubes with the axis parallel to the plane of rotation.

The flow in a rectangular channel of high aspect ratio, with long sides, parallel to the rotation axis, is mainly dependent on the effect of destabilization and stabilization in the regions with different signs of the velocity derivative in the direction of the Coriolis force [2].

The development of the secondary flow is a defining factor in channels where the long sides are parallel to the axis of rotation [3].

The interaction of the aforementioned effects results in a specific interest in researching turbulent flows in channels with an approximately square cross-section.

Experimental data on the distribution of mean longitudinal

velocity were included in Ref. [4-6].

2.   TURBULENT FLOW

Experimentally, by means of a hot-wire anemometer, the authors measured the characteristics of an almost developed air stream in a channel with a cross-section of 20 x 20 mm at a distance of 35 hydraulic diameters from a honeycomb set up at the inlet. The measuring system consisted of a wire probe with tungsten sensor of 5 μm diameter and approximately 1 mm length perpendicular to the probe axis, an unlinearized rotating constant temperature hot-wire anemometer and an automatic series of registering devices with a microprocessor. The Reynold's number

$$Re = U_m h/\nu$$

changed from 3,000 to 11,000 and the Rossby number

$$Ro = U_m/\omega h$$

from 5 to 40; U_m = the mass mean velocity of the stream, ω = the angular velocity of rotation, and h = the size of cross-section of the channel. The total relative error in the measurements of the mean longitudinal velocity component, U, did not exceed ± 2% and the error in turbulence intensity $\frac{\sqrt{(\overline{u^2})}}{U}$ was estimated at ± 6%.

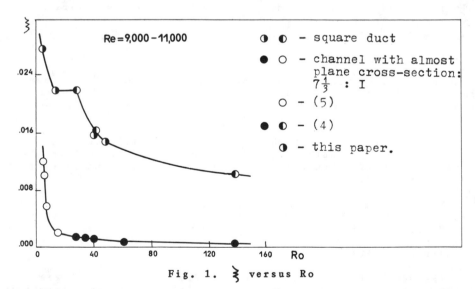

Fig. 1. $\gtrless$ versus Ro

where $\gtrless = \int_{-1}^{1} \left| 1 - \frac{U}{U(\omega=0)} \right| d\overline{x}$ = velocity profile deformation parameter along the line $\overline{x}, \overline{y} = 0$ (see Fig. 2).

The results obtained enable us to draw the following conclusions about certain peculiarities of turbulent flow in a rotating channel.

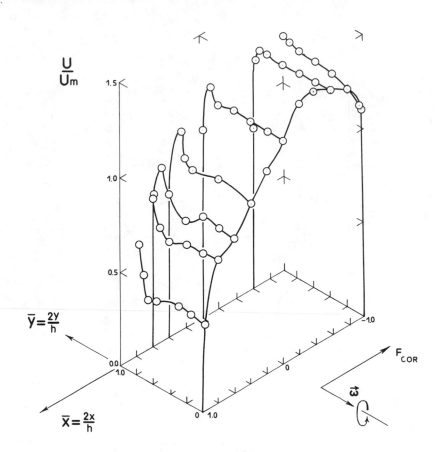

$\dfrac{U}{U_m}$

$\overline{y} = \dfrac{2y}{h}$

$\overline{x} = \dfrac{2x}{h}$

F_{COR}

$\vec{\omega}$

Fig. 2. Nondimensional longitudinal velocity profiles in
half the cross-section of a rotating channel
at Re = 9,000 and Ro = 5

The maximum in the longitudinal velocity distribution moves to
the side with the higher pressure. Up to Ro > 30, field deformation is defined by the effects of secondary flow, and beginning
from Ro ≃ 12, by the joint effects of secondary flow and stabilizing action. (Fig. 1). With Ro decreasing, velocity equalization
may be observed in the central region along lines parallel to the
rotational axis. (Fig. 2). Owing to the symmetry of distribution
of flow characteristics relating to the plane of rotation, the
given measurements are taken from one half of the cross-section
channel. In the region by the walls perpendicular to the
rotational axis, a typical velocity distribution for a non-linear
Ekman layer occurs. On the side of low pressure, the effects of
inertia in these layers results in the highlighting of a well-
defined maximum of longitudinal velocity and the fluctuation of its
value perpendicular to the wall.

Measurements carried out to estimate secondary velocity
components have proved under all conditions that the angle between

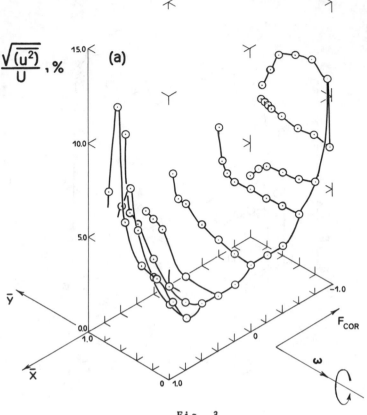

$$\frac{\sqrt{(\overline{u^2})}}{U}, \%$$

Fig. 3.

(a)   Distribution of turbulent intensity in a half of rota-
      ting channels cross section at Re = 9,000 and Ro = 5.

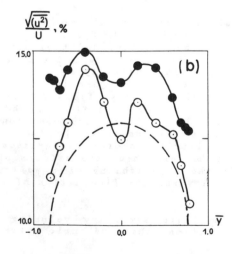

$$\frac{\sqrt{(\overline{u^2})}}{U}, \%$$

(b)   Distribution of $\sqrt{(\overline{u^2})}/U$
      near the pressure side
      caused by Taylor Görtler-
      like cells at the same
      values of Re and Ro:

○ – $\overline{x} = -0.97$,

● – $\overline{x} = -0.95$,

      (dotted line – for non-
      rotating channel)

the velocity vector and the longitudinal axis of the channel in the centre of the stream does not exceed 7°.

With an increase of rotation, the turbulent intensity in the centre of the stream is greatly lessened in the half of the cross-section adjoining the side of low pressure (Fig. 3a). In the region of destabilization, the changes in values of $\sqrt{(\overline{u^2})}$ are not so radical compared with a stationary channel. This shows the redistribution of turbulent energy between longitudinal and normal components, resulting in an increase in the intensity of fluctuations in the direction of the main x-component of the Coriolis force. Moreover, the character of the value changes of $\sqrt{(\overline{u^2})}$ on the pressure side illustrates the formation of two pairs of large-scale Taylor Görtler-like cells at Ro $\simeq$ 7 (Fig. 3b). A similar phenomenon in the close range of values Re and Ro has been discovered by a flow visualization technique in channels of high aspect ratio [2].

A tendency for flow characteristics to equalize along lines parallel to the axis of rotation also becomes apparent in the distribution of the value $\sqrt{(\overline{u^2})}$ (Fig. 3a). Defining the conditions in regions where the influence of the effects of inertia in the Ekman layer is insignificant, the turbulent intensity even decreases in proportion to the nearness of the wall, which proves the absence of turbulent production in the wall layer. Furthermore, near the low pressure side, the local maximum in the distribution of $\sqrt{(\overline{u^2})}$ can be observed, caused by the intense generation of turbulent motion in the shearing layer at the outer edge of the Ekman layer.

3. HEAT TRANSFER

It goes without saying that the reconstruction of the flow under the influence of rotation is apparent in a general change of the intensity of heat exchange. The non-uniformity of the characteristics of turbulent flow along the perimeter of a cross section of a rotating channel is apparent in the irregular distribution of local heat transfer coefficients.

The quantitative definition of these effects was investigated by the authors using a model of a circular tube with an inner diameter of d = 9.7 mm and a length of 25 diameters. For the measurement of local heat transfer, the authors added a wall with a practically constant heat flow on the tube surface. Note that the value of the Grashof number created by the centrifugal and Coriolis acceleration did not exceed 4 x 10^{-2} x Re^2. The temperature was measured by thermocouples at eleven points along the perimeters of corresponding cross section during the rotation of tube on its longitudinal axis.

In all experimental conditions, $5,500 \leq Re = \dfrac{U_m \, d}{\nu} \leq 50,000$, $50 \leq Re_\omega = \dfrac{\omega d^2}{\nu} \leq 400$, visible non-uniformity of local Nusselt coefficients $Nu = \dfrac{\alpha d}{\lambda}$ has been noted along the perimeter in contradiction to Ref. [8]. (Fig. 4a,b).

In the region of high pressure with the superimposition of rotation heat transfer is intensified. In the opposite region (for

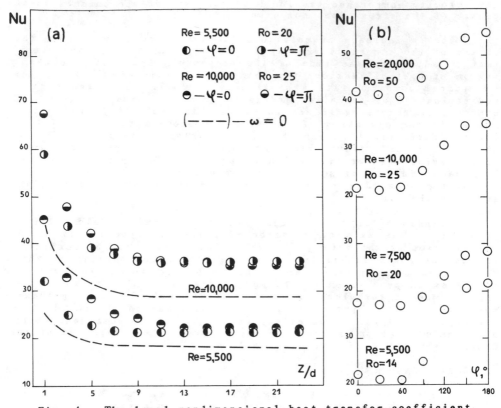

Fig. 4. The local nondimensional heat transfer coefficient
         distribution in rotating tube:
         (a) along the longitudinal lines φ = 0 and π ;
         (b) along the perimeter of cross section at z/d=20.

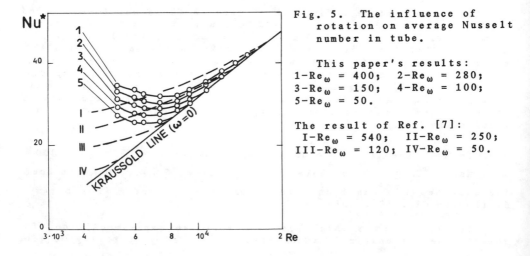

Fig. 5. The influence of
         rotation on average Nusselt
         number in tube.

         This paper's results:
$1-Re_{\omega}$ = 400; $2-Re_{\omega}$ = 280;
$3-Re_{\omega}$ = 150; $4-Re_{\omega}$ = 100;
$5-Re_{\omega}$ = 50.

The result of Ref. [7]:
 $I-Re_{\omega}$ = 540; $II-Re_{\omega}$ = 250;
$III-Re_{\omega}$ = 120; $IV-Re_{\omega}$ = 50.

Re $>$ 7,000) it is decreased in comparison with a stationary tube. With decreasing Ro = Re/Re$_\omega$ the relative irregularity of heat transfer coefficients increases. In the section x/d $>$ 12, the ratio of measured heat transfer coefficients for diametrically opposite points in the plane of rotation is approximated by the formula

$$\frac{Nu\ (\varphi = \pi)}{Nu\ (\varphi = 0)} = 0.88 + 3.6\ Ro^{-1/2}, \quad Re \geq 10^4, \quad 25 \leq Ro \leq 250$$

For Ro $<$ 100 in the region of low pressure, some increase in the heat transfer ($\varphi \to 0$) can be noticed, (Fig. 4b), which may be connected with the local maximum of turbulent intensity in the area of low pressure, mentioned earlier, and also with the twisting of flow at the merging of secondary streams near the wall. These data were obtained on a model with a sharp edge inlet. It is interesting to notice that on a model where a section of ten diameter length begins with a smooth inlet nozzle with a ratio of 4:1 and is positioned above the warmed area, results practically coincide.

In certain aspects, the dependence of the mean Nusselt number Nu* for the whole surface of the tube (Fig. 5) conforms to the results obtained in Ref. [7] (shown as a dotted line). The continuous line, approximating the author's data for a stationary tube, coincides with the Kraussold formula.

REFERENCES

1. Greenspan, H.P. 1968. The theory of rotating fluids. Cambridge Univ. Press.

2. Johnston, J.P., Halleen, R.M. and Lezius, D.K. 1972. Effects of spanwise rotation on the structure of two-dimensional fully developed turbulent channel flow. Journ. Fluid Mech., Vol. 56, pp. 533-557.

3. Wagner, R.E. and Velkoff, H.R. 1974. Measurements of secondary flows in rotating duct. Trans. ASME, Ser. A, No. 4, pp. 31-39.

4. Moore, J. 1967. Effects of Coriolis on turbulent flow in rotating rectangular channels. Massachusetts, Cambridge, MIT, Gas turbine laboratory, rep. 89.

5. Halleen, R.M. and Johnston, J.P. 1967. The influence of rotation on flow in a long rectangular channel - an experimental study. Stanford University, Department of Mech. Eng., rep. MD-18.

6. Dobner, E. 1959. Über den Strömungswiderstand in einem rotierenden Kanal. Darmstadt, Technische Hochschule, Diss.

7. Zysina-Molojen, L.M., Dergatch, A.A. and Kogan, G.A. 1976. Experimental study on heat transfer in a radial rotating pipe. Acad. Sci. U.S.S.R., TVT, Vol. 14, pp. 1108-1111.

8. Mori, J., Fukada, T. and Nakayama, W. 1971. Convective heat transfer in a rotating radial circular pipe. Int. Journ. Heat and Mass Transfer, Vol. 14, pp. 1807-1824.

The Effect of Mass Forces on Heat Transfer in Turbine Rotor Blades and on Hydraulic Resistance in Cooling Channels

V. P. POCHUEV, M. E. TSARLIN, and V. F. SHERBAKOV
Central Aeroengine Institute
Moscow, USSR

ABSTRACT

The results of an investigation of local heat transfer between gas and turbine blades are given in the paper. The method of unsteady heat regime of thin wall is used to determine local heat flows. The heat transfer coefficients obtained both for rotating and fixed blades are compared. The effect of non-isothermal flow in rotating rotor systems on heat transfer and hydraulic resistance is analysed. The qualitative coincidence of heat transfer coefficients and hydraulic resistance in rotating radial and fixed vertical ducts is acknowledged.

1. LOCAL BLADE SURFACE HEAT TRANSFER

There are many well known studies of local heat transfer at the surface of turbine blades. But only few of them deal with heat transfer on rotating blades and they give contradictory results. Some of those investigations show 1.2-1.4 (and even 1.8) times increase of the mean heat transfer coefficient α_r on the rotating blade surface [1]. It is noted in [2] that heat transfer rates on air cooled blades in static and rotating conditions are the same.

The research of local heat transfer on blade surfaces at close to real conditions is of great practical interest. Covered in this paper are experimental investigations on local heat transfer coefficients between gas and rotating blade external surface that were conducted on a singe-stage experimental turbine. The mean turbine diameter is 0.618 m. The circumferential velocity is 220-275 mps. Turbine entry gas temperature, T_r^*, is in the range 840-1200K. The local heat fluxes on the blade surfaces were determined by the unsteady thin wall method [3].

The investigations were conducted on two rotor-mounted uncooled thin wall cast blades with a wall thickness of 0.7-2.1 mm. The blade chord b = 33.33 mm, blade pitch t = 24.2 mm, inlet incidence angle β_1 = 50° (the angle between tangent to the centre line of an airfoil at inlet and cascade front), outlet angle β_2 = 28° (the angle between tangent to the centre line of an airfoil at exit and cascade front), leading edge diameter 4.6 mm, trailing edge diameter 1.6 mm, blade height at exit section 62 mm, and airfoil aspect ratio 0.21.

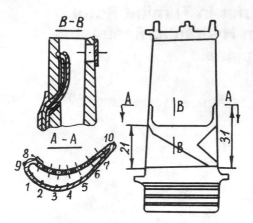

Fig. 1.   Rotor blade preparation

The experimental blades were instrumented with 19 chromel-alumel thermocouples with 0.2 mm diameter wires. The scheme of the thermocouple installation is shown in Fig. 1.   The thermocouples were installed on the inner sides of the concave and convex surfaces of the blades.   Thermocouples on leading and trailing edges were installed on the external sides.   The temperature measurements in relative movement T_w^* were carried out by two leading edge mounted thermocouples.

The averaged heat transfer coefficients for parts of the airfoil equal to the distance between two thermocouples (~ 5-6 mm) were calculated from a heat balance equation for the blade wall area using the results of wall and gas temperatures measurements in unsteady conditions.

The unsteady regime of blade heating was produced by suddenly raising the gas temperature. T_w^* rose by 270-280 K while rotor speed increased from 110 to 140 rps.   The gas flow incidence angle at this condition is i ≈ +15°.

Analysis shows that the determination error of thin-wall blade local heat transfer coefficients for Bi = $\alpha_r \cdot \delta_{cm}/\lambda_{cm}$ < 0.1 (where δ_{cm}-wall thickness, λ_{cm}-wall heat conductivity) does not exceed 10-15%.

Fig. 2 shows the distribution of local heat transfer coefficients along the airfoil at gas flow Reynolds number Re ≈ 2×10^5, calculated by row exit flow velocity, blade chord and gas viscosity at temperature T_w^*.   Obtained values of local heat transfer coefficients α_r were compared with those obtained in static conditions. The static conditions experiment was carried out on the row of five blades with flow inlet incidence angles i = 0, i = +15°, i = -15°. The row inlet gas temperature was ~ 990 K, gas pressure ~ 2×10^5 Pa, Mach number M_2 ≈ 0.7.

Fig. 2 shows the variation of $Nu_b = \alpha_{rib}/\lambda_r$ along the airfoil in static conditions for shockless flow (i = 0) and incidence angle i = +15°.

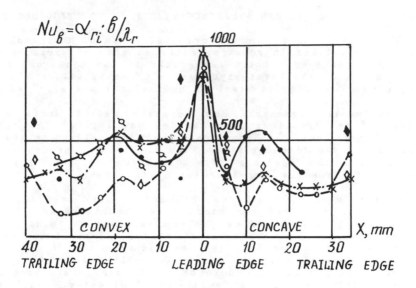

Fig. 2.   The comparison of local values of Nu number
for rotating and static cascades

- • ——— $Re_2 = 1.96 \times 10^5$
- ◐ ——— $Re_2 = 2.0 \times 10^5$ } for rotating
- ◆ ——— $Re_2 = 2.0 \times 10^5$ the results of [4]
- ○ - - - $Re_2 = 1.86 \times 10^5$, $i = 0°$
- ◇ ——— $Re_2 = 1.83 \times 10^5$, $i = +15°$ } static cascade
- ✕ —·—· $Re_2 = 2.0 \times 10^5$, $i = 0°$ the results of [4]

As a rule the value of heat transfer on the blade in rotary conditions (at $i = +15°$) is increased compared with that for static conditions. The local values of heat transfer coefficient on the concave blade surface in rotary conditions are 1.2–1.8 times higher than those for static conditions. This may be explained, it seems, by early flow transition to a turbulent state. On the convex blade surface the heat transfer coefficient, α_r, increase due to rotation is negligible. The flow over the blade with positive incidence angle results in increase of α_r on this surface.

Fig. 2 also contains the experimental data of [4] on local heat exchange in rotary conditions obtained by regular regime technique on the blade row having a configuration close to the investigated one (in the mean blade section $b = 37$ mm, $t = 23.5$ mm, $\beta_1 = 55°$, $\beta_2 = 33°$). The tests were performed for the same values of Re_2. The rotation speed changes from 100 to 150 rps. In the present investigation the observed influence of rotation on heat transfer was stronger at concave blade surface and somewhat weaker at the convex side in the leading edge region. The results show that the distribution of local heat transfer coefficients on the turbine blade surface is strongly dependent on mass forces.

2. HEAT TRANSFER AND HYDRAULIC RESISTANCE IN ROTATING DUCTS

The investigation of mass force influence on heat exchange and hydraulic resistance in rotating ducts is also of great interest. The problem has been treated repeatedly in the technical literature. In one of the first experimental investigations [5] (of an S-shaped duct) a large increase of hydraulic resistance was found.

Experimental researches on rotating radial ducts with different flow directions were carried out [6,7]. The authors observed large heat exchange rate and hydraulic resistance increases for the flow direction from the axis to periphery, while the data for the opposite flow direction do not differ significantly from the corresponding static ones.

A large number of other experimental works are also known [8-10], in which either the increase of heat exchange rate on rotating ducts is observed or such an increase is found small or even decreasing. The authors of the above papers while preparing their experiments proceeded from the assumption that radial duct rotation influence is mainly due to Coriolis acceleration leading to vortex pair formation. Therefore such experiments were usually performed with the ducts of small inner diameter, small wall heating rate and at high flow velocities. This physical model does not fully correspond to the real phenomenon, which is why with the same values of kinematic and hydraulic similarity criteria different results were obtained.

The flow in rotating radial ducts exists in the presence of a force field generated by both Coriolis and centripetal accelerations and is accompanied by vortex system formation.

The case is further complicated if the flow is nonisothermal and this is specific to cooling ducts. Under such conditions if the coolant moves in the direction of centrifugal forces the heat transfer and hydraulic resistance are influenced by the fluid density difference across the duct. This influence results in convective streams counteracting the main flow. Similar counteraction also exists in fixed vertical pipes when the thermal (free) convection is directed against the forced flow.

It was shown in [11] for multidirectional convection conditions that may arise in vertical pipes (depending on relative Reynolds and Grashof numbers) that it is possible to obtain both decreases and increases of heat exchange in comparison with that for forced convection only.

In modern turbines centripetal acceleration of $8 \times 10^4 g$ may be reached on the mean diameter. In such conditions a more suitable physical model will be obtained with the assumption that the flow non-isothermality, resulting in convection current generation, has a great influence on the heat exchange and the hydraulic resistance. The generation of these currents explains all the known facts of the influence of rotation on heat exchange and hydraulic resistance both when the flow direction is coincident with the direction of centrifugal forces and when the flow direction is opposite to centrifugal forces.

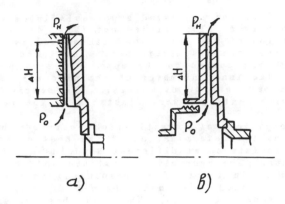

$$a) \qquad \qquad \textit{6)}$$

Fig. 3. Experimental configurations for heat transfer
investigations in (a) closed and (b) open
radial ducts

To confirm this assumption experimental research on three
model facilities was carried out over a wide range of Reynolds and
Grashof numbers, which were used to control the similarity of
interaction of secondary convective currents with the main flow.
Two of the facilities are cantilever rotors with duralumin disks
having heating elements with total power of 3 kW (Fig. 3). In one
configuration there were 48 radial 6 mm ducts drilled inside the
disk, and in the other configuration there were 24 radial grooves
with rectangular cross-section cut on the side of the disk,
rotating close to the fixed restricting wall. In both configurations the disk diameter was equal to 400 mm and the air supply
duct radius was 75 mm. The rotation speed was varied from 2000 to
6000 rpm, and the rate of duct heating was also varied.

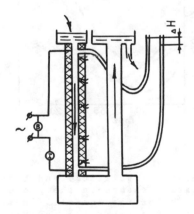

Fig. 4. Experimental configuration for investigation
of mass force field influence on heat transfer
and hydraulic resistance in a vertical pipe

To confirm the suggested hypothesis investigations in a fixed vertical pipe were also carried out (Fig. 4), where the similarity on interaction of the main flow with convective currents caused by thermal convection was maintained. The tests were conducted with water which was pumped to the upper end of the pipe and directed downward. The inner diameter of the pipe was 39 mm, the length of the heated part was 40 diameters. The heating was done by AC current, with heating power adjustable in the range of 0 – 30 kW.

The experimental investigations carried out on this latter facility enabled determination of the heat exchange and change in hydraulic resistance at different Reynolds and Grashof number combinations. The hydraulic resistance and heat transfer relative changes given in terms of corresponding values for nonrotating ducts or for fixed duct in isothermal flow at the same Reynolds number are shown in Fig. 5. The data obtained show that one of the factors influencing heat transfer and hydraulic resistance is flow non-isothermality, which in the presence of mass force fields may lead to significant heat transfer intensification and hydraulic resistance increase.

The same effect of mass forces may take place in cooling systems of axial and centripetal turbines. In the latter for this reason the penalty in aerodynamic performance may be caused by aerodynamic loss increases from flowing gas about the cold impeller.

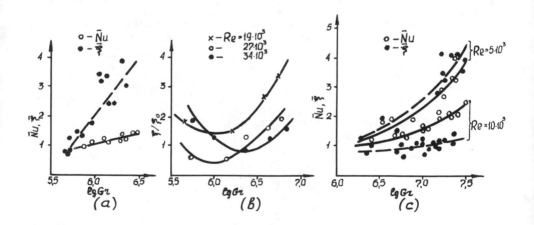

Fig. 5.   Hydraulic resistance and heat transfer – the relative change in the presence of mass force fields for the installations shown in Fig. 3a – (a), Fig. 3b – (b) and Fig. 4 – (c)

REFERENCES

1. Ivanov, V.M., Manushin, E.A. and Lapin, Yu.D., ''Some Results of the Experimental Investigation of a Cooled Turbine,'' Izv. V.U.Z., sec. Aviation Enginering, 1966, No. 2, pp. 143–150.

2. Kopelev, S.Z., Gurov, S.V., and Avilova-Shulgina, M.V., ''Heat Transfer in Cooling Exposed Flow Parts of a Turbine,'' Izv. A.N. USSR, sec. Energetics and Transport, 1971, No. 4, pp. 105–111.

3. Pochuev, V.P., ''Determination of Local Heat Transfer Coefficients on the Surface of Turbine Blades,'' Energy Equipment, 1975, No. 15, pp. 4–8.

4. Trushin, A.V. and Zakirov, M.V., ''Concerning Heat Transfer on Rotor Blades,'' Izv. V.U.Z., sec. Aviation Engineering, 1973, No. 2, pp. 151–154.

5. Seeling, W., ''On the Phenomenon of Rotational Turbulence in Rotating Tubes and Channels,'' Proc. Int. Congress for Technical Mechanics, Vol. 1, 1930.

6. Perelman, R.G. and Polinovsky, V.I., ''Hydraulic Resistance of Rectilinear Channels in a Centrifugal Force Field,'' Izv. A.N. USSR, sec. Techn. Sci., 1958, No. 10, pp. 150–153.

7. Kovalevskaya, A.E. and Perelman, R.G., ''Heat Transfer Under the Influence of Centrifugal Forces,'' Izv. A.N. USSR, sec. Techn. Sci., 1958, No. 11, pp. 92–94.

8. Zysina-Molojen, L.M., Dergatch, A.A. and Kogan, G.A., ''Experimental Study on Heat Transfer in a Radial Rotating Pipe,'' Acad. Sci., USSR, TVT, 1976, Vol. 14, pp. 1108–1111.

9. Mori, Y., Fukada, T. and Nakayama, W., ''Convective Heat Transfer in a Rotating Radial Circular Pipe,'' Int. J. Heat and Mass Transfer, 1971, Vol. 14, pp. 1807–1824.

10. Shukin, V.K., ''Heat Transfer and Hydraulics of Internal Flows in the Field of Mass Forces,'' in ''Mechanical Engineering,'' Moscow, 1970, p. 330.

11. Brown, W.G., ''Superposition of Forced and Natural Convection in a Vertical Tube at Low Flow,'' VDI-Forschungsheft 480, B, Vol. 26, 1960, p. 31.

ROTATING SURFACES AND ENCLOSURES

Fluid Flow and Heat Transfer in Rotating Disc Systems

J. M. OWEN
School of Engineering and Applied Sciences
University of Sussex
Falmer, Brighton, Sussex BN1 9QT, England

ABSTRACT

The paper provides a review of fluid flow and heat transfer in the rotating disc systems that are relevant to designers of turbomachinery. Starting with the free disc, the review includes rotor-stator systems (which are used to simulate turbine discs rotating near stationary casings) and rotating cavities (which are used to simulate corotating turbine discs). Although there are many papers devoted to these systems, and some design information does exist, there are still a number of important areas that need further theoretical and experimental research. These areas include the study of mainstream gas ingress (in a rotor-stator system) and turbulent buoyancy-driven flows (inside rotating cavities).

NOMENCLATURE

a	inner radius of disc
b	outer radius of disc
C	constant
C_m	moment coefficient ($\equiv M/\frac{1}{2}\rho\Omega^2 b^5$)
C_p	specific heat at constant pressure
C_w	flow coefficient ($\equiv Q/\nu b$)
G	gap ratio ($\equiv s/b$)
G_c	shroud-clearance ratio ($\equiv s_c/b$)
Gr_θ	rotational Grashof number ($\equiv \beta\Delta T\,Re_\theta^2$)
Gr_Ω	rotational Grashof number ($\equiv \beta\Delta T\,Re_\Omega^2$)
k	thermal conductivity
M	frictional moment on one side of the disc
n	power-law index
Nu	local Nusselt number ($\equiv qr/k\Delta T$)
Nu_Ω	local Nusselt number ($\equiv qs/k\Delta T$)
Nu_{av}	average Nusselt number ($\equiv q_{av}b/k\Delta T_{av}$)
Pr	Prandtl number
q	heat flux from disc surface to coolant
Q	volumetric flow rate
r	radial location
Re_r	radial Reynolds number ($\equiv C_w/2\pi(r/b)$)
Re_z	axial Reynolds number ($\equiv 2\overline{W}a/\nu$)
Re_θ	rotational Reynolds number ($\equiv \Omega b^2/\nu$)
Re_Ω	rotational Reynolds number ($\equiv \Omega s^2/\nu$)
s	axial gap between rotor and stator and between corotating discs
s_c	clearance between shroud and rotor
T	temperature

	tangential component of velocity relative to a stationary frame
	bulk-average axial velocity in inlet tube
	volume expansion coefficient
ΔT	boundary-layer thickness
ε_z	appropriate temperature difference
ν	axial Rossby number ($\equiv \bar{W}/\Omega a$)
ξ	kinematic viscosity
ρ	dimensionless parameter in source-sink flows
σ	density
Ω	dimensionless buoyancy parameter ($\equiv Pr\beta\Delta T\ Re_{\Omega}^{\frac{1}{2}}$)
	angular speed of the rotating disc

Subscripts

av	radially-weighted average value
c	value at the cold-disc surface
ent	entrained value
h	value at the hot-disc surface
i	value at the edge of the inner layer
I	value at inlet to the system
ℓ	laminar
min	minimum value
max	maximum value
ref	appropriate reference value
s	value at the disc surface
t	turbulent
∞	'free-stream' value

1. INTRODUCTION

A rotating disc can often serve as the model for the flow and heat transfer that occurs inside turbomachinery. As shown in Fig. 1, the gas turbine provides many examples of this: an air-cooled turbine disc rotating near a casing can be modelled by a simple rotor-stator system; two corotating discs can be modelled by a rotating cylindrical cavity. In fact, many rotating flows of practical importance can be considered in terms of the rotor-stator system or of the rotating cavity, both of which have been the subject of numerous experimental and theoretical studies.

The designer who is involved with stressing a turbine disc, and with estimating its fatigue life, needs to know the temperature distribution under steady-state and transient conditions. In order to calculate these temperatures, he requires a knowledge of the flow structure near the disc and of the heat transfer coefficients between the surrounding air and the disc surface. He wants to optimize the coolant flow rate to an air-cooled turbine disc: too much coolant is inefficient and expensive; too little can lead to catastrophic failure. The designer has also to calculate how the coolant flow affects the windage power dissipated by the disc, and he needs to predict the pressure distribution between the disc and its casing or between two corotating discs.

In the real engine, the geometry is usually complicated and many of the design parameters are not well specified. Measurements on the engine itself are always difficult and expensive, and the data obtained are usually limited in both accuracy and universality. Rotating disc systems can be more readily modelled, both experimentally and theoretically, and they can be employed to give insight into the nature of the flow and heat transfer processes that occur inside an engine. Data obtained from the study of these systems can then be used to provide answers to some of the questions posed by the designer.

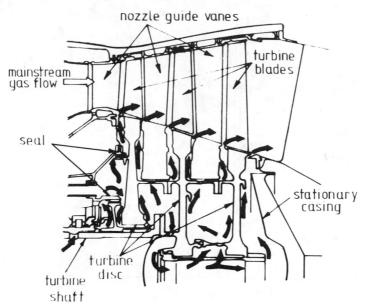

nozzle guide vanes

mainstream
gas flow

turbine
blades

seal

stationary
casing

turbine
disc

turbine
shaft

Fig 1 Air-cooled gas-turbine discs

➜ cooling air flow

The object of this paper is to provide a review of the rotating disc systems that are relevant to the above-stated problems. There are two excellent books devoted to rotating systems: Dorfman (1)[*] is concerned mainly with rotor-stator systems; Greenspan (2) includes a large amount of material on rotating cavities. Each year, a number of papers on the subject of rotating disc systems is published; very few, however, throw much light on the problems facing the designer of turbomachinery. The topics selected below and the papers cited are the ones that, in this writer's opinion, provide either design data or an improved understanding of the nature of a particular flow.

Section 2 is devoted to the 'free disc': this is the genesis of all rotor-stator systems and also has relevance to rotating cavities. Enclosed, open and shrouded rotor-stator systems are discussed in Section 3; and Section 4 is concerned with sealed rotating cavities and with cavities that have superimposed flows.

2. THE FREE DISC

Fig. 2 shows a schematic representation of a free disc (that is, a disc rotating in an infinite quiescent environment). This simple model serves as the first 'building block' with which to begin the synthesis of more complex rotating disc systems. It has been the subject of experimental and theoretical studies too numerous to discuss here, and the salient features are described below.

---

[*] Reference numbers will be used the first time that an author appears in each Section or Sub-Section.

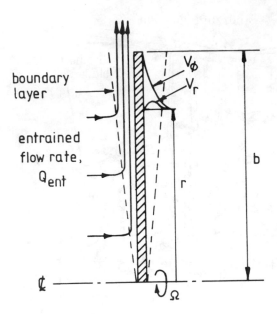

Fig. 2   Schematic diagram of boundary - layer flow on the free disc

2.1   Isothermal Flow

Flow is entrained into a boundary layer on the rotating disc and the tangential component of velocity, V_ϕ, (relative to a stationary frame of reference)is sheared from a value of Ωr on the disc surface to zero outside the layer. Transition from laminar to turbulent flow occurs at $(r/b)^2 Re_\theta \simeq 3 \times 10^5$ (where $Re_\theta \equiv \Omega b^2 / \nu$).

For laminar flow, the boundary layer thickness, δ, is independent of radial location, and

$$\delta \simeq 4(\nu /\Omega)^{\frac{1}{3}} \tag{2.1}$$

where δ is the axial location (according to Cochran's (3) exact solution) where $V_\phi / \Omega r = 0.026$. The moment coefficient, C_m (where $C_m \equiv M/\frac{1}{2}\rho\Omega^2 b^5$, and M is the frictional moment generated on <u>one</u> side of the disc) is

$$C_m = 1.935 \ Re_\theta^{-\frac{1}{2}} \tag{2.2}$$

The dimensionless flow rate, $C_{w,ent}$, (where $C_{w,ent} \equiv Q_{ent}/\nu b$, and Q_{ent} is the volumetric flow rate entrained, or 'pumped', by one side of the disc) is

$$C_{w,ent} = 2.78 \ Re_\theta^{\frac{1}{2}} \tag{2.3}$$

For turbulent flow (according to Karman's (4) solutions of the integral momentum equations),

$$\delta/r = 0.525((r/b)^2 Re_\theta)^{-0.2} \tag{2.4}$$

$$C_m = 0.073 \, Re_\theta^{-0.2} \tag{2.5}$$

$$C_{w,ent} = 0.219 \, Re_\theta^{0.8} \tag{2.6}$$

A more accurate expression for C_m, which is in good agreement with measured values up to $Re_\theta \simeq 7 \times 10^6$, is that given by Dorfman (1) where

$$C_m = 0.491 \, (log_{10} \, Re_\theta)^{-2.58} \tag{2.7}$$

2.2 Heat Transfer

The local Nusselt number ($Nu \equiv qr/k(T_s - T_{ref})$, where the subscripts s and ref refer to the disc surface and a suitable reference value, respectively) and the average value ($Nu_{av} \equiv q_{av}b/k(T_s - T_{ref})_{av}$, where the subscript av refers to the radially-weighted average value) can be calculated from the Reynolds analogy. Dorfman (1) has shown that

$$Nu^*_{av} = Re_\theta C_m/\pi \tag{2.8}$$

where * signifies that $Pr = 1$ and $(T_s - T_{ref}) \propto r^2$. For arbitrary Prandtl numbers and power-law temperature profiles, where $(T_s - T_{ref}) \propto r^n$, the Nusselt number can be calculated from

$$Nu_{av} = f(Pr)g(n)Nu^*_{av} \tag{2.9}$$

$f(Pr)$ and $g(n)$ are functions of Prandtl number and the temperature index n, respectively.

The Reynolds analogy was extended by Owen (5) to include frictional heating effects, which can be significant at large Reynolds numbers. Eqn (2.8) is still valid if T_{ref} is calculated from

$$T_{ref} = T_\infty + \tfrac{1}{2}R\Omega^2 r^2/c_p \tag{2.10}$$

where T_∞ is the temperature of the surrounding fluid and R is a recovery factor. It is suggested that $R = Pr^{1/3}$ for fluids with 'moderate' Prandtl numbers ($Pr = 0(1)$).

For laminar flow, Dorfman's solution gives (for moderate Pr and $-2 < n < 3$)

$$(b/r)Nu = Nu_{av} = 0.308Pr^{\frac{1}{2}} \, (n + 2)^{\frac{1}{2}} \, Re_\theta^{\frac{1}{2}} \tag{2.11}$$

which agrees well with most available experimental data (e.g. Cobb and Saunders (6)) for which gravitation-induced buoyancy effects are negligible. For turbulent flow,

$$(b/r)^{1.6} Nu = (n + 2.6)Nu_{av}/(n + 2) = 0.0197 \, Pr^{0.6} (n + 2.6)^{0.2} Re_\theta^{0.8} \tag{2.12}$$

For $Re_\theta < 10^6$, $Pr = 0.72$ and $n = 0$, eqn (2.12) is in good agreement with the data of Cobb and Saunders.

3. ROTOR-STATOR SYSTEMS

Schematic diagrams of enclosed, open (or 'partially enclosed') and shrouded rotor-stator systems are shown in Fig. 3. These systems can be used to model the type of flow that occurs when a disc rotates close to a stationary casing,

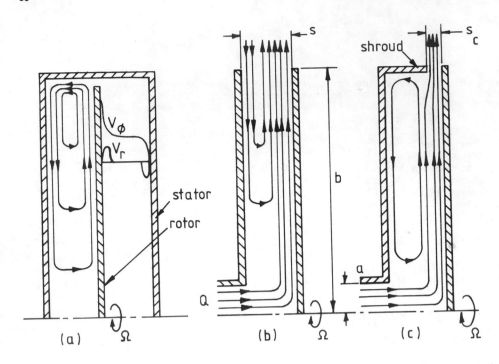

Fig. 3 Schematic diagram of flow in rotor - stator systems:(a) enclosed disc;
(b) open rotor-stator; (c) shrouded rotor-stator.

with or without a superimposed flow. In the limiting case where the superimposed flow is zero and the casing is removed, the systems reduce to the free disc.

3.1   The Enclosed Rotating Disc

Unlike the free disc, the enclosed disc (see Fig. 3a) must entrain fluid from a finite reservoir. At sufficiently large values of the axial clearance ratio, G (where $G \equiv s/b$), separate boundary layers form on the rotor and stator. Fluid moves radially outward near the rotating disc, axially across the cylindrical casing and radially inward on the stator. In the central core, which rotates at a speed less than that of the rotor, there is an axial flow of fluid from the stator to the rotor. As the axial clearance is reduced, the boundary layers merge and, for small clearances, Couette flow (in which inertial terms are negligible) occurs.

The earliest theoretical and experimental work for laminar and turbulent isothermal flow was conducted by Schultz-Grunow (7), and an improved theoretical estimate of the measured moment coefficients was obtained by Okaya and Hasegawa (8); neither analysis included the effect of the cylindrical casing. The experimental data of Pantell (9) showed that there is an optimum gap ratio ($G \simeq 2.97 \, Re_\theta^{-0.34}$) for which the moment coefficient, C_m, is a minimum (approximately 47% of the free disc value); for larger values of G, the moment coefficient tends to that of the free disc.

In a combined theoretical and experimental study, Daily and Nece (10) delineated the flow into four regimes (laminar or turbulent flow, merged or separate boundary layers); they produced correlations for C_m in each regime. A laminar analysis, for the separate boundary layer case, was conducted by Ashiwake and Nahayama (11), who solved the momentum and energy integral equations. Their calculations of the radial pressure distribution were in good agreement with the data of Daily and Nece. Cooper and Reshotko (12) conducted a combined laminar and turbulent flow analysis for the flow in an infinite rotor-stator system. Despite the fact that they did not include the outer cylindrical surface in the model, their predicted moment coefficients and velocity profiles were in good agreement with the results of Daily and Nece.

For heat transfer in an enclosed disc system, Dorfman (1) showed that the Reynolds analogy is valid provided (for the case of separate boundary layers on the rotor and stator) that account is taken of the core rotation. For laminar flow, this results in a 50% increase in local Nusselt numbers (using the core temperature as the reference value) over the free disc case; for turbulent flow, there is a 6% decrease.

3.2   The Open Rotor-Stator System

In the open (or 'partially-enclosed') system, see Fig. 3(b), fluid can be supplied at the centre with a volumetric flow rate Q (to simulate an air-cooled gas turbine disc). For the case of zero superimposed flow rate and an infinite gap between the rotor and stator, the flow will be the same as that for the free disc. However, as the gap is reduced (the superimposed flow still being zero), the fluid entrained into the boundary layer on the rotor is supplied by inflow down the stator.

At large values of G, the boundary layers will remain separate; if the gap is reduced below a certain value of G (G = G_{max}, say) the layers will start to interact. This interaction will restrict the flow entrained by the rotor and thereby reduce the tangential momentum efflux in its boundary layer. Thus, for G < G_{max}, there will be a tendency for the moment coefficient (and, by analogy, the Nusselt number) to be reduced, as was mentioned in Section 3.1 for the enclosed disc.   However, if the gap is further reduced below a limiting value of G (G = G_{min}, say), Couette flow will occur and the moment coefficient and Nusselt number will start to increase with decreasing G.

The above effect is shown in Fig. 4 which indicates the variation of C_m and Nu_{av} (T_∞, in eqn (2.10) is replaced by T_I, the coolant inlet temperature) with G for a range of values of C_w at Re_θ = 9 x 10^5. The experimental data and numerical solutions (of the boundary layer equations) were obtained by Owen, Haynes and Bayley (13). It can be seen that for C_w < 1.3 x 10^4, C_m and Nu_{av} drop below the free disc value; for $C_w \geq$ 2.2 x 10^4, this effect does not occur. It should be noted that for these tests the flow was turbulent, and coolant entered through the stator at a radius ratio of a/b = 0.133.

An estimate of G_{max}, referred to above, may be made by considering the results of Section 2. Eqn (2.4) for the free disc can be rewritten, for r = b, as

$$\delta/s = 0.525 \, G^{-1} \, Re_\theta^{0.2} \tag{3.1}$$

Now, the fluid entrained by the rotor must be supplied in the space between the stator and the edge of the rotor boundary layer. It is postulated that the stator will not influence the rotor if the space available for inflow, at r = b, is at least as great as the space available for outflow. Using this argument,

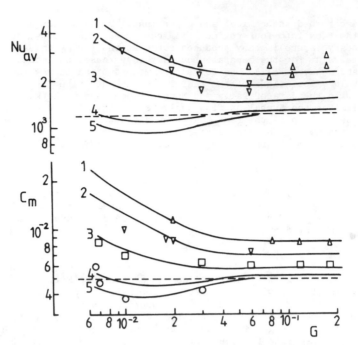

Fig.4. Effect of gap ratio on average Nusselt numbers and moment coefficients for an aircooled rotating disc, $Re_\theta = 9 \times 10^5$. Numerical solutions: curve 1, $C_w = 7.4 \times 10^4$; curve 2, $C_w = 4.7 \times 10^4$; curve 3, $C_w = 2.2 \times 10^4$; curve 4, $C_w = 1.3 \times 10^4$; curve 5, $C_w = 0.53 \times 10^4$. Experimental data: ▵ $C_w = 7.4 \times 10^4$; ▽ $C_w = 4.7 \times 10^4$; □ $C_w = 2.2 \times 10^4$; ○ $C_w = 0.53 \times 10^4$. ––––– Free disc

the flow over the rotor will be affected (that is, $G = G_{max}$) when $\delta/s = 1/2$; hence, from eqn (3.1),

$$G_{max} = 1.05 \, Re_\theta^{-0.2} \qquad\qquad (3.2)$$

For $Re_\theta = 9 \times 10^5$, $G_{max} = 0.068$, which is consistent with the data of Fig. 4. Eqn (3.2) is also consistent with the mass transfer data of Kreith, Doughman and Kozlowski (14) and with the heat transfer data of Metzger (15).

Using a similar argument, but employing the boundary layer thicknesses calculated by Okaya and Hasegawa (8) for the enclosed disc, Haynes (16) showed that

$$G_{min} = 0.23 \, Re_\theta^{-0.2} \qquad\qquad (3.3)$$

For $Re_\theta = 9 \times 10^5$, $G_{min} = 0.015$, which is also consistent with the data of Fig. 4. It is also interesting to observe that, for $Re_\theta = 9 \times 10^5$, eqn (2.6) for the free disc gives $C_{w,ent} = 1.27 \times 10^4$. Referring to Fig. 4, it would appear that if $G < G_{max}$ and $C_w < C_{w,ent}$ then both C_m and Nu_{av} can drop below the free disc value.

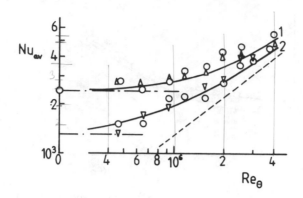

Fig. 5. Average Nusselt numbers for an aircooled rotating disc at a small gap ratio, $G = 0.02$. Numerical solutions: curve 1, $C_w = 7.5 \times 10^4$; curve 2, $C_w = 3.2 \times 10^4$ Experimental data: △ enthalpy balance method; ○ conduction method; ▽ Reynolds analogy.————— , Asymptotic solution;——————, Free disc.

Fig. 5 shows the variation of Nu_{av}, for an air-cooled rotating disc, with Re_θ for $C_w/10^4$ = 3.2 and 7.5 and G = 0.02. The numerical solutions and exper-imental data were obtained by Owen, Haynes and Bayley. The data were based on three independent techniques: (i) the solution of Laplace's conduction equation for the rotor; (ii) an enthalpy balance on the cooling air; (iii) the Reynolds analogy applied to measured moment coefficients. It can be seen that the effect of Re_θ is weak at the smaller values of the rotational Reynolds numbers, and the effect of C_w is weak at the higher values.

At this smaller value of G, there is an asymptotic expression for large C_w ($C_w/2\pi GRe_\theta > 1.7$) which, for air (Pr = 0.72), can be expressed as

$$Nu_{av} = 0.0145 \, (C_w/G)^{0.8} \tag{3.4}$$

The absence of any effect of Re_θ on heat transfer was also observed by Kreith, Doughman and Kozlowski and by Mitchell (17). At the other end of the range (small values of $C_w/2\pi GRe_\theta$), there is little effect of C_w (as observed by Kapinos (18)), and the results tend to those of the free disc.

Fig. 6 shows the variation of Nu_{av} with Re_θ for large gap ratios (G > 0.055) at $C_w/10^4$ = 7.4 and 9.8. It can be seen that the effect of gap ratio is weak (the numerical solution was conducted for G = 0.055). Above this value of G, 'separation' occurred in the numerical solution implying that the boundary layer equations were no longer valid.

Design formulae were produced by Owen and Haynes (19) from the above res-ults. Correlations of the mean and local Nusselt numbers and moment coef-ficients were obtained for a wide range of rotational Reynolds numbers (Re_θ < 4 x 10^6), coolant flow rates (C_w < 10^5) and gap ratios (0.01 < G < 0.12).

Metzger, Mathis and Grochowsky (20) conducted experiments on a rotating disc, the heated rim of which was cooled by impinging jets of air. In order to simulate the types of rotor geometry that are used in gas turbines, four dif-

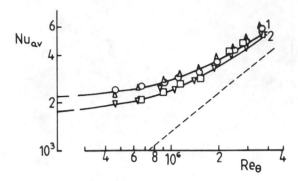

Fig. 6. Average Nusselt numbers for the air-cooled rotating
disc at large gap ratios, G > 0.055. Numerical solutions:
curve 1, $C_w = 9.8 \times 10^4$, G = 0.055; curve 2, $C_w = 7.4 \times 10^4$,
G = 0.055. Experimental data: o $C_w = 9.8 \times 10^4$, G = 0.12;
△ $C_w = 9.8 \times 10^4$, G = 0.08; □ $C_w = 7.4 \times 10^4$, G = 0.12; ▽ $C_w = 9.8 \times 10^4$,
G = 0.12. -----, Free disc.

ferent disc profiles were tested. For each profile, the geometry of the heated
disc rim was unchanged, and either one or two air jets was used. It was found
that the disc profile had no significant effect on heat transfer from the disc
rim. Without jet impingment, the measured Nusselt numbers were (for
$0.1 < Re_\theta/10^5 < 2.4$) consistent with free disc correlations. With a single jet,
no effect of impingment was observed for coolant flow rates less than 10% of the
free disc entrainment value: for jet flow rates equal to the entrainment value,
heat transfer rates were two or three times those of the free disc.

3.3   The Shrouded Rotor-Stator System

      A shrouded rotor-stator system, see Fig. 3(c), can be used to model the
flow that occurs near a sealed air-cooled gas turbine disc. In the turbine, a
peripheral seal is used to minimize the coolant flow rate (Q_{min}, say) necessary
to prevent the penetration (or ingress) of hot mainstream gas into the cavity
between the rotor and its casing. In the shrouded rotor-stator system, the seal
can be modelled by cylindrical shrouds attached to the stator and/or the rotor.

      Bayley and Owen (21) studied the fluid dynamics in a system which had a
stationary shroud, with an axial clearance, s_c, between the shroud and the rotating disc. Pressure distributions and moment coefficients were measured, on a
rig with a/b = 0.133, for $0.06 < G < 0.18$ and $G_c = 0.0033$ and 0.0067 ($G_c \equiv s_c/b$)
over a range of C_w ($C_w < 5 \times 10^4$) and Re_θ ($Re_\theta < 4 \times 10^6$). It was found that the
gap ratio, G, had a small effect compared with the shroud clearance ratio, G_c,
and C_m increased with decreasing G_c.

      The pressure distribution is calculable by the superposition of the pres-
sure drop due to the shroud onto the unshrouded pressure distribution. The
latter can be calculated from a solution of the boundary-layer equations; the
former can be estimated from an approximation of the inviscid radial momentum
equation over the shroud. The pressure drop across the shroud increases with
increasing C_w, decreasing Re_θ and decreasing G_c. Using the argument that ingress

of external fluid into the rotor-stator cavity can only occur if the pressure in the cavity is negative (with respect to the external pressure), Bayley and Owen produced the correlation

$$C_{w,min} = 0.61 \, G_c \, Re_\theta \tag{3.5}$$

(where $C_{w,min} \equiv Q_{min}/\nu b$).

Owen and Phadke (22) used pressure measurements and flow visualization to evaluate $C_{w,min}$ for a similar geometry to that studied by Bayley and Owen. For a gap ratio of G = 0.1 and 0.0025 < G_c < 0.04, they obtained the correlation

$$C_{w,min} = C \, G_c^n \, Re_\theta \tag{3.6}$$

From the flow visualization results, values of C = 0.14 and n = 0.66 were found to give the best fit to the data for 2×10^5 < Re_θ < 10^6.

The ingress problem was also studied, for a shrouded disc system with a radial clearance between the cylindrical shroud and the rotor, by Phadke and Owen (23). Unlike their axial-clearance counterparts, radial-clearance seals can exhibit a 'pressure-inversion' effect where the pressure inside the rotor-stator cavity increases, rather than decreases, with increasing rotational speed. As a consequence, the radial-clearance seal can be more effective at preventing ingress than the axial-clearance seal. It should be pointed out that all the above ingress tests were conducted in the absence of the external axial flow that occurs in actual gas turbines; this external flow is likely to have a significant effect on the ingress problem.

Heat transfer measurements were made in a stationary shrouded system with an axial clearance by Haynes and Owen (24). For a range of shroud clearance ratios, coolant flow rates and rotational Reynolds numbers, they obtained the mean Nusselt numbers, under turbulent conditions, on the air-cooled rotating disc. Using a modified form of the boundary-layer equations, their numerical solutions were in good agreement with the data, which showed an increase in Nu_{av} with decreasing G_c. Their results show that the effect of the shroud was negligible for G_c > G_{min}, where the latter is calculated from eqn (3.3).

Sparrow et al (25-28) conducted turbulent heat transfer experiments in a shrouded rotor-stator system with a/b = 0.111, G $\leq$ 2, $C_w \leq 10^4$, $Re_\theta \leq 10^6$. Yu, Sparrow and Eckert (25) measured Nusselt numbers on the shroud and on the rotor for the case where cooling air entered through the centre of the rotor; it left through a radial clearance between the insulated stator and the heated stationary shroud. Sparrow, Buszkiewicz and Eckert (26) carried out similar measurements for the case where the coolant entered through the centre of the rotor and left through a radial clearance between a stationary shroud and the rotor. The latter work was repeated by Sparrow and Goldstein (27) but with the coolant entering through the centre of the stator; Sparrow, Shamsundar and Eckert (28) obtained data for the axial throughflow case.

The latter case was studied by De Socio, Sparrow and Eckert (29), who obtained numerical solutions of the turbulent recirculating flow equations. Gosman et al (30,31) also solved these equations numerically; they obtained good agreement with some of the data obtained in refs. 21, 24 and 25.

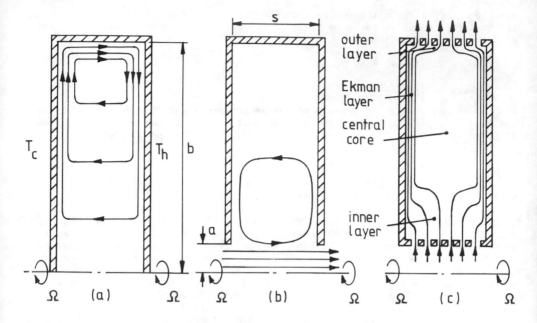

Fig. 7 Schematic diagram of flow in rotating cavities:(a)sealed cavity;
(b) cavity with axial through flow;(c) cavity with radial outflow

4   ROTATING CAVITIES

Rotating cavities, see Fig. 7, can be used to model the flow between corotating turbine or compressor discs. For the limiting case of an isothermal
cavity with no superimposed flow, solid-body rotation (with zero velocity relative to the cavity) occurs. It is, therefore, convenient to refer flows to
this basic datum by the use of a rotating frame of reference. For isothermal
source-sink flows in which the tangential component of velocity (relative to the
rotating frame) is small, the nonlinear inertial terms can be neglected. The
resulting linear equations represent a balance between Coreolis, pressure and
viscous forces; the boundary layers which form on the discs at large Reynolds
numbers are referred to as Ekman layers.

For a non-isothermal cavity, the large centripetal accelerations (in turbines, accelerations 20,000 times that due to gravity are common) give rise to
buoyancy forces that can, for the case of zero or relatively small superimposed
flow rates, control the flow. An example of this is the flow that occurs in the
sealed rotating cavity described below.

4.1   The Sealed Rotating Cavity

Fig. 7(a) shows a schematic diagram of the buoyancy-induced flow for the
case where the right-hand and left-hand discs have isothermal temperatures T_h
and T_c, respectively, ($T_h > T_c$). At large rotational Reynolds numbers, fluid
moves radially outward in an Ekman layer on the cold disc and then flows, via a

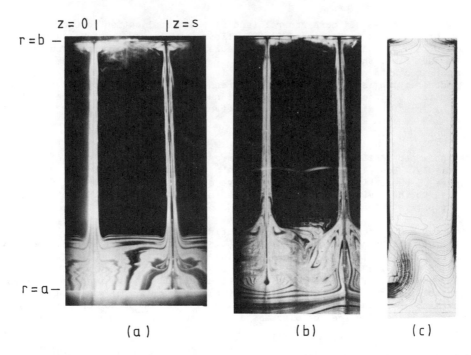

Fig. 8   Flow structure in a rotating cavity with radial outflow: $a/b = 0.1$, $G = 0.267$, $C_w = 79$, $Re_\theta = 2.5 \times 10^4$. (a)radial inlet: visualization (b)axial inlet: visualization (c)axial inlet: computation.

boundary layer on the cylindrical shroud, inward near the hot disc. In the central part of the cavity, there is an 'axial wind' in which fluid migrates from the hot to the cold Ekman layer.

Using the Boussinesq approximation for the temperature-dependent density, Ostrach and Braun (32) appear to be the first to have attempted an analysis of laminar free convection in the sealed cavity. Dorfman (33), Hudson (34) and Chew (35) considered the related problem of laminar buoyancy-driven flows between infinite corotating discs. Dorfman's solution, which uncouples the momentum and energy equations, is only valid when conduction dominates over convection; Hudson's solution, which assumes that the temperature throughout the Ekman layer equals that of the adjacent disc, is only valid in convection dominated flows where

$$Re_\Omega^{-\frac{1}{2}} \ll \sigma^{-1} \ll 1$$

Here

$$\sigma \equiv Pr \, \beta \Delta T \, Re_\Omega^{\frac{1}{2}}, \tag{4.1}$$

and

$$Re_\Omega \equiv \Omega s^2 / \nu = G^2 \, Re_\theta, \tag{4.2}$$

Pr and β are the Prandtl number and volume expansion coefficient, respectively, and $\Delta T = T_h - T_c$.

Chew's solution is more general (and also has application to the source-sink flows discussed in Section 4.3) and allows for a power-law temperature distribution ($T \propto r^n$) on each disc. If the radial temperature distribution is the same on each disc, then there is no effect of buoyancy in the solutions. If $r^2 \Delta T$ increases with radius, the axial wind is from the hot to the cold disc; the wind direction is reversed if $r^2 \Delta T$ decreases with radius. For the isothermal case ($n = 0$), Chew's calculated Nusselt numbers ($Nu_\Omega \equiv qs/k\Delta T$, q being the heat flux from the disc to the fluid) are given by

$$Nu_\Omega = \pm \tfrac{1}{4}\sigma/(1 - \exp(\mp\sigma)) \tag{4.3}$$

where the upper positive and negative signs refer to the hot and cold discs, respectively. Eqn (4.3) reduces to Dorfman's and Hudson's solutions for small and large σ, respectively. In all the above solutions, the tangential velocity (relative to the cavity) is sheared, across the core between the Ekman layers, from a positive value near the hot disc to a negative value near the cold disc.

Homsy and Hudson (36) extended the infinite disc analysis to include the effect of a cylindrical shroud, and Hudson, Tang and Abell (37) conducted experiments on an oil-filled rotating cavity. Two different silicone oils were used ($Pr \simeq 7$, $\beta \simeq 1.3 \times 10^{-3} K^{-1}$, and $Pr \simeq 3100$, $\beta \simeq 10^{-3} K^{-1}$) in a cavity with $G = 0.07$ and 0.14. The temperature difference between the two isothermal discs was variable up to $35^\circ C$, and the cavity (of radius b = 0.14 m) was rotated up to 860 rev/min. For the low Pr oil, Ekman layer flow was produced for $\sigma > 1$, and the Nusselt number for the hot disc was correlated by

$$Nu_\Omega = 0.533 \ (\beta\Delta T)^{0.822} \ Re_\Omega^{0.449} \ G^{-0.173} \tag{4.4}$$

For the high Pr oil, non-Ekman layer flows were produced for $Gr_\Omega G^3 > 2$, and

$$Nu_\Omega = 1.13 \ Gr_\Omega^{0.267} \ G^{0.855} \tag{4.5}$$

where Gr_Ω is the 'rotational Grashof number' ($Gr_\Omega \equiv \beta\Delta T \ Re_\Omega^2$).

For turbulent flow, Kapinos (38) obtained approximate solutions of the integral boundary-layer equations. With air (Pr = 0.72), his results simplify to

$$Nu_\Omega = C \ G^{-0.6} \ Gr_\Omega^{0.4} \tag{4.6}$$

where C = 0.0259 and 0.0195 for the hot and cold discs, respectively. It should be noted that Kapinos assumed that the tangential velocity (relative to the cavity) was zero in the core, and, for this case, ΔT is the disc-to-core temperature difference. It is not obvious that the assumption of zero tangential velocity in the core is valid; it is certainly not true for laminar flow. In view of this questionable assumption (and bearing in mind that there is no accepted criterion for transition from laminar to turbulent flow in a sealed rotating cavity), eqn (4.6) should be used with caution.

4.2   The Rotating Cavity with Axial Throughflow

A schematic diagram of the isothermal structure inside a rotating cavity with a central axial flow (or jet) of fluid is shown in Fig. 7(b). The central jet generates secondary flow in the cavity; under 'ideal' conditions, the secondary flow is an axisymmetric toroidal vortex which increases in strength with increasing flow rate and decreasing rotational speed. Under some conditions, as discussed below, vortex breakdown (an abrupt change in the structure of the flow) occurs, and the flow in the jet, and in the cavity, can become non-axisymmetric and unsteady. For non-isothermal flow, buoyancy effects are usually significant; at sufficiently large rotational Grashof numbers, the flow structure should be similar to that found in the sealed cavity.

Hennecke, Sparrow and Eckert (39) obtained numerical solutions of the laminar Navier-Stokes and energy equations, and Owen and Bilimoria (40) and Owen and Onur (41) made heat transfer measurements, under turbulent flow conditions, in an air-cooled cavity in which the downstream disc was heated. Owen and Pincombe (42) used flow visualization and laser-doppler anemometry (LDA) to observe the flow structure and measure the velocity distribution inside the cavity under laminar and turbulent conditions.

In the cavity tested by Owen and Pincombe, the inner to outer radius ratio, a/b, was 0.1, the gap ratio, G, was 0.53, and measurements were limited to $Re_\theta < 4 \times 10^5$ and $Re_z < 10^5$ (where $Re_z \equiv 2\bar{W}a/\nu$, $\bar{W}$ being the bulk-average axial velocity in the rotating central-inlet tube). Transition from laminar to turbulent flow inside the cavity was associated with transition in the central jet. Turbulent flow was found to start at $2 \times 10^3 < Re_z < 2 \times 10^4$, depending on conditions in the inlet tube. Regimes of spiral and axisymmetric vortex breakdowns were delineated by the axial Rossby number, ε_z ($\varepsilon_z \equiv W/\Omega a$). The most dramatic effect was found during turbulent spiral breakdown, $21 \leq \varepsilon_z \leq 100$, and the jet precessed about central axis and destroyed the axisymmetry of the flow inside the cavity. A similar effect was also observed by Yu et al (25) in their rotor-stator system where the flow entered through a rotating tube in the rotor and left through a small clearance between the stator and a stationary shroud.

For heat transfer tests, the presence of vortex breakdown can make measurements extremely difficult to interpret. For small values of Rossby number, $\varepsilon_z < 10$, where vortex breakdown appears to have a negligible effect on heat transfer, Owen and Onur correlated the average Nusselt number for the heated downstream disc (using the coolant inlet temperature as the reference value) by

$$Nu_{av} = 0.50 \ G^{0.38} \ Re_z^{0.15} \ Gr_\theta^{0.31} \qquad (4.7)$$

where Gr_θ is the rotational Grashof number ($Gr_\theta \equiv \beta\Delta T \ Re_\theta^2$, and ΔT is the maximum temperature difference between the heated disc and the coolant). The measurements were conducted in a cavity with a/b = 0.1, and the correlation was obtained for G = 0.133, 0.267 and 0.4, $1.8 \times 10^4 < Re_z < 1.6 \times 10^5$, $0.14 < \beta\Delta T < 0.27$, $2.3 \times 10^4 < Re_\theta < 2 \times 10^6$ and $\varepsilon_z < 10$ (that is, 50 $Re_z/Re_\theta < 10$). It should be noted that, unlike the experiments conducted by Owen and Bilimoria, a restriction was placed in the downstream outlet tube (the ratio of the inlet tube diameter to that of the outlet tube was 1.25). Without the restriction, only the data of Owen and Bilimoria for G = 0.267 are correlated by eqn (4.7); for the other two gap ratios, no such correlations could be obtained. For low flow rates, as discussed in Section 4.3, heat transfer is dominated by free convection.

4.3   The Rotating Cavity with Radial Source-Sink Flow

Fig. 7(c) shows a simplified representation of the isothermal flow structure inside a rotating cylindrical cavity with radial outflow from a uniform source, at r = a, to a uniform sink, at r = b. Considerable insight into the laminar flow structure is provided by the analysis by Hide (43), and by the numerical studies of Bennetts and Jackson (44) and Chew (45).

Fluid enters the cavity from the source and forms an inner (or source) layer which distributes the flow equally into an Ekman layer on each rotating disc. From the Ekman layers, the flow leaves the sink via an outer (or sink)

layer. Between the inner, outer and Ekman layers is a central core of rotating
fluid in which the radial and axial components of velocity are zero. From the
solution of the laminar 'linear equations' (that is, the Navier-Stokes equations, referred to a rotating frame of reference, in which the nonlinear inertial terms are neglected), the tangential component in the central core
(referred to a stationary frame of reference) is given by

$$V_\phi/\Omega r \;=\; 1 \pm \xi_\ell^{-1} \tag{4.8}$$

where

$$\xi_\ell \;\equiv\; 2\pi(r/b)^2 C_w^{-1} Re_\theta^{\frac{1}{2}} \tag{4.9}$$

The positive and negative signs in eqn (4.8) refer to radial inflow and outflow,
respectively.

Owen and Pincombe (46) used flow visualization and LDA to observe and
measure the flow structure in a rotating perspex cavity in which air entered,
either axially or radially, at the centre (r = a) and left through a perforated
outer shroud (r = b). The air was seeded with micron-sized oil particles, and
flow visualization was achieved by illuminating an r - z plane, through the axis
of rotation. Using an argon-ion laser, photographs were taken with the optical
axis of the camera normal to the plane of illumination.

Photographs obtained in this way are shown in Fig. 8 for laminar flow in a
cavity with a/b = 0.1, G = 0.267, C_w = 79 and Re_θ = 2.5 x 10^4. It should be
noted that reflections from each perspex disc gave rise to mirror images at
z = 0 and z = s. For Fig. 8(a), the radial-inlet case, the flow structure is
similar to that indicated schematically in Fig. 7(c). Close inspection of the
Ekman layers reveals signs of spatially-periodic cells that are attributed to
Ekman layer instabilities; at sufficiently large values of C_w, these instabilities are associated with transition from laminar to turbulent flow.

Fig. 8(b) is for the axial-inlet case where the flow enters through the
centre of the upstream (left-hand-side) disc. Whilst the flow structure in the
outer part of the cavity is similar to that for the radial-inlet case, the inner
layers are clearly different. In particular, the jet in Fig. 8(b) shows the
non-axisymmetric behaviour associated with spiral vortex breakdown (which was
discussed in Section 4.2).

Chew's numerical solutions are, for both the radial- and axial-inlet
cases, in good agreement with the measurements and flow visualization of Owen
and Pincombe. An example of the computed streamlines are shown in Fig. 8(c)
where, bearing in mind that the numerical method is for steady axisymmetric
flow, the agreement between these streamlines and the smoke pattern shown in
Fig. 8(b) is very good.

For the axial-inlet case, Chew has shown that under conditions where vortex breakdown is likely to occur, the flow close to each disc in the cavity is
similar to that which occurs in the radial-inlet case. However, if the flow is
sufficiently high, a wall jet forms on the downstream disc. Near the upstream
disc (and near both discs for the radial-inlet case), the radial-component of
velocity in the inner layer is similar to that found near the free disc.   To a
good approximation, the inner layer extends to a radius at which the flow rate
entrained by the free disc equals that supplied at the centre of the cavity.
Using this argument, and eqn (2.3), the radius of the inner layer, r_i, can be
approximated by

$$r_i/a \;=\; 1.5(b/a)(C_w/4\pi Re_\theta^{\frac{1}{2}})^{\frac{1}{2}} \tag{4.10}$$

which is consistent with the measurements of Owen and Pincombe.

The integral momentum equations for the flow in an Ekman layer have been derived, and approximate solutions of the linear equations have been obtained, by Owen and Rogers (47). The solutions, for the tangential velocity in the central core and the Ekman layer thickness, are applicable to laminar and turbulent source-sink flows in which buoyancy effects may be significant. For laminar isothermal flow, the solution for V_ϕ agrees with eqn (4.8); for turbulent isothermal flow, it is shown that

$$V_\phi / \Omega r \; = \; 1 \pm \xi_t \qquad\qquad (4.11)$$

where

$$\xi_t \; \equiv \; 0.450(r/b)^{13/8} \; C_w^{-5/8} \; Re_\theta^{\frac{1}{2}} \qquad\qquad (4.12)$$

and the positive and negative signs in eqn (4.11) refer to radial inflow and outflow, respectively.

Rogers and Owen (48) obtained numerical solutions of the nonlinear integral equations. For sufficiently large values of ξ_ℓ or ξ_t, the solution approaches an asymptotic curve which differs only slightly from the linear solution. At smaller values of ξ_ℓ or ξ_t, the results are affected by the choice of initial conditions (for example, the value of swirl at inlet to the cavity). A significant buoyancy-induced effect occurs for turbulent outflow when the discs are hotter than the fluid; at a critical value of $\xi_t (\beta \Delta T \xi_t \simeq 2.7)$, the Ekman layer thickness increases significantly with increasing ξ_t. This effect, which also occurs in the linear solutions, is thought to signal the onset of buoyancy-induced instability, which is discussed below.

Pincombe, Rogers and Owen (49) show good agreement between velocities predicted from the integral equations and those measured in rotating cavities with source-sink flows. Fig. 9 shows the comparison between the theoretical values of $V_\phi / \Omega r$, according to eqns (4.8) and (4.11), and the measured values at $r/b = 0.767$ for an isothermal cavity with a/b = 0.1 and G = 0.133. It is interesting to observe, from Fig. 9, that transition from laminar to turbulent flow appears to occur at $C_w \simeq 860$. In fact, eqns (4.8) and (4.11) produce equal values of V_ϕ when $Re_r = 180$ (where $Re_r \equiv C_w/2\pi(r/b)$), which, for r/b = 0.767, occurs at $C_w = 860$. It is therefore suggested that this value of Re_r can be used to estimate transition from laminar to turbulent flow in a rotating cavity with source-sink flow.

For turbulent outflow, as for the laminar case, it is possible to estimate the size of the inner layer by assuming that Ekman layers start when the flow supplied at the centre of the cavity equals that entrained by a turbulent free disc. Making this assumption, and using eqn (2.6), gives

$$r_i/b \; = \; C \, C_w^{5/13} \, Re_\theta^{4/13} \qquad\qquad (4.13)$$

where C = 1.38. In practice, a value of C = 1.80 gives better agreement with measured values.

Heat transfer measurements in a rotating cavity with a radial outflow of cooling air were made by Owen and Bilimoria (40) and Owen and Onur (41). The average Nusselt numbers were determined for the heated downstream disc in a cavity with an axial inlet at a/b = 0.1, with G = 0.133, 0.267 and 0.4, for a range of turbulent flows and rotational speeds. Owen and Bilimoria identified three flow regimes: (i) a source-flow regime, which occurs at large flow rates where the source and sink layers fill the entire cavity and no Ekman layers are present; (ii) a 'developing Ekman layer regime' in which Ekman layers begin to form; (iii) a 'fully-developed Ekman layer regime', which occurs at large

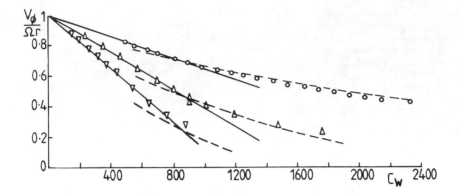

Fig.9 The effect of flow rate on the tangential velocity in a rotating
cavity with radial outflow for G = 0·133, r/b = 0·767, z/s = 0·5

$\triangledown$, $Re_\theta = 10^5$; $\triangle$, $Re_\theta = 2 \times 10^5$; $\circ$, $Re_\theta = 6 \times 10^5$; —— eqn(4·8)— —eqn(4·11)

rotational speeds when Ekman layers extend throughout most of the cavity. The
heat transfer results were also found to fall into three regimes.

Owen and Onur obtained the following correlations.

Regime I   : Nu_{av} = 1.94 $G^{1/6}$ $C_w^{2/3}$ (4.14)

        II   : Nu_{av} = 0.07 $C_w^{1/3}$ $Re_\theta^{1/2}$ (4.15)

        III : Nu_{av} = 4.11 $G^{1/9}$ $C_w^{1/2}Re_\theta^{1/9}$ (4.16)

The correlations were based on measurements for G = 0.133, 0.267 and 0.400,
970 < C_w < 79000, 2 x 10^4 < Re_θ < 2.2 x 10^6, and $\beta\Delta T \simeq 0.27$. Only one disc of
the cavity was heated, and $\beta\Delta T$ was equal to the difference between the maximum
disc temperature and the coolant inlet temperature; for the Nusselt number, the
reference temperature was taken to be the coolant inlet temperature. The
boundaries between the three regimes are given by the intercepts of eqns (4.14)
to (4.16).

Flow visualization in the heated cavity, for $\beta\Delta T \simeq 0.27$, revealed that
buoyancy-induced instability occurred when

    $Gr_\theta^{1/2}/C_w > C$ (4.17)

At a value of C $\simeq$ 180, the source layer started to oscillate; at C $\simeq$ 330, the
isothermal flow structure (described above) broke down into 'chaotic' flow. It
was considered that for C > 530, the flow and heat transfer inside the cavity
was controlled by free convection.

Thus, in addition to the above-mentioned three, forced-convection, regimes
there is a fourth, free-convection, regime. Owen and Onur were able to correlate the data from the radial-outflow experiments (and the axial-throughflow

experiments described in Section 4.2) for $10^9 < Gr_\theta < 1.3 \times 10^{12}$ in this fourth regime by

$$Nu_{av} = 0.267 \, Gr_\theta^{0.286} \qquad\qquad (4.18)$$

It would therefore appear that, at sufficiently large rotational Grashof numbers, the flows inside a rotating cavity with radial outflow or axial throughflow will tend to the buoyancy-driven flows that occur inside a sealed rotating cavity. This limiting free-convection case has obvious application to gas turbines (where Grashof numbers in excess of 10^{13} are common); yet it is a problem that is not well understood. For free convection in a rotating cavity, there is no accepted criterion for transition from laminar to turbulent flow, and there is, to this writer's knowledge, no proven turbulence model. In view of the limitations of the available theoretical models and experimental data, buoyancy-induced rotating flows present an important challenge to theoreticians and experimenters alike.

5   CLOSURE

The free disc, which provides the first 'building block' for synthesizing rotating disc systems, has been the subject of numerous theoretical and experimental studies. Moment coefficients and Nusselt numbers have been measured and predicted for laminar and turbulent flow; there is no shortage of reliable data for this problem (but this will not deter future research workers!).

Rotor-stator systems, which have practical relevance for the designer of turbomachinery, can be classified into three areas: (i) the enclosed rotating disc; (ii) the open rotor-stator system; (iii) the shrouded rotor-stator system. For the isothermal enclosed disc, there are a number of experimental and theoretical studies from which moment coefficients and pressure distributions can be estimated; for the non-isothermal case, comparatively few studies have been conducted (this might reflect the limited practical application of this problem).

The open rotor-stator has been used to model the flow and heat transfer that occurs near air-cooled gas-turbine discs. The effect of rotor-stator axial gap and the effect of coolant flow rate are complex: below a certain flow rate, and at a gap ratio smaller than a limiting value, the frictional moments and heat transfer rates can be less than those for the free disc; at large flow rates, the free disc levels are exceeded and the effects of rotational speed are negligible; at large rotational speeds, the moment and heat transfer approaches that of the free disc. These effects, which appeared contradictory a decade or so ago, are now reasonably well understood, thanks largely to the numerical solution of the parabolic boundary layer equations and a large amount of reliable experimental data.

By contrast, the shrouded rotor-stator system (which can be used to model the flow that occurs near a sealed air-cooled gas turbine disc) is an elliptic rather than parabolic system. Nevertheless, the flow in the centre of the disc (away from the peripheral shroud) is similar to that found in the open system. One of the most important problems facing the turbine designer is the estimation of the minimum coolant flow rate to prevent the ingress of hot mainstream gas into the rotor-stator cavity. Data exist, to estimate this minimum coolant flow rate, for a number of seal geometries in the absence of an external axial flow ; but little data are currently available on the effect of external flow. This is an important area that is open to experimenters and theoreticians alike.

The flow structure in a rotating cavity (which can be used to model conditions between corotating turbine or compressor discs) is significantly different to that found in rotor-stator systems. For the sealed heated cavity, where centrifugal buoyancy forces create free convection, a number of laminar flow analyses have been conducted and there are some experimental data available. There is, however, no accepted criterion for transition from laminar to turbulent flow, and there are no satisfactory theoretical models available to predict the turbulent flow structure.

There are some design data available for the case of a heated rotating cavity with an axial throughflow or radial outflow of coolant. For the latter case, and for rotating source-sink flows in general, there have been a number of theoretical and experimental laminar flow studies. In view of the complexity of the flow structure, an understanding of laminar source-sink flow is necessary before the turbulent case can be effectively studied. To date, turbulent flow analysis has been limited to the solution of the integral momentum equations; the predictions of velocity obtained from these equations are, however, in good agreement with available experimental data.

Experimental heat transfer data for the rotating cavity with axial throughflow or radial outflow suggest that, at sufficiently high rotational Grashof numbers, the flow structure is dominated by buoyancy effects. It would therefore appear that free convection is the 'final state' for all non-isothermal flow in a rotating cavity. It is certainly a problem of great relevance to the turbine designer, and much more theoretical and experimental research (particularly in the turbulent flow case) is required.

It is apparent, from the above, that there are design data available in some areas and a shortage in others. Of necessity, in this brief review, it has not been possible to include many design formulae, or to present theoretical and experimental details. However, if enthusiastic readers wish to pursue the subject further then the writer directs their attention to ref. (50)!

ACKNOWLEDGEMENTS

I should like to thank the colleagues whose work has contributed to this review and to apologise to those authors whose work has been overlooked. The following companies and institutions have, over a number of years, sponsored much of the work at the University of Sussex, and have suggested many fascinating (if intractable!) problems: GEC Gas Turbines Limited; Ministry of Defence; Motoren- und Turbinen-Union; Rolls Royce Limited; Ruston Gas Turbines Limited; Science and Engineering Research Council; Sulzer Brothers.

REFERENCES

1.   Dorfman, L.A. 1963 Hydrodynamic resistance and heat loss of rotating solids. Oliver and Boyd, Edinburgh.

2.   Greenspan, H.P. 1968 The theory of rotating fluids. CUP, Cambridge.

3.   Cochran, W.G. 1934 The flow due to a rotating disc. Proc. Cambridge Phil. Soc. 30, 365.

4.   Kármán, Th.v. 1921 Über laminare und turbulente Reibung. Z. angew. Math. Mech. 1, 233.

5.   Owen, J.M. 1971 The Reynolds analogy applied to flow between a rotating and a stationary disk. Int. J. Heat Mass Transfer 14, 451

6.   Cobb, E.C. and Saunders, O.A. 1956 Heat transfer from a rotating disk. Proc. Roy. Soc. A236, 343.

7.   Schultz-Grunow, F. 1935 Der Reibungswiderstand rotierender Scheiben in Gehäusen. Z. angew. Math. Mech. 15, 191.

8.   Okaya, T. and Hasegawa, M. 1939 On the friction to the disc rotating in a cylinder. Japanese J. Phys 13, 29.

9.   Pantell, K. 1950 Versuche über Scheibenreibung. Forsch. a.d. Geb.d.Ing. Wes. 16, 97.

10.   Daily, J.W. and Nece, R.E. 1960 Chamber dimension effects on induced flow and frictional resistance of enclosed rotating discs. J. Basic Eng. 82, 217.

11.   Ashiwake, N. and Nahayama, W. 1979 Laminar convective heat transfer from a rotating disc in an enclosed cylindrical container. Heat Transfer Japanese Research 8, 1.

12.   Cooper, P. and Reshotko, E. 1975 Turbulent flow between a rotating disc and a parallel wall. AIAA J. 13, 573.

13.   Owen,J.M., Haynes, C.M. and Bayley, F.J. 1974 Heat transfer from a air-cooled rotating disk. Proc. Roy. Soc.A336. 453.

14.   Kreith,F. Doughman, E. and Kozlowski, H. 1963 Heat and mass transfer from an enclosed rotating disk with and without source flow. J. Heat Transfer 85, 153.

15.   Metzger, D.E. 1970 Heat transfer and pumping on a rotating disc with freely induced and forced cooling. J. Engng Power 92, 342.

16.   Haynes, C.M. 1973 Heat transfer in rotating disc systems. D.Phil. thesis, University of Sussex, England.

17.   Mitchell, J.W. 1963 A study of the fluid dynamics and heat transfer behaviour for radially inward flow over a shrouded rotating disk. Report No.TR57. Stanford University, U.S.A.

18.   Kapinos, V.M.1965 Heat transfer from a disk rotating in a housing with a radial flow of coolant. J. Engng Phys.8, 35.

19.   Owen,J.M. and Haynes,C.M. 1976 Design formulae for the heat loss and frictional resistance of air-cooled rotating discs. See: Improvements in fluid machines and systems for energy conversion, 4, Hoepli, Milan.

20.   Metzger, D.E., Mathis, W.J. and Grochowsky, L.D. 1979 Jet cooling at the rim of a rotating disk. J. Engng Power 101, 68.

21.   Bayley, F.J. and Owen, J.M. 1970 The fluid dynamics of a shrouded disk system with a radial outflow of coolant. J. Engng Power 92, 335.

22. Owen, J.M. and Phadke,U.P. 1980 An investigation of ingress for a simple shrouded rotating disc system with a radial outflow of coolant. 25th ASME Gas Turbine Conf., Paper No. 80-GT-49.

23. Phadke, U.P. and Owen, J.M. 1982 An investigation of ingress for an air-cooled shrouded rotating disc system with radial-clearance seals. 27th ASME Gas Turbine Conf., Paper No. 82-GT-145.

24. Haynes, C.M. and Owen,J.M. 1975 Heat transfer from a shrouded disk system with a radial outflow of coolant. J. Engng Power 97, 28.

25. Yu, J.P., Sparrow,E.M. and Eckert,E.R.G. 1973 Experiments on a shrouded parallel disk system with rotation and coolant throughflow. Int.J.Heat Mass Transfer 16, 311.

26. Sparrow, E.M., Buszkiewicz, J.C. and Eckert, E.R.G. 1975 Heat transfer and temperature field experiments in a cavity with rotation, recirculation and coolant throughflow. J. Heat Transfer 97, 22.

27. Sparrow,E.M. and Goldstein,L. 1976 Effect of rotation and coolant throughflow on the heat transfer and temperature field in an enclosure. J. Heat Transfer 98, 387.

28. Sparrow, E.M., Shamsundar, N. and Eckert, E.R.G. 1973 Heat transfer in rotating cylindrical enclosures with axial inflow and outflow of coolant. J. Engng Power 95, 278.

29. De Socio, L.M., Sparrow, E.M. and Eckert, E.R.G. 1976 Analysis of rotating recirculating turbulent flow and heat transfer in an enclosure with fluid throughflow. Int. J. Heat Mass Transfer 19, 345.

30. Gosman, A.D., Lockwood, F.C. and Loughhead, J.N. 1976 Prediction of recirculating swirling turbulent flow in rotating disc systems. J. Mech. Engng Sci. 18, 142.

31. Gosman, A.D., Koosinlin, M.L., Lockwood,F.C. and Spalding,D.B. 1976 Transfer of heat in rotating systems. 21st ASME Gas Turbine Conf. Paper No. 76-GT-25.

32. Ostrach, S. and Braun, W.H. 1958 Natural convection inside a flat rotating container. NACA TN4323.

33. Dorfman, L.A. 1968 Laminar thermal convection in the rotating cavity between two discs. Izv.AN.SSSR.Mekhanika Zhidkusti i Gaza 3, 40.

34. Hudson, J.L. 1968 Non-isothermal flow between rotating discs. Chem. Engng Sci. 23, 1007.

35. Chew,J.W. 1981 Similarity solutions for non-isothermal flow between infinite rotating discs. Report No. 81/TFMRC/38, School of Engng & Appl.Sciences, University of Sussex, England.

36. Homsy, G.M. and Hudson,J.L. 1969 Centrifugally driven thermal convection in a rotating cylinder. J. Fluid Mech. 35, 33.

37. Hudson, J.L., Tang, O. and Abell, S. 1978 Experiments on centrifugally driven thermal convection in a rotating cylinder. J. Fluid Mech. 86, 147.

38.   Kapinos, V.M. 1966 Convective heat transfer in the closed cavity between two rotating discs in a turbulent flow regime. Izv. vuz. aviat. tekh. No.1, 123.

39.   Hennecke, D.K., Sparrow, E.M. and Eckert, E.R.G. 1971 Flow and heat transfer in a rotating enclosure with axial throughflow. Warme-und Stoffubertragung $\underline{4}$, 222.

40.   Owen, J.M. and Bilimoria, E.D. 1977 Heat transfer in rotating cylindrical cavities. J.Mech.Engng Sci. $\underline{19}$, 175.

41.   Owen, J.M. and Onur, H.S. 1982 Convective heat transfer in a rotating cylindrical cavity. 27th ASME Gas Turbine Conf., Paper No. 82-GT-151.

42.   Owen, J.M. and Pincombe, J.R. 1979 Vortex breakdown in a rotating cylindrical cavity. J. Fluid Mech. $\underline{90}$, 109.

43.   Hide,R. 1968 On source-sink flows in a rotating fluid. J.Fluid Mech. $\underline{32}$, 737.

44.   Bennetts, D.A. and Jackson, W.D.N. 1974 Source-sink flows in a rotating annulus: a combined laboratory and numerical study. J. Fluid Mech. $\underline{66}$, 689.

45.   Chew, J.W. 1980 Computation of the flow in rotating cavities. Part 1b: Isothermal laminar source-sink flows. Report No. 80/TFMRC/12, School of Engng and Appl. Sciences, University of Sussex, England.

46.   Owen, J.M. and Pincombe, J.R. 1980 Velocity measurements inside a rotating cylindrical cavity with a radial outflow of fluid. J. Fluid Mech. $\underline{99}$, 111.

47.   Owen, J.M. and Rogers, R.H. 1980 Solution of the integral momentum equations for an Ekman layer in a heated rotating cavity. Part 1: The full equations and the linear approximation. Report No. 80/TFMRC/15, School of Engng and Appl. Sciences, University of Sussex, England.

48.   Rogers, R.H. and Owen, J.M. 1981 Solution of the integral momentum equations for an Ekman layer in a heated rotating cavity. Part 2: The non-dimensional form of the equations and the numerical solution. Report No. 81/TFMRC/16, School of Engng and Appl. Sciences, University of Sussex, England.

49.   Pincombe, J.R., Rogers, R.H. and Owen, J.M. 1982 Solution of the integral momentum equations for an Ekman layer in a heated rotating cavity. Part 3: Comparison between theory and experiment. Report No. 82/TFMRC/17, School of Engng and Appl. Sciences, University of Sussex, England.

50.   Owen, J.M. and Rogers, R.H. Fluid flow and heat transfer in rotating disc systems. Research Studies Press, Chichester (Monograph in preparation).

Turbulent Flow between a Rotating Disk and a Stationary Wall with Heat Transfer

JANUSZ W. POLKOWSKI
Gas Turbines Division
Brown, Boveri & Co.
Baden, Switzerland

ABSTRACT

This paper presents an extension of the two known classic solutions: one deals with the flow over a disk rotating in a free space; the other is for an enclosed system with no through flow, where the core flow is assumed to rotate like a solid body.

These two solutions are the special cases of the more general one presented here, in which the concept of a core flow is retained and e.g. a free disk corresponds to a zero core rotation. In the solution of the integral momentum equations the velocity profiles in the boundary layers satisfy a $1/n$ power law, where n is arbitrary number; this makes a particular solution adjustable to any range of Reynolds numbers.

The moment coefficient has been found to be a function of rotational Reynolds number, coolant flow rate coefficient and geometry (the rotating shroud being a part of the rotor and/or stationary shroud fixed to the wall).

On that basis, Dorfman's solution of heat transfer from a disk rotating in a free space has been extended to the actual situation of the turbine rotor. In the solution obtained, the Nusselt number is a function of geometry (rotating and/or stationary shroud), coolant flow rate coefficient, Reynolds and Prandtl numbers; it is valid for the coolant flow rates from zero to about two times the flow rate induced by a free disk.

NOMENCLATURE

r, θ, z	radial, circumferential and axial directions in cylindrical coordinate system	$Re_r = \omega r^2/\nu$	Reynolds number
		S	axial distance between a disk and a wall
$C_M = M/(0.5\rho\omega^2 r^5)$	moment coefficient	S_d	length of the rotating shroud
$C_m = \dot{m}/(\rho R^3 \omega)$	mass flow coefficient	S_w	length of the stationery shroud
C_p	specific heat		
h	static enthalpy	$S'd = S_d/R, S'_w = S_w/R$	dimensionless length of shroud
H	total enthalpy		
M	friction moment	T	temperature
$\dot{m}$	coolant mass flow rate	V	velocity
$Nu = qr/[\lambda(T_d - T_{dad})]$	Nusselt number	β	core flow angular velocity
p	pressure		
$Pr = \mu C_p/\lambda$	Prandtl number	δ	boundary layer thickness
R	disk radius		

λ	thermal conductivity	ρ	density
μ	dynamic viscosity	ω	disk angular velocity
ν	kinematic viscosity	Φ	dissipation function

Subscripts: cool - coolant; d - disk; w - stationary wall; sh - shroud; in - inlet; out - outlet; f.d. - free disk; av or - mean value; ad - adiabatic;

1. INTRODUCTION

Flow between a plane, rotating disk and a stationary wall constitutes a model of the flow of coolant over a gas turbine rotor. An extensive literature on the subject represents two essentially different ways of attacking the problem. One approach is based on semi-empirical relationships developed for turbulent flow with some assumptions based on experimental results, e.g. the concept of a core flow rotating like a solid body in a closed system without through flow. Another approach is to solve the set of governing equations over the whole flow between the rotor and the wall without any a priori assumptions about velocity profiles etc. A promising attempt has been made by Owen et-al [3], [4]. They concentrated, however, on systems with large coolant flow rates, exceeding the free disk flow rates many times. The numerical method of solution, they used, was not applicable to recirculating flows. Recirculation takes place within the system (inward flow along the stationary wall) for the coolant flow rates between zero and that of a free disk. Enclosed system with no through flow corresponds to the maximum intensity of recirculation and maximum rotation of the core. When the through flow rate becomes equal to the free disk flow rate, the whole process takes place in one only (disk) boundary layer; there is no recirculation and no core rotation. With the further increase of the flow rate, the influence of the rotating disk becomes steadily less "felt" and the flow conditions tend to those in a diffuser.

This paper deals with small and moderate flow rates. In many turbines the amount of coolant in the cooling system of the rotor is of the order of (and often less than) the flow rate of the free disk. The concept of the core flow is retained,hence the solution is valid if the gap between the rotor and the wall is greater than the sum of thicknesses of the boundary layers. This condition is satisfied in most of the actual situations.

2. SOLUTION OF THE MOMENTUM EQUATION

Velocity profiles in a disk boundary layer are assumed as follows:

$$V_r = \alpha r \omega \eta^{1/n} (1-\eta) \quad (1) \qquad V_\Theta = r\omega (1-k\eta^{1/n}) \quad (2)$$

where $\eta = z/\delta$ and $(1-k) \omega = \beta$ is a core angular velocity.

Boundary conditions:

$$z = 0 : V_r = 0, V_\Theta = r\omega \quad (3) \qquad z = \delta : V_r = 0, V_\Theta = r\omega (1-k) \quad (4)$$

For very small z, $V_r \sim \alpha r \omega \eta^{1/n}$.   In the relative coordinates system, attached to the rotor

$$V_{\Theta rel} = V_\Theta - r\omega = -kr\omega \eta^{1/n} \quad (5) \qquad V_{r\,rel} = V_r \sim \alpha r\omega \eta^{1/n} \quad (6)$$

The resultant relative, reference velocity can be written as

$$V_{ref} = (V_r^2 + V_{\Theta rel}^2)^{0.5} = r\omega (k^2 + \alpha^2)^{0.5} \quad \text{as } \eta = 1 \quad (7)$$

According to (1), V_r first increases from zero at the wall to a certain max. value at $\eta = 1/(n + 1)$ and then decreases to zero at the outer edge of the boundary layer. Eq. (7) is obtained, extrapolating V_r profile, defined by (6), up to the outer edge of the boundary layer. In order to apply the formula for a pipe, so obtained reference velocity (V_{ref}) and thickness of the boundary layer (δ) are considered to be equivalent to the V_{max} at the axis of the pipe and the radius of the pipe (R), respectively.

The empirical 1/7 power law for a pipe has the form

$$U_{max}/\upsilon_* = 8.74 \ (\upsilon_* R/\nu)^{1/7}, \qquad \text{where} \qquad \upsilon_*^2 = \tau_w/\rho \qquad (8)$$

An exponent 1/8 or even 1/10 may give better agreement with experimental data for very high Reynolds number; therefore the expression $A(\upsilon_* R/\nu)^{1/n}$, instead of $8.74 \ (\upsilon_* R/\nu)^{1/7}$, has been applied in the present analysis.

Setting $V_{ref} = U_{max}$ and $\delta = R$ we obtain the expression for a wall shear stress along the disk

$$\tau_o = \bar{A}^{-2n/(n+1)} \rho \ V_{ref}^{2n/(n+1)} (\nu/\delta)^{2/(n+1)} \qquad (9)$$

setting $A' = A^{-2n/(n+1)}$ and substituting (7) into (9) we obtain

$$\tau_o = A' \ \rho \ (\nu/\delta)^{2/(n+1)} \ (r\omega)^{2n/(n+1)} \ (\alpha^2 + k^2)^{n/(n+1)} \qquad (9'')$$

From (5) and (6) we have $V_r/V_{\Theta rel} = \tau_r/\tau_\Theta = -\alpha/k$, hence

$$\tau_\Theta = \tau_o \ [1 + (\alpha/k)^2]^{-0.5} \qquad (10)$$

Substituing (9'') into (10) we obtain

$$\tau_\Theta = A'\rho \ (\nu/\delta)^{2/(n+1)} \ (r\omega)^{2n/(n+1)} \ k(\alpha^2 + k^2)^{(n-1)/(2n+2)} \qquad (11)$$

Equations of motion in radial and circumferential directions can be written as

$$V_r \frac{\partial V_r}{\partial r} + V_z \frac{\partial V_r}{\partial z} - \frac{V^2_\Theta}{r} = -\frac{1}{\rho} \frac{\partial p}{\partial r} + \frac{1}{\rho} \frac{\partial \tau_{zr}}{\partial z} \qquad (12)$$

$$V_r \frac{\partial V_\Theta}{\partial r} + V_z \frac{\partial V_\Theta}{\partial z} + \frac{1}{r} V_\Theta V_r = \frac{1}{\rho} \frac{\partial \tau_{z\Theta}}{\partial z} \qquad (13)$$

for $\rho = $ const., continuity eq. can be written

$$V \cdot \bar{V} = \frac{\partial V_r}{\partial r} + \frac{V_r}{r} + \frac{\partial V_z}{\partial z} = 0 \qquad (14)$$

All other derivatives of the stress tensor components on the right hand side of equations (12) and (13) can be omitted in this case, and for simplicity the following notation is adapted:

$$\tau_{z\Theta} = \tau_\Theta \ , \ \tau_{zr} = \tau_r$$

Equations (12) and (13) can be integrated along z from 0 to the outer edge of the boundary layer (δ). It is assumed that the pressure is constant along z and it is only a function of r. In some papers, e.g. Dorfman in [2], terms containing V_z in equations (12) and (13) are considered to be negligible and are omitted in the solution of the problem of the enclosed disk. This is unnecessary simplification, as the set of equations can easily be solved retaining those terms.

As a result of integration of (12) and (13), we obtain

$$\frac{d}{dr}\left(r\int_0^\delta v^2_r\, dz\right) - \int_0^\delta v^2_\Theta\, dz = -\frac{r}{\rho}\tau_r - \frac{r}{\rho}\delta\frac{dp}{dr} \qquad (15)$$

$$\frac{d}{dr}\left(r^2\int_0^\delta v_r v_\Theta\, dz\right) = -\frac{r^2}{\rho}\tau_\Theta \qquad (16)$$

After substituting equations (1) and (2) into equations (15) and (16), and after integration and rearrangement of terms, we get

$$\alpha^2\omega^2 a\,(3r^2\delta + r^3\frac{d\delta}{dr}) - r^2\omega^2\delta\,(bk^2 - 2kc + 1) + r^2\omega^2\delta\,(1-k)^2 = -\frac{r}{\rho}\tau_r \qquad (17)$$

$$\alpha\omega^2 e\,(4r^3\delta + r^4\frac{d\delta}{dr}) = -\frac{r^2}{\rho}\tau_\Theta \qquad (18)$$

where $a = \dfrac{n}{n+2} + \dfrac{n}{3n+2} - \dfrac{n}{n+1}$, $b = \dfrac{n}{n+2}$, $c = \dfrac{n}{n+1}$, $d = \dfrac{n}{2n+1}$

$$e = c - d + k\,(0.5c - b)$$

Equations (17) and (18), which describe the flow in the disk boundary layer have two unknowns: α and δ; k can be considered as a parameter, to be determined later. Equations (17) and (18) can easily be solved by anticipating that (r/δ) $(d\delta/dr)$ = constant. This leads to δ = constant $\cdot r^m$, where m can be determined from dimensional analysis, resulting in m = $(n-1)/(n+3)$. The final form of the solution of (17) and (18) is given by

$$\delta_d = A^{\frac{-2n}{n+3}}\,(\frac{5n+11}{n+3})^{-\frac{n+1}{n+3}}\,(\alpha e)^{-\frac{n+1}{n+3}}\,k^{\frac{n+1}{n+3}}\,(\alpha^2 + k^2)^{\frac{n-1}{2n+6}}\,(\nu/\omega)^{\frac{2}{n+3}}\,r^{\frac{n-1}{n+3}} \qquad (19)$$

$$\alpha_d = k\,[\frac{2/(n+1) - k2/(n+2)}{ka(4n+8)/(n+3) + e\,(5n+11)/(n+3)}]^{0.5} \qquad (20)$$

Substituting (19) into (11) we obtain

$$\tau_{\Theta d} = B\rho\nu^{\frac{2}{n+3}}\,\omega^{\frac{2(n+2)}{n+3}}\,r^{\frac{2(n+1)}{n+3}} \qquad (21)$$

where $B = A^{\frac{-2n}{n+3}}\,(\frac{5n+11}{n+3})^{\frac{2}{n+3}}\,k^{\frac{n+1}{n+3}}\,(\alpha^2_d + k^2)^{\frac{n-1}{2(n+3)}}\,(\alpha_d e)^{\frac{2}{n+3}}$

moment is given by

$$M_d = \int_0^R 2\Pi r^2\tau_\Theta\, dr = \frac{2\Pi}{\frac{2(n+1)}{n+3}+3}\,B\rho\nu^{\frac{2}{n+3}}\,\omega^{\frac{2(n+2)}{n+3}}\,R^{\frac{2(n+1)}{n+3}+3} \qquad (22)$$

and moment coefficient

$$C_{Md} = M_d\,/\,(0.5\rho\omega^2 R^5) = \frac{4\Pi}{\frac{2(n+1)}{n+3}+3}\,B\,Re^{\frac{-2}{n+3}} \qquad (23)$$

From the mass flow rate in a disk boundary layer given by $\dot{m}_d = 2\Pi\rho r\int_0^\delta V_r\, dz$, the mass flow coefficient can be found in the form

$$C_{\dot{m}d} = 2\Pi\,(\frac{n}{n+1} - \frac{n}{2n+1})A^{\frac{-2n}{n+3}}\,(\frac{5n+11}{n+3})^{-\frac{n+1}{n+3}}\,(\frac{k}{e})^{\frac{n+1}{n+3}}\,(k^2+\alpha^2)^{\frac{n-1}{2n+6}}\,\alpha^{\frac{2}{n+3}}\,Re r^{\frac{-2}{n+3}} \qquad (24$$

Similar procedure can be applied to the stationary wall boundary layer. Assuming velocity profiles

$$V_r = -\alpha_w r \, \omega (1-k) \, \eta^{1/n}(1-\eta) \quad (25) \qquad V_\theta = r\omega(1-k)\eta^{1/n} \quad (26)$$

we finally arrive at the following relationships

$$\delta_w = A^{\frac{-2n}{n+3}} f \alpha_w^{-\frac{n+1}{n+3}} (1+\alpha_w^2)^{\frac{n-1}{2(n+3)}} [\frac{\nu}{\omega(1-k)}]^{\frac{2}{n+3}} \cdot r^{\frac{n-1}{n+3}} \quad (27)$$

$$\alpha_w = [\frac{4(n+3)(n+1)(3n+2)}{n^2(5n+11)(3n+2) - 8n^3(n+2)}]^{0.5} \quad (28)$$

$$\tau_{\theta w} = A^{\frac{-2n}{n+3}} f^{\frac{2}{n+3}} \rho \nu^{\frac{2}{n+3}} [\omega(1-k)]^{\frac{2(n+2)}{n+3}} \alpha_w^{\frac{2}{n+3}} (1+\alpha_w^2)^{\frac{n-1}{2(n+3)}} r^{\frac{2(n+1)}{n+3}} \quad (29)$$

$$C_{Mw} = \frac{4\Pi}{\frac{2(n+1)}{n+3}+3} B' (1-k)^{\frac{2(n+2)}{n+3}} Re^{\frac{-2}{n+3}} \quad (30)$$

where $\quad f = (\frac{n}{n+2} - \frac{n}{2(n+1)})(\frac{5n+11}{n+3})$, $B' = A^{\frac{-2n}{n+3}} f^{\frac{2}{n+3}} \alpha_w^{\frac{2}{n+3}} (1+\alpha_w^2)^{\frac{n-1}{2(n+3)}}$

The effect of shrouds is accounted for in the following way

$$M_{rot} = M_d + M_{dsh} \quad (31) \qquad M_{cas} = M_w + M_{wsh} \quad (32)$$

where M_d and M_w are moments at the plane part of the disk and the parallel wall, while M_{dsh} and M_{wsh} are corresponding additional moments at the rotating shroud, being a part of the rotor, and at the stationary shroud, respectively.

The shear stress at the shroud ($\tau_{\theta dsh}$ or $\tau_{\theta wsh}$) is assumed to be the same as at the disk or the wall at the corresponding radius. This seems to be an acceptable approximation on the basis of the following reasoning: the circumferential velocity, which has a predominant effect on τ_θ, is the same at the shroud and at the disk, or wall; the thickness of the boundary layer can change, but shear stress is a weak function of δ ($\tau \sim \delta^{-(2/(n+1))}$) and e.g. for $n = 7$ an increase of δ by 100 % results in the change in τ by about 15 %. Consequently M_{dsh} and M_{wsh} can be written as

$$M_{dsh} = \tau_{\theta dsh} \, 2\Pi R^2 S_d \quad (33) \qquad M_{wsh} = \tau_{\theta wsh} \, 2\Pi R^2 S_w \quad (34)$$

where $\tau_{\theta dsh}$ and $\tau_{\theta wsh}$ are assumed to be equal to $\tau_{\theta d}$ and $\tau_{\theta w}$ at $r = R$, respectively. Moments at the rotor (disk and shroud) and at the casing (wall and shroud) satisfy the following relationship

$$M_{rot} - M_{cas} = \dot{m}_{cool} (r_{out} \bar{V}_{\theta out} - r_{in} \bar{V}_{\theta in}) \quad (35)$$

where $\bar{V}_{\theta out}$ and $\bar{V}_{\theta in}$ are mean values of the circumferential velocity of the coolant. It is assumed that in the range $0 < \dot{m}_{cool} \le \dot{m}_{f.d.}$, an outflow of coolant takes place in the disk boundary layer, and that

$$\bar{V}_{\theta out} = (1-k) \, r\omega + L [r\omega - (1-k) \, r\omega] = r\omega (Lk + 1-k) \quad (36)$$

where L is a const. Using (21) to (24), (29) to (34) and (36), (35) can be written as

$$2\Pi B (\frac{n+3}{5n+11}+S'_d) - 2\Pi B'(1-k)^{\frac{2n+4}{n+3}} (\frac{n+3}{5n+11}+S'_w) - C_m R_e^{\frac{2}{n+3}} [D - (\frac{r_{in}^2}{R})(\frac{\omega_{in}}{\omega})] = 0 \quad (37)$$

where $D = Lk + 1-k$. Const. L can be found from the free disk conditions ($S'_d = S'_w = 0$; $k = 1$). For a given geometry (shrouds) and coolant flow rate k can be found from (37) and then the moment, moment coefficients or shear stresses can be determined from (21) to (23) and from (29) to (34). For the free disk conditions ($k = 1$, $S_d = S_w = 0$) and for $n = 7$ and $A = 8.74$ we obtain from (23) and (24)

$$C_M Re^{0.2} = 0.07288 \qquad (38) \qquad C_{\dot{m}} Re^{0.2} = 0.2188 \qquad (39)$$

It is assumed that pressure remains const. in θ and z directions; from (12) pressure gradient can be written as

$$\frac{dp}{dr} = \rho \ (v_\theta^2 /r - \bar{V}_r \ \partial \bar{V}_r /\partial r) \tag{40}$$

where $\bar{V}_r = \dot{m}_{cool} / (2\Pi rs\rho)$. Terms $V_z \ \partial V_r/\partial z$ and $\partial \tau_{zr}/\partial z$ are omitted, being considered negligible.
From (40) we obtain

$$\Delta p = 0.5\rho \ [(1-k)^2 \omega^2 (r_2^2 - r_1^2) + (\frac{\dot{m}_{cool}}{2\Pi s\rho})^2 \frac{1}{r_1^2} - \frac{1}{r_2^2}] \tag{41}$$

In most of actual situations and for $0 < \dot{m}_{cool} \le \dot{m}_{f.d.}$, only the first term on the r.h.s. of eg. (40) or (41), which determines the effect of the core rotation, is significant. In one of BBC turbine, where $\dot{m}_{cool} \sim 0.7 \ \dot{m}_{f.d.}$, the second term of eq. (41) represented only 1 % of the entire Δp between r_{in} and R.

The experimental results, obtained just recently in the Turbines Institute of the Tech. University in Aachen indicate clearly, that for $0 < \dot{m}_{cool} < \dot{m}_{f.d.}$ Δp depends only on the core rotation. Δp decreases with the increase of $\dot{m}$, becomes almost zero for $\dot{m} = \bar{m}_{f.d.}$, and then Δp increases again with the further increase of $\dot{m}$, but this time the shape of the curve p (r) clearly indicates that Δp is due to the diffusor effect (2-nd term of eq. 41).

Free disk and enclosed disk with $\dot{m}_{cool} = 0$ constitute the limiting cases of the solution presented here. For $k = 1$, equations from (19) to (24) constitute a solution for a free disk, and for $n = 7$, moment coefficient corresponds to that quoted in the literature. For an enclosed disk with negligible effect of shroud and with absence of through flow ($\dot{m}_{cool} = 0$), the core rotates at $\beta/\omega = 0.487$ ($k = 0.513$), and the corresponding value of moment coefficient $C_M Re^{0.2} = 0.0371$; ($2 \times C_M Re^{0.2} = 0.0742$). Schlichting quotes 0.0622 for $C_M Re^{0.2}$ in this case, with a comment that it is about 17 % less than the experimental results. Taking into account this correction $2 \times C_M Re^{0.2}$ would be equal to 0.075. On the other hand it is difficult to imagine an arrangement of experiment, in which both through flow and shroud are absent, and the stationary shroud always results in an increase of C_M (see Fig. I). Consequently such a situation ($\dot{m}_{cool} = 0$, $s_w = 0$) should only be considered as a reference system. Dorfman in [2] obtained $C_M Re^{0.2} = 0.0357$, probably due to the fact that he omitted terms containing V_z in equations (12) and (13), while solving this case.

A comparison of the present calculation ($n = 7$) with the experimental results of Daily and Nece [1] for an enclosed disk are presented in Table 1 ($\dot{m}_{cool} = 0$, $S_d = 0$).

Fig. 1 presents an influence of shrouds (S_d or S_w) on the moment coefficient and core rotation for the enclosed system ($\dot{m}_{cool} = 0$). When S_w increases, the core rotation (β) decreases, resulting in an increase of the moment coefficient (C_{Md}). When $S_w \to \infty$, then $\beta \to 0$ and $C_{Md} \to C_{Mf.d.}$. The core rotation increases with an increase of S_d, resulting in a decrease of C_{Md}, but the prevailing effect of C_{Mdsh} results in an increase of C_{Mrot}.

The relationship between the mass flow coefficient and the moment coefficient for $S_d = S_w = 0$ is presented in Fig. 2. The hitherto published experimental results imply, that an increase of $\dot{m}_{cool}$ from $\dot{m}_{f.d.}$ up to 2 to 3 x $\dot{m}_{f.d.}$, results in a very small increase of C_{Md} over $C_{Mf.d.}$, and the greater is S, the smaller is an influence of $\dot{m}$ greater than $\dot{m}_{f.d.}$ on C_{Md}. Consequently the method of computing C_M presented here, is valid for coolant flow rates from zero to at least 2 x $\dot{m}_{f.d.}$. In the first rotor of one of BBC turbines $\dot{m}_{cool}$ is 2 to 3 kg/s, while $\dot{m}_{f.d.}$ = 3 kg/s (Re_R = 2, 3 x 10^7). Reasonable results could be expected up to $\dot{m}_{cool} \sim 7$ kg/s.

3.   HEAT TRANSFER FROM A ROTATING DISK

3.1 Reynolds Analogy

Energy equation for a steady process can be written as

$$\rho \frac{dH}{dt} = \Phi + \rho \vec{V} \cdot \vec{f} + \nabla \cdot (\lambda \nabla T) \tag{42}$$

Equations (12), (13), (14) and (42) describe the problem. The following simplifying assumptions are necessary in order to arrive at the identity of the momentum and energy equation (Reynolds analogy). Equation (12) is reduced to $\rho V_\theta^2/r =$ = dp/dr, considering the remaining terms negligible; it is assumed that

$$\Phi = \tau_{z\theta} \, \partial V_\theta/\partial z, \quad \rho \vec{V} \cdot \vec{f} = V_\theta \partial \tau_{z\theta}/\partial z, \quad \nabla \cdot (\lambda \nabla T) = \partial/\partial z (\lambda \partial T/\partial z),$$

$T_T = T + V_\theta^2/(2Cp)$, Cp = const and Pr = 1. As a result, (42) takes on the form

$$\rho (V_z \frac{\partial T_T}{\partial z} + V_r \frac{\partial T_T}{\partial r}) = \frac{\partial}{\partial z} (\mu_{eff} \frac{\partial T_T}{\partial z}) \tag{43}$$

(43) becomes identical to (13) if $T_T = C_1 + C_2 \, rV_\theta$ \hfill (44)

(44) implies T = const. = C_1 at z = s (stationary wall) and

$$T = T_w + crV_\theta - V_\theta^2 / (2Cp) \qquad (45) \qquad \text{at } 0 \leq z < S.$$

At the disk (45) takes on the form

$$T_d = T_w + r^2 \omega \, (C - \frac{\omega}{2Cp}) \tag{46}$$

Heat flux can be written as q = $\tau_{z\theta}$ (V_θ - C Cp r) \hfill (47)

(47) determines the heat flux at the solid boundaries of the system, as $\tau_{z\theta}$ is known at z = 0 (disk) and z = s (stationary wall). Setting q_d = 0 one obtains from (47) C = C_{ad} = ω/Cp and then from (46) the adiabadic disk temperature can be written as

$$T_{dad} = T_w + (\omega^2 r^2) / (2 Cp) \tag{48}$$

and the heat flux at the disk

$$q_d = C_p \frac{\tau_d}{r\omega} (T_d - T_{dad}) = C_p r \tau_d (C - C_{ad}) \tag{49}$$

Local Nusselt number can be expressed as Nu = $\tau_d/(\mu\omega)$ or, using (21) for n = 7 we obtain

$$N_u = B \, R_{er}^{0.8} \qquad (50) \qquad \text{or} \qquad Nu = B \, (r/R)^{1.6} \, R_{eR}^{0.8} \qquad (51).$$

For r = R, equations (46) to (51) are valid for the rotating shroud as well.

Using (21) for n = 7, the amount of heat transferred at the surface of the disk between 0 and r can be written as

$$Q_d = 2\Pi \int_0^r q_d \, rdr = 0.4348 \, \Pi \, Cp \, (C - C_{ad}) \, B\rho \, \nu^{0.2} \omega^{1.8} r^{4.6} \tag{52}$$

Similar expression can be found for Q, transferred at the shroud, and the total amount of heat, exchanged between the rotor and coolant can be expressed as

$$Q_{rot} = Q_d + Q_{sh} = \Pi Cp(C-C_{ad})B\rho\nu^{0.2}\omega^{1.8}R^{4.6}(0.4348 + 2 S'_d) \tag{53}$$

Q_d and Q_{rot} can be expressed by means of friction moment in the following way

$$Q_d = Cp \ (C-C_{ad}) \ M_d \tag{54} \qquad Q_{rot} = C_p \ (C-C_{ad}) \ M_{rot} \tag{55}$$

Mean Nusselt number can be defined as

$$\bar{N}_u = \frac{\bar{q}r}{\lambda \ (T_d-T_{dad})_{av}} \tag{56}$$

where $\bar{q} = Q/(\Pi r^2) = (2/r^2)\int_0^r qr \ dr$

and $(T_d - T_{dad})_{av} = (2/r^2)\int_0^r (T_d - T_{dad}) \ rdr = 0.5 \ \omega r^2 \ (C - C_{ad})$

Using (54) and expressing moment by means of moment coefficient we obtain from (56)

$$\bar{N}_u = \frac{1}{\Pi} Cp/\lambda \rho C_M \omega r^2 \tag{57}, \text{ for Pr = 1,} \qquad \bar{Nu} = \frac{1}{\Pi} C_M \ Rer \tag{58}$$

Substituting (23) for n = 7 into (58) we obtain

$$\bar{Nu} = 0.0567 \ k^{0.8} \ (\alpha_d^2 + k^2)^{0.3} \ (\alpha_d e)^{0.2} \ Rer^{0.8} \tag{59}$$

Fig. 2, showing moment coefficient as a function of mass flow coefficient can also be interpreted as the presentation of the mean Nusselt number as a function of mass flow coefficient; it follows form (58) that $\Pi Rer^{-0.8} \ \bar{Nu} = C_M \ Rer^{0.2}$

Figure 3 presents the local Nusselt number (Nu) as a function of radius (r/R) for Rer = 2.312 x 10[7] with shrouds (S'$_w$ or S'$_d$) and $\dot{m}_{cool}$ as parameters. It is evident that the influence or the shroud decreases with the increase of $\dot{m}_{cool}$. In particular the stationary shroud becomes less "felt" with the increase of $\dot{m}_{cool}$. E.g. for $\dot{m}$ = 1, there is some difference (though very small) between S'$_d$= = 0.05 and 0.1, while Nu is almost constant (at the given r) for a wide range of S'$_w$. It is also evident that the presence of the stationary shroud (S$_w$) results in an increase of Nu, while the rotating shroud (S$_d$) acts in the opposite manner. This is clear, bearing in mind the relationship between the moment coefficient and the shrouds.

Starting from $\dot{m}_{cool} \simeq \dot{m}_{f.d.}$, the further increase of $\dot{m}_{cool}$ affects Nusselt number (like moment coefficient) only very slightly. This is illustrated in the table 2, where the results of this analysis are compared with Owen's experiments (p. 145-147 in [5]) for unshrouded system and for Re$_R$ = 4 x 10[6]. More than two fold increase of $\dot{m}$ gives only 10 % increase in $\bar{Nu}$, and in one case $\bar{Nu}$ remains constant (with the accuracy of reading small diagram in the log. scale). The numerical values of $\bar{Nu}$ in [5] are, however, by about 10% higher (in one case, see last column, even more) than computed here. One of the reasons for the difference could be a temperature distribution along the disk, which was not a quadratic one in Owen's experiments (Figure 2 in [4]).

3.2 Solution With an Arbitrary Radial Temperature Distribution at the Disk

Dorfman found a solution for the heat transfer from the freely rotating disk with an arbitrary disk temperature distribution [2].

Considering the heat balance for an annular element of the thermal boundary layer of thickness δ_T the following integral relation was obtained

$$qr = \rho \, C_p \frac{d}{dr} \left(\int_0^{\delta_T} rV_r T dz \right) \tag{60}$$

Setting $\int_0^{\delta_T} V_r T dz = r\omega T_d \delta_T^{**}$ one obtains $\delta_T^{**} = \frac{1}{r\omega T_d} \int_0^{\delta_T} V_r T dz$

Thermal boundary layer Reynolds number was defined as $\overset{**}{Re_T} = (r\omega/\nu) \, \overset{**}{\delta_T}$. As a solution the following formula was obtained

$$(\overset{**}{Re_T})^{m+1} = \frac{m+1}{A_T} Re r T_d^{-(m+1)} r^{-(m+3)} \int_0^r T_d^{(m+1)} r^{(m+2)} \, dr \tag{61}$$

where $T_d(r)$ is an arbitrary function.
Having given $\overset{**}{Re_T}$, $\bar{Nu}$ can be found from the relation

$$\bar{Nu} = 2 \, Pr \, \overset{**}{Re_T} \frac{T_d(r)}{\bar{T}_d(r)} \tag{62}$$

Local Nusselt number can be evaluated from the equation

$$Nu = \bar{Nu} + \frac{1}{T_d} \left(\int_0^r T_d r dr \right) \frac{d}{dr} \left(\frac{\bar{Nu}}{r} \right) \tag{63}$$

Constants m and A_T in (61) can be found from the known, particular solution. The known solution for $T_d = cr^2$ yields $\bar{Nu} = 4 \, Pr \, \overset{**}{Re_T}$ (64)

Dorfman found m = 0.25 and A_T = 135.8 $[Pr/\phi(Pr)]^{1.25}$ where

$$\phi(Pr) = 0.72/[1-2.73\sqrt{N_{uPr} = 1/Rer}] \text{ for } Pr_{lam} = 0.72. \tag{65}$$

A_T = 135.8 for Pr = 1.

If frictional terms are omitted in the energy equation and p and Cp are assumed to be const., then (42) is reduced to $\rho \, CpdT/dt = \nabla \cdot (\lambda \nabla T)$; this equation identifies Dorfman's formulation of the problem. As a consequence the concept of adiabatic temperature is not existent. It means, referring to the disk case, that the relation $q \sim (T_d - T_w)$ holds rather than $q \sim (T_d - T_{dad})$. It turns out that simply replacing $(T_d - T_w)$ by $(T_d - T_{dad})$ and redefining Nusselt number accordingly, Dorfman's solution becomes valid for the actual flow with friction. For $T_d = T_w + ar^2$ and $T_{dad} = T_w + br^2$ we have $T_d - T_{dad} = (a - b)r^2$. Substituting the last expression into (61) we obtain

$$\overset{**}{Re_T} = (\frac{1}{A_T} \frac{m+1}{3m+5})^{1/(m+1)} Re r^{1/(m+1)} \tag{66}$$

From (59) for k = 1 (free disk) we have

$$\bar{N}_u = 0.0232 \, Re r^{0.8} \tag{67}$$

From (64), (66) and (67) for Pr = 1 we obtain m = 0.25 and A_T = 135.8. The convenient form of the approximation of the given $T_d(r)$ is the expression of the type

$$T_d = T_{dad} + Cr^\gamma = a + br^2 + Cr^\gamma \quad (68) \qquad \text{It follows that} \qquad T_d - T_{dad} = Cr^\gamma \quad (69)$$

Replacing T_d by $(T_d - T_{dad})$ and assuming temperature distribution according to (69), we obtain from (61), (62) and (63), for Pr = 1, the following relations

$$\overset{**}{Re_T} = (1.25/A_T)^{0.8} (1.25\gamma + 3.25)^{-0.8} Re r^{0.8} \tag{70}$$

$$\bar{Nu} = 2\overset{**}{Re_T} \frac{T_d - T_{dad}}{(T_d - T_{dad})_{av}} = (1.25/A_T)^{0.8} (\gamma+2)(1.25\gamma+3.25)^{-0.8} Re r^{0.8} \tag{71}$$

$$Nu = A_T^{-0.8} \; (\gamma+2.6)^{0.2} Re_r^{0.8} = \frac{\gamma+2.6}{\gamma+2} \; \overline{Nu} \tag{72}$$

where $(T_d - T_{dad})/(T_d - T_{dad})_{av} = 0.5 \; (\gamma + 2)$. Constant $m = 0.25$ remains the same for any configuration, while A_T can be found from the known solution for $\gamma=2$. From (71) and (59) we obtain

$$A_T = 44.45 \; k^{-1} \; (\alpha^2_d + k^2)^{-0.375} \; (\alpha_d e)^{-0.25} \tag{73}$$

For $Pr \neq 1$, the mean Nusselt number can be written as

$$\overline{N}_u = (\gamma+2) \; (\frac{1.25}{1.25\gamma+3.25})^{0.8} \; A_T^{-0.8} \phi \; (Pr) Re_r^{0.8} = \phi(Pr) \overline{Nu}_{Pr = 1} \tag{74}$$

If $Pr = 0.72$, $\phi \; (Pr)$ is defined by equation (65), or for $Re_r=10^6$ the simple approximation $\phi \; (Pr) \cong Pr^{0.6}$ can be used. Assuming $\phi \; (Pr)$ constant along radius, local Nusselt number can be found from (74) using (72). If $Pr \neq 1$, the adiabatic disk temperature can be defined as $T_{dad} = T_w + \mathcal{R} \; (\omega r)^2 / (2 \; C_p)$, where the recovery factor $\mathcal{R}$ is a function of Pr_{turb} and Pr_{lam}. It can be approximated by a simple formula $\mathcal{R} = (Pr_{lam})^{1/3}$.

The results of computation for a free disk ($k = 1$) for two different temperature distributions are presented in Table 3. The computation has been performed for the following data: $\rho = 4.15 \; kg/m^3$, $\nu = 8 \times 10^{-6} \; m^2/s$, $\omega = 663.7/s$, $C_p= 1100 \; J/(kgK)$, $T_w = 700 \; K$, $T_{dR} = 850$, $T_{dad} = 700 + 200.23r^2$ ($Pr = 1$), $T_{dad} = 700 + 179.46r^2$ ($Pr = 0.72$), $T_d = 700 + 538.05r^2$ ($\gamma = 2$), $T_d = 700 + 1019.04r^3$, approximated by $T_d - T_{dad} = 1028.93r^{3.65}$ ($\gamma = 3.65$, $Pr = 0.72$). It is evident that Q from 0 to 0.5R amounts to only 3.8 % of Q_{0-R} for $\gamma = 2$, and even much less for $\gamma > 2$. Nusselt number increases with γ, but the actual amount of heat, transferred from the disk for the same temperature range (700, 850) decreases.

4. CONCLUSIONS

1. The moment coefficient, as well as local and mean Nusselt numbers, have been determined as functions of geometry (stationary and/or rotating shroud) coolant flow rate, Reynolds and Prandtl numbers.

2. The solution is basically valid for $0 \leq \dot{m}_{cool} \leq \dot{m}_{f.d.}$. Its validity can be extended up to about $2 \times \dot{m}_{f.d.}$, as both, C_M and Nu do not exceed $C_{Mf.d.}$ and $Nu_{f.d.}$ (in this range of $\dot{m}_{cool}$) by more than a few percent. The solution is valid for any range of Reynolds numbers, as the velocity profiles in the boundary layers satisfy a $1/n$ power law, where n is an arbitrary number.

3. If $S > \delta_d + \delta_w$, the further increase of the gap itself does not affect C_M or Nu, if shrouds (S_d and/or S_w) remain unchanged or are not present (Fig. 5, p. 464 in [4]. The system is, however, usually bounded by the shroud, which affects the flow and heat transfer phenomena. The influence of shroud is decreasing with the increase of $\dot{m}_{cool}$, and it is almost negligible if $\dot{m}_{cool} \geq \dot{m}_{f.d.}$, especially the influence of a stationary shroud.

4. In a heat transfer problem the boundary condition at the disk, $T_d = T_w + Cr^\gamma$, is "flexible" enough to approximate well the actual possible temperature distribution, whereas the condition at the stationary wall, $T_w = const$. is a rather restrictive one. If $\dot{m}_{cool} \geq \dot{m}_{f.d.}$, the condition $T_w = const$. coincides with the condition $\partial T/\partial z = 0$ at the wall. This makes the free disk solution an acceptable approximation of the actual situation. Preliminary measurements in one of BBC turbines showed surprisingly, that temperature along the stationary wall is really almost const.

5.   The whole analysis is valid for the full disk, from r = 0 to r = R, and for constant physical properties of the coolant. Inaccuracy resulting from the actual r_{in} > 0 is relatively minimal, as heat transferred from the inner part of the disk (0 ≤ r ≤ 0.5 R) amounts to only 4 % (for γ = 2) or less (for γ > 2) of the total amount of heat, transferred between r = 0 and r = R. The same applies to the friction moment. Although the flow parameters at any given radius depend on r_{in}, but bearing in mind that all flow phenomena are predominantly dependent on circumferantial velocity component (V_Θ), the dependence on r_{in} (at least for r_{in} ≤ 0.5) is negligible. On that basis, an error arising from constant physical properties can be minimized, assuming these properties (μ, Cp, ρ) equal those at certain radius close to R; e.g. for r_{in}= 0.5 R at R = 0.85 R.

5. TABLES AND DIAGRAMS

Table 1:

S'_w	β/ω	$C_M Re^{0.2}$ x 2 [1]	β/ω	$C_M Re^{0.2}$ x 2 present calc.
0,025	0,490	0,0755	0,470	0,0776
0,0636	0,464	0,0772	0,447	0,0788
0,10	0,445	0,080	0,428	0,0826
0,2145	0,411	0,0873	0,382	0,0894

Influence of the stationary shroud on the core rotation and moment coefficient (S_d=0; $\dot{m}_{cool}$=0). Comparison of the present calculation with Daily and Nece[1] results.

Table 2:

$\dot{m}/\dot{m}_{f.d.}$	0.76	1.12	1.77	2.34
Nu	4260	← 4437 →		
Nu,S'=.02	4470	4480	5000	
Nu,S'=.03		5000	5000	5000
Nu,S large			5000	5000

[5]

Nu as a function of $\dot{m}_{cool}/\dot{m}_{f.d.}$ and S'. Comparison of the present calculation with Owen and Haynes [5] results.

Table 3:

r	r/R	Rer x10^{-6}	N̄u	Nu	qx10^{-4}	Q_{o-r} x10^{-4}	Pr	γ
			5780	6647	2.12	0.19	1.0	2
0.259	.49	5.56	4746	5458	2.57	0.24	0.72	2
			5246	5803	0.84	0.057		3.65
			18067	20777	13.53	5.15	1.0	2
0.528	1.0	23.12	14835	17060	16.38	6.24	0.72	2
			16397	18138	17.42	4.88		3.65

Heat transfer computation for a free disk for 2 different temperature distribution and R_{er} = 23.12 x 10^6.

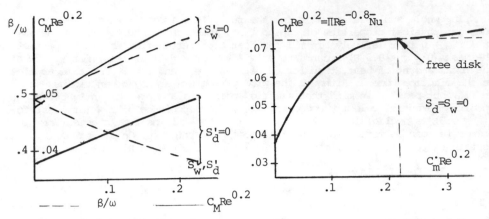

Fig.1. Moment coefficient and core rotation as functions of shrouds for $\dot{m} = 0$

Fig.2. Moment coefficient and Nusselt number as functions of mass flow rate

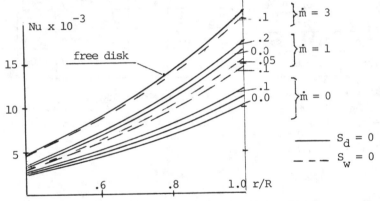

Fig.3. Influence of shrouds and mass flow rate on local Nusselt number for $Re_R = 23.12 \times 10^6$

6. REFERENCES

[1] Daily J.W., Nece R.E.: "Chamber Dimension Effects on Induced Flow an Frictional Resistance of Enclosed Rotating Disks", J. of Basic Eng., March 1960.

[2] Dorfman L.A.: "Hydrodynamic resistance and heat loss of rotating solids", Oliver and Boyd, Edinburgh, 1963.

[3] Bayley F.J., Owen J.M.: "Flow Between a Rotating and a Stationary Disc", Aeronautical Quarterly, Vol. 20, 1969.

[4] Owen J.M., Haynes C.M., Bayley F.J.: "Heat Transfer from an Air-Cooled Rotating Disk", Proc. R. Soc., London A.336, 1974.

[5] Owen J.M., Haynes C.M.: "Design formulae for the heat loss and frictional resistance of air-cooled rotating disks", Improvements in Fluid Machines and Systems for Energy Conversion, 4, Haepli, Milan, 1976.

Experiments on Friction, Velocity and Pressure Distribution of Rotating Discs

G. DIBELIUS, F. RADTKE, and M. ZIEMANN
Institute for Steam and Gas Turbines
of the Technical University Aachen
Aachen, FRG

ABSTRACT

Frictional effects and related flow mechanisms in gaps between a housing and a rotating disc have been investigated experimentally. The rotational Reynolds number varied from $2 \cdot 10^6$ to $3 \cdot 10^7$ for a disc radius of 0.4 m. Different axial clearances and different radial limitations of the gaps as well as screw heads mounted at the disc were studied. The effect of superimposed air flow through the gaps in both centrifugal and centripetal directions on velocity field, pressure distribution, axial thrust, and frictional torque is reported.

1. INTRODUCTION AND NOMENCLATURE

The secondary flow of fluid in the gap between a housing and an enclosed rotor is induced by frictional effects. Actually in turbomachinery this motion is superimposed by a throughflow out of disc sealing and cooling systems. This throughflow has a considerable effect on the flow mechanisms. In particular massflow as well as the swirl at the entrance of the gap have a considerable effect on friction losses, pressure distribution and axial thrust.

The following nomenclature in alphabetic arrangement and according to Fig. 1 is used.

b thickness of disc

c_r absolute radial velocity component

c_u absolute tangential velocity component

$C_{F1} \equiv \dfrac{\dot{m}_{F1}}{R\,\mu}$ throughflow number

$C_M \equiv \dfrac{M}{\frac{1}{2}\,\rho\,\omega^2\,R^5}$ friction torque number on one face of the disc

$C_p \equiv 1 - \dfrac{P_{(R)} - P_{(r)}}{\rho\,\omega^2\,R^2}$ pressure number

117

$$C_{Th} \equiv \frac{\int_{r_i}^{R} (p_{(R)} - p_{(r)}) \, 2 \pi r \, dr}{\rho \, \omega^2 \, R^2 \, (R^2 - r_i^2) \, \pi} \qquad \text{axial thrust number}$$

$$C_V \equiv \frac{\dot{V}_r}{\omega \, R^3 \, \pi} \qquad \text{volume flow number}$$

$f \, ()$ function of

$K \quad \equiv \frac{\beta}{\omega}$ core rotation factor

l_i all lengths for describing the geometrie of the gap

$\dot{m}_{F1}$ superimposed mass flow

M frictional torque on one face of the disc

p pressure

r radius

r_i inner radius

r_w radius of the cylindrical wall in the gap

R radius of the disc

$Re \quad \equiv \frac{\rho \, \omega \, R^2}{\mu}$ rotational Reynolds number

s axial clearance between disc and housing

t radial tip clearance of the disc

$\dot{V}_r \quad \equiv r \, \pi \int_{0}^{s} |c_r| \, dz$ total volume flow radially transported by secondary flow only

z distance normal to the disc

α exit angle of the swirl vanes, opposed to circumferential direction

β angular velocity at $z = s/2$

μ dynamic viscosity

ρ density of the fluid at the radius R

ω angular velocity of the disc

     For ready application of the experimental results to different cases occuring in turbomachines they have to be expressed in dimensionless terms. The dimensionless expressions used for the correlation are

$$\left. \begin{array}{l} \dfrac{c_u}{\omega \, r} \\[4pt] C_V \\[2pt] C_p \\[2pt] C_{Th} \\[2pt] C_M \end{array} \right\} \quad = \quad f(Re, \; C_{F1}, \; \alpha, \; \tfrac{l_i}{R}) \qquad\qquad (1)$$

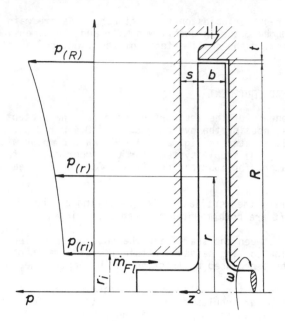

Fig. 1.   Notation

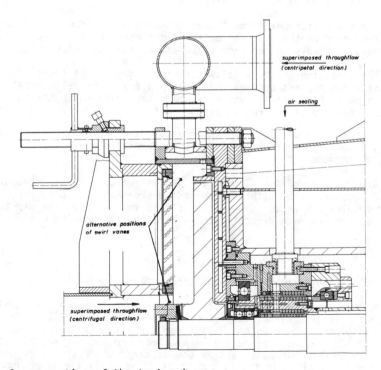

Fig. 2.   Cross section of the test set up

The different variables on the left hand side are mainly a function of the Reynolds number, the throughflow number and the dimensionless geometry of the gap, neglecting the small effects of the isentropic coefficient and Mach number |1|.

2. EXPERIMENTAL EQUIPMENT

Main components of the test set up used for the present work are (1) the test apparatus, housing the overhung disc of 0.4 m radius, mounted in two bearings as illustrated in Fig. 2, (2) a 65 kW d-c motor with a speed range of 0 - 6000 rpm, (3) supply systems for the superimposed throughflow of air, for cooling water, for lubricating oil and for the air sealing, and (4) the instrumentation.

For measuring the ventilation torque acting on the disc the disc shaft and the motor shaft are both mounted in cradle bearings.

The duct between the shafts and the housing is sealed off by a complex air barrier. The cooling passages within the side walls of the disc housing transfer the heat produced by friction in the gap to the cooling water.

A superimposed throughflow e.g. in centripetal direction is supplied by 8 pipes to the outer annulus torus and then through 100 holes of 15 mm diameter before entering the gap and leaving at the axis back into the closed-circuit system. For centrifugal throughflow the air is supplied axially and leaves the casing at its circumference.

For the present experiments the geometry of the gap and throughflow were modified only on the side facing the front cover plate, keeping the other side of the disc always in the same shape and at the same clearance.

The variation of the axial width of the gap on the front side was effected by different spacers between the cover plate and the cylindrical casing. As indicated in Fig. 2 swirl vanes can be attached at the outer angle ring for centripetal and at the radius r_i for centrifugal superimposed throughflow.

The front cover plate has an arrangement of holes in three radial directions, 11 positions each for velocity and pressure measurements and 6 positions to hold thermocouples.

The velocity of the flow in the gap is measured with a conventional hook shaped pressure probe with three bores. Whereas the velocity probe and the thermocouples can be shifted axially across the width of the gap the pressure measurements are made directly at the bores at the housing wall.

3. EXPERIMENTAL RESULTS

3.1 Velocity distribution without throughflow

As measure for the circumferential component of the core velocity half-way of the axial clearance the core rotation factor is plotted in Fig. 3 as function of the dimensionless radius for a rotational Reynolds number Re = $4.2 \cdot 10^6$. Parameters are the axial clearance and the radial extension of the gap both made dimensionless by the disc radius. With the extension of the gap beyond the disc rim the circumferential velocity component decreases in radial direction as compared with an almost constant factor measured for the gap

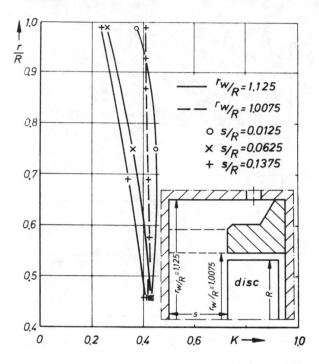

Fig. 3.   Core rotation factor as a function of the radius (Re = $4.2 \cdot 10^6$, $C_{F1} = 0$)

limited roughly at the disc rim. In the latter case the core rotates as a solid body in agreement with results of Daily and Nece [2]. The effects of radial extension become smaller while decreasing the axial clearance. This is taken into account by the equation of Zilling [3]

$$K = \frac{1}{1 + (\frac{r_w}{R})^2 \sqrt{\frac{r_w}{R} + 5 \cdot \frac{s}{R}}} \tag{2}$$

which is independent of the dimensionless radius and gives an averaged core rotation of K = 0.37 for r_w/R = 1.125   and   s/R = 0.1375 while from the experiments (Fig. 3) K varies from 0.41 at the inner radius to 0.24 at the rim.

The radial component of the velocity is represented by the volume flow number plotted in Fig. 4 in the same way as the core rotation factor. With the radial extension of the gap the volume flow number is doubled. The observed shear flow at the gap width s/R = 0.0125 gives the smallest volume flow number.

3.2 Velocity distribution with throughflow

Fig. 5, 6 and 7 are representative of results obtained for constant Reynolds number and a centripetal throughflow with a swirl at the entrance. They

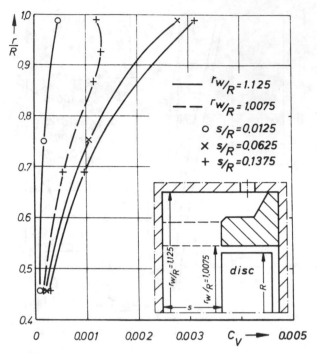

Fig. 4. Volume flow number as a function of the radius (Re = $4.2 \cdot 10^6$, C_{F1} = 0)

indicate clearly how the superimposed throughflow affects the flow mechanisms induced by frictional effects.

In Fig. 5 the swirl introduced with the throughflow at the rim is relatively small in comparison to the circumferential velocity of the disc raising the core rotation factor on Fig. 6 from 0.4 to 0.6.

However since the swirl of the throughflow is almost maintained along the way to the center the circumferential velocity of the fluid increases and surpasses the circumferential velocity of the disc as the flow approaches the center.

The core rotation factor increases following closely to the law of constant swirl plotted as dash - and - dot line in Fig. 6. Towards the center the centrifugal field becomes small. Then the radial flow inward may even extend to the disc embodying two small secondary eddies as sketched in Fig. 7 according to velocity traverses at three different radii.

3.3 Radial pressure distribution and axial thrust

Variation of the pressure in axial direction of the gap is very small as measured with special static pressure probes. Therefore, the radial distribution measured by static wall taps in the casing is taken valid also for the rotating disc.

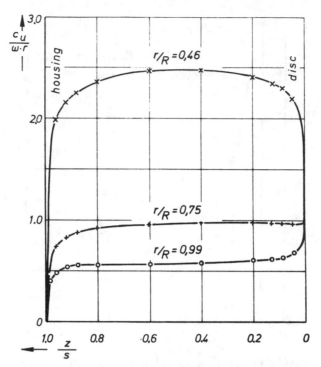

Fig. 5. Velocity profiles of tangential component in the gap at different radii
(Re = $4.2 \cdot 10^6$, C_{F1} = $14 \cdot 10^4$, centripetal, α = 12^0, s/R = 0.0625)

In Fig. 8 and 9 radial pressure distribution expressed by the pressure number
is plotted for two different axial clearances. It depends substantially on the
direction and the amount of the superimposed throughflow.

Neglecting the friction and assuming a homogenous rotating core as a
solid body as well as an incompressible fluid the reduced Navier - Stokes
equation for the radial direction has the simple form

$$- \frac{1}{\rho} \frac{dp}{dr} = c_r \frac{dc_r}{dr} - \frac{c_u^2}{r} \tag{3}$$

Then, the radial pressure distribution may be calculated as

$$C_{p(r/R)} = 1 - \frac{1}{8\,\pi^2} \left(\frac{C_{F1}}{Re} \frac{R}{s}\right)^2 \left(\left(\frac{R}{r}\right)^2 - 1\right) - \frac{1}{2} K^2 \left(1 - \left(\frac{r}{R}\right)^2\right) \tag{4}$$

This represents a centrifugal flow within a radial diffusor described by the
throughflow number superimposed by a rotating core identified by its rotation
factor K.

Evaluating both terms of the sum within Eq. (3) on the basis of the experimental results the first term is of two orders of magnitude smaller than
the second one. Neglecting the first term of Eq. (3) the radial pressure distribution can be calculated by a method of differences for the complicated flow
mechanisms demonstrated at Fig. 5 to 7. In Fig. 8 this calculation is compared

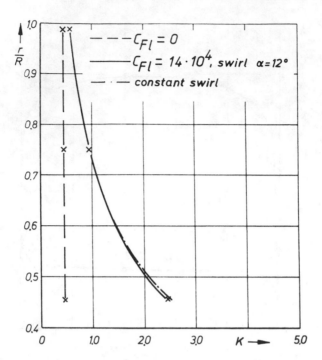

Fig. 6.   Core rotation factor as a function of the radius (Re = $4.2 \cdot 10^6$, cen-
          tripetal, α = 12°, s/R = 0.0625)

to the experimental results. The comparatively good agreement shows that the
pressure distribution depends mainly on the core rotation while the diffusor
effect is very small.

   With the throughflow number the pressure number first increases to a
certain value and then decreases. This is gathered from the axial thrust number
plotted in Fig. 10 and 11 resulting in smaller thrust for higher pressure
numbers. Velocity measurements with the hook shaped probe disclosed that the
minimum of the thrust number for throughflow without swirl is experienced close
to the critical flow rate. This is defined by the minimum amount of radial
throughflow necessary to prevent inflow of fluid at the rim radius of the disc
throughout the total axial clearance. The axial thrust number increases with
the swirl.

   The effect of screw heads on the thrust number is also shown in Fig. 10.
The eight socket-head cap screws of 20 mm height were mounted on the disc at a
radius r/R = 0.75. Increasing throughflow gives a decreasing influence of the
screw heads.

   The dash - and - dot line in Fig. 10 is derived from an approximate calcu-
lation according to Daily, Ernst and Asbedian |4|. From their experiments they
derived a formula for calculating the core rotation factor, depending on the
centrifugal throughflow without swirl at the entrance of the gap. As the geometry at the outer part of their gap deviates from the present there is a
difference in the axial thrust number. Only for throughflow numbers out of their
experimental range this curve is near to our experimental results.

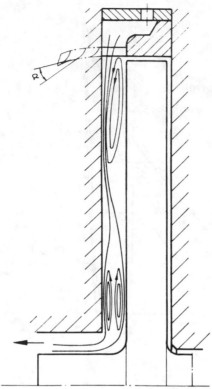

Fig. 7.   Flow pattern (Re = $4.2 \cdot 10^6$, C_{F1} = $14 \cdot 10^4$, centripetal, $\alpha = 12^0$,
s/R = 0.0625)

3.4 Frictional torque

The frictional torque acting on the front side of the disc is determined
as the difference between the measured torque and that part of the torque
acting on the back side and on the cylindrical surface of the disc. The torque
for the back side has been taken from experiments with the same conditions on
both sides of the disc. The torque acting on the cylindrical surface was deter-
mined by variation of the disc width in agreement with other experimenters |5|.
These parts of the total torque vary only slightly with the operating conditions.

In Fig. 12 and 13 the experimental results are plotted versus the Reynolds
number. The outward flow affects more the boundary layer at the disc and the
inward flow more the boundary layer at the housing. Hence, the frictional
torque number for outward flow exceeds that for inward flow. A positive swirl
in the forced throughflow reduces the frictional torque. At the same swirl
angle α and throughflow number C_{F1} the frictional torque number is influenced
more by inward flow than by outward flow. For the gap geometry with radial
limitation at the disc rim and s/R = 0.1375 there is good agreement with the
results of Daily and Nece |2|. Radial extension of the gap in combination with
decreasing core rotation factor enlarges the frictional torque number.

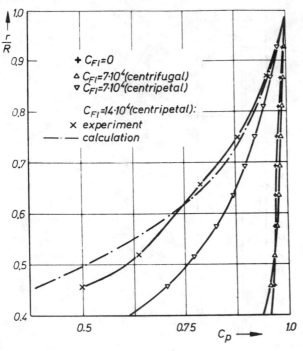

Fig. 8. Pressure number as
         a function of the
         radius and the
         throughflow number
         (Re = $4.2 \cdot 10^6$,
         $\alpha = 12^0$, s/R = 0.0625)

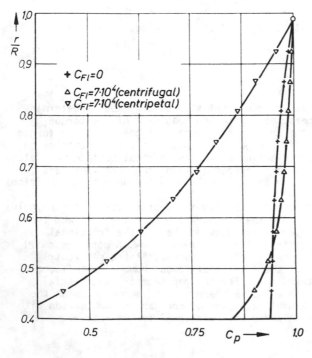

Fig. 9. Pressure number as
         a function of the
         radius and the
         throughflow (Re =
         $4.2 \cdot 10^6$, $\alpha = 12^0$,
         s/R = 0.0125)

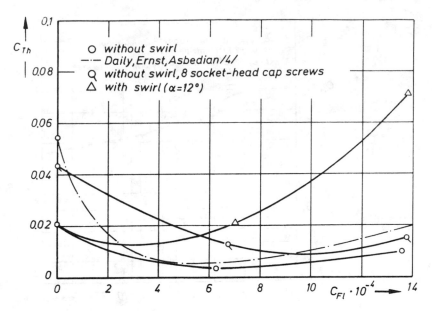

Fig. 10.   Axial thrust number versus throughflow number (Re = 4.2·10⁶, s/R = 0.0625, outward flow)

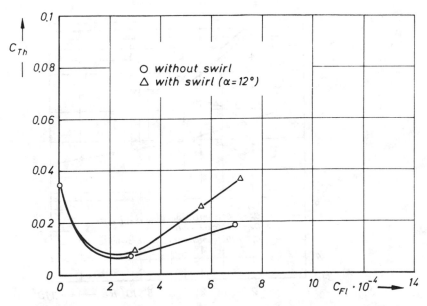

Fig. 11.   Axial thrust number versus throughflow number (Re = 4.2·10⁶, s/R = 0.0125, outward flow)

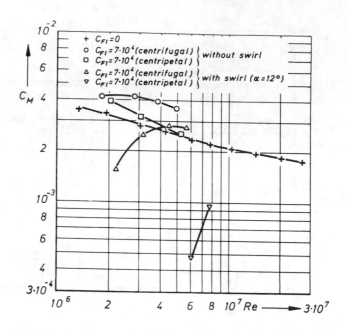

Fig. 12.   Frictional torque number as a function of Reynolds number
           (s/R = 0.0625)

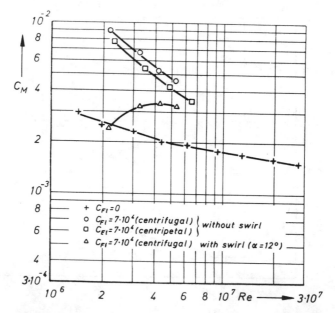

Fig. 13.   Frictional torque number as a function of Reynolds number
           (s/R = 0.0125)

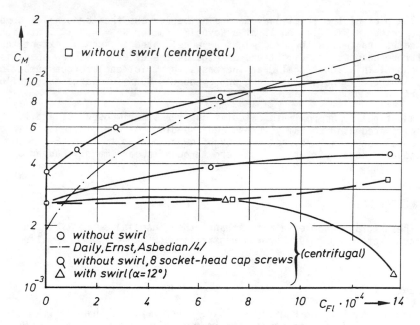

Fig. 14. Frictional torque number versus throughflow number (Re = $4.2 \cdot 10^6$, s/R = 0.0625)

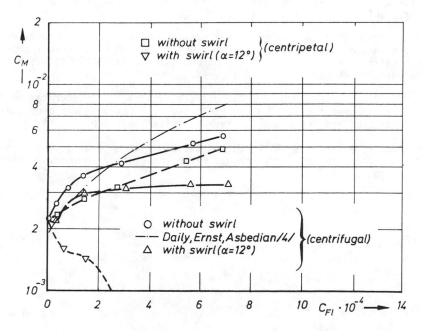

Fig. 15. Frictional torque number versus throughflow number (Re = $4.2 \cdot 10^6$, s/R = 0.0125)

The plot of the frictional torque number versus the throughflow number is presented in Fig. 14 and 15 at a Reynolds number Re = $4.2 \cdot 10^6$. At high flow rates and swirls the disc might be accelerated to such an extent that the torque changes its direction. In Fig. 14 the effect of screw heads is also shown. This additional drag increases the frictional torque number by 40 %, if there is no throughflow, and by about 100 % for large throughflow numbers.

Additionally approximate calculations for superimposed outward flow without swirl according to Daily, Ernst and Asbedian |4| are plotted. Their experimental range only covered a throughflow number up to about $3 \cdot 10^4$ and was extrapolated here. Differences at zero throughflow are mainly caused by the deviating geometry of the gap.

ACKNOWLEDGEMENT

The autors acknowledge with thanks the financial support of the Forschungsvereinigung für Verbrennungskraftmaschinen e.V.

REFERENCES

1. Theodorsen, Th. and Regier, A. 1944. Experiments on drag of revolving disks, cylinders, and streamline rods at high speeds. NACA Rep. No. 793.

2. Daily, J.W. and Nece, R.E. 1960. Chamber dimension effects on induced flow and frictional resistance of enclosed rotating disks. J. Basic Eng., Vol. 82, pp.217-232.

3. Zilling, H. 1973. Untersuchung des Axialschubs und der Strömungsvorgänge in den Radseitenräumen einer einstufigen radialen Kreiselpumpe mit Leitrad. Mitt. des Inst. für Strömungslehre und Strömungsmaschinen, Universität Karlsruhe, Heft 15, p. 48.

4. Daily, J.W., Ernst, W.D. and Asbedian, V.V. 1964. Enclosed rotating disks with superposed throughflow: mean steady and periodic unsteady characteristics of induced flow. MIT Hydrodynamics Lab., Rep. No. 64.

5. Bilgen, E. and Boulos, R. 1973. Functional dependence of torque coefficient of coaxial cylinders on gap width and Reynolds number. J. Fluids Eng., pp. 122-126.

Convection in a Rotating Annular Cavity

G. de VAHL DAVIS, E. LEONARDI, and J. A. REIZES
School of Mechanical and Industrial Engineering
The University of New South Wales
Kensington, N.S.W., Australia, 2033

ABSTRACT

A numerical study has been made of natural and forced convection in a vertical annular cavity formed between two concentric cylinders and two horizontal planes. The inner cylinder and either of the horizontal surfaces are rotating and heated; the outer cylinder and the other horizontal surface are stationary and cooled. Solutions have been obtained for an air-filled cavity of aspect ratio 1 and radius ratio 2 for a range of values of Reynolds number ($0 \leqslant Re \leqslant 1\,000$) and Rayleigh number ($0 \leqslant Ra \leqslant 200\,000$).

1. INTRODUCTION

The problem of cooling rotating electric machinery increases in difficulty with increasing size. This is because the heat developed is roughly proportional to the volume of the machine, i.e. to the cube of a linear dimension, whereas the surface area available for cooling only increases with the square of the dimension. For medium to large machines, therefore, an elaborate external cooling system must be provided.

In small machines, however, this is neither practicable nor necessary, and reliance is placed on natural cooling processes. In such situations, heat is transferred from the interior, where it is generated both by bearing friction and by iron and copper losses, to the external surfaces of the machine, where it is transmitted to the environment. This heat transfer takes place by conduction through the structure of the machine and by convection through fluid-filled cavities. When the surfaces of such cavities are stationary, the heat transfer calculations are fairly simple. In electric machinery, on the other hand, there may be cavities in which one or more of the surfaces rotate and in which the various surfaces may be at different temperatures. For example, the annulus between the outer casing of a motor, which is cooled by heat transfer to the environment, and the rotor and shaft, which are heated, presents a difficult design problem. Heat transfer will occur in such cavities both by natural convection as a result of gravity and of the centrifugal and Coriolis forces caused by rotation, and by forced convection as a result of relative motion

131

between some of the walls of the cavity.   The mechanism of heat transfer is then much more complex.

In this paper we describe a numerical study of convection in such a cavity. The results are strictly applicable only to motors with a vertical axis, because we have assumed axial symmetry. However it is shown that, for sufficiently high rotational speeds, the results can also be applied to the more usual case of a machine with a horizontal axis. Use of the laminar flow equations limits the Reynolds number to a value of the order of 10^3 and the Rayleigh number to a value of the order of 10^6. Above these values, turbulent flow can be expected. Fortunately, this nevertheless permits a useful range of parameter values to be examined. For example, a motor containing a cavity with a characteristic dimension of 10 mm rotating at 1000 rpm and experiencing a temperature rise of 40°C in the winding will operate at Re = 700 and Ra = 5 000.

2.   MATHEMATICAL FORMULATION

Consider a vertical annulus of inner radius R'_i, outer radius R'_o and height L' filled with a homogeneous Newtonian fluid. (A prime denotes a dimensional quantity.) The origin of a cylindrical coordinate system (r, ϕ, z) is located at the centre of the upper boundary. The geometry is determined by the radius ratio $R = R'_o/R'_i$ and the aspect ratio $L = L'/(R'_o-R'_i)$.

All fluid properties are assumed to be constant except the density in the body force term. Since the flow is rotating, the body force term includes the forces due to centrifugal and Coriolis accelerations as well as the acceleration due to gravity [1]. This is an extension of the usual Boussinesq approximation.

We use $R'_o-R'_i$, $1/\Omega'$ (where Ω' is the angular velocity of the rotating boundary), $(R'_o-R'_i)\Omega'$ and $\rho'(R'_o-R'_i)^2\Omega'^2$ as scale factors for length, time, velocity and pressure. (At Re = 0 a different scheme must be used; $(R'_o - R'_i)^2/\nu$ and $\nu/(R'_o-R'_i)$ then serve as scale factors for time and velocity.) The non-dimensional temperature is defined as $\Theta = (T'-T'_c)/(T'_h-T'_c)$, in which the subscripts c and h denote the cold and hot boundaries respectively. The non-dimensional equations representing the conservation of mass, momentum and energy then become

$$\nabla.\underline{V} = 0 \tag{1}$$

$$\frac{\partial u}{\partial t} + \underline{V}.\nabla u = (1 - \frac{RaFr}{Re^2Pr}\theta)\frac{v^2}{r} - \frac{\partial p}{\partial r} + \frac{1}{Re}(\nabla^2 u - \frac{u}{r^2}) \tag{2a}$$

$$\frac{\partial v}{\partial t} + \underline{V}.\nabla v = -(1 - \frac{RaFr}{Re^2Pr}\theta)\frac{uv}{r} + \frac{1}{Re}(\nabla^2 v - \frac{v}{r^2}) \tag{2b}$$

$$\frac{\partial w}{\partial t} + \underline{V}.\nabla w = (\frac{1}{Fr} - \frac{Ra}{Re^2Pr}\theta) - \frac{\partial p}{\partial z} + \frac{1}{Re}(\nabla^2 w) \tag{2c}$$

$$\frac{\partial \theta}{\partial t} + \nabla \cdot (\underline{V}\theta) = \frac{1}{RePr}\nabla^2 \theta \tag{3}$$

in which $\underline{V} = u\hat{\underline{r}} + v\hat{\underline{\phi}} + w\hat{\underline{z}}$ is the velocity vector, $Re=(R_o'-R_i')^2\Omega'/\nu$, $Ra=Pr.g\gamma(T_h'-T_c')(R_o'-R_i')^3/\nu^2$ and the Froude number $Fr=(R_o'-R_i')\Omega'^2/g$.

Since the motion is axisymmetric, the momentum equations can be rewritten in terms of a Stokes stream function, defined by

$$u = -\frac{1}{r}\frac{\partial \xi}{\partial z}, \qquad w = \frac{1}{r}\frac{\partial \xi}{\partial r}; \tag{4}$$

a vorticity

$$\zeta = \frac{\partial u}{\partial z} - \frac{\partial w}{\partial r}; \tag{5}$$

and a swirl velocity

$$\Gamma = rv. \tag{6}$$

Eqns. (2a) and (2c) yield the vorticity transport equation

$$\frac{\partial \zeta}{\partial t} = \frac{u\zeta}{r} + \frac{2\Gamma}{r^3}\frac{\partial \Gamma}{\partial z} - \frac{1}{r}\frac{\partial ru\zeta}{\partial r} - \frac{\partial w\zeta}{\partial z} + \frac{Ra}{Re^2Pr}\frac{\partial \theta}{\partial r}$$

$$- \frac{RaFr}{Re^2Pr}\frac{\Gamma}{r^3}(\Gamma\frac{\partial \theta}{\partial z} + 2\theta\frac{\partial \Gamma}{\partial z}) + \frac{1}{Re}(\frac{1}{r}\frac{\partial \zeta}{\partial r} + \frac{\partial^2 \zeta}{\partial r^2} + \frac{\partial^2 \zeta}{\partial z^2} - \frac{\zeta}{r^2}). \tag{7}$$

while (6) with (2b) leads to the swirl velocity equation

$$\frac{\partial \Gamma}{\partial t} = -\frac{1}{r}\frac{\partial ru\Gamma}{\partial r} - \frac{\partial w\Gamma}{\partial z} + \frac{RaFr}{Re^2Pr}\frac{\theta u\Gamma}{r^2} + \frac{1}{Re}(\frac{\partial^2 \Gamma}{\partial r^2} + \frac{\partial^2 \Gamma}{\partial z^2} - \frac{1}{r}\frac{\partial \Gamma}{\partial r}). \tag{8}$$

The relationship between ξ and ζ is obtained from (4) and (5):

$$\zeta = -\frac{1}{r}(-\frac{1}{r}\frac{\partial \xi}{\partial r} + \frac{\partial^2 \xi}{\partial r^2} + \frac{\partial^2 \xi}{\partial z^2}) \tag{9}$$

Equations (3), (7), (8) and (9), with suitable boundary conditions, describe the flow and thermal fields.

3. BOUNDARY CONDITIONS

Each boundary has been assumed to be impermeable and isothermal (different boundaries have different temperatures), and either moving at a constant angular velocity or stationary. The two cases considered are described in Table 1. Case U is chosen to simulate a region between the upper end of a vertical rotor and the casing of the motor; the inner and lower surfaces of the cavity are heated and rotating, while the other two surfaces are cooled and stationary. Case L represents a region at the lower end of a motor; the inner and upper surfaces are heated and rotating, and the other surfaces are cooled and stationary.

CASE	BOUNDARY			
	$r=R_i$	$r=R_o$	$z=0$	$z=L$
U	$\Omega=1$	$\Omega=0$	$\Omega=0$	$\Omega=1$
	$\Theta=1$	$\Theta=0$	$\Theta=0$	$\Theta=1$
L	$\Omega=1$	$\Omega=0$	$\Omega=1$	$\Omega=0$
	$\Theta=1$	$\Theta=0$	$\Theta=1$	$\Theta=0$

TABLE 1.   Boundary conditions for temperature Θ
and angular velocity Ω

At any boundary

$$\xi_b = \text{constant} \quad \text{and} \quad (\partial\xi/\partial n)_b = 0; \tag{10}$$

n is the normal coordinate. In a closed system the value of ξ_b is arbitrary and is taken as zero.

The boundary condition for the swirl velocity Γ is

$$\Gamma_b = r^2\Omega_b. \tag{11}$$

From equations (9) and (10) the boundary vorticity ζ_b is

$$\zeta_b = -\frac{1}{r}\left(\frac{\partial^2\xi}{\partial n^2}\right)_b \tag{12}$$

The numerical implementation of this is an extension of the Woods formula [2]. At $r=R_i$, for example, it becomes

$$\zeta_b = -\frac{3\xi_2}{\Delta r^2(R_i+\Delta r/4)} - \frac{(R_i/2) + (\Delta r/4)}{(R_i+\Delta r/4)}\,\zeta_2$$

in which the subscript 2 refers to the first internal mesh point.

4.   METHOD OF SOLUTION

Advantage was taken of the fast convergence rate of the false transient technique [3]. This consists of adding a fictitious time derivative term $\partial\xi/\partial t$ to (9) and multiplying the time derivative terms in (3), (7), (8) and (9) by factors which allow individual adjustment of the four time steps.

The finite difference approximations (FDA) of these equations were derived by replacing the time derivatives with forward differences and the spatial derivatives with second order central differences. The FDAs were solved with the Samarskii-Andreyev ADI scheme [4]. Most solutions were obtained in a few hundred time steps, depending in part upon the initial condition chosen, using a uniform 31x31 mesh. The computations took about two seconds per iteration on a CDC Cyber 171 with this mesh. A finer mesh (41x41) was used for some high Re solutions.

5. RESULTS AND DISCUSSION

The annulus considered here has a radius ratio R=2 and aspect
ratio L=1.   These values were chosen because they permit some
interesting features of the flow to be examined while being not too
different from values likely to be encountered in practice.   In due
course, a full parametric study of this problem will be made,   the
results from which will be published elsewhere. Solutions have
been obtained for 0≤Re≤1 000 and 0≤Ra≤200 000; Pr = 0.7 in all
solutions, corresponding to air.

For an annulus with dimensions of the order of a centimetre
rotating at a speed of the order of 1 000 rpm, Fr is of the order
of 10. Numerical experiments showed that the flow is independent
of Fr for Fr≤10, and the results presented are all for Fr = 0.

Only a small portion of the vast amount of computer output can
be described here.   The stream function, the temperature and the
circumferential velocity are probably the most informative of the
variables,   and contour plots of them for some of the solutions are
shown in Figures 1 and 2.   The contour levels drawn for ξ are
mostly 0.1,   0.5, 0.9 and 0.98 of ξ_{max}; occasionally, different or
additional contours are used to illustrate a particular feature of
the flow.   The values of ξ_{min} and ξ_{max} are given adjacent to each
diagram; a positive value of ξ is associated with anticlockwise
flow and _vice_ _versa_.   The contour levels for θ are 0.2, 0.4, 0.6
and 0.8, and for Γ they are 0.25, 0.5, 0.75 and 1.0.

5.1 The Upper Annulus

Figure 1 describes the flow in the annular region at the upper
end of the motor: Case U of Table 1.   The left hand border of each
diagram is the inner, rotating boundary of the annulus and the axis
of rotation is off each diagram to the left.

At low Re, the flow is dominated by natural convection.   When
Ra is also low (Figure 1a), the isotherms show that heat transfer
takes place largely by conduction.   The motion consists of a single
toroidal cell the displayed portion of which rotates in a clockwise
direction.   This motion is weak;   the maximum radial and axial
velocity components u and w (values of which are not shown) are
only about one quarter of the maximum value of the circumferential
velocity v. The contours of v show that the circumferential motion
is not greatly affected by the motion in the azimuthal plane.

At low Re and large Ra, Figure 1b, the azimuthal motion is
considerably stronger: the maximum stream function has increased by
a factor of about 20 for a 50-fold increase in Ra, and so have the
radial and axial velocities.   As a consequence, the isotherms show
considerable distortion; hot air is drawn across the upper, cooled
boundary of the annulus, while cold air is drawn across the lower,·
heated boundary.

The velocity v is affected in a similar manner.   It has a
maximum value of 2 at the lower right hand corner of the annulus,
decreases linearly across the lower boundary to a value of 1 on the
inner wall and is zero on the upper and outer walls.   As a result

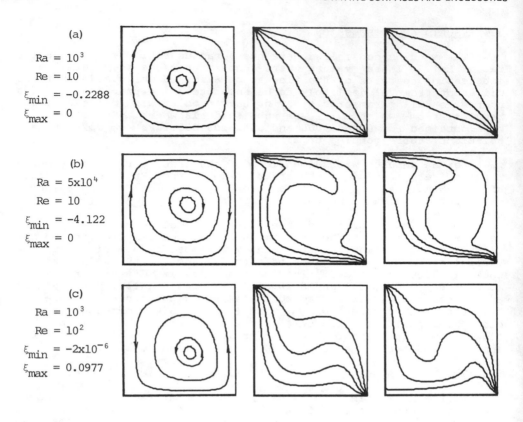

(a)

Ra = 10^3

Re = 10

ξ_{min} = -0.2288

ξ_{max} = 0

(b)

Ra = 5×10^4

Re = 10

ξ_{min} = -4.122

ξ_{max} = 0

(c)

Ra = 10^3

Re = 10^2

ξ_{min} = -2×10^{-6}

ξ_{max} = 0.0977

Figure 1.   Streamlines, Isotherms and Isovels: Case U

of the motion in the azimuthal plane, fluid with high v is transported from the rotating left hand wall to the upper region of the cavity, while fluid with low v is carried from the outer stationary wall to the lower region of the cavity.

The motion induced by the rotating lower boundary causes fluid near it to move radially outwards, opposing the buoyancy motion. At Re = 100 and Ra = 1 000 (Figure 1c), this centrifugal motion dominates the flow, the stream function contours showing a single anti-clockwise cell. The upper thermal layer has thickened as a result of cold fluid being drawn across it from the outer, unheated zone. Similarly, hot fluid has been drawn down onto the lower surface, where the thermal layer has also thickened. As Ra increases (Figure 1d), the buoyancy flow strengthens, and a more complex flow structure forms. A clockwise cell generated by fluid rising near the heated surface occupies the upper portion of the annulus, and an anti-clockwise cell generated by centrifugal forces near the lower boundary occupies the lower portion. The isotherms and isovels reflect this flow pattern. A further increase in Ra leads to growth and eventual dominance of the buoyancy flow (Figure 1e). The flow is similar to that shown in Figure 1b, which is for the same Ra but a lower Re.

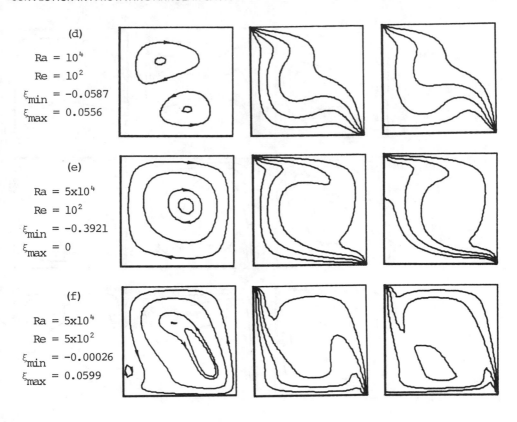

(d)

Ra = 10^4

Re = 10^2

ξ_{min} = -0.0587

ξ_{max} = 0.0556

(e)

Ra = 5×10^4

Re = 10^2

ξ_{min} = -0.3921

ξ_{max} = 0

(f)

Ra = 5×10^4

Re = 5×10^2

ξ_{min} = -0.00026

ξ_{max} = 0.0599

Figure 1.   Streamlines, Isotherms and Isovels: Case U (Cont.)

At Re = 500, the flow at low and moderate values of Ra is almost entirely dominated by centrifugal forces. The interaction of these with buoyancy forces at high Ra, however, causes a complex flow structure to form, Figure 1f. Anti-clockwise flow occupies most of the cavity. Near the lower end of the heated vertical boundary, a weak buoyancy-driven cell has formed which will, at higher Ra, repeat the flow development seen at Re = 100 (Figures 1d and e).

At all except the lowest Re, then, centrifugal forces have a significant effect on the flow. When these forces dominate, an anti-clockwise flow is established. The radial and axial velocities are relatively weak - of the order of v_{max}/8 or less. This contrasts with the buoyancy-driven flow, the velocities in which can be many times v_{max} at high Ra and low Re. Although the maximum <u>non-dimensional</u> value of u or w is of the order of 0.25 at all Re (compared with v_{max} which is 2 at all Re), the <u>dimensional</u> velocities are proportional to Re and therefore to the rate of rotation Ω' of the moving walls of the annulus.

(a)

$Ra = 2 \times 10^5$

$Re = 10$

$\xi_{min} = -2.268$

$\xi_{max} = 0$

(b)

$Ra = 10^3$

$Re = 5 \times 10^2$

$\xi_{min} = -0.0612$

$\xi_{max} = 0$

(c)

$Ra = 2 \times 10^5$

$Re = 5 \times 10^2$

$\xi_{min} = -0.0573$

$\xi_{max} = 0$

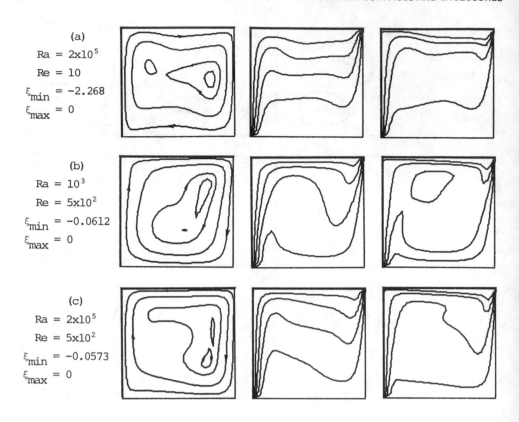

Figure 2.   Streamlines, Isotherms and Isovels: Case L

5.2 The Lower Annulus

In the annular region at the lower end of the motor, Case L, centrifugal forces and buoyancy forces act together to generate clockwise motion. The flow consists of one or a pair of clockwise cells. The two-cell flow at low Re, high Ra (Figure 2a) is similar to that in a stationary cavity at high Ra. As Re increases (Figure 2b), these cells merge but a new pair of clockwise cells forms near the outer boundary, possibly illustrating an instability similar to that observed in rectangular cavities at high aspect ratio and high Ra. The pattern persists as Ra increases (Figure 2c). This behaviour is not exhibited in the Case U solutions, at least over the present range of values of Re.

At low Re, the flow is driven almost entirely by the buoyancy forces, and when also Ra is high the velocities u and w are an order of magnitude greater than the maximum centrifugal velocity. At high Re, the velocities relative to v_{max} are much smaller, although somewhat larger than the corresponding values in the upper annulus.

This change with increasing Re in either annulus is due to a change in the relative importance of the centrifugal and buoyancy forces. The coefficient of the buoyancy term $\partial\theta/\partial r$ in Equation (7) is proportional to Ra/Re^2; thus as Re increases, the buoyancy term is reduced relative to the centrifugal terms (the first terms on the right hand side of Equation (7)).

5.3   Heat Transfer

The parameter of practical importance in the thermal design of electric motors is the Nusselt number. We use subscripts i, o, t and b on Nu to denote the average Nusselt numbers on the inner, outer, top and bottom boundaries respectively. Nu_i and Nu_t are

$$Nu_i = -\frac{1}{L}\int_o^L \frac{\partial\theta}{\partial r}\, dz$$

and
$$Nu_t = -2(\theta_b-\theta_t)\frac{R-1}{R+1}\int_{R_i}^{R_o}\frac{\partial\theta}{\partial z}\, r\, dr.$$

Similar expressions define Nu_o and Nu_b.

For L=1 and R=2, the heat transfer across each annulus is

$$Nu_{tot,U} = Nu_i + 3Nu_b/2 \qquad \text{and} \qquad Nu_{tot,L} = Nu_i + 3Nu_t/2.$$

The total rate of heat transfer is the sum of these.

Figure 3 shows the variation of Nu_i with Ra at the heated inner surface of the upper annulus for various values of Re. (Broken lines in this and the following figures indicate estimates made in the absence of solutions for some combinations of parameter values.) Re has no effect on Nu_i at very low Re and a diminishing effect at high Ra. Otherwise, Nu_i increases with Re. However, Nu_i decreases with Ra up to a critical value which increases with Re. This is a reflection of the change, with increasing Ra, from centrifugal to buoyancy flow, with a consequent thickening of the thermal boundary layers during the transition. At the lower surface of the upper annulus, the effect of flow reversal is also to be seen. As Figure 4 shows, Nu_b initially decreases with increasing Re. At sufficiently high Re, Nu_b starts to increase and

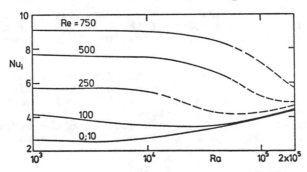

Figure 3.   Nu at Inner Surface of Upper Annulus

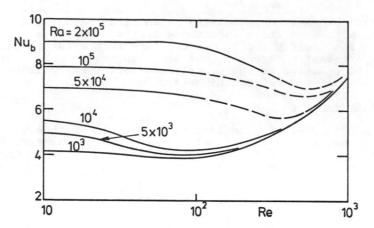

Figure 4.   Nu at Bottom Surface of Upper Annulus

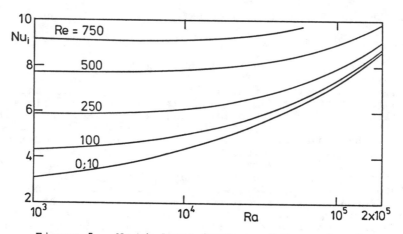

Figure 5.   Nu at Inner Surface of Lower Annulus

becomes independent of Ra as the centrifugal motion becomes dominant over the lower boundary of the annulus. The total rate of heat transfer across the upper annulus, $Nu_{tot,U}$, shows a similar behaviour.

The practical significance of this finding is that the rate of cooling of a motor across the upper air gap initially <u>decreases</u> with increasing rotational speed. It is only when the speed exceeds a certain critical value, which depends upon Rayleigh number (and therefore, for a given motor, upon the temperature rise in the rotor), that the rate of cooling starts to increase with speed.

Across the lower annulus, the Nusselt number increases with Re at all Ra. Figure 5 shows Nu_i. The lines are converging, showing the diminishing effect of Re with increasing Ra. Nu_i increases

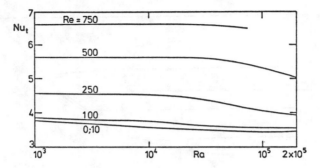

Figure 6.   Nu at Top Surface of Lower Annulus

with increasing Ra. The reverse is true for Nu_t, on the heated top
surface of the lower annulus, as shown in Figure 6. The reason for
this is that, with increasing Ra, heat transfer from the inner
boundary causes the fluid temperature adjacent to the upper
boundary to increase, thereby reducing the rate of heat transfer
from it. The effect is, however, small, and $Nu_{tot,L}$ increases with
both Ra and Re.

At low values of Re and high values of Ra, the rate of heat
transfer from the upper end of the motor is substantially greater -
up to about 30% for the parameter values studied - than from the
lower end. This is potentially troublesome, since it can lead to a
non-symmetrical temperature distribution, unequal rates of thermal
expansion and possible thermal stress problems. For intermediate
values of Re (i.e., when Case U flow reversal occurs), the rate of
cooling from the upper end is less than from the lower end. With
increasing Re, the Nusselt numbers at the two ends of the motor
approach each other, this requiring progressively higher values of
Re as Ra increases. Figure 7 shows $Nu_{tot,U}$ and $Nu_{tot,L}$ as
functions of Re for two different values of Ra. The decrease in Nu
with Re up to a critical value can be seen, as can the fact that,
at sufficiently high Re, Nu becomes less dependent upon Ra and also
upon the differences between Case U and Case L.

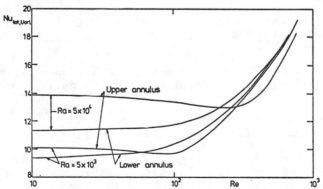

Figure 7.   Total Heat Transfer across Upper and Lower Annulus

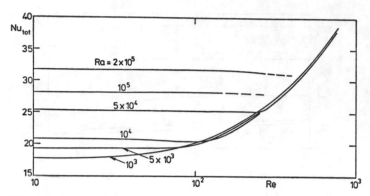

Figure 8.   Total Heat Transfer across Both Annuli

For Re of the order of 500 or more, natural gravity becomes negligible compared with the centrifugal forces. Under these conditions, the effect of motor orientation disappears and the results can be applied to a motor with a horizontal axis.

The total heat transfer rate from the motor is shown in Figure 8. Nu_{tot} increases with Ra and decreases very slightly with Re until a particular Re is reached (the value of which increases with Ra) above which Nu_{tot} increases strongly with Re and becomes independent of Ra.

6.   ACKNOWLEDGEMENT

We are grateful to the National Energy Research, Development and Demonstration programme administered by the Commonwealth Department of National Development and Energy which provided financial support during the early stages of this work.

7.   REFERENCES

1.   Homsy, G.M. and Hudson, J.L., Centrifugally Driven Thermal Convection in a Rotating Cylinder, J. Fluid Mech., Vol. 35, Pt. 1, pp.33-52, 1969.

2.   Woods, L.C., A Note on the Numerical Solution of Fourth Order Differential Equations, Aero. Q., Vol. 5, pp.176-184, 1954.

3.   Mallinson, G.D. and de Vahl Davis, G., The Method of the False Transient for the Solution of Coupled Elliptic Equations, J. Comp. Phys., Vol. 12, pp.435-461, 1973.

4.   Samarskii, A.A. and Andreyev, V.B., On a High Accuracy Difference Scheme for Elliptic Equations with Several Space Variables, USSR Comp. Math. & Math. Phys., Vol. 3, pp.1373-1382, 1963.

Mass Transfer in the Annulus between Two Coaxial Rotating Cylinders

KUNIO KATAOKA, YASUHIKO BITOU, KEIJI HASHIOKA,
TOSHIYA KOMAI, and HIDEKI DOI
Department of Chemical Engineering
Kobe University
Rokkodai, Nada, Kobe 657, Japan

ABSTRACT

Mass transfer is observed on the stationary outer cylinder by using an electrochemical technique and discussed with the aid of the observations of the dynamical instabilities in Taylor Couette flow. Seven distinct modes of mass transfer have been confirmed to exist owing to the cascade transitions of flow as Taylor number is increased: the spectral evolution to a chaotic flow does not occur gradually as an infinite sequence of instabilities but with one or two predominant fundamental frequencies as a small number of transitions; local mass fluxes are time-independent in laminar Couette flow and steady Taylor vortex flow and time-dependent in wavy vortex, quasiperiodic wavy vortex, weakly turbulent vortex, and turbulent vortex flows as well as turbulent flow; over the wide range of Taylor number beyond the Taylor instability, local mass transfer coefficients on the time average show an axially periodic distribution, indicating the existence of a cellular structure.

1. INTRODUCTION

Convective mass transfer is studied experimentally taking into account the dynamical instabilities and transitions of a viscous incompressible fluid between two coaxial cylinders as the rotation speed of the inner cylinder is raised.

The modes of flow in this flow system can be primarily characterized as a function of Taylor number, i.e. $Ta = (Ri\Omega d/\nu) \sqrt{d/Ri}$. When Ta exceeds a critical value Ta_c, the instability of purely laminar Couette flow leads to a steady axisymmetric Taylor vortex flow which has a periodic cellular structure in the axial direction. It is generally recognized (e.g.[1]) that several stable modes of flow can exist owing to a sequence of transitions as Ta increases: purely laminar Couette flow $\longrightarrow$ laminar vortex flow (singly periodic) $\longrightarrow$ wavy vortex flow (doubly periodic) $\longrightarrow$ quasiperiodic wavy vortex flow $\longrightarrow$ turbulent vortex flow $\longrightarrow$ turbulent flow. Recent work [2,3] on the dynamical instabilities revealed that the transition to turbulence occurs as a definite sequence of hydrodynamical instabilities and that the Taylor vortex flow has only one or two main characteristic frequencies before the final transition to chaotic Taylor vortex flow.

The present authors [4] and Mizushina *et al.* [5] successfully measured local variation of mass fluxes on the inside wall of the stationary outer cylinder with and without an axial flow and found a sinusoidal-like, axially periodic distribution of mass transfer, corresponding to the axially periodic cellular structure. The present experiment was designed to measure local time-dependent mass

143

fluxes and velocity gradients on the inside wall of the outer cylinder by using an electrochemical technique [6].

2. EXPERIMENT

2.1 Apparatus and Electrochemical Technique

Figure 1 shows the experimental apparatus consisting of two vertically mounted concentric cylinders of copper. The inner cylinder was 58.0 mm outer diameter and 380 mm long. The outer cylinder was 94.0 mm inner diameter and 400 mm long. The radius ratio $R_i/R_o = 0.617$ and the aspect ratio $H/d = 21.1$. The upper and lower fluid surfaces were fixed. The critical Taylor number was experimentally confirmed to be $Ta_c = 60$.

The experiment was performed using a diffusion-controlled electrode reaction of Cu^{++} ions. The inner cylinder served as the anode and the outer one as the main cathode for mass transfer measurement. Local measurement of the mass transfer was accomplished using 36 circular point cathodes (K-electrode) which were fabricated by inserting 0.4 mm diameter copper wires concentrically into 0.6 mm inner diameter holes drilled in the outer cylinder wall. The K-electrodes, mounted flush with the surrounding main cathode, were located at every 4 mm in the axial direction. The limiting current density Id on each K-electrode gives the local mass flux or mass-transfer coefficient k at its position:

$$k = Id/F\ Cb \tag{1}$$

The mass transfer measurements can be considered as those of heat transfer to a constant-temperature wall from a high Prandtl number fluid.

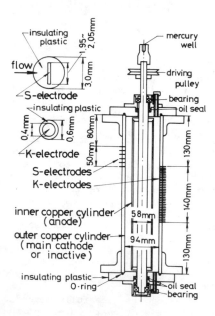

Fig.1. Experimental Apparatus and Details of Test Electrodes.

Table 1. Working fluid and experimental conditions

Electrolytic aqueous solution

$CuSO_4$ (kg-equiv./m^3)	0.001 ~ 0.019
H_2SO_4 (kg-equiv./m^3)	4
Glycerin (wt %)	0 ~ 58.4

Number of rotation (RPM)
                  12 ~ 425

Taylor number (—)
                  50 ~ 19,500

Schmidt number (—)
                  $3 \times 10^3 \sim 7.7 \times 10^5$

Local measurement of the velocity gradient on the outer cylinder wall was performed using 6 rectangular cathodes (S-electrode) which were fabricated by inserting about 2 mm wide and 0.20 mm thick copper sheets into 3 mm inner diameter holes drilled in the outer cylinder wall. The longer sides of each rectangular cathode were fixed perpendicular to the main rotating flow, i.e. to the azimuthal direction. The S-electrodes were located at every 10 mm in the axial direction. The limiting current density Id from each S-electrode with the main cathode inactive gives local velocity gradient:

$$s = 1.90 \ (Id/F \ Cb)^3 (L/D^2) \qquad\qquad (2)$$

The working fluid and the experimental conditions are listed in Table 1. At a given Taylor number there were several distinct stable spatial states, depending upon the Taylor number history. However, there was no appreciable difference of mass transfer except in axial wavelength among those accessible spatial states. Therefore, the present experiment was performed by rapidly accelerating the inner cylinder from rest to a given Taylor number.

2.2 Data Acquisition and Statistical Analysis

The data-processing equipment is diagrammed with the electrode circuit in Figure 2. Two limiting currents from an arbitrarily chosen pair of S-electrodes were simultaneously transformed into the corresponding voltage drops by DC resistors (5 kΩ), amplified by DC amplifiers, and then recorded by a multi-channel data recorder. By means of an AD converter, the analog voltage signals were transformed at every 10 milliseconds for 200 seconds into the digital signals and stored as a time history into the memory of a digital computer. According to the work of Swinney and his coworkers [2], it was expected that the spectral power of the second fundamental would be two or more orders of magnitude less than that of the first fundamental. As shown in Figure 3, two time intervals were used for time averaging in the autocorrelation analysis so as to observe the second fundamental in distinction from the first one.

The shorter time interval Δt' (usually 1 second) was used to observe higher frequency components:

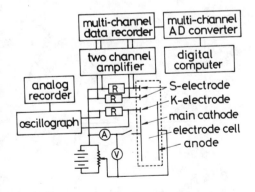

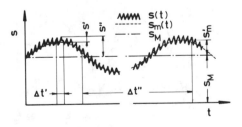

Fig.2. Data-processing Equipment and Electrode Circuit.

Fig.3. Schematic Diagram of Time-averaging with Two Time Intervals.

$$s'(t) = s(t) - s_m(t) \tag{3}$$

where
$$s_m(t) = \frac{1}{\Delta t'} \int_t^{t+\Delta t'} s(t) \, dt$$

The autocorrelation of higher frequency components was defined as

$$R_{EH} = \frac{\overline{s'(t) \, s'(t + \tau)}}{\overline{s'^2}} \tag{4}$$

The longer time interval $\Delta t''$ (usually 100 seconds) was used to observe lower frequency components:

$$s_m'(t) = s_m(t) - s_M \tag{5}$$

where
$$s_M = \frac{1}{\Delta t''} \int_t^{t+\Delta t''} s_m(t) \, dt$$

The autocorrelation of lower frequency components was defined as

$$R_{EL} = \frac{\overline{s_m'(t) \, s_m'(t + \tau)}}{\overline{s'^2}} \tag{6}$$

The power spectrum was computed from 8,192 or 16,384 points of the velocity-gradient vs. time record:

$$s''_j = s_j - s_M \qquad (j = 0,1, \text{-----}, N-1) \qquad N = 8,192 \text{ or } 16,384$$

The power spectrum is

$$P_k = \frac{2}{N\Delta t} \left[(\Delta t \, Re(A_k))^2 + (\Delta t \, Im(A_k))^2 \right] \qquad (k = 0,1, \text{---}, N/2-1) \tag{7}$$

where Δt: sampling time (10 ms)

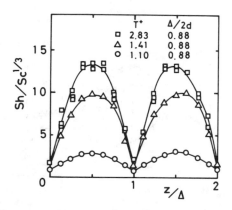

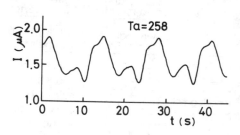

Fig.4. Axial Distribution of Mass
Transfer Coefficients in Laminar
Vortex Flow.

Fig.5. Oscillogram of S-electrode
Current in Wavy Vortex Flow.

$Re(A_k)$, $Im(A_k)$: real and imaginary parts of the Fourier transform of velocity-gradient fluctuations $s_k"$, respectively.

3. EXPERIMENTAL RESULTS AND DISCUSSION

(i) <u>Purely laminar flow</u>   ($T^+ < 1$)   Local mass fluxes show neither time-dependency nor axial variation.

(ii) <u>Laminar vortex flow</u> ($1 < T^+ < 3$)   In a very short supercritical range of Taylor number, the instability of Couette flow leads to a steady laminar axisymmetric vortex flow. Local mass fluxes do not show any time-dependency but the axial distribution of mass transfer, shown in Figure 4, indicates cycloidal-like periodicity which results from the axially-arrayed, laminar axi-

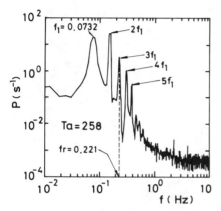

Fig.6. Power Spectrum of Velocity-gradient Fluctuations in Wavy Vortex Flow.

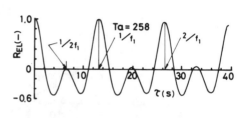

Fig.7. Autocorrelation Coefficient in Wavy Vortex Flow.

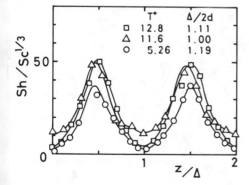

Fig.8. Axial Distribution of Mass Transfer Coefficients in Wavy Vortex Flow.

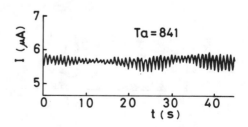

Fig.9. Oscillogram of S-electrode Current in Quasiperiodic Wavy Vortex Flow.

symmetric Taylor vortices. Each peak in the mass transfer distribution corresponds to the outflow boundary where a jet-like outward flow impinges on the
outer cylinder wall and each valley corresponds to the inflow boundary. (The
upper limit of Taylor number for this flow regime remains to be verified.)

(iii) <u>Wavy vortex flow</u> $(3 < T^+ < 13.3)$ Figure 5 is an oscillogram of
S-electrode current I (μA) obtained at the middle of a Taylor cell. There appear two kinds of very slow oscillations in the velocity gradient. Figures 6
and 7 show the power spectrum and autocorrelation. It is seen from those figures that there exist a single fundamental frequency component ($f_1 = 0.0732$ Hz
or $\omega_1 = 0.331$) and its harmonics. The smaller peak in the autocorrelation curve
can be considered as the second harmonic. The third harmonic on the power spectrum is coincident with the rotation of the inner cylinder ($fr = 0.221$ Hz).

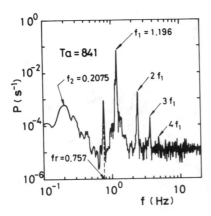

Fig.10. Power Spectrum of Velocity-
gradient Fluctuations in Quasiperiodic Wavy Vortex Flow.

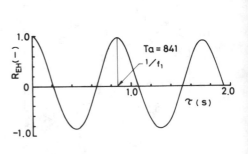

Fig.11. Autocorrelation Coefficient
of Higher-frequency Components in
Quasiperiodic Wavy Vortex Flow.

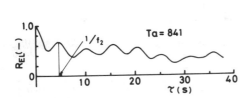

Fig.12. Autocorrelation Coefficient
of Lower-frequency Components in
Quasiperiodic Wavy Vortex Flow.

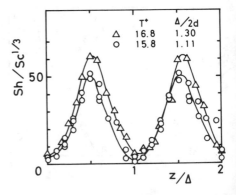

Fig.13. Axial Distribution of Mass
Transfer Coefficients in Quasiperiodic Wavy Vortex Flow.

This suggests that the velocity of the azimuthally travelling waves is 1/3 that of the inner cylinder. However, local mass fluxes still remain time-independent and have a sinusoidal-like axially periodic distribution, as shown in Figure 8. The frequency of the fundamental seems too low to affect the overall resistance to mass transfer, distributed over the whole annular gap. It is difficult to perceive the transition from laminar vortex to wavy vortex flow by the mass transfer observation.

(iv) Quasiperiodic wavy vortex flow   ($13.3 < T^+ < 18$)   Figure 9 shows that the oscillogram of S-electrode current obtained near the outflow boundary has polarogram-like, regular oscillations. It is seen from the power spectrum of Figure 10 that there appear two incommensurate fundamental components ($f_1 = 1.196$ Hz or $\omega_1 = 1.58$ and $f_2 = 0.2075$ Hz or $\omega_2 = 0.272$). The successive peaks except for $fr = 0.757$ Hz are the harmonics of the first fundamental. The two incommensurate fundamentals are confirmed by two kinds of autocorrelations (Figures 11 and 12). In this flow regime, local mass fluxes begin to show similar polarogram-like wavy oscillations, the frequency of which coincides with that of the first fundamental. Figure 13 is an axially-periodic distribution of the time-averaged mass transfer coefficients, showing the existence of an axially periodic cellular structure.

(v) Weakly turbulent vortex flow   ($18 < T^+ < 33$)   Figure 14 is an oscillogram of S-electrode current obtained near the inflow boundary. The autocorrelation and power spectrum are shown in Figures 15 and 16. There appear wavy but slightly irregular oscillations, which can be considered as a sign of turbulence. It is seen from Figures 15 and 16 that there still exists the second fundamental component ($f_2 = 0.226$ Hz or $\omega_2 = 0.278$) but that the first fundamental component ($f_1 = 1.135$ Hz or $\omega_1 = 1.394$) is about to disappear. The first fundamental component could not be confirmed from the autocorrelation of higher frequency components. It should be noted that the background continuum level begins to form the power spectrum of turbulent flow. In this flow regime, local mass fluxes have similar wavy but slightly irregular oscillations. As shown in Figure 17, the axially periodic distribution of the time-averaged mass transfer coefficients indicates the presence of the cellular vortex structure. It should be noted that there appears a small peak at every inflow boundary owing to the separation of boundary layer at the higher Taylor number.

(vi) Turbulent vortex flow   ($33 < T^+ < 160$)   The S-electrode current obtained at the outflow boundary and the corresponding autocorrelation and power spectrum are shown in Figures 18 to 20. There appears chaotic turbulence and both fundamental frequency components disappear from the power spectrum. The continuous power spectrum suggests the transition to turbulent flow. As shown in Figure 21, however, the axially periodic distribution of the time-averaged mass transfer coefficients indicates the persistent presence of cellular structure.

(vii) Turbulent flow   ($T^+ > 160$)   In this flow regime, both the currents of S-electrode and K-electrode have irregularily oscillating fluctuations but do not show any periodic structure in the axial direction.

In spite of many dynamical transitions, the Sherwood number obtained by time- and spatial-averaging of local mass transfer coefficients can be well correlated with Taylor number by the equation

$$Sh = 6.04 \, T^{+ \, 1/2} \, Sc^{1/3} \qquad\qquad T^+ > 1 \qquad\qquad (8)$$

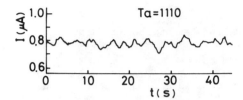

Fig.14. Oscillogram of S-electrode
Current in Weakly Turbulent Vortex
Flow.

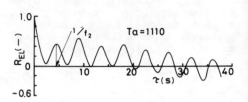

Fig.15. Autocorrelation Coefficient
of Lower-frequency Components in
Weakly Turbulent Vortex Flow.

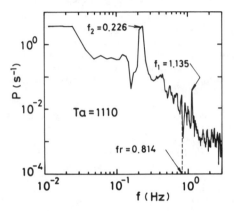

Fig.16. Power Spectrum of Velocity-
gradient Fluctuations in Weakly
Turbulent Vortex Flow.

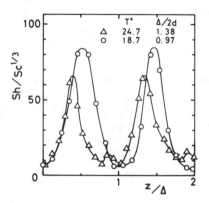

Fig.17. Axial Distribution of Mass
Transfer Coefficients in Weakly
Turbulent Vortex Flow.

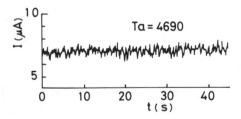

Fig.18. Oscillogram of S-electrode
Current in Turbulent Vortex Flow.

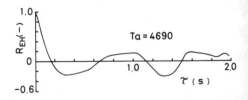

Fig.19. Autocorrelation Coefficient
of Higher-frequency Components in
Turbulent Vortex Flow.

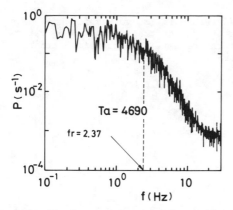

Fig.20. Power Spectrum of Velocity-gradient Fluctuations in Turbulent Vortex Flow.

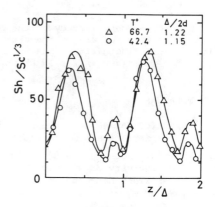

Fig.21. Axial Distribution of Mass Transfer Coefficients in Turbulent Vortex Flow.

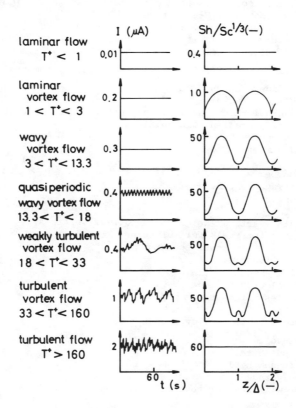

Fig.22. Classification of Mass Transfer Modes.

4. CONCLUSION

The experimental study on the mass transfer in Taylor vortex flow has been successful with the aid of the observations of the spectral evolution of velocity-gradient fluctuations. As a summary of the results, the classification of the modes of mass transfer is shown in Figure 22.

NOMENCLATURE

A   : surface area of test electrode
A_k : Fourier transform of velocity-gradient fluctuations
Cb   : bulk concentration of Cu^{++} ion
D   : diffusivity of Cu^{++} ion
d   : gap width between two coaxial cylinders
F   : Faraday constant
f   : frequency
fr   : frequency of inner cylinder rotation
H   : height of vertical annulus
I   : test electrode current
Id   : current density on test electrode
k   : mass transfer coefficient on the internal surface of outer cylinder
L   : streamwise length of S-electrode
N   : number of points of digitalized data record
P   : power spectrum of velocity-gradient fluctuations
R_{EH} : autocorrelation coefficient of higher-frequency components
R_{EL} : autocorrelation coefficient of lower-frequency components
Ri   : inner cylinder radius
Ro   : outer cylinder radius
s   : velocity gradient on the internal surface of outer cylinder
s' and s" : velocity-gradient fluctuations
Sc   : Schmidt number = ν/D
Sh   : Sherwood number = $2kd/D$
t   : time
Ta   : Taylor number = $(Ri\Omega d/\nu)\ \sqrt{d/Ri}$
T^+ : relative Taylor number = Ta/Ta_c
z   : axial length

Δ : wavelength (vertical height of a pair of vortices)
$\Delta t'$ and $\Delta t"$: time intervals for averaging
Δt : sampling time for power spectrum
ν : kinematic viscosity
τ : time difference
Ω : angular velocity of inner cylinder
ω : dimensionless frequency = f/fr

Subscripts, superscripts, and overlines

c   : critical value
1   : first fundamental
2   : second fundamental
m and M : time-averaged
───── : average with respect to time

REFERENCES

1. DiPrima, R.C. 1981. Transition in flow between rotating concentric cylinders. in TRANSITION AND TURBULENCE, Academic Press, NY, pp.1 - 23.

2. Fenstermacher, P.R., Swinney, L.H., and Gollub, J.P. 1979. Dynamical instabilities and the transition to chaotic Taylor vortex flow. *J. of Fluid Mechanics*, Vol.94, pp.103 - 128.

3. Bouabdallah, A. and Cognet, G. 1979. Laminar - turbulent transition in Taylor Couette flow. in I.U.T.A.M. Symposium on Laminar - Turbulent Transition, Stuttgart, F.R.G., September 16 - 22.

4. Kataoka, K., Doi, H. and Komai, T. 1977. Heat/mass transfer in Taylor vortex flow with constant axial flow rates. *International J. of Heat and Mass Transfer*, Vol.20, pp.57 - 63.

5. Mizushina, T., Ito, R., Kataoka, K., Yokoyama, S., Nakashima, Y. and Fukuda, A. 1968. Transition of flow in the annulus of concentric rotating cylinders. *Kagaku-Kogaku (Chemical Engineering, Japan)*, Vol.32, pp.795 - 800.

6. Mizushina, T. 1971. The electrochemical method in transport phenomena. in ADVANCES IN HEAT TRANSFER, Academic Press, NY, Vol.7, pp.87 - 161.

Analysis of Heat Transfer from a Rotating Cylinder with Circumferential Fins

SUHAS V. PATANKAR and JAYATHI Y. MURTHY
Department of Mechanical Engineering
University of Minnesota
Minneapolis, Minn. 55455, USA

ABSTRACT

The paper presents a numerical study of laminar flow and heat transfer in the situation where a cylinder with circumferential fins rotates within a stationary shroud. The flow is created by only the cylinder rotation; no axial throughflow is considered. The results are presented for a range of the values of the Taylor number, the clearance between the fins and the shroud, and the aspect ratio of the inter-fin space. It is shown that, at high Taylor numbers, the secondary flow significantly augments the heat transfer. Also the torque required to rotate the cylinder increases. The heat transfer augmentation is found to be more pronounced at higher Prandtl numbers.

1. INTRODUCTION

Heat transfer from rotating cylindrical bodies occurs in many practical applications. Electric motors and generators are the primary examples of such applications, in which the internally generated heat must be dissipated to the surroundings. The surface of the rotor in these machines is seldom smooth; it has ribs, grooves, or other fin-like projections. Therefore, a study of the heat transfer from a finned rotating cylinder is directly relevant to the heat transfer problems in electrical machines. Such a study is also useful in the analysis of rotating stirrers and agitators used in chemical industry.

The physical configuration considered in this paper is shown schematically in Fig. 1. Circumferential fins are uniformly spaced on the rotating cylinder, which is enclosed in a stationary shroud. Heat transfer from the rotating cylinder to the shroud takes place through the rotating fluid between them. The fins not only increase the surface area available for heat transfer but also increase the rate of heat transfer by creating secondary flows sketched in Fig. 1. These secondary vortices are similar to the well-known Taylor vortices (see, e.g. [1]), but are not formed due to a hydrodynamic instability. They result from the nonuniformity of the centrifugal force caused by the presence of the fins. In this sense, the secondary flows indicated in Fig. 1 are similar to the buoyancy-induced flows encountered in [2].

As in most heat transfer augmentation devices, the increased heat transfer from the finned cylinder is accompanied by a penalty in terms of the increased torque required to rotate it. Thus the knowledge of both the heat transfer and fluid flow characteristics is desirable.

The purpose of this paper is to present a numerical study of the flow and

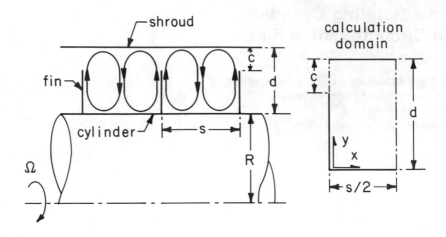

Fig. 1   The physical situation and the calculation domain

heat transfer for the shrouded finned rotor shown in Fig. 1. The results apply
to laminar flow and cover a range of geometrical configurations and rotation
speeds.

It appears that no previous analysis for this problem is available in the
literature. References [2] and [3] deal with a <u>stationary</u> shrouded fin array
with and without the effects of buoyancy.

2. MATHEMATICAL FORMULATION

To limit the number of parameters to be considered, some assumptions are
made about the nature of the geometry. First, the radius R of the rotating
cylinder is regarded to be much larger than the distance d between the cylinder
and the shroud. Most electric motors and generators do have this characteristic. As a result, the equations governing the flow in the calculation domain
(shown in Fig. 1) are the same as those in the Cartesian coordinates x, y ,
except that a centrifugal force must be included. Second, the thickness of
the fins will be considered to be negligible in comparison with the fin spacing
s.

The flow between the finned cylinder and the shroud is considered to be
generated only by the cylinder rotation; no axial throughflow is included in
this study. For a laminar, constant-property, axisymmetric flow, the governing
equations can be written as follows.

continuity:
$$\frac{\partial u}{\partial x} + \frac{\partial v}{\partial y} = 0 \tag{1}$$

momentum:
$$u\,\frac{\partial u}{\partial x} + v\,\frac{\partial u}{\partial y} = -\frac{1}{\rho}\frac{\partial p}{\partial x} + \nu\left(\frac{\partial^2 u}{\partial x^2} + \frac{\partial^2 u}{\partial y^2}\right) \tag{2}$$

$$u \frac{\partial v}{\partial x} + v \frac{\partial v}{\partial y} = -\frac{1}{\rho} \frac{\partial p}{\partial y} + \nu \left(\frac{\partial^2 v}{\partial x^2} + \frac{\partial^2 v}{\partial y^2} \right) + \frac{w^2}{R} \tag{3}$$

$$u \frac{\partial w}{\partial x} + v \frac{\partial w}{\partial y} = \nu \left(\frac{\partial^2 w}{\partial x^2} + \frac{\partial^2 w}{\partial y^2} \right) \tag{4}$$

energy:

$$u \frac{\partial T}{\partial x} + v \frac{\partial T}{\partial y} = \alpha \left(\frac{\partial^2 T}{\partial x^2} + \frac{\partial^2 T}{\partial y^2} \right) \tag{5}$$

Here w is the circumferential velocity. The last term in Eq. (3) repre-sents the centrifugal force acting in the y direction. There is no pressure gradient in Eq. (4), since the circumferential velocity does not experience a pressure force in an axisymmetric flow. The boundary conditions for the velo-city w are provided by the fact that the cylinder and the fins rotate at a con-stant angular velocity Ω . For heat transfer, the cylinder surface and the fins are assumed to be at a uniform temperature T_1. (This implies that the fins are made of a high-conductivity material.) The shroud is supposed to be at a different uniform temperature T_2.

The governing equations can be cast into appropriate dimensionless forms by means of the following dimensionless variables.

$$X = x/d \quad , \quad Y = y/d \tag{6}$$

$$U = ud/\nu \quad , \quad V = vd/\nu \quad , \quad W = w/(R\Omega) \tag{7}$$

$$P = (p/\rho)d^2/\nu^2 \quad , \quad Re = R\Omega d/\nu \quad , \quad Ta = Re(d/R)^{1/2} \tag{8}$$

$$\theta = (T-T_2)/(T_1-T_2) \quad , \quad Pr = \nu/\alpha \tag{9}$$

Here, Ta is the Taylor number, which is the main governing parameter for the flow field. The dimensionless differential equations are

$$\frac{\partial U}{\partial X} + \frac{\partial V}{\partial Y} = 0 \tag{10}$$

$$U \frac{\partial U}{\partial X} + V \frac{\partial U}{\partial Y} = -\frac{\partial P}{\partial X} + \frac{\partial^2 U}{\partial X^2} + \frac{\partial^2 U}{\partial Y^2} \tag{11}$$

$$U \frac{\partial V}{\partial X} + V \frac{\partial V}{\partial Y} = -\frac{\partial P}{\partial Y} + \frac{\partial^2 V}{\partial X^2} + \frac{\partial^2 V}{\partial Y^2} + Ta^2 W^2 \tag{12}$$

$$U \frac{\partial W}{\partial X} + V \frac{\partial W}{\partial Y} = \frac{\partial^2 W}{\partial X^2} + \frac{\partial^2 W}{\partial Y^2} \tag{13}$$

$$U \frac{\partial \theta}{\partial X} + V \frac{\partial \theta}{\partial Y} = \frac{1}{Pr} (\frac{\partial^2 \theta}{\partial X^2} + \frac{\partial^2 \theta}{\partial Y^2}) \tag{14}$$

The boundary conditions for these equations can be given with reference to the calculation domain shown in Fig. 1. They are

shroud $U = 0 \quad , \quad V = 0 \quad , \quad W = 0 \quad , \quad \theta = 0$ \tag{15} cylinder and fin $U = 0$, $V = 0$, $W = 1$, $\theta = 1$ (16)

other boundaries $U = 0$ and $\partial/\partial X = 0$ for V, W, θ (17)

The flow problem is completely determined by the values of the Taylor number Ta and two geometrical parameters S (= s/d) and C (= c/d). The heat transfer problem requires one additional parameter, namely the Prandtl number Pr.

An overall outcome of the flow field solution can be expressed in terms of a dimensionless shear force F on the shroud. F is defined as

$$F = \frac{\int \tau dx}{(\mu R\Omega/d) \int dx}$$ (18)

where τ is the shear stress on the shroud and the integrals extend over the x-direction length of the calculation domain. In terms of the dimensionless variables, the expression for F is

$$F = - (2/S)\int_0^{S/2} (\partial W/\partial Y) \, dX$$ (19)

The overall Nusselt number Nu is defined as

$$Nu = \frac{\overline{q} \, d}{k(T_1 - T_2)}$$ (20)

where $\overline{q}$ is the average heat flux at the shroud surface. It then follows that Nu can be calculated from

$$Nu = - (2/S)\int_0^{S/2} (\partial\theta/\partial Y) \, dX$$ (21)

The set of governing differential equations was solved by the numerical method described in [4]. In particular, the coupling between Eqs. (10-12) was handled by the SIMPLER procedure of [5]. At first, a converged solution was obtained for the flow equations (10-13) for a given value of the Taylor number. Then the temperature equation (14) was solved for two different values of the Prandtl number. The range of the Taylor number was limited to 2000, since the flow is expected to be turbulent for higher values of this parameter.

The results reported in this paper have been obtained with a 20x25 grid in the xy coordinates. The grid point distribution was made nonuniform with a fine spacing near the solid surfaces and near the fin tip. Exploratory calculations were performed by varying the number of grid points and the nonuniformity of the grid. They indicate that the reported results are accurate to at least 1 percent.

3. RESULTS AND DISCUSSION

A wealth of information is produced by the numerical solutions of the governing equations. It includes the distributions of the three velocity components and pressure, the temperature distribution, and the viscous stresses and heat transfer rates at the boundaries. Here, the main overall results will be presented first. Then some details of the velocity and temperature distributions will be shown.

3.1 Results for Vanishing Taylor Number

The effect of the centrifugal force is to induce the secondary flows sketched in Fig. 1. However, at very small values of the Taylor number, the secondary flows are very weak, and as a result, both the velocity w and the temperature T are little influenced by convection terms. It can be seen that, for the case of Ta → 0 , the dimensionless variables W and θ both satisfy the Laplace equation. Further, since they have the same boundary conditions, their solutions are identical. This may be termed as the conduction solution for the problem. It should also be noted that the conduction solution for θ is independent of the Prandtl number.

In the present problem, computations for Ta = 0.1, 1, and 10 gave virtually identical results. These results are used as a reference for the solutions at higher Taylor numbers. The values of the dimensionless force F and the Nusselt number Nu for Ta → 0 are denoted by F_0 and Nu_0 . They are listed in Table 1 for different values of the geometrical parameters. Since the solutions for W and θ are identical for Ta → 0, it follows that F_0 equals Nu_0.

The physical significance of the values in Table 1 can be best discussed by treating them as heat transfer rates. The values decrease with increasing C; this is the result of the decrease in the height and area of the fins as the clearance increases. The decrease in the fin spacing S implies that more fins are placed per unit axial length of the cylinder; hence, the heat transfer increases. A limiting behavior can be visualized for which S → 0 ; then the value of Nu_0 (or F_0) will approach 1/C.

3.2 Overall Nusselt Number

The influence of the secondary flow on the Nusselt number is shown in Figs. 2, 3, and 4, where the ratio Nu/Nu_0 is plotted as a function of the Taylor number. Each figure is drawn for a particular value of the dimensionless spacing S and shows the results for three values of the clearance C and two values of the Prandtl number.

The secondary flow induced by the centrifugal force is, in general, seen to enhance the heat transfer; augmentation factors of the order of 10 are encountered. Obviously, the Nusselt number increases with the Taylor number.

The heat transfer enhancement is more pronounced for large clearances, i.e. small fin heights. This behavior can be understood by considering the role of the circulating fluid. The fluid transports the low temperature on the shroud to the hot surface of the cylinder; it then flows along the fin and returns to the shroud. Thus, the heat transfer is enhanced on the cylinder̄ surface and near the root of the fin. However, since the fluid flowing over the remainder of the fin is already preheated, the enhancement is not significant. Therefore, the shorter fins seem to benefit more from the secondary flow.

Table 1   Values of F_0 or Nu_0

S	C=0	C=0.25	C=0.5
2.0	3.61	1.57	1.20
1.0	6.66	2.08	1.39
0.5	12.09	2.67	1.60
0.0	∞	4.00	2.00

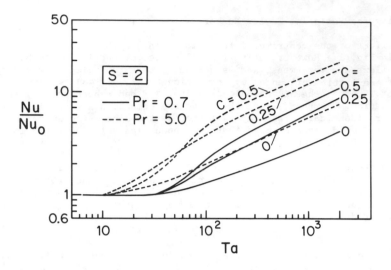

Fig. 2   Variation of Nu/Nu_0 for S = 2

The effect of fin spacing is also interesting.   Closely spaced fins tend to provide greater resistance to the secondary flow.   Therefore, the onset of appreciable secondary motion and the associated heat transfer augmentation are delayed as S decreases.   However, once the secondary flow is established, the augmentation is substantial.

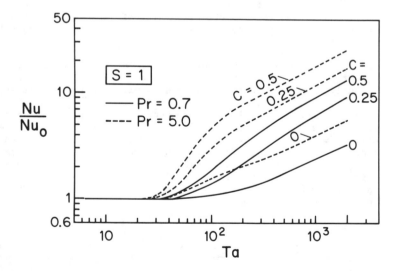

Fig. 3   Variation of Nu/Nu_0 for S = 1

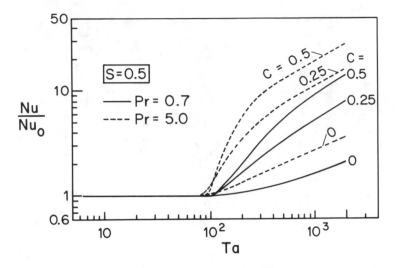

Fig. 4   Variation of Nu/Nu_0 for S = 0.5

    In various augmentation devices, it is found that the secondary motion is
particularly beneficial for the large Prandtl number fluid. This is confirmed
by the behavior seen in Figs. 2-4. A large Prandtl number can be considered
to be the result of low conductivity or high specific heat. Since conduction
heat transfer is weak, even a small amount of recirculating flow contributes
significantly to the heat transfer augmentation. The values of Nu/Nu_0 in
Figs. 2-4 for Pr = 5 are appreciably greater than the Pr = 0.7 values for corresponding conditions.

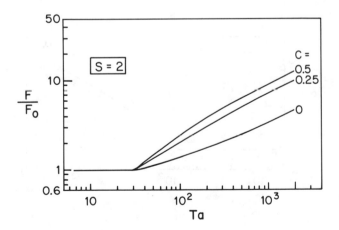

Fig. 5   Variation of F/F_0 for S = 2

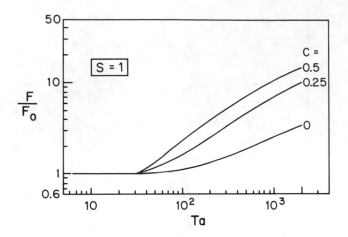

Fig. 6 Variation of F/F_0 for $S = 1$

3.3 Shear Force on the Shroud

The torque required to rotate the finned cylinder can be calculated from the shear force exerted on the shroud. In general, heat transfer augmentation is accompanied by increased shear force. The ratio F/F_0 is plotted as a function of the Taylor number in Figs. 5, 6, and 7 for the different geometrical configurations.

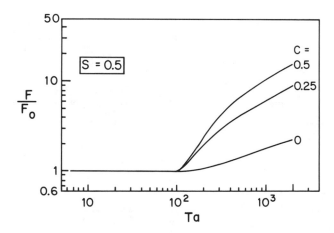

Fig. 7 Variation of F/F_0 for $S = 0.5$

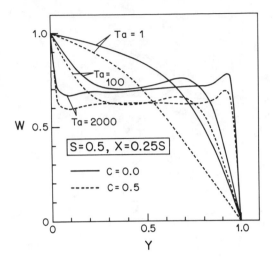

Fig. 8 Circumferential velocity profiles for different clearances

From Eqs. (13) and (14) and their boundary conditions, it can be concluded that the solutions for W and θ are identical for Pr = 1. Then it follows that the F values are indeed equal to Nu for Pr = 1. Thus, the plots in Figs. 5-7 can be thought of as the results for Nu/Nu_0 at Pr = 1.

Because of this interpretation of the values of F/F_0, the discussion in the previous subsection about the influence of Ta and the geometrical parameters on Nu/Nu_0 is also applicable to the variation of F/F_0 in Figs. 5-7. No further elaboration, therefore, seems necessary.

3.4 Profiles of Circumferential Velocity

Some idea of the variation of w in the domain of interest can be obtained from the profiles shown in Figs. 8 and 9, where W is plotted as a function of Y for a location corresponding to X = 0.25 S. Incidentally, the profiles of W can also be interpreted as profiles of θ for Pr = 1.

The influence of the Taylor number on the velocity profiles is to make them more uniform. Another notable feature is the local peak near the shroud, which is formed due to the convection of high-w fluid by the secondary flow.

In Fig. 8 the clearance parameter C does not show any striking influence on the profiles. However, the effect of spacing S shown in Fig. 9 is significant. For the closely spaced fins (S = 0.5), the distortion due to the secondary flow does not begin even at Ta = 100. On the other hand, for S = 2, the profile for Ta = 100 is appreciably distorted from its shape at Ta = 10. Also, the values of W are in general higher for the closely spaced fins, as the rotation of the fins is more effectively transmitted to the fluid.

3.5 Secondary Flow and Temperature Field

Further insight into the pattern of the secondary flow and the resulting

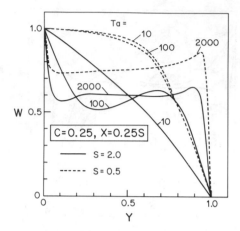

Fig. 9   Circumferential velocity profiles for different spacings

temperature field can be obtained from Fig. 10.   The contours shown correspond
to Ta = 500 and S = 1.   The left half of each diagram displays the streamlines,
while temperature contours for the two Prandtl numbers are drawn in the right
half.

The pattern of circulation is almost the same for the three values of C,
although some slight evidence of the viscous resistance at the fin surface can
be detected.

The value of ψ_m given for each case is the maximum value of the dimensionless function ψ defined by

$$\psi = - \int\limits_0^Y UdY \tag{22}$$

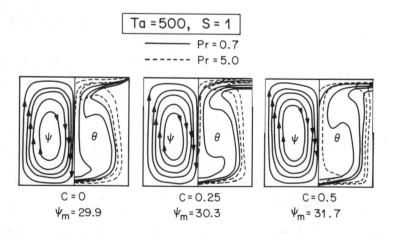

Fig. 10   Streamlines and temperature contours for Ta = 500

The magnitude of ψ_m is a measure of the mass flow rate of the secondary motion. The value of ψ_m does not seem to be significantly influenced by the clearance parameter C. This is because two opposing factors are at work. A long fin does extend the circumferential velocity of the cylinder further into the fluid, thus encouraging stronger secondary flow. On the other hand, the friction at the extended fin surface inhibits the secondary motion.

The temperature contours for Pr = 0.7 are more evenly spread throughout the domain, while those for Pr = 5 indicate thin thermal boundary layers near the heated and cooled surfaces and a uniform temperature core in the bulk of the fluid. The distortion of the isotherms due to the secondary flow is clearly in evidence.

4. CLOSING REMARKS

This paper has described a numerical study of a problem of practical importance. The results for heat transfer and shear force should be useful for design calculations. The discussion of the trends of the results is provided to develop physical insight into the flow and heat transfer behavior. Possible extensions of the work include the calculations for turbulent flow and the inclusion of an axial throughflow.

ACKNOWLEDGMENT

This research was performed under the auspices of NSF Grant CME 8007476.

NOMENCLATURE

C	dimensionless clearance, c/d
c	clearance space, Fig. 1
d	distance between cylinder and shroud, Fig. 1
F	dimensionless shear force, Eq. (18)
F_0	value of F for Ta $\rightarrow$ 0
k	thermal conductivity of the fluid
Nu	Nusselt number, Eq. (20)
Nu_0	value of Nu for Ta $\rightarrow$ 0
P	dimensionless pressure, Eq. (8)
p	pressure
Pr	Prandtl number of the fluid
$\bar{q}$	average heat flux at the shroud surface
R	radius of the rotating cylinder, Fig. 1
Re	Reynolds number, Eq. (8)
S	dimensionless spacing, s/d
s	spacing between adjacent fins, Fig. 1
T	temperature
T_1	temperature of the finned cylinder
T_2	temperature of the shroud
Ta	Taylor number, Eq. (8)
U	dimensionless axial velocity, Eq. (7)
u	axial velocity
V	dimensionless radial velocity, Eq. (7)
v	radial velocity
W	dimensionless circumferential velocity, Eq. (7)
w	circumferential velocity
X	dimensionless axial coordinate, Eq. (6)
x	axial coordinate, Fig. 1

Y	dimensionless radial coordinate, Eq. (6)
y	radial coordinate, Fig. 1
α	thermal diffusivity of the fluid
θ	dimensionless temperature, Eq. (9)
μ	dynamic viscosity of the fluid
ν	kinematic viscosity of the fluid
ρ	density of the fluid
τ	shear stress on the shroud surface
ψ	dimensionless stream function, Eq. (22)
ψ_m	maximun value of ψ
Ω	angular velocity of the rotating cylinder

REFERENCES

1. Schlichting, H., <u>Boundary Layer Theory</u>, seventh edition, McGraw-Hill, 1979.

2. Acharya, S. and Patankar, S. V., Laminar Mixed Convection in a Shrouded Fin Array, <u>J. Heat Transfer</u>, Vol. 103, pp. 559-565, 1981.

3. Sparrow, E. M., Baliga, B. R., and Patankar, S. V., Forced Convection Heat Transfer from a Shrouded Fin Array with and without Tip Clearance, <u>J. Heat Transfer</u>, Vol. 100, pp. 572-579, 1978.

4. Patankar, S. V., <u>Numerical Heat Transfer and Fluid Flow</u>, McGraw-Hill-- Hemisphere, 1980.

5. Patankar, S. V., A Calculation Procedure for Two-Dimensional Elliptic Situations, <u>Numerical Heat Transfer</u>, Vol. 4, pp. 409-425, 1981.

Experimental Studies on Diabatic Flow in an Annulus with Rough Rotating Inner Cylinder

K. V. C. RAO and V. M. K. SASTRI
Heat Transfer and Thermal Power Laboratory
Department of Mechanical Engineering
Indian Institute of Technology
Madras-600 036, India

ABSTRACT

Heat transfer in an annulus with the inner rotating cylinder was studied experimentally. Investigations were carried out on smooth as well as rough inner cylinders to study the heat transfer particularly at large Taylor numbers. The effect of surface roughness on heat transfer from a rotating surface was studied. Water was used as the axial fluid. The percentage increases in heat transfer for the rotating rough surfaces were evaluated.

1. INTRODUCTION

In many practical applications such as electrical machines, cyclone chambers, journal bearings, heat is to be removed from a rotating surface. A good approximation to such a system is the fluid motion in a concentric annulus with an inner rotating core. Dorfman [1] gave an excellent review of the state of art of heat transfer and fluid flow in rotating systems. His treatment is mostly theoretical. Over the past several years the number of articles dealing with flow and heat transfer in rotating systems has increased rapidly indicating the importance of this problem. One of the major problems facing the designer is the prediction and measurement of the rate of heat transfer from the rotor to the fluid flowing in the rotor stator gap of an electrical machine.

It is highly desirable not only to predict to a reasonable degree of accuracy, the heat transfer coefficient from the rotating surface of the electrical machine but augment the heat transfer from the surface which has a direct bearing on the cooling and, consequently, on the size of the machine. Alternatively it is essential to determine the rate of heat transfer if such a provision already exists on the surface.

The general parameters which effect the heat transfer from a rotating surface include: the gap between the rotor and stator, speed of rotation of the rotor, nature of the coolant fluid, configuration of the path of the cooling fluid, the geometry and nature of the surface. Of the variables, the influence of all except the nature of the surface seems to have been studied fairly completely. In many practical systems such as a typical electrical machine, the nature of the rotating surface cannot be assumed to be smooth. For example in high capacity generators, the tendency is to use the so called gap-pickup cooling system. In this system the coolant fluid is scooped through the wedges which project slightly outside the rotor surface. The flow through the diagonal inner passages is discharged through the slots in the wedges which are in a direction opposite to that of inlet wedge slots. Under such conditions the rotor is no longer smooth. Since a rough surface generally gives rise to augmentation in heat transfer, it is all the more important that the actuel heat transfer coefficient be determined for such a system.

Numerous techniques are available for augmenting the heat transfer from a stationry surface by convection to the surrounding fluid, such as surface roughness, displaced turbulent promotors, swirl flow, surface tension devices, liquid additives etc. Bergles and Webb [2] presented a bibliography of world literature on techniques to agument convective heat transfer. However, little or no effort appears to have been made to examine the augmentation techniques in rotating systems. It is with this view an attempt is made to investigate the effect of roughness on heat transfer from a rotating surface.

2. PREVIOUS WORK

If one simplifies the rotor of an electrical machine to a small model consisting of two concentric cylinders, with the inner cylinder rotating and outer cylinder stationary, it gives a good approximation for finding out the heat transfer coefficient from the rotating surface of an electrical machine. The first major aspect to be considered is the nature of flow: laminar, turbulent, vortex or a combination of any of these. Axial flow superimposed with rotational flow affects the stability of flow and it has considerable influence on heat transfer. An important parameter recognised by many authors [3,4,5,6,7,8] but not investigated is the geometry and the nature of the rotating surface through which heat transfer takes place. However a few [5,6,7,8] studies on slotted surface heat transfer are available.

Considerable information is available for restricted cases. Luke [9] seems to have been the first to investigate such problems. He studied the heat transfer rate from an inner rotating cylinder to air flowing through the annulus. It was difficult to draw any general conclusions from Lukes experiments due to the dimensions of the experimental setup and the nature of the flow. Gazley [5] in his extensive study measured the heat transfer in an annular gap for various combinations of axial flow and speeds of rotation. A variety of rotor and stator surfaces were provided so as to give both smooth and various combinations of slot configurations covering the range typical of an electrical machine. Kaye and Elgar [3] reported data on modes of adiabatic and diabatic fluid flow in an annulus with an inner rotating cylinder. The main emphasis was on adiabatic flow and only a few preliminary results were reported for diabatic flow. The work was subsequently extended by Becker and Kaye [4], particularly for the case of diabatic flow. They clearly indicated the demarcation lines for various flow regimes and compared their data with those of earlier authors Fig. 1.

A more recent paper on this topic is that of Gardiner and Sabersky [7]. They measured the heat transfer coefficient for the flow in the annular space between an inner rotating cylinder and an outer stationary one with superimposed axial flow. They considered Taylor numbers upto 10^6 and flow Reynolds numbers upto 7000. Experiments were performed for three different Prandtl numbers. They conducted experiments on smooth and slotted rotors.

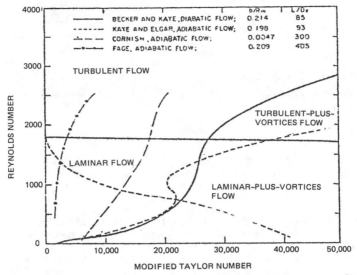

Fig. 1   Demarcation lines for flow regions in annulus [4]

They concluded that for the smooth rotor, the Ta for the onset of vortex flow first increases with increasing axial flow and after reaching a maximum seems to decrease slightly with further increase of axial flow. This tendency towards decreasing Ta is very definate for the slotted rotor. They felt that free convection may have a strong influence on the transition point.

It can be seen from the literature that not much effort was directed to investigate the effects of surface conditions and geometry. There is extensive literature [2] on the effects of surface roughness on heat transfer characteristics on stationary surfaces.

2.1   EFFECT OF ROUGHNESS ON HEAT TRANSFER

There are different roughness forms which increase heat transfer such as ribs, grooves, threads etc. The different roughness forms can be classified according to their shape and type of roughness. In general the roughness can be classified in two ways viz. integral roughness and overlapped roughness. The former has a good thermal contact with the parent surface while latter will act mainly as a turbulence promotor. Fig. 2 shows different forms of roughness. Transverse ribs are either formed by a helical coil or multiple rings, of round or square wire bonded to the surface of the circular cross section, or by helical threads machined in the inner surface. Discrete-rib-type is the term used to denote transverse ribs which have a pitch "s" much larger than the rib height (s/e = 10), where as thread type is the term used to denote transverse ribs which have relatively small pitch (s/e = 1.4-2.0).

3.   EXPERIMENTAL SETUP

The general layout of the experimental setup is shown in Fig. 3. The test apparatus is designed for the purpose of measuring the heat transfer from the rotating surface to the fluid flowing through the annulus. A sectional view of the test setup is shown in Fig. 4. The rotor is of mild steel and is made hol-

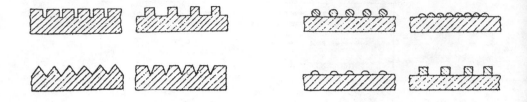

Fig. 2 (a) Integral Roughness   (b) Overlapped Roughness

low to permit the housing of the cylindrical heating element. At either end of
the rotor two steps are provided to seat the insulating discs to minimise axial
conduction. The rotor is held by two flanges and hollow shaft at either end
with an insulating hylam sheet in between rotor-side and flange. The whole assembly consisting of the test section and the hollow shaft and flanges are
supported on two self aligning ball bearings. The stator is also made up of
mild steel of the same length and when supported it is concentric with rotor,
providing an annular gap of 11 mm. Both the outer surface of the rotor and
the inner surface of the stator are chrome-plated to prevent the formation of
rust and maintain a clean surface. The stator is well insulated to minimise
radial losses and provide adiabatic boundary at the out-side of stator. The
rotor and stator are sealed by packing material. By proper arrangement and
pressure on the seals the friction between the seal and the rotating surface
is maintained and leakage prevented.

   The rotor is driven by a variable speed motor through a pulley and belt
drive. The speed of the motor is controlled by a rectifier unit. In the present
experiments the maximum speed of the rotor was about 400 rpm which gave Ta
above 10^8.

   Water is used as the coolant fluid. The water is pumped from a sufficiently large sump to an overhead tank, in which constant level is maintained. The
water inlet of the test section is connected to the outlet of the water tank
directly. The outlet of the test section is connected to drain through a regu-
lating valve to ensure that the test section is always full with water and no air
is trapped. The flow rate is measured by collecting water for a known time and
by weighing. Two surge tanks are provided at the inlet and outlet of the test

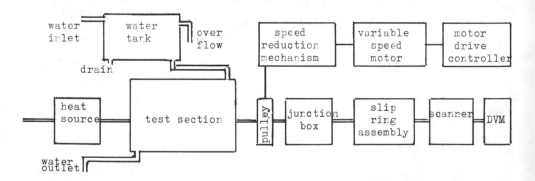

Fig. 3 Layout of the Experimental Set-up

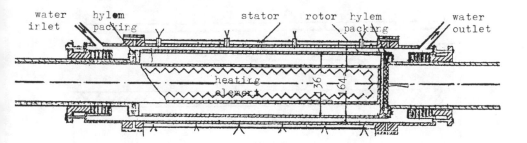

Fig. 4 Sectional View of the Experimental Set-up

section to minimise the disturbances and ensure uniform flow in the annulus. From the surge tanks three outlets are taken and connected to the annulus circumferentially 120° apart. Similarly at the exit of the annulus three outlets are taken and connected to the surge tank.

Five rotors, all of same length and diameter, were made from mild steel with a wall thickness of 3 mm. One was smooth and four had integral roughness on surface. The different kinds of roughness considered are shown in Fig. 5. After machining to the required condition they were chrome-plated.

Double insulated iron-constantan thermocouples of 0.5 mm dia were fixed to the rotor, stator and in water space. The location of thermocouples is shown in Fig. 6. Thermocouple beads were made in an inert atmosphere and axial grooves were made to seat the thermocouple properly. At the point of fixing, a slightly enlarged hole was made to house the bead and filled with Technical-G, copper cement to ensure good thermal contact. A slipring mechanism specially designed for transmission of signals from sensors on rotating machinary was used. This slipring assembly consists of 24 rings and they are gold plated to ensure minimum contact resistance. The assembly is capable of measuring at speeds upto 15000 rpm. A more detailed description of the temperature measurement is given in reference [10].

The heating element is a cylindrical chamber with resistance coils inserted in it. The heater element can be inserted and removed easily even while the rotor is moving. The inner race of the ball bearing is such that it permits the entry of the heating element into the rotor section. The heater is held with a hollow pipe with insulation in between to minimise losses by heat conduction along the pipe. However two thermocouples are fixed on the heating element holder at known distance to determine the conduction losses.

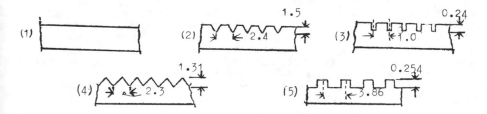

Fig. 5 Kinds of Roughness Considered. (1) Smooth (2) Triangular Grooves
(3) Square Threads (4) Whitwoth Threads (5) Rectangular Ribs.

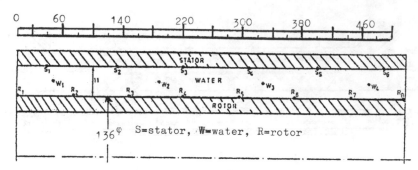

Fig. 6   Location of Thermocouples

When placed inside the rotor, the heater element is concentric with the rotor
and capable of delivering a maximum of 2 KW. The heater is of the same length
as the test section and has a diameter of 62 mm. The heater is stationry and
the heat transfer to the rotor body is by both convection and radiation. This
type of heating arrangement is found to be very convenient when compared to
the conventional methods of heating the rotor by giving the supply to the heating element through sliprings.

4.   TEST RESULTS

The Prandtl number is determined from the film temperature using the mid
rotor temperature and the average bulk temperature of the water. The change
in the value of Pr due to the change in temperature ($\mp 2^{\circ}C$ for the entire range
of readings) is neglected and the average Pr is taken as 4.5.

All the test results apply to the mid section normal to the axis of the
inner cylinder. The point considered is midway between the inlet and outlet
of the test section. All the temperatures are measured with reference to the
inlet water temperature. The axial temperature variation in the mid-section
concentric to the cylinders is fitted by the method of least squares and the
value of "ΔT" is computed from the rotor-water temperature. Typical graph is
shown in Fig. 7. The heat transfer coefficient from rotor surface to axial
fluid is expressed in the nondimensional form

$$Nu_R = q\, b/(\Delta T)\, k \quad .$$

The heat transfer rate is calculated based on the heat carried away by
axial fluid by measuring the inlet and outlet temperatures of water and its
mass flow rate. The stator is well insulated and assumed adiabatic as no heat
is transfered through the stator surface. All the heat supplied is assumed to
be transfered to the axial fluid. However a heat balance is made between the
heat supplied and heat carried away by the axial fluid. The difference is attributed to the losses due to axial conduction and radiation. The losses are
calculated and found to be generally constant for a given heat input and they
are estimated to be of the order of 5% of the heat supplied. The Nusselt number is expressed in terms of Nu_R/Nu_C, where Nu_C is the conduction Nusselt
number given by

$$Nu_C = (b/R_1)/\ln(R_2/R_1) \quad .$$

This is normally done to camouflage the explict dependence of heat transfer coefficient on gap ratio. However in the present investigation the gap

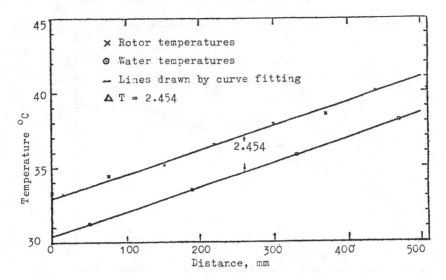

Fig. 7   Rotor Surface and Water Temperature along the
Length of the Test Section. Typical Plot.

ratio is constant and the above difinition is useful for comparing the present
data with other published work.

The Reynolds number is given by

$$Re = 2\,W.b/\nu$$

in terms of the axial velocity component and the hydraulic diameter of the
annulus.

The modified Taylor number is given by

$$(Ta)_m = \frac{2\Omega^2\,R_1^2\,b^3}{\nu^2\,(R_1+R_2)}\left(\frac{1697}{4}\,P\right)$$

Where P is the geometric factor given by

$$P = 0.0571\left(1 - 0.652\,\frac{b}{R_1}\right) + 0.00056\left(1 - 0.652\,\frac{b}{R_1}\right)^{-1}$$

Experiments were carried out at one Pr and three Re for various Ta. The
range of Re considered is low due to the limitation on the heater capacity.
Ta considered are of the order of 10^8. The flow in the present situation is
considered to be laminar-vortex flow. For low speeds it is considered to be
thermally developing and for higher rpm it is thermally developed flow. All
the tests were conducted in the same test set-up keeping geometric parameters
unaltered. Results were obtained on one smooth and four rough surfaces. The
experiments on smooth rotor were conducted mainly in order to be able to compare the present results with those of the earlier investigations and subsequently serve as a base for comparing the results obtained on rough cylinders.
For the purpose of comparison, the data of Gardiner and Sabersky [7] is taken.

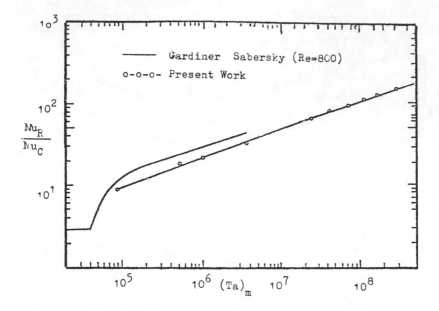

Fig. 8   Heat Transfer Coefficient Vs. Modified Ta.

They reported experiments for both smooth and slotted rotors upto Ta of 10^6.
The results of the results of the present investigation and those of [7] are
plotted in Fig. 8. The comparision may be seen to be satisfactory. The slight
deviation is attributed to the method of evaluating "ΔT". Measurements were
made at low values of Ta to compare with the results of [7]. The Ta range is
extended to values of 10^8 and larger for high Pr.

Fig. 9 shows the variation of Nu with respect to Ta for various Re. For
the Ta range of 10^5 to 10^8, the Nu increases with increase in Ta as expected.
It is also seen that the variation of Nu with Re, for the range considered, is
not significant up to a Ta of 5×10^7, whereas beyond this number a considerable
difference is observed. Further the Nu decreases with increasing Re. This can
be explained by the stability criteria. The graph further indicates that the
values correspond to the region of laminar-plus-vortex flow, since the Nu
continuously increases with Ta. As the speed increases the formation of vortices creates turbulence and heat transfer will incresea. It is evident from
Fig. 9 that for the range of Re considered, the effect of rotation, particularly at low speeds is negligible. As speed increase the difference in Nu also
increases. The effect of Re is shown in Fig. 10. It may be noted that Nu remains practically constant whereas, for large Ta there is slight variation.
Similar results were obtained with rough cylinders. Fig. 11 shows the variation of Nu along the length of the rotor for different Ta. For the case of
smooth cylinder at Re = 360, these Nu are calculated based on the temperature
difference evaluated at different locations or from the curves obtained by curve
fitting the data by least squares. In most of the cases the variation along
the length is negligible indicating that the flow is practically fully developed thermally, particularly at high Ta.

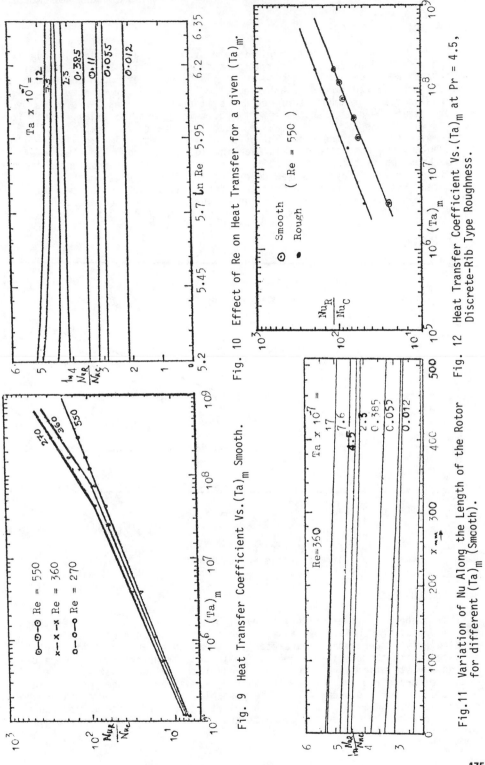

Fig. 10 Effect of Re on Heat Transfer for a given (Ta)$_m$.

Fig. 12 Heat Transfer Coefficient Vs.(Ta)$_m$ at Pr = 4.5, Discrete-Rib Type Roughness.

Fig. 9 Heat Transfer Coefficient Vs.(Ta)$_m$ Smooth.

Fig.11 Variation of Nu Along the Length of the Rotor for different (Ta)$_m$ (Smooth).

4.1 EFFECT OF ROUGHNESS

Fig. 12 shows the effect of rib type of roughness (s/e = 15.2) on heat transfer coefficient compared to that on a smooth surface. For the purpose of comparison the values corresponding to Re = 550, both for rough and smooth surfaces are plotted. There is enhancement of heat transfer of about 70%, compared to that for smooth surface. For stationary surfaces it is normally expected that the heat transfer effect due to roughness for axial flow Re considered will be negligible. It may be noted that due to superimposition of rotation, considering the resultant of axial and rotational velocity the flow is turbulent. A composite plot is made in Fig. 13 to show the relative effects of roughness on heat transfer. It is seen that, in general, roughness increases heat transfer to a varying degree depending upon the type of roughness. It is estimated that the increase in heat transfer is about 40% for Whitworth thread type of roughness, 9 - 10 % for the square thread type of roughness and 4 - 5 % for V-groove roughness. The discrete rit type of roughness has shown considerable improvement over thread type of roughness, obviously also due to the fin effect.

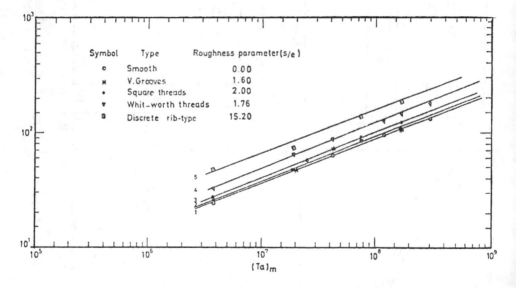

Fig. 13   Heat Transfer Coefficient Vs. $(Ta)_m$ for
Different Kinds of Roughness.

5.   CONCLUSIONS

From the present experimental investigations it can be concluted that the nature of the surface has a significant influence on heat transfer. Particularly, the roughness parameter s/e seems to be an important quantity in determining the possible augmentation in heat transfer. However more detailed investigations are required to predict the influence of s/e as a parameter. Discrete rib type of roughness is more desirable since increase in heat transfer of the order of 70% are possible. Since such increases are possible even at low axial flow Re, high heat transfer rates for low pressure drop values can be achieved. High heat transfer rates can be obtained even for smaller axial flow conditions.

NOMENCLATURE

b	=	gap between rotor and stator	m
c_p	=	specific heat of the fluid	kcal/kg-$^\circ$C
D_e	=	equivalent diameter	m
e	=	height of roughness	m
h	=	heat transfer coefficient	kcal/hr-m^2-$^\circ$C
k	=	thermal conductivity	kcal/hr-m-$^\circ$C
L	=	length of the rotor	m
Nu_C	=	conduction Nusselt number	-
Nu_R	=	rotor to axial fluid Nusselt number hb/k	-
Pr	=	Prandtl number $\mu c_p/k$	-
q	=	heat transfer rate per unit area	kcal/hr-m^2
R_m	=	mean radius $(R_1 + R_2)/2$	m
R_1	=	outer radius of inner cylinder	m
R_2	=	inner radius of outer cylinder	m
Re	=	axial flow Reynolds number	-
ΔT	=	temperature difference between rotor - axial fluid	$^\circ$C
Ta	=	Taylor number	-
$(Ta)_m$	=	modified Taylor number	-
s	=	pitch of ribs	m
W	=	axial velocity of the fluid	m/hr
Ω	=	angular velocity of the rotor	rad/hr
μ	=	dynamic fluid viscosity	kg/hr-m
ν	=	kinematic fluid viscosity	m^2/hr

REFERENCES

1. Dorfman, L.A. 1963. Hydrodynamic resistance and heat loss of rotating solids, Oliver and Boyd Publishing Co., Edinburg and London.

2. Bergles, A.E. and Webb, R.L. 1970. Bibliography on augmentation of convective heat and mass transfer, presented at the seminar on "Augmentation of convective heat and mass transfer", sponsored by the heat transfer division of ASME at the winter annual meeting of the ASME, New York, December 2, pp. 1-15.

3. Kaye, J. and Elgar, E.C. 1958. Modes of adiabatic and diabatic fluid flow in an annulus with inner rotating cylinder, J. of Heat Transfer, Trans. ASME, Vol. 80, No. 3, pp. 753-765.

4. Becker, K.M. and Kaye, J. 1962. Measurements of diabatic flow in an annulus with an inner rotating cylinder, J. of Heat Transfer, Trans. ASME, Ser. C, Vol. 84, No. 2, pp. 97-105.

5.   Gazley, G. jr. 1958. Heat transfer characteristics of rotational and axial flow between concentric cylinders, J. of Heat Transfer, Trans. ASME, Vol. 80, No. 1, pp. 79-90.

6.   Hoseason, D.B. 1931. The cooling of electrical machines, Proceedings of IEE J., No. 409, Vol. 69, pp. 121-155.

7.   Gardiner, S.R.M. and Sabersky, R.H. 1978. Heat transfer in a annular gap, International Journal of Heat and Mass Transfer, Vol. 21, No. 12, pp. 1459-1466.

8.   Tachibana, F. and Fukui, S. 1964. Convective heat transfer of rotational and axial flow between two concentric cylinders, J. Soc. Mech. Engr. Bul. Vol. 7, No. 26, pp. 385-391.

9.   Luke, G.E. 1923. The cooling of Electrical machines, Trans. AIEE., Vol. 42, pp. 636.

10. Rao, K.V.C. 1981. Heat transfer in an annulus with a rotating rough inner cylinder, Ph.D. Thesis, Dept. of Mech. Engg., Indian Institute of Technology, Madras, India.

Numerical Study of Friction on Enclosed Rotating Perforated Discs with Forced Through-Flow

WEN–JEI YANG
Department of Mechanical Engineering
and Applied Mechanics
University of Michigan
Ann Arbor, Mich., USA

J. H. WANG
Babcock and Wilcox
Alliance, Ohio, USA

ABSTRACT

The power requirement for a rotating perforated disc in a housing with through-flow is theoretically studied by a numerical method. The enclosure consists of a cylindrical and two conical elements. The full Navier-Stokes equations are integrated using finite-difference techniques. The results are utilized to determine the disc torque, i.e. power consumption. Both laminar and turbulent flow are treated. The disc perforation is modelled as a ring slot with an equivalent opening. The effects of through-flow rate, disc rotational speed and housing geometry on the power requirement are determined. The results for the no through-flow case are compared with the existing theoretical and experimental data.

1. INTRODUCTION

A rotating perforated-disc type contactor (RDC) consists of one cylindrical and two conical elements as illustrated in Fig. 1. In applications, two phases, for example water and air, flow in countercurrent in a longitudinal direction. A rotating perforated disc disperses the gas phase in the liquid for radial mixing. Perforations are provided for breaking up gas bubbles into smaller sizes to increase the phase interfacial area for mass transfer. A multi-stage construction accomplishes a periodic interruption of the longitudinal transport for intensive radial mixing. Thus, the interfacial area is periodically renewed, resulting in mass transfer enhancement.

Two dimensionless parameters pertinent to the operating conditions are the Reynolds number Re and the Taylor number Ta which are defined as

$$Re = \frac{2\dot{m}}{\pi\mu(R_s + R_o)} \; , \quad Ta = \frac{\Omega R_d^2}{\nu} \tag{1}$$

Here, $\dot{m}$ denotes the through-flow rate of the liquid phase; μ, the liquid absolute viscosity; R_s, R_o, and R_d, the radii of the shaft, the cylindrical wall, and the disc, respectively; Ω, the rotational speed of the disc; and ν, the liquid kinematic viscosity. Re is defined based on the hydraulic diameter of the cylindrical section, $2(R_o - R_s)$.

The device is a mass exchanger developed for mass transfer operations such as absorption, extraction and fermentation. It is especially applicable for gas-liquid systems with high Henry's constants, i.e. low gas solubility in liquid. When used for a liquid-gas system, the unit can be operated with a wide range of flow rates for both phases. It is characterized by two special features:

(i) Very strong secondary flow, equal in strength to primary flow when Ta exceeds approximately 500; it promotes mixing resulting in the mass transfer rate which is better by a factor of 24 to 48 than that in plate columns [1]. Equally favorable results are expected in extraction processes.

(ii) High gas holdup in each stage results in less power requirement. The operating range of gas holdup may be designed at about 15% which is much higher than in conventional mixing equipment. The corresponding power requirement in an aerated system is reduced to about 50 to 70% of that to pure liquids [2, 3].

Brauer [3], an inventor and pioneer himself, provided a thorough review on the decade-long research achievement of his Institute in the development of this RDC. So far, only experimental studies on flow behavior, power requirement, gas holdup and mass transfer performance (to be reported at a later date) are fairly complete. However, theoretical approach is still at its infancy, i.e. liquid-phase behavior in the absence of through-flow, countercurrent air stream and perforations.

Glaeser et al [4] disclosed that the perforated-disc type device performed better than propellers and turbines as gas dispersion equipment. Through numerical integration of the full Navier-Stokes equations, Thiele [5] obtained a theoretical expression for the power requirement valid in the range of very small Ta, less than approximately 10. Huber [6], Jain [7] and Igwemezie [8] conducted experiments on the power requirement at high values of Ta. Results indicated a substantial power reduction that was associated with the presence of gas phase. Correlating these test results in the form of characteristic power curves, Yang [2] provided the explanation for the physical causes of the large power reduction. Dylag and Brauer [9] conducted experiments on the RDC in an open tank with and without flow baffles.

The present work concerns theoretical study of the power requirement to non-aerated systems in the rotating perforated-disc contactors. Both laminar and turbulent flow are considered. The full Navier-Stokes equations are numerically integrated using a finite-difference technique [10, 11]. The velocity distributions in the flow field are used to evaluate the power requirements with and without through-flow. Results for the through-flow case are compared with the existing theoretical and experimental data [3, 4].

2.   THEORETICAL APPROACH

The physical system to be studied is shown in Fig. 1. The disc is perforated with circular holes at uniform circumferential spacing. A liquid flows longitudinally downward in the RDC. The cylindrical coordinates (r, θ, z) are employed with the origin fixed at the disc center. z measures the axial distance in the direction against the liquid flow. The flow through perforations is assumed to be uniformly distributed in the circumferential direction such that a ring slot of an equivalent opening may be used to replace the perforations in the analysis. Both laminar and turbulent flow are considered. In the turbulent flow case, a K-ε two-equation model is employed to determine the turbulent viscosity.

With the use of the vorticity ω and the stream function ψ defined as

$$\omega = \frac{\partial U}{\partial z} - \frac{\partial W}{\partial r} \; ; \quad U = -\frac{1}{\rho r}\frac{\partial \psi}{\partial z} \; ; \quad W = \frac{1}{\rho r}\frac{\partial \psi}{\partial r} \tag{2}$$

the full Navier-Stokes equations for quasi-steady, axi-symmetric flow can be expressed as [10]

$$a = \frac{\partial}{\partial z}\left(\phi\,\frac{\partial\psi}{\partial r}\right) - \frac{\partial}{\partial r}\left(\phi\,\frac{\partial\psi}{\partial r}\right) \; - \frac{\partial}{\partial z}\; br\,\frac{\partial}{\partial z}\,(c\phi)$$

$$- \frac{\partial}{\partial r}\; br\,\frac{\partial}{\partial r}\,(r\phi) \; + rd = 0 \tag{3}$$

Here, (U, V, W) denote the velocity components in the (r, θ, z) directions, respectively; ρ, liquid density; and ϕ, general dependent variable. The coefficients a, b, c and d together with ϕ are listed in Table 1

TABLE 1: Definition of a, b, c, d and ϕ

ϕ	a	b	c	d
ψ	0	$1/\rho r^2$	1	$-\omega/r$
ω/r	r^2	r^2	μ_e	$-(\rho v^2)/\partial z$
rV	1	$\mu_e r^2$	$1/r^2$	0
κ	1	μ_e/σ_k	1	$\mu_t G - \rho\varepsilon$
ε	1	μ_e/σ_e	1	$C_1\varepsilon\mu_t G/\kappa - C_2\rho\varepsilon^2/\kappa$

Equation (3) can be applied for both laminar and turbulent flows. In the case of laminar flows, $\mu_e = \mu$ and only the first three variables in Table 1 are needed; for turbulent flows, all five equations must be solved $\mu = \mu + \mu_t$, where μ and μ_t are the laminar and turbulent viscosity, respectively. μ_t is defined as $^tC_D\,\rho\ell\kappa^{1/2}$ which is equal to $\rho\kappa^2 C_u/\varepsilon$. The constants C_1, C_2^t, C_u, C_D, σ_e and σ_k are 1.45, 2.0, 0.082, 1.0, 1.3 and 1.0, respectively [10]. G signifies the turbulent viscous dissipation function.

The appropriate boundary conditions are

(a) at inlet: U = V = 0, W - f(r); $\qquad\qquad$ (4a)

(b) on stationary walls: U = V = W = 0; $\qquad\qquad$ (4b)

(c) on rotating surfaces: U = W = 0, V = Ωr; $\qquad\qquad$ (4c)

(d) at exit: U = V = 0, $\partial W/\partial z = 0$ $\qquad\qquad$ (4d)

For the near-wall layers where viscous effects predominate over turbulent ones, the "wall function method" [10] is incorporated in the computer program. A fast-converging line-iterative technique is employed to solve the above equations in finite difference form for the distribution of flow velocities (U, V, W). Details of the numerical computational procedure are available in [10, 11] and will not be repeated here.

The circumferential component of the shearing stress for a disc wetted on both sides is

$$\tau = \mu \left(\frac{\partial V_u}{\partial z} + \frac{\partial V_b}{\partial z} \right) \tag{5}$$

In the case of turbulent flow, the effective viscosity should be used for μ. V_u and V_b represent the circumferential velocities above and below the disc, respectively. The moment or torque of the disc wetted on both sides

$$T_d = \int_o^{R_d} (2\pi r^2 \tau) dr \tag{6}$$

is then calculated followed by evaluating the power number N_p and disc-moment coefficient C_m which are defined as

$$N_p = \frac{T_d}{\rho n^3 (2R_d)^5} , \quad C_m = \frac{2T d}{\Omega^2 \rho R_d^5} = (2/\pi)^3 N_p \tag{7}$$

n denotes the number of disc revolutions per unit time. The power number is also referred to as the Newton number Ne.

3.   RESULTS AND DISCUSSION

The mixing performance as well as power requirement of RDC's vary depending upon eight geometric parameters R_s, R_o, R_d, B, b, L and $\propto$ or R_i; two operating factors Ω and $\dot{m}$, and fluid properties. In design application, L and R_o determine the overall size or compactness of an RDC and may be considered fixed, while R_s and b are specified by material strength and load. The range of B/L most favorable to flow distributions is between 1/4 and 1/2. This study was conducted with L, R_o, R_s, B and b fixed at the value shown in Fig. 1 which was employed in references 2 through 8. In other words, only R_d, $\propto$, Re and Ta were varied to determine their influence on the performance of an RDC.

Numerical results were obtained for two housing geometries, in which $\propto$ = 45° has been found to be the most efficient for mixing while $\propto$ = 90° has been commonly employed in liquid-liquid extraction. In each housing geometry, the results for different combinations of Re and Ta were obtained. R_d was also varied to determine its influence on the RDC performance. Results are graphically presented in Figs. 2, 3 and 4.

3.1 Laminar Flow Regime, $0 \le Re \le 58$ and $0 \le Ta \le 1 \times 10^4$

Numerical results for the RDC of $\propto$ = 90° [10, 11] disclosed four very extensive recirculating zones in the upper half of the vessel, leaving only a narrow street from the inlet to the exit for the main stream. This flow pattern exhibited a large dead zone in the flow field. Therefore, the vessel of $\propto$ = 90° seems to be much less effective for flow mixing than that of $\propto$ = 45°. Nevertheless, Fig. 2 shows that these two RDC's share the same curves in the Np versus Ta plot. The results reveal that : (a) Log Np

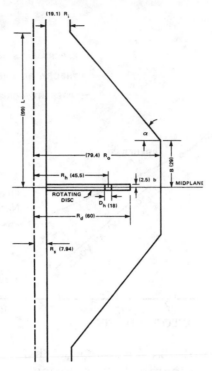

Fig 1. A Schematic of Rotating Perforated Disc Contactor (with all dimensions in mm).

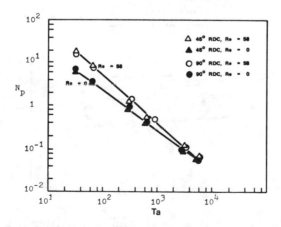

Fig 2. Theoretical Prediction of Power Number versus Taylor Number for 45° and 90° RDC's in Laminar Flow Regime.

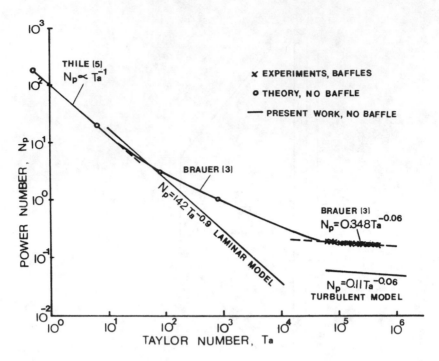

Fig. 3. Comparison of Present Work with Existing Data for
45° RDC of Identical Dimensions.

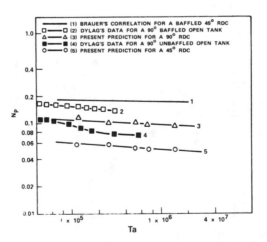

Fig. 4. Theoretical and Experimental
Results of Power Number versus
Taylor Number for 45° and 90°
RDC's in Turbulent Flow Regime.

varies linearly with log Ta and can be expressed as

$$Np = 142 \ Ta^{-0.9 \pm 0.02} \quad \text{for Re} = 0 \tag{8a}$$

$$Np = 691 \ Ta^{-1.1 \pm 0.01} \quad \text{for Re} = 58 \tag{8b}$$

(b) For Ta less than 300, the through-flow Re is critical in determining the power requirement. As shown in Fig. 2, Np increases with Ta. However, this effect diminishes with an increase in Ta. When Ta exceeds 1000, Np is practically independent of Re. (c) In the laminar flow range, the power requirement for a 90° RDC is about the same as that for a 45° RDC. It is postulated that at such a low disc speed, the housing wall of the 45° RDC is under more frictional effect than that of the 90° RDC, but if the 90° RDC carries a larger amount of fluid in motion these two factors cancel each other out. (d) Thiele [5] solved the full Navier-Stokes equations and obtained the theoretical result for the no through-flow case:

$$Np = \frac{2\pi^4}{Ta} \ \frac{0.95(b/R_o)^{0.8}}{(R_d/R_o) \ 1-(R_d/R_o)^2} + \frac{0.95(L/R_o)^{1.4}}{(R_s/R_o) \ 1-(R_s/R_o)^{1/4}} \ (R_s/R_d)^3$$

$$+ \frac{\sigma}{\pi^3} \ (R_d/R_o)^{1/3} \ 2-(R_s/R_d)^4 \tag{9}$$

which was valid for Ta less than 10. Later, Thiele's numerical technique was modified by Brauer [3] to determine the theoretical power number for higher values of Ta up to 1,000. Figure 3 compares their results with the present study equation (8a) in the absence of through-flow. The slope of Thiele's power performance curve, -1.0 is close to that of equation (8a), -0.9. Equation (8a) intersects with Brauer's curve. One may conclude that the present work agrees satisfactorily with the two previous studies [3, 5].

A numerical experiment revealed that the flow on both sides of the disc and immediately adjacent to the slot edges were characterized by high radial and tangential velocities, resulting in a small net flow through the slot. Consequently, the presence of the slot caused negligible change in power consumption. Even doubling the slot width produced little difference in power. Thus, it is concluded that the effect of disc perforations on the power consumption can be ignored.

3.2 Turbulent flow

In the absence of through-flow, Yang [2] obtained the critical Ta value of approximately 6.9×10^4 for transition from laminar to turbulent flow in the RDC. The present study predicted that for turbulent flow in the range of $0 \leq Re \leq 580$ and $5.7 \times 10^4 \leq Ta \leq 1.37 \times 10^6$,

$$Np = 0.11 \ Ta^{-0.06 \pm 0.005} \quad \text{for } \alpha = 45° \tag{10a}$$

$$Np = 0.30 \ Ta^{-0.08 \pm 0.005} \quad \text{for } \alpha = 90° \tag{10b}$$

Equations (10) were correlated from Fig. 4 which plotted numerical results of Np as a function of Ta. Also displayed in the figure are experimental results [6, 7, 8] for a perforated type 45° RDC with flow baffles (correlated

in [3]); and Dylag's test data for a perforated disc in an open tank with and
without flow baffles [9]. The dimensions of the systems used by Dylag were
h_b/R_o = 0.664, h/R_o = 2.0, S_p/R_o = 0.2, R_d/R_o = 0.333, 0.444, 0.5 and D_h/R_o =
0.146, 0.156, 0.21° in which R_d and R_o denote the radii of the disc and tank,
respectively; h, the head of the fluid; h_p, the height of the disc over the
tank bottom; and S_p, the width of the baffles. Results from the present study
indicate that: (a) The power number was very sensitive to the near-wall
velocity profile. A 10% increase in the von Karman's constant resulted in
an enhancement of the power number by 20%, due to the dependence of the wall
shear stress on the near-wall velocity gradient. The values of von Karman
and von Driest constants were derived from a one-dimensional Couette flow
and were assumed universal for any flow. For high swirling flows, the accuracy of these universal constants becomes questionable. However, rather
than adjusting their values to fit experimental data, these constants were
fixed throughout the study.   (b) The axial through-flow failed to influence
the power number due to an overwhelming effect of the swirling velocity.
(c) While the laminar case exhibited similar power requirements for both
α = 45° and 90° systems, the power number for the 90° RDC was almost twice
that of the 45° RDC in the turbulent flow range.   (d) For a 45° RDC with a
perforated disk (of identical dimensions as Fig. 1) and four vertical flow
baffles, Brauer [3] obtained the correlation equation using the test data of
[5, 6, 7]:

$$Np = 0.348 \ Ta^{-0.06} \tag{11}$$

The empirical curve had the same slope but was about 300% higher in magnitude
than that predicted by the present study for an unbaffled 45° RDC. This
discrepancy is mainly due to the effect of the baffles.   (e) In the case
of open tanks [8], the power requirement for a disc in a baffled open tank
with free surface was much higher than that of the same system without
flow baffles. The value of Np for the unbaffled open tank was slightly
lower than the present prediction for a 90° RDC. This discrepancy resulted
from the presence of a free surface which tended to reduce the power requirement.   (f) A numerical experiment disclosed that the presence of a slot
(with the same open area as the perforations) on a disc produced only slightly
different distribution patterns of velocity, vorticity, streamline, integral
length scale and kinetic energy from those of a solid disc. The power
number was found to increase by approximately 1% in the presence of the slot.
Even the slot size was doubled, only a small net flow passed through the
slot. Therefore, the effect of perforations on the power consumption is
quite small.

4.   CONCLUSIONS

A ring-slot model is proposed to simulate the disc perforations. The
effect on power consumption is found to be negligible. In laminar flow
range $0 \leq Re \leq 58$ and $0 \leq Ta \leq 1 \times 10^4$, equation (8) predict the power
consumption. At low disc speed, the power requirement increases with a
decrease in Ta. This effect diminishes at high disc speed. The power requirements of 45° and 90° RDC's are essentially the same. In the case of
turbulent flow, equations (10) for the power consumption are valid in the
range of $0 \leq Re \leq 580$ and $5.7 \times 10^4 \leq Ta \leq 1.37 \times 10^6$. The 90° RDC consumes
up to 50% more power than the 45° unit. Equation (8) for the no through-
flow case is in good agreement with the existing theoretical results [3, 5].
Equation (10) for the 45° RDC has the same slope as the experimental data
for no through-flow. However, due to the absence of flow baffles, the

for no through-flow. However, due to the absence of flow baffles, the equation predicts a power number that is approximately 1/3 that required in the experiments.

5.   NOMENCLATURE

a, b, c, d	Constants
B_b	
C_1, C_2, C_D, C_u	Constants
C_m	disc-moment coefficient as defined by equation (7)
G	turbulent viscous dissipation function
K	turbulent kinetic energy
L	one-half length of RDC
1	integral length scale in turbulent flow
$\dot{m}$	through-flow rate of liquid phase
Np	power number as defined by equation (7)
Ne	Newton number
n	number of disc revolution per unit time
R	radius; R_d, of disc; R_i, of R_o, of cylindrical wall; R_s of shaft (see Fig. 1)
(r, θ, z)	cylindrical coordinates
Re	Reynolds number as defined by equation (1)
Ta	Taylor number as defined by equation (1)
T_d	disc torque as defined by equation (6)
U, V, W	velocity components in (r, θ, z) direction near-wall
V	circumferential flow velocity, V_u and V_b, over and under the disc, respectively

GREEK SYMBOLS

∝	angle of conical sections (see Fig. 1)
ε	rate of dissipation energy
μ	liquid absolute viscosity; μ_e, effective value $= \mu + \mu_t$; μ_t turbulent component
ν	liquid kinematic viscosity
ρ	liquid density
σ_k, σ_e	constants
τ	local shear stress on both sides of the disc at r, as defined by equation (5)
φ	general dependence function
ψ	stream function
Ω	angular speed of disc rotation

ω   vorticity

REFERENCES

1. Stichlmair, J. 1978. Grundlagen der Dimensionierung des Gas Flussigkeit Kontaktapparates Bodenkolonne. Verlag Chemie, Weinheim, New York.

2. Yang, Wen-Jei, 1978. Mechanism of power dissipation in liquid-gas mixing in a perforated-disk type stirring cascade. Heat Transfer 1978, Vol. 4, Hemisphere, Washington, D.C., pp. 7-12.

3. Brauer, H. 1980. Fluid dynamics and mass transfer in a three-stage rotating disc reactor. German Chem. Eng., Vol. 3, pp. 66-78.

4. Glaeser, H., Biesecker, B.O. and Brauer, H. 1973. Begassung von Flussigkeiten mit Propeller und Lockscheidenruhrer. Verferenstechnik Vol. 2, pp. 31-49.

5. Thiele, H. 1972. Stromung und Leistungsbedarf beim Ruhren Newtonscher Flussigkeiten mit Anker-, Blatt, und Turbinenruhrern im Laminaren bereich. Diss. Technische Universitat Berlin.

6. Huber, A. 1971/72. Flussigkeitsbegasung in Erner Mehrstufigen Ruhrkaskade. Studienarbeit Matr. -Nr. 34578, Lehrstul fur Verfahren-stechnik, Technische Universitat Berlin.

7. Jain, S.H. 1973. Einfluss geneigter Strombrecher auf die Flussigkeitsbegasung in einem Mehrphasenruhrreaktor. ibid, A4219.

8. Igwemezie, L. 1974. Leistungsbedarf und Gasinhalt in Einer Mehrstufigen, Begasten Ruhrkaskade, ibid, A4780.

9. Delag, M. and Brauer, H. 1976. Leistungsbedarf bei exzentrischer Anordnung des Ruhrers. Verfahrenstechnik Vol. 10, p. 637.

10. Wang, J.H. and Yang, Wen-Jei 1980. Turbulent flows in a disc-type stirrer with cone-shaped housing. Numerical Methods in Non-Linear Problems (ed. by C. Taylor, E. Hintron and D.R.J. Owen), Vol. 1, Pineridge Press, Swansea, U.K. pp. 885-893.

11. Wang, J.H. and Yang, Wen-Jei 1981. Numerical Solutions for laminar flows in disc-type stirrers. Numerical Methods in Laminar and Turbulent Flow (ed. by C. Taylor & B.A. Schrefler), Pineridge Press, Swansea, U.K. pp. 129-140.

Study of the Turbulent Field inside the Blade to Blade Channel of a Radial Compressor

J. P. BERTOGLIO, H. SALAUN, and D. JEANDEL
Laboratoire de Mecanique des Fluides
Ecole Centrale de Lyon
69130 Ecully, France

ABSTRACT

In previous papers, a spectral study of the action of Coriolis forces on a homogeneous turbulence has been made. The main equations and results of the study are summarized here . The validity of this preceding study was limited by the fact that non linear and inhomogeneous effects were not taken into account. In the present paper, we use an equation for the effective strain, recently proposed by Maxey, to extend our model to more complex non linear and non homogeneous flows. A comparison with experiments is made in the case of a rotating boundary layer.

1. INTRODUCTION

It is known that, in rotating machines, turbulence is strongly modified by the presence of Coriolis forces. In particular in centrifugal compressors the action of rotation strongly affects the stability of turbulence and experiments have shown the existence of a coupling effect between Coriolis forces and production of turbulence by the mean shear. Either increase or decrease of turbulence may result, depending on which side of the blade to blade channel is considered. Consequences concerning the behaviour of flows inside compressors are strong : on the suction side, the decrease of turbulence stresses due to Coriolis forces may result in a tendancy to separation of the boundary layer.

In previous papers, (Bertoglio [1]), the effect of Coriolis forces on turbulence has been studied in the very idealized frame of rapid distortion theory. Computations of a homogeneous turbulent field submitted to a uniform shear flow in a rotating frame were carried out without taking into account the non linear action of the turbulent motion on itself. The main results of this study will be summarized in section II.

In fact, in a real machine, the condition of a rapid distortion is rarely satisfied. It will only be the case if the residence time of turbulence inside the blade to blade channel is short, when compared to a characteristic time of the interaction between turbulent eddies, a condition which is usually not strictly satisfied. Therefore rapid distortion theory in itself will not provide a practical turbulence model. However, several authors (Townsend [2], [3]) have shown that the structure of turbulence and its anisotropy were fairly well described by rapid distortion theory, even though the condition of a rapid distortion was not satisfied, on the condition to limit the action time of the mean gradient to an asymptotical value. This value can be considered as a memory time of turbulence and is directly connected to the presence of non linear effects. This memory time limits the stain to an effective value.

In the particular case of the action of Coriolis forces, we have already pointed out that assumptions of this kind can make sense [4], [1].

Recently, Maxey [5] has proposed an equation for the effective strain which furthermore takes inhomogeneous effects into account. In the present paper, our rapid distorsion computation, which includes the effects of Coriolis forces, is used together with the equation of Maxey, an equation for the turbulent kinetic energy (as suggested by Mathieu [6]), and an equation for a length scale. We compute rotating turbulent boundary layers.

2. HOMOGENEOUS TURBULENCE IN A ROTATING FRAME : RAPID DISTORTION THEORY

We summarize in the present section equations and results more extensively presented in ref. [1] and [4].

2.1 Equations

The two-point correlation equations in a frame rotating at the angular speed Ω can easily be obtained :

$$
\begin{aligned}
& \frac{\partial}{\partial t}\overline{u_i u'_j} + \overline{U}_1 \frac{\partial \overline{u_i u'_j}}{\partial X_1} + \overline{U'}_1 \frac{\partial \overline{u_i u'_j}}{\partial X_{1'}} \\
& + (\frac{\partial \overline{U}_i}{\partial X_1} + 2\,\varepsilon_{ikl}\,\Omega_k)\,\overline{u_l u'_j} \\
& + (\frac{\partial \overline{U}'_j}{\partial X'_1} + 2\,\varepsilon_{jkl}\,\Omega_k)\,\overline{u'_l u_j} \\
& + \frac{\partial}{\partial X_1}\,\overline{u_i u_l u'_j} + \frac{\partial}{\partial X'_1}\,\overline{u'_j u'_l u_i} \\
& + \frac{1}{\rho}\,\overline{(u'_j \frac{\partial p}{\partial X_i}} + \overline{u_i \frac{\partial p'}{\partial X'_j})} \\
& - \nu\,(\frac{\partial^2 \overline{u_i u'_j}}{\partial X_k{}^2} + \frac{\partial^2 \overline{u_i{}' u_j}}{\partial X'_k{}^2}) = 0
\end{aligned} \tag{1}
$$

For a typical situation encountered inside a rotating radial machine, we have in particular studied the case :

$$
\lambda_{ij} = S\,\delta_{i1}\,\delta_{j2}
$$
$$
\Omega i = \Omega\,\delta_{i3}
$$

2.2 Results

In ref. [4], we have used a numerical approach to solve eq (2). Results concerning one point correlations have been found by integrating over the values of the wave vector $\vec{K}$. They show that the destabilizing or stabilizing effects of Coriolis are taken into account : for $\Omega/S > 0$, all the components of the Reynolds stress tensor are found to be amplified by the rotating motion of the frame, whereas they are decreased when $\Omega/S < 0$.

One of the major results of ref. [4] concerns the behaviour of the mixing-length ratio $1/1_o$, in which 1 is defined by

$$1 = \left| \overline{u_1 \, u_2} \right|^{1/2} / S$$

and 1_o is the same quantity when $\Omega = 0$. This ratio is plotted in figure 1 versus the Richardson number :

$$R_i = \frac{- 2 \, \Omega \, (S - 2 \, \Omega)}{S^2}$$

It is relevant to point out that figure 2 provides a support to the assumption usually made for turbomachinery applications [7, 8, 9]: $1/1_o$ is nearly a linear function of R_i :

$$1/1_o = 1 - \beta R_i \tag{2}$$

Applying a three dimensional Fourier transform, we can rewrite Eq. (1) in spectral form as :

$$(\frac{\partial}{\partial t} + 2 \, \nu K^2) \, \phi_{ij} \, (\vec{K},t) - \lambda_{lm} \{ K_1 \, \frac{\partial \phi_{ij}}{\partial K_m}$$

$$+ \, 2 \, \frac{K_i K_1}{K^2} \, \phi_{mj} \, + \, 2 \, \frac{K_j K_1}{K^2} \, \phi_{im}$$

$$- \, \delta_{i1} \, \phi_{mj} \, - \, \delta_{j1} \, \phi_{im} \, \}$$

$$+ \, 2 \, \varepsilon_{mkl} \, \Omega_k \, \{ \, \phi_{1j} \, (\delta_{im} - \frac{K_i K_m}{K^2})$$

$$+ \, \phi_{i1} \, (\delta_{jm} - \frac{K_j K_m}{K^2}) \, \} \, = 0 \tag{3}$$

In the above equation, turbulence is regarded as homogeneous and is submitted to a mean velocity gradient uniform and steady in the rotating frame, defined as :

$$\lambda_{ij} = \frac{\partial \overline{U}_i}{\partial x_j}$$

In this section, we have omitted the non-linear terms, assuming for the moment, that we shall only consider evolution times shorter than the non-linear characteristic time of turbulence : τ_{NL}. This assumption, usually called rapid distortion, will be discussed later.

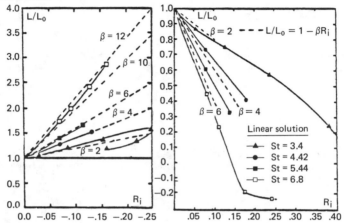

Fig. 1 - Mixing length ratio vs Richardson number. Comparison between rapid distorsion theory and the relation used in turbomachinery.

2.3 Discussion

The main advantage of the two-point description of turbulence is that the role played by pressure is exactly taken into account. No modeling of pressure has been introduced in our approach, which is particularly fitting to study the effects of Coriolis forces, since the validity of the classical one-point models for the pressure-strain correlation may in fact be very questionable in rotating frames (see ref. [1]).

In return, the validity of our linear model is restricted by the lack of non linear terms : the transfer of energy between eddies of various sizes is not taken into account. In the case of a rotating machine, in order to strictly satisfy the condition of validity of rapid distortion, the residence time of turbulence inside the blade to blade channel τ_R, must be short when compared to a characteristic time of the non-linear terms τ_{NL}

$$\tau_R \ll \tau_{NL} \qquad\qquad\qquad (4)$$

We may choose a local expression for τ_{NL}, for example :

$$\tau_{NL} \sim L/(\overline{q^2})^{1/2}$$

in which

$$\frac{\overline{q^2}}{2} = \frac{1}{2} (\overline{u_1^2} + \overline{u_2^2} + \overline{u_3^2})$$

is the turbulent kinetic energy, and L an integral length scale.

Condition (4) is a first limitation to the validity of our results. A second limitation is imposed by the fact that in eq. (2), we have assumed that the turbulent field was homogeneous, which is far to be satisfied inside a machine.

An attempt to extend our model to non linear and inhomogeneous flows, such as boundary layers, will be made in the next section.

3. EXTENSION TO NON-LINEAR AND INHOMOGENEOUS FIELDS

In the preceding section, we have restricted the validity of our rapid distortion model to times shorter than τ_{NL}. However, several authors, and in particular Townsend [2, 3], have pointed out that, in fact, for larger values of time, the rapid distortion theory can still provide useful informations, on the condition to limit the action time of the linear processes to a fictious value : t_*^* which can be considered as a memory time of the turbulent motion. It is suggested that the history of the distortion over the life-time of an eddy must be used, rather than its complete action since the beginning of the shearing process. The effective strain, defined as

$$\alpha = St^*,$$

is then introduced as a crucial parameter.

We may then consider that some non dimensional quantities should be regarded as being fairly well predicted by rapid distortion : for example, quantities characterizing the anisotropy of turbulence such as :

$$b_{ij}(t) = \overline{u_i u_j} (t)/\overline{q^2} (t) - \delta_{ij}/3$$

or other ratios like :

$$\overline{u_1 u_2} (t)_{(\Omega \neq 0)} / \overline{u_1 u_2} (t)_{(\Omega = 0)}, \text{ etc...}$$

As for the characteristic length and velocity scales, L and $\overline{q^2}$, they are considered to be too much affected by the lack of energy transfer between eddies, for their estimates by rapid distortion theory to be of any use as soon as (4) is not strickly satisfied . Townsend [4] and Maxey [5] have suggested that useful prediction models may use the results of rapid distortion to specify b_{ij}, and a separate equation to specify q^2, following the proposal made by Mathieu [6].

In the particular case of the action of Coriolis forces on a turbulent boundary layer, we have already pointed out that a good estimate of some parameters could be derived from the results of our rapid distortion study, by replacing t^* by $\tau'_{NL} \sim 2 \, \delta / \overline{U}_{ex}$, a memory time which can be regarded as a turnover time depending on the boundary layer thickness δ . For example, the parameter β appearing in eq. (3) was fairly well predicted in that way [4].

A more general limit for t^* may be provided by $L/(\overline{q^2})^{1/2}$. Maxey [5] has recently proposed an equation for α .

$$\frac{\partial \alpha}{\partial t} = S - \alpha / T_D \tag{5}$$

in which $T_D = L/d(\overline{q^2})^{\frac{1}{2}}$, d being a constant.

This equation limits the value of $\alpha = St^*$ to the asymptotical value $SL/d(\overline{q^2})^{\frac{1}{2}}$ for large time, and preserves $\alpha \sim St$ for small values of t.

In order to take into account the effects of inhomogeneity, Maxey has furthermore introduced a diffusion term in the equation for α , as proposed by Townsend [2], who argued that, at a fixed point of the flow, eddies with different strain histories would be observed.

In the present paper we adopt the model proposed by Maxey to use it in rotating frame : an equation for α provides values which are thereafter used as input for the rapid distortion computation described in section II. Like in Maxey's model a separate equation for q^2 is needed . Unlike Maxey who specified the length scale as a function of the distance to the wall, we use the equation for the length proposed by Jeandel [10].

The final set of equations :

$$\frac{\partial \overline{U}_1}{\partial x_1} + \frac{\partial \overline{U}_2}{\partial x_2} = 0 \tag{6}$$

$$\overline{U}_1 \frac{\partial \overline{U}_1}{\partial x_1} + \overline{U}_2 \frac{\partial \overline{U}_1}{\partial x_2} = \frac{\partial}{\partial x_2} (-\overline{u_1 u_2}) + \frac{\partial}{\partial x_2} (\nu \frac{\partial \overline{U}_1}{\partial x_2}) \tag{7}$$

$$\overline{U}_1 \frac{\partial \overline{q^2}/2}{\partial x_1} + \overline{U}_2 \frac{\partial \overline{q^2}/2}{\partial x_2} = -\overline{u_1 u_2} \frac{\partial \overline{U}_1}{\partial x_2} - C_{DIS} (\overline{q^2}/2)^{3/2}/L +$$
$$\partial/\partial x_2 \{(\frac{(\overline{q^2}/2)^{1/2} L}{\sigma_2} + \nu) \frac{\partial \overline{q^2}/2}{\partial x_2}\} \tag{8}$$

$$\overline{U}_1 \frac{\partial \overline{q^2} L/2}{\partial x_1} + \overline{U}_2 \frac{\partial \overline{q^2} L/2}{\partial x_2} = C_P \, b_{12} \overline{q^2} \, L \frac{\partial \overline{U}_1}{\partial x_2} - C_M (\frac{\overline{q^2}}{2})^{3/2} \tag{9}$$
$$+ \partial/\partial x_2 \{(\frac{(\overline{q^2}/2)^{1/2} L}{\sigma_1} + \nu) . \frac{\partial \overline{q^2} L/2}{\partial x_2}\} + 2(C_4 \, L/x_2)^n \, b_{12} \frac{\overline{q^2} \, L}{2} \frac{\partial \overline{U}_1}{\partial x_2}$$

$$\overline{U}_1 \frac{\partial \alpha}{\partial x_1} + \overline{U}_2 \frac{\partial \alpha}{\partial x_2} = \frac{\partial \overline{U}_1}{\partial x_2} - C_D \, \alpha (\frac{\overline{q^2}}{2})^{1/2}/L + \partial/\partial x_2 \{(\frac{(\overline{q^2}/2)^{1/2} L}{\sigma_3} + \nu) \frac{\partial \alpha}{\partial x_2}\}$$
$$\tag{10}$$

is closed when supplemented by equation (2) giving $\overline{u_1u_2}/q^2$ at each point as function of the mean velocity gradient $S = \partial\overline{U_1}/\partial X_2$ acting during an effective straining time t * defined by

$$t^* = \alpha/S$$

Considering the above model, we must note that eq. (6) (8) and (9) are not modified by the rotating motion of the frame. The terms involving Ω in (7) have been neglected since it is known that the Coriolis effect on turbulence is one order of magnitude larger that the direct effect of Ω in (7) (see ref. [7]). The pressure terms have been omitted in (7).

4. RESULTS IN THE CASE OF A BOUNDARY LAYER - COMPARISON WITH EXPERIMENTS

The values of the constants appearing in eq. (6) to (9) are the values proposed by Jeandel (11).

$$C_{Dis} = 0.09, \quad C_P = 0.98, \quad C_M = 0.058, \quad C_4 = 4.3,$$

$$n = 6, \quad \sigma_1 = 1., \text{ and } \quad \sigma_2 = 1.$$

The value of the diffusion coefficient for α, which does not have much effect on the distortion strain equation (5), has been chosen in order to fit with the model of Maxey :

$$\sigma_3 = 0.5$$

The value of C_D deduced from Maxey was leading to a ratio $\overline{u_1u_2}/\overline{q}^2$ constant in a large part of the boundary layer : C_D has been optimized in order to adjust the value of this constant ratio to the experimental result : - 0.15 in the steady frame case.

The initial data for $\overline{q^2}$, L and $\overline{U}$ in the first section are determined by the method proposed by Jeandel. α is equal to zero in the first section. The initial data for equation (2) are the spectrum measured by Comte-Bellot [12] and correspond to isotropic turbulence.

At the wall, the values of the different parameters are chosen to be in agreement with the classical boundary layers experimental results. The results presented here correspond to $\alpha = 2$ at the wall.

The numerical code used to solve eq. (6) (7), (8) and (9) is the one proposed by Jeandel [11]

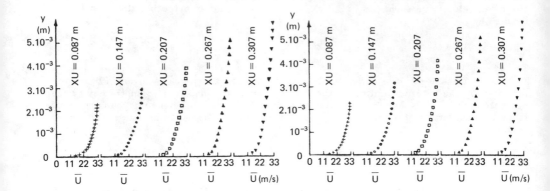

Fig. 2 - Evolution with time of the mean velocity profiles. Comparison between the present model and the results of Jeandel ($U_{ex} = 33$ ms^{-1} ; $\Omega = 0$).

In the case of a steady frame boundary layer, figure 2 shows that our results agree fairly well with Jeandel's method. The evolution of $-\overline{uv}/q^2$ is given in figure 3.

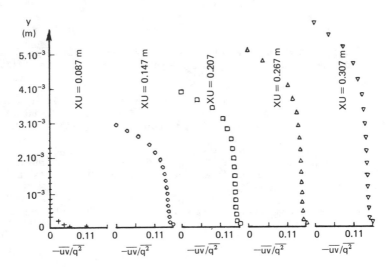

Fig. 3 - Evolution of the ratio $-\overline{uv}/q^2$ ($U_{ex} = 33$ ms^{-1} , $\Omega = 0$).

In order to study the effect of rotation, conditions corresponding to the experiment of Koyama et Al [13] have been adopted : $U_{ex} = 5$ m/s, $\Omega = \pm 32$ rd s^{-1}, initial boundary thickness 5 10^{-3} m. Results corresponding to the last testing section are given (X = .555 m).

The stabilizing and destabilizing effects of Coriolis forces appear in figure 4 and 5 : - $\overline{uv}$ and $\overline{q^2}$ are increased when Ω/S is positive and decreased for negative Ω/S.

The behaviour of the components of the Reynolds stress tensor are plotted in figure 6. The mean velocity profile (Fig. 7) clearly exhibits the same trends as the experimental results (Fig. 8) :

- when Ω/S is positive (pressure side), existence of a flat outer portion of the boundary layer

- when Ω/S is negative (suction side), reduction in the slope of the velocity profile near the wall.

The discrepancy between the two cases, Ω positive and Ω negative, is however smaller than in the experiment.

CONCLUSION

Rapid distortion theory is particularly fitting to study the coupling action of rotation and shear on homogeneous turbulence. The model proposed by Maxey is a very simple way to extend the results of rapid distortion to inhomogeneous flows in which non linear effects are not negligible. In the present paper we have pointed out that our linear analysis, the equation of Maxey for the effective strain, together with the computation code of Jeandel are providing a method that predicts the general trends of the action of rotation on turbulent boundary layers.

The present work can also be considered as a test of the model proposed by Maxey. We have shown that the model can be used together with an equation for a lenght scale, which can be helpful for computing turbulent flows away from boundary. We have pointed out that, in the case of a steady frame boundary layer, the results of the model are in very good agreement with the results obtained by Jeandel. We have furthermore shown that the model predicts the experimental trends exhibited by a turbulent flow submitted to the action of Coriolis forces, which can be considered as a particular case of action of an external force field on turbulence.

More complete two-point closures [14] could be used to test or improve the model of Maxey. Some comparisons with a non-linear homogeneous closure already support the model in the case of a steady frame shear flow. A study could be made for rotating fields. Nevertheless the effective strain model provides a very suitable and efficient computation method, as it is far less expensive than any method based on more sophisticated two point closures.

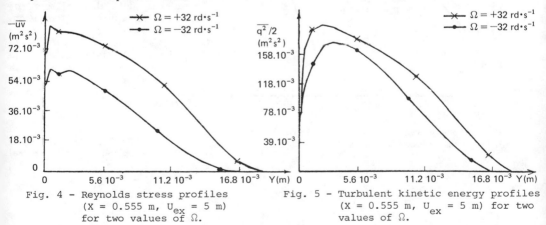

Fig. 4 - Reynolds stress profiles (X = 0.555 m, U_{ex} = 5 m) for two values of Ω.

Fig. 5 - Turbulent kinetic energy profiles (X = 0.555 m, U_{ex} = 5 m) for two values of Ω.

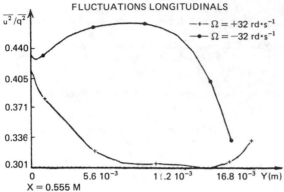

FLUCTUATIONS LONGITUDINALS

$\overline{u^2}/q^2$

$-\!+\!-\ \Omega = +32\ \text{rd}\cdot\text{s}^{-1}$
$-\!\bullet\!-\ \Omega = -32\ \text{rd}\cdot\text{s}^{-1}$

0.440

0.405

0.371

0.336

0.301

0   5.6 10^{-3} 1(.2 10^{-3} 16.8 10^{-3} Y(m)

X = 0.555 M

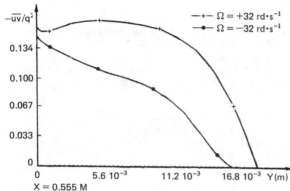

$-\overline{uv}/q^2$

$-\!+\!-\ \Omega = +32\ \text{rd}\cdot\text{s}^{-1}$
$-\!\bullet\!-\ \Omega = -32\ \text{rd}\cdot\text{s}^{-1}$

0.134

0.100

0.067

0.033

0

0   5.6 10^{-3} 11.2 10^{-3} 16.8 10^{-3} Y(m)

X = 0.555 M

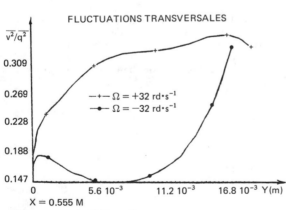

FLUCTUATIONS TRANSVERSALES

$\overline{v^2}/q^2$

0.309

0.269

0.228

$-\!+\!-\ \Omega = +32\ \text{rd}\cdot\text{s}^{-1}$
$-\!\bullet\!-\ \Omega = -32\ \text{rd}\cdot\text{s}^{-1}$

0.188

0.147

0   5.6 10^{-3} 11.2 10^{-3} 16.8 10^{-3} Y(m)

X = 0.555 M

Fig. 6 - Profiles of $\overline{u^2}/q^2$, $\overline{v^2}/q^2$, $\overline{uv}/q^2$
for two values of Ω

REFERENCES

1. Bertoglio, J.P. 1982, Homogeneous turbulent field within a rotating frame, AIAA Journal, vol. 20, n° 9, pp. 1175-1181.

2. Townsend, A.A. 197O, Entrainment and the structure of turbulent flow. J. Fluid Mech., Vol. 41, pp. 13-46.

3. Townsend, A.A. 1980, The response of sheared turbulence to additional distortion. J. Fluid Mech. Vol. 98, pp. 171-191.

4. Bertoglio, J.P. 1980, Influence des forces de Coriolis sur une turbulence soumise à des gradients. Thèse de Doct. Ing., Univ. C. Bernard, Lyon.

5. Maxey, M.R. 1982, Distorsion of turbulence in flows with parallel streamlines. Submitted to J. Fluid Mech.

6. Mathieu, J. 1971. Réflexions sur les écoulements turbulents à surface libre. Von Karman Institute, Lect. Series 36.

7. Bradshaw, P. 1973, Effects of streamline curvature on turbulence flow. Agadograph, N° 169, pp. 1-8O

8. Johnston, J.P. and Eide, 1976, Turbulent boundary layers on centrifugal compressor blades : Prediction of the effects of surface curvature and rotation. ASME 76-Fe-10

9. Papailiou, K.D. 1978, Effet de la force de Coriolis et effet de la courbure de paroi sur une couche limite. Rapport D.G.R.S.T. 76/7/1298

10. Jeandel, D., Brison, J.F. and Mathieu, J. 1978, Modeling methods in physical and spectral space. Physics of Fluids, vol. 21, pp. 169-182.

11.   Jeandel, D. 1975, Une approche phénoménologique des écoulements turbulents inhomogènes. Thèse Doct. Etat, Univ. C. Bernard, Lyon

12.   Comte-Bellot, G. and Corrsin, S. 1971, Simple Eulerian time correlation of full and narrow band velocity signals in grid-generated isotropic turbulence. J. Fluid Mech. vol. 48, pp. 273-337.

13.   Koyama, H., Masuda, S. Ariga, I. and Wanatabe, I. 1979, Stabilizing and destabilizing effects of Coriolis force on two-dimensional laminar and turbulent boundary layers. J.
Eng. Power, vol. 101. pp. 23-32

14.   Bertoglio, J.P., 1981, A model of three-dimensional transfer in non-isotropic homogeneous turbulence. Third Sym. Turb. Shear Flows, Davis, to appear in Springer-verlag.

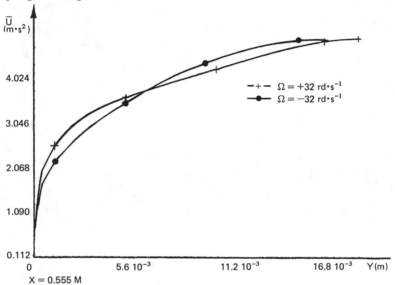

Fig. 7 - Mean velocity profiles, at X = 0.555 m, computed with the present model (U_{ex} = 5 m/s).

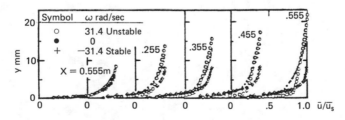

Fig. 8 - Mean velocity profiles. (Measurements of Koyama, et al.)

An Experimental Investigation and some Analytical Considerations Concerning the Vaporous/Gaseous Cavity Characteristics of an Eccentic Shaft Seal or Bearing

MINEL J. BRAUN
University of Akron
Akron, Ohio 44304, USA

ROBERT C. HENDRICKS
NASA Lewis Research Center
MS23-2
Cleveland, Ohio 44135, USA

ABSTRACT

This paper describes the experimental pressure and temperature results obtained when rotating a shaft in an eccentric lucite casing at velocities ranging from 209 to 628 rad/sec (2000 to 6000 RPM). The results are presented in terms of three-dimensional plots and contour maps. Photographic evidence (1) is presented to illustrate how the downstream and upstream regions of the cavity evolve into the well known finger patterns. The Swift-Stieber, separation and Floberg boundary conditions are discussed in the light of their applicability to the experiments which show the development of sub-atmospheric pressures within the cavity region. A simplified thermal analysis in conjunction with the bubble nucleation theory is used to determine the nature of the cavity content.

1. INTRODUCTION

1.1 Cavitation Boundary Conditions. Short Theory Review

The basic theory of hydrodynamic lubrication was set in 1886 when Reynolds [1] derived the differential formulation for pressure build-up in a thin lubricating film. The equation,

$$\frac{\partial}{\partial x}\left(\frac{\rho h^3}{12\eta}\frac{\partial P}{\partial x}\right) + \frac{\partial}{\partial y}\left(\frac{\rho h^3}{12\eta}\frac{\partial P}{\partial y}\right) = \frac{\partial}{\partial x}\left[\frac{\rho(U_1 + U_2)}{2}h\right] + \frac{\partial h}{\partial t} \tag{1}$$

can accurately describe the fluid film pressure distribution - under loading - in the portions where the pressure gradient $\left(\frac{\partial P}{\partial x}\right) > 0$ i.e., in the convergent portion and also a limited part of the unruptured divergent film. In 1904, Sommerfeld [2], published his now famous solution of the pressure distribution for an infinitely wide, 360°, full clearance oil film journal bearing. He recognized cavitation as a limitation of his theory and predicted the existence of sub-atmospheric pressures in the diverging section of the bearing surfaces, as shown in Fig. 1A. Sommerfeld also postulated the disruption of the continuum liquid phase due, above all, to the incapability of the oils to withstand

---

(1) A motion picture supplement is available as experimental supporting evidence.

negative (sub-atmospheric) pressures. Gumbel [3] made the first attempt to model the film rupture, followed by Swift [4] and Stieber [5]. All three authors neglected the sub-atmospheric pressure loop, and Swift and Stieber solved Eq. (1) using as boundary condition for the inception of the cavitation region,

$$\partial P/\partial x = 0 \; ; \; P = P_{cavity} = P_{atm} \tag{2}$$

This film rupture condition yielded a pressure distribution as the one shown in Fig. 1B. Hopkins [6] in 1957, Bretherton [7] and Taylor [8] in 1960, and Coyne and Elrod [9] in 1970 observed the experimental occurance of the sub-cavity pressure loop in bearings and suggested a mechanism whereby the flow separation effect may be instrumental in film rupture. Separation would occur when

$$\partial u/\partial y = 0 \; , \tag{3}$$

a condition equivalent to the inception of secondary flows, resulting in a region of flow reversal and conglomeration of the dissolved gases at the point of separation. This implies an adverse pressure gradient at the cavitation boundary, which if one surface is stationary, is given by

$$\partial P/\partial x = 2\eta U/\delta^2 \tag{4}$$

The separation theory results in a pressure distribution profile which can accommodate the sub-atmospheric pressures, Fig. 1C. While the separation theory [7-10], considers that the lubricant moves under and/or over the cavity, Fig. 2A, Floberg [11,12], postulates that: (i) the flow is carried between the cavities, Fig. 2B, and (ii) no net mass flow penetrates into the cavity. In the light of (i) and (ii) he proposes a boundary condition which connects the oil film region to the cavitation region,

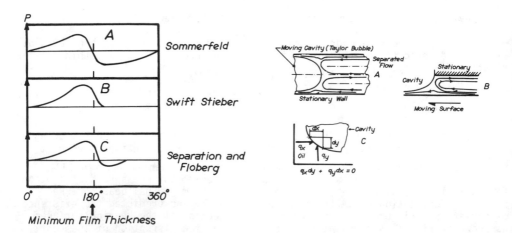

Figure 1 Possible pressure configurations for cavitation boundary
conditions

Figure 2 Separation versus Floberg
Case

$$(\partial P/\partial x) - (\partial P/\partial y)(dx/dy) = 6\eta(U_1 + U_2)/\delta^2 \qquad (5)$$

The Reynolds Eq.(1) and Eq. (5) when solved together for a specified number of streamers (n) can predict the pressure field with subcavity loop pressures and the location of the cavitation bubble.

1.2 Pseudo, Gaseous and Vaporous Cavitation: Some New Arguments

The content of the cavitation bubble has been a matter of argument ever since Reynolds [1] derived the basic equation of lubrication theory. Vapor versus liberation of dissolved gases--that is the question! One can distinguish between three types of mechanisms which can create and cause a bubble to grow. Let us assume that there exists a homogeneous film mixture of oil and dissolved gases at a given pressure P_f.

(a) If suddenly the pressure is reduced to P_g (larger than P_v) where $P_g < P_f$, the dissolved gas nuclei will merge into a separate gaseous phase and, at a critical cluster radius rupture the homogeneity of the oil film. This is the essence of gaseous cavitation. The lowest pressure in the oil film depends on mass fraction, solubility and type of gases dissolved in the oil. The sub-cavity pressure represents the pressure drop which the oil can withstand right before the film is broken due to the release of dissolved gases. This pressure drop has to be equivalent with the maximum surface tension, σ, which the liquid can tolerate. For a bubble of arbitrary shape

$$P_{sc} - P_f = \sigma[1/R_1 + 1/R_2] \qquad (6)$$

In Eq. (6), the sub-cavity pressure is, in most cases, the sum of the vapor pressure and the pressure of the non-condensable impurities,

$$P_{sc} = P_v + P_g \qquad (7)$$

By introducing Eq. (7) into Eq. (6) one obtains,

$$P_v + P_g - P_f = \sigma[1/R_1 + 1/R_2] \qquad (8)$$

and gaseous cavitation ensues as a function of the surface tension, film pressure and geometry.

(b) If the pressure is lowered to $P < P_v$, then volatile hydrocarbon nuclei will merge (a process akin to nucleate boiling) and, either form a new cavity or be released into an existing gaseous cavity, forming a mixture of dissolved gases and vapors. For the case where no non-condensable impurities are present in the oil, the lowest pressure in the film just before rupture will be the oil-vapor pressure, P_v, and in Eq. (6) we simply replace P_{sc} by P_v.

Both the mixture and the separate components are considered perfect gases, and according to Dalton's law one can write for the mixture

$$P_{sc} V = nRT ; \quad (n = n_v + n_g) \qquad (9)$$

and for the components

---

[1] vapor/gas release is often triggered by sites on the boundaries, following the active site criteria of Hsu.

$$P_i \Psi = n_i RT \qquad (i = n,g) \tag{10}$$

In order to determine the ratio of partial pressures

$$P_v/P_g = n_v/n_g \tag{11}$$

one has to know from the onset the molar ratio of the dissolved gases, volatility and concentrations of hydrocarbons in the oil and their respective vapor pressures. It is the authors' opinion that only by connecting Eqs. (9-11) to the nucleation (ebullition) theory can one decide whether vaporous or gaseous cavitation has occured (See Section 4.3).

(c) Finally, the so-called pseudo-cavitation can occur when the size of the bubble changes due to a variation of P_f rather than due to an adiabatic/isothermal mass exchange from the oil to the cavity. If the mass stays constant, for a bubble of arbitrary shape undergoing a transformation from state (i) to state (j),

$$\Psi_i \rho_i = \Psi_j \rho_j \tag{12}$$

Considering that for an adiabatic/isothermal transformation

$$P_{sc,i}/(\rho_{sc,i})^m = P_{sc,j}/(\rho_{sc,j})^m \qquad (m = 1,k) \tag{13}$$

and that $R_1 = R_2 = R$ in Eq. (6), one can relate the film pressure to the cavity pressure by

$$P_f/P_{sc,j} = 1 - \frac{1}{P_{sc,j}} (P_{sc,j}/P_{sc,i})^{1/3m} (2\sigma/R_i) \tag{14}$$

Where $P_{sc,j}$ is given by Eq. (7). While Eq. (6) gives for a combination (P_{sc}, P_f) the radius at which cavitation can start, Eq. (14) shows that for an adiabatic or isothermal transformation what may seem as an increase/decrease in the mass amount of the gaseous phase, may be in reality an illusion due to the change in P_{sc} and P_f.

Thus, great care should be exerted in differentiating and identifying the various types of cavitation mechanisms.

2. SCOPE

This paper's main endeavor is to present and explain experimental data concerning the pressure and temperatures characterizing the behavior of the journal bearing interface film when operating in the fully flooded mode. The data obtained are of significance for both seal and bearing technology.

For the first time the attempt has been made (to the best of the authors knowledge) to obtain three dimensional (3-D) digitized images of the temperature and pressure envelopes associated with both the full oil film and gaseous and/or vaporous cavitation zones. The resulting experimental data are presented and analyzed. A thermal model for ebullition incipience in the cavitation tone, is postulated, thus allowing to establish the limits within which vaporous cavitation can occur.

3.   EXPERIMENTAL EVIDENCE

3.1   The Experimental Facility

A schematic of the test section and oil flow diagram of the experimental rig is shown in Fig 3. The detail B presents a sketch of the temperature-pressure transducer (TPT) used in the experiment. The test section is formed out of a journal 50.8 mm in diameter with an L/D ratio of 0.75 which is rotated by a spindle whose velocity can be varied from 193 to 524 rad/sec (1843 to 10000 RPM). The eccentricity ratio between the journal and the housing center-line was fixed at 0.4, with a minimum clearance of 0.00684 cm, [13] and located approximately 70-degrees from the bottom of the configuration. The journal was enclosed by an acrylic transparent housing which was connected to a motor which permitted a 360° rotation of the housing at low but variable speeds. The housing-motor configuration was mounted on a motorized linear slide which allowed the axial displacement of the housing, Fig 3.

To accurately follow and record the position of the TPT an electronic linear and angular indexing system was used in conjunction with two x-y plotters, recording concomitantly both pressures and temperatures at specific locations.

The turbine flowmeters FM1 and FM2 located on each of the oil exit lines together with the valves V1 and V2 were used to measure and control the oil flow on each side of the submerged rotating journal. The temperature at the test section oil inlet line was controlled by means of the water cooled heat exchanger, HE. A thermocouple located at the heat exchanger exit measured the temperature of the oil feeding the test section.

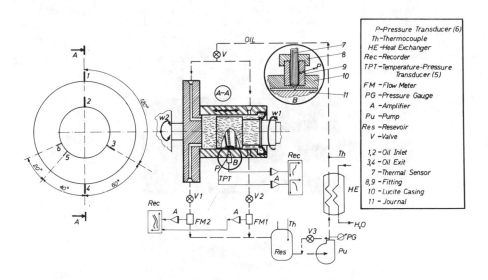

Figure 3   The experimental facility

3.2 The Experimental Procedure

Prior to data taking, oil was pumped with the by-pass fully open to flood and slightly pressurize the test section until all the visible air bubbles were eliminated from the oil contained within the transparent reservoir. Special attention was given to the elimination of air bubbles in and around the TPT access orifice. After the test regime speed was reached, (2000, 4000, 6000 rpm), the test section oil inlet temperature was allowed to stabilize and reach a steady state value. Then the mechanized TPT would perform sequential, axial or circumferential sweeps with the pressure and temperature recorded concurrently on the x-y plotters. The oil mass flows through FM1 and FM2 were monitored and recorded during each sweep in order to assess the influence of axial flow through the journal clearance.

For the circumferential sweeps, the motion of the TPT is initiated at $\theta = 0$, (Fig 3, Fig 6AA), and proceeds clockwise 360°. Axially, the origin Z = 0 of the TPT is located in the oil bath at 4 mm to the left of the journal vertical wall, and the longitudinal sweep is 45.7 mm to the right of the origin, (see also Fig 6AA).

4. EXPERIMENTAL RESULTS

4.1 The Start-Up Regime

The study and discussion of the start-up behavior of the journal is of a qualitative nature rather than quantitative. A Fastax movie camera was used to record the incipient formation of the cavitation zone in both the downstream (convergent recompression zone) and upstream (incipient nucleation zone) portions of the journal. Figures 4A-C show the cavity in the process of developing from individual gas (or vapor) nuclei which spread in the oil, to a cluster of bubbles, which congregate to reach a critical radius and break the film homogeneity. As can be seen from the overlays, new small buble nuclei form at the periphery of the main bubble cavity and then are either engulfed in the larger gaseous (vaporous) structure or are carried downstream by the oil.

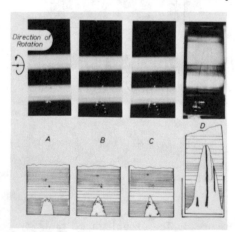

Figure 4   Downstream Section Cavitation Inception and Development During Journal Start-Up TRansients.

The cavitation zone can not develop properly during the initial stage of the transient and we shall clasify the phenomenon as pseudo-cavitation. As the journal velocity increases, the pressure drops further below P_g at the location of minimum clearance while the downstream regions now drop to P_g. Thus, a sustained process of gas release is triggered throughout the divergent section, causing the bubbles to grow through a process of mass addition and coalescence. The apex of the bubble extends into the film up to the point where the pressure in the oil film P_f is larger than P_g. The process of true gaseous (and possibly vaporous) cavitation replaces the pseudo-cavitation at the end of the transient. Figure 4D, shows the form of the cavity during the steady state regime for a journal angular speed of 2000 RPM.

Figures 5A through 5F show the appearance and development of the gaseous phase in the upstream region of the cavitation zone. This region penetrates up to the level of the minimum clearance where it joins the downstream zone. The rather different shape of the bubbles, is probably due to a combined effect of the flow and pressure patterns which characterize the geometry in the vicinity of the minimum clearance. Figures 5A through 5E show the development of the transient from the fern leaf structure, associated with the incipient stages, (Figs. 5A-D) to the more structured finger-like pattern (Figs. 5E,F) of the final stages of the transient.

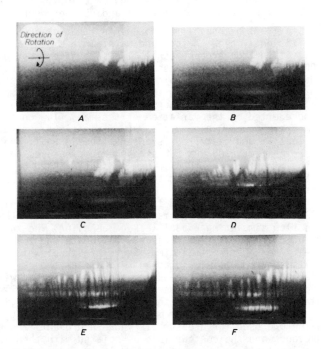

Figure 5 Upstream Section Cavitation Inception and Development During Journal Start-Up Transients.

4.2 The Steady State Regime

For this regime the experiments have been run at 2000,4000 and 6000 RPM first with air dissolved in the oil at atmospheric pressure, then at 4000 RPM with

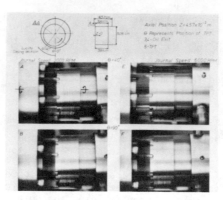

Figure 6   Steady State Downstream Cavitation Zone Spread (2000,6000 RPM)

carbon dioxide saturated oil at a reservoir pressure of .122 to .136 MPa (3 to 5 psig).

Figure 6 shows the downstream cavitation zone (gaseous air) at 2000 and 6000 RPM respectively, while the TPT performs a rotation around the journal. It is evident from the photograph that as the angular velocity increases, the low pressure (P_g less than P_f) penetrates further and further downstream allowing the extension of the cavitation zone. It is noteworthy that at 6000 RPM the shear forces become sufficiently high to form and maintain a cloud of gas bubbles at the tip of the cavity(or convergence zone). The same effect can be observed also at 2000 and 4000 RPM, but with much less intensity. The photographic evidence provides thus data supporting the physical aspects of the separation theory [6-9].

The pressure measurements. During the entire set of experiments the total mass flow of oil has been maintained constant, thus assuring that little or no influence will be exerted by the variation of the mass flow on the pressure or temperature behavior. Figure 7 presents a set of digitized pressure experimental data, as it develops around the circumference of the journal. From Figs. 7A1,A2 and 7A3 it appears that at 2000 RPM sub-atmospheric pressures were not reached anywhere along the cicumference; the minimum pressure ranged between .137 and .130 MPa. As the velocity is increased to 4000 and 6000 RPM the regions of minimum pressure drop from above atmospheric to ranges between .08 and .034 MPa, Figs. 7C1,C2 and 7C3. The pressure variation takes place both longitudinally and circumferentially. The explanation seems to reside in the fact that we have a small axial mass flow simulating a seal leakage behavior. The three dimensional graphs and the contour maps of Fig. 7 clearly delineate the zones of maximum and minimum pressures. The Z-θ plane (at atmospheric pressure) which intersects the tree dimensional pressure dome, isolates and visualizes the cavitation region. Maximum pressure occurs immediately ahead of the 90° location, with the minimum clearance at approximately θ=70°. These results corroborate well the pressure profiles postulated by the separation theory and by Floberg [11,12].

The temperature measurements. The methodology for recording the temperatures was similar to the one used for pressures. Figure 8 presents the resulting three dimensional envelopes and the contour maps associated with them. It can be deduced by simple inspection,that angular velocity, as it is varied from 2000 to 4000 and 6000 RPM contributes to the increase in the temperature of the oil film and that of the gaseous(vaporous) phase. The reference plane Z-θ is located at 20°C and the contour maps show the lines of constant pressure and help form a complete image of the development of the temperature envelope around the journal. Since mass flows are maintained constant and viscous dissipation increases with angular speed,it was expected that temperatures would increase too. It is our belief that

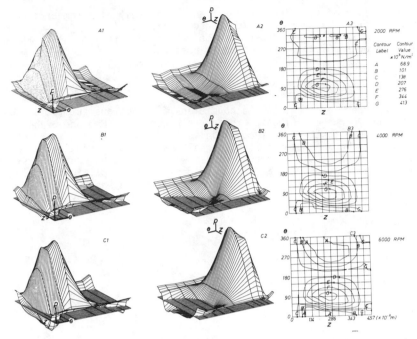

Figure 7   Three Dimensional Steady State Experimental Pressure Measurements.

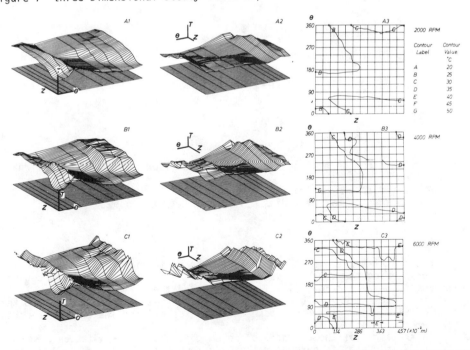

Figure 8   Three Dimensional Steady State Experimental Temperature Measurements

a combination of very low pressures and relatively high temperatures can result in hydrocarbon vapor release (de facto boiling).

The next section proposes a model which can predict the limits within which such hydrocarbon vapor production is possible.

4.3 Ebullition Model For Vaporous Cavitation.

In order to establish the conditions for vaporous cavitation to occur, we shall consider the nucleation theory as proposed by Hsu and Graham [14] in combination with the energy equation as it is applied to a layer of lubricant separating the bearing from the journal. As a first approximation we shall consider the bearing circumference unwrapped and the lubricant as a slab with internal heat generation. The concept provides an upper(optimistic) bound in the lubricant temperature profile and accounts for the energy generated in the fluid due to frictional shear. For the configuration detailed by Fig. 9, the governing energy equation is

$$d^2T/dy^2 + \dot{q}'''/k = 0 \; ,$$

(15)

with the following boundary conditions

$$\text{at} \quad x = 0 \qquad\qquad T = T_s$$

$$\text{at} \quad x = \delta \qquad\qquad dT/dy\big|_\delta = -\frac{U}{k_{oil}}(T_\delta - T_\infty)$$

(16)

where

$$U = 1/(L_p/k_p) + (1/h_a)$$

(17)

Solving Eq. 16 in conjunction with the boundary conditions of Eq. 17, one obtains.

$$T-T_s=(-\dot{q}'''/2k_{oil})x^2+[\dot{q}'''\delta(1+U\delta/2k_{oil})-U(T_s-T_\infty)][1/(k_{oil}+\delta U)]x$$

(18)

where

$$\dot{q}''' = R_j \eta U^2/[(R_j+\delta/2)\delta^2]$$

(19)

Eq. 20 calculates the heat generated inside the lubricant due to frictional heat dissipation.

According to the nucleation theory advanced by Hsu [14] and Hsu and Graham, [15], for a surface pit of radius R_c, a superheat given by

$$T - T_{sat} = 2\sigma T_{sat}/1.6 \; h_{fg}\rho_v y$$

(20)

is necessary at the top of the bubble for the nucleation to occur and develop.

Considering the vapor pressure curve for a typical oil [16], in conjunction with Eqs. (10-12) and Eqs. (16-21) we developed the curves of Fig. 9. It becomes clear that nucleation can occur only when the curves describing the film temperature will intersect and overtake the curve describing the bubble superheat requirements.

In the light of these findings, we suggest that the cavitation observed in Figs. 6, 7 and 8 is of gaseous rather vaporous nature. The fern-leaf structure characteristic to the start up regime observed by Jacobson and Hamrock (17) and

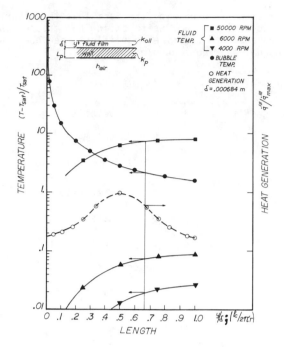

Figure 9   Lubricant Film and Bubble Temperature Profiles for Various Angular
          Velocities

Braun and Hendricks [18] must also be of gaseous nature. The model presented
herein constitutes an upper bound for the fluid temperatures. Under flow conditions the calculated temperatures are likely to be lower removing even more
the vaporous cavitation possibility.

5. CONCLUSION

     The experimental information presented has been reduced and analyzed in
order to obtain three-dimensional plots and contour maps of pressure and temperatures around a non-loaded eccentric journal bearing. To complement these
results, a visual study was carried out to describe the inception and development
of the cavity around the non-loaded journal (the only load being gravity) from
the start-up to the steady state regime. We have discussed the dilemma of the
gaseous vs pseudo or vaporous cavity formation, and attempted to determine which
of the three cavitation condition models (Swift-Stieber, separation, Floberg)
are closer to the reality. It seems from the data gathered that the Floberg
pressure profile, Fig. 1, comes closer in shape to the profiles we are presenting.
The development of sub-atmospheric pressures in the cavity are confirmed and, it
is quite apparent that these pressures drop as the angular velocity is increased.
     An analytical model for the determination of the inception of vaporous
cavitation has been proposed. The model involves the energy equation of the
lubricant film and the ebullition criterion postulated by Hsu and Graham [15].
As a result of this model we learned that the cavity can contain hydrocarbon
vapors only when the pressures and the temperatures of the bubble and fluid are
compatible for the production of vapors.

Finally, it is worthwhile stressing the fact that the slight axial flow occuring between the journal and the lucite casing has created the simulated conditions of seal leakage, and thus the results obtained are beneficial to both the seal and journal bearing technology.

SYMBOLS

h_a - heat transfer coefficient for air

h_{fg} - latent heat

δ - clearance height

K - isentropic expansion coefficient

k - conductivity

l - wall thickness

n - number of moles

P - pressure

$\dot{q}'''$ - heat flux

$\bar{R}$ - gas constant

R - radius

T - temperature

t - time

U - velocity

Ψ - volume

x, y, z, Θ - spatial coordinates

η - viscosity

ρ - density

σ - surface tension

Subscripts

f - film

g - gas

j - journal

p - plastic

sc - sub-cavity

v - vapor

REFERENCES

1. Reynolds, O. "On the Theory of Lubrication and its Application to Mr. Beauchamps Tower's Experiments Including and Experimental Determination of the Viscosity of Olive Oil", Phil. Trans. Roy. Soc., Vol. 177, 1886.

2. Sommerfield, A. "Zur Hydrodynamische Theorie der Schmiermittelreibung", A. Math. Phys., Vol. 50, 1904.

3. Gumbel, L. Monatsblatter Berlin Bezirksver, V.D.I., Vol. 5, 1914.

4. Swift, H.W. "The Stability of Lubricating Films in Journal Bearings", Proc. Instit. of Civil Engrs., (London) Vol. 233, 1932.

5. Stieber, W. "Das Schwimmlager", V.D.I. 1933, Berlin

6. Hopkins, M.R. "Viscous Flow Between Rotating Cylinders and a Sheet Moving Between Them" Brit. J. Appl. Phys., Vol. 8, 1957.

7. Bretherton, F.P. "The Motion of Long Bubbles in Tubes", Jourl. of Fluid Mech., Vol. 9, 1960, pp. 218-224.

8. Taylor, G.I. "Cavitation of Viscous Fluid in Narrow Passages", J. of
    Fluid Mech., Vol. 16, 1963, pp. 595-619.

9. Coyne, J.C. and Elrod, H.G. "Conditions for the Rupture of a Lubricating
    Film, Part I: Theoretical Model", J. of Lub. Tech., Vol. 92, pp. 451-456.

10. Cox, B.G. "An Experimental Investigation of the Stream Lines in Viscous
    Fluid Expelled form a Tube",   J. of Fluid Mechanics, Vol. 20, 1964,
    pp. 193-200.

11. Floberg, L. "On Hydrodynamic Lubrication with Special Reference to Sub-
    cavity Pressures and Number of Streamers in Cavitation Regions", Acta
    Polytechnica Scandivanica, ME, Series 19, 1965, pp. 1-35.

12. Floberg, L. "Sub-cavity Pressures and Number of Oil Streamers in
    Cavitation Regions with Special Reference to the Infinite Journal Bearing",
    Acta Polytechnica Scandinavica, ME, Series 37, 1968, pp. 1-36.

13. Etsion, I. and Ludwing, L.P. "Observation of Pressure Variation in the
    Cavitation Region of Submerged Journal Bearings", NASA TM 81582, 1981.

14. Hsu, Y.Y. "On the Site Range of Active Nucleation Cavities on a Heating
    Surface", J. Heat Transfer, 84 C(3), 1962, pp. 207-216.

15. Hsu, Y.Y., Graham, R.W. "An Analytical and Experimental Study of the
    Thermal Boundary Layer and Ebullition Cycle in Nucleate Boiling, NASA TND-
    594, 1961.

16. "Data Book for Designers, Fuels Lubricants and Hydraulic Fluids, Marketing
    Technical Services, Humble Oil and Refinery Company, 1969.

17. Jacobson, B.O., Hamrock, B.J. "High Speed Motion Picture Camera
    Experiments of Cavitation in Dynamically Loaded Journal Bearings",
    TM-82789, 1981.

18. Braun, M. J., Hendricks, R.C., "An Experimental Investigation of the
    Vaporous Gaseous Cavity Characteristics", Submitted to ASME/ASLE Lub-
    rication Conference, October 1982.

Temperature and Thermal Stress Field in a Rotating Water Spray Cooled Roller Supporting a Hot Slab

BRUNO LINDORFER and GERHARD HOFER
Voest-Alpine AG
Linz, Australia

1. INTRODUCTION

The main application of water spray-cooled rollers is in hot rolling mills as well as in continuous casting plants, where the rollers are not only stressed by thermal stresses due to the temperature distribution, but also by mechanical loads.

The principal reasons why it is necessary to know the temperature distribution in a roller are:

- Insufficient cooling can lead to a shortened life time of the roller due to surface cracks or fatigue caused by thermal stresses.

- The temperature field significantly affects the shape of the roller which, in turn, influences the loading conditions and, in case of rolling mills, the result of the rolling process, i. e. profile and residual stress distribution.

- The material properties, especially the yield stress, and therefore the load carrying capacity are strongly temperature dependent.

In the examples mentioned above - hot rolling mills and continuous casting plants, heat transfer into the rollers occurs within a very small local contact zone and water-spray cooling also takes place only over a small portion of the surface of the roller. The arrangement of the cooling and heating zones determines the thermal boundary conditions of the problem. The following heat transfer mechanisms are examined:

- Local hot spot contact between roller and hot slab. Extremely high heat transfer coeffizient (HTC) (it may be reduced by scale layer).

- Surface to surface radiation from the hot slab to the roller.

- Water film cooling. The HTC mainly depends on the specific water flow rate and the surface temperature.

- Water-spray-cooling. The HTC essentially depends on the flow rate of the cooling water and on its velocity on leaving the nozzle. The distribution of the water varies with the type of nozzle used.

- Heat transfer to the surroundings by convection and radiation.

These boundary conditions alter during rotation of the roller.

2. THERMAL MODEL OF THE ROLLER

Since the circumferential temperature gradients are significantly smaller than those for radial heat flow, a one-dimensional model, neglecting the circumferential heat flow was developed. This offers both the advantage of a sufficiently accurate but not too expensive transient solution of the problem and the possibility of extensive parametric studies.

3. NUMERICAL PROCEDURE

The numerical procedure is a lumped parameter method. To solve the transient temperature distribution of an irregular body, the body is devided into a number of cells (Fig. 1). Each cell is associated with its geometric centre which shall be called an interior node.

Surfaces of cells lying on the boundary of the body are associated with one of the surface points which shall be called a surface node. Thus a three-dimensional mesh or grid represents the body.

The computer program solves the general heat conduction equation:

$$\frac{\partial}{\partial x}(k\,\frac{\partial T}{\partial x}) + \frac{\partial}{\partial y}(k\frac{\partial T}{\partial y}) + \frac{\partial}{\partial z}(k\frac{\partial T}{\partial z}) + Q = \rho\ c_p\ \frac{\partial T}{\partial \tau} \quad (1)$$

by solving the heat balance equation

$$\sum_j H_{ij}(T_j - T_i) + Q_i = C_i\,\frac{\Delta T_i}{\Delta \tau} \qquad\qquad (2)$$

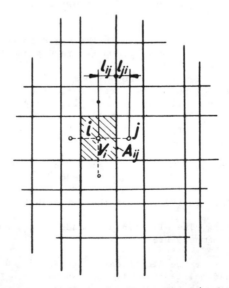

Fig.1. Element grid for mathematical model

for each node in each time step iteratively.
j ranges over the neighbours of i.

H_{ij} .... admittance between node i and j

T_i .... temperature of node i

Q_i .... internal heat generation during $\Delta\tau$ at node i

$\Delta\tau$ .... time increment

where:

$$H_{ij} = \frac{A_{ij}}{\dfrac{l_{ij}}{k_i} + \dfrac{l_{ji}}{k_j}}$$

A_{ij} .... Contact area between cells i and j

$l_{ij} + l_{ji}$ .. length of heat path between nodes ij

k_i .... conductivity of material in cell i

$$C_i = c_i \, \rho_i \, V_i \quad .... \text{ capacity}$$

c_i .... specific heat of material of cell i

ρ_i .... mass density in cell i

V_i .... volume of cell i

For a number of nodes n a system of n linear equations (2) must be solved. The Gauss-Seidel procedure is used and the nonlinearities due to temperature dependence of the material properties and nonlinear boundary conditions are taken into consideration by updating the coefficients of the system of equations. The iteration for the time step k + 1 for the i th temperature T_i^{k+1}, is as follows:

$$^{\tau+\Delta\tau}T_i^{k+1} = \frac{\displaystyle\sum_{j<i} H_{ji}\,^{\tau+\Delta\tau}T_j^{k+1} + \sum_{j>i} H_{ji}\,^{\tau+\Delta\tau}T_j^{k} + Q_i + \frac{C_i}{\Delta\tau}\,^{\tau}T_i}{\displaystyle\sum_j H_{ji} + \frac{C_i}{\Delta\tau}} \quad (3)$$

This iteration continues until all internal temperatures meet the following criterion:

$$\left| ^{\tau+\Delta\tau}T_i^{k} - ^{\tau+\Delta\tau}T_i^{k+1} \right| < \text{CRIT } (= 0.001)$$

4. PRACTICAL RELEVANCE - DEMONSTRATED BY CONSIDERATION OF A SUPPORTING ROLLER
    IN A CONTINUOUS CASTING PLANT

Figure 2 shows a cross-section of the considered hollow supporting roller and the arrangement of the cooling and heating zones.

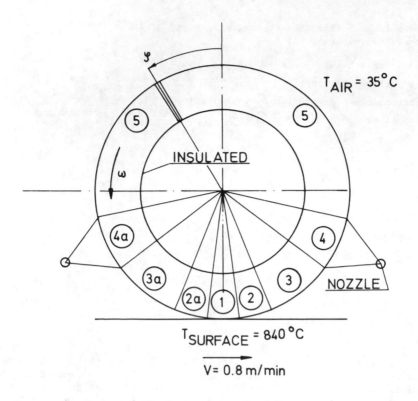

Fig.2. Boundary conditions

Supporting roller:
external diameter   D = 175 mm
inside diameter   d = 108 mm
casting speed   v = 0.8 m/min
(velocity of the strand)

ZONE	angle/deg	boundary condition
1	2°	hot spot contact zone
2,2a	9°	radiation
3,3a	35°	waterfilm cooling
4,4a	30°	water spray (nozzle) cooling
5	210°	convection and radiation

4.1 Boundary Conditions (Figure 2)

    Since the hollow space of the roller is filled with grease without any
cooling mechanism, the inside surface of the supporting roller can be con-

sidered to be insulated.

The cooling and heating zones and the determination of the heat-transfer-coefficients (HTC) in each zone are explained as follows:

Zone 1. Local hot spot contact between roller and hot strand: The length b of the contact zone is calculated from the Hertz-formula: (for explanation of notations see Figure 3)

$$b = 3.04 \sqrt{\frac{F\ D}{2\ E_m\ \ell}} \qquad\qquad (4)$$

where:

$$E_m = \frac{2\ E_1\ E_2}{(E_1 + E_2)}$$

F .... force between roller and strand normal to the strand surface;

D .... external diameter of the roller

E_1 .... Young's modulus of the roller

E_2 .... Young's modulus of the strand

ℓ .... axial bearing length of the roller

The value of the angle 2β of the contact zones results from:

$$2\beta^\circ = \frac{180}{\pi}\ \frac{2b}{D} \qquad\qquad (5)$$

In the considered example:

$$F \approx 150\ kN$$
$$D = 0.175\ m$$

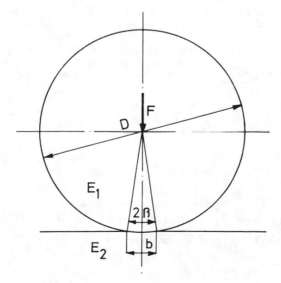

Fig.3. Contact zone

$$\ell = 0.695 \text{ m}$$
$$E_1 = 2 \times 10^{11} \text{ N/m}^2$$
$$E_2 = 1.25 \times 10^{11} \text{ N/m}^2$$
$$E_m = 1.5 \times 10^{11} \text{ N/m}^2$$

These values yield a circumferential bearing length b:

$$b = 1.1 \times 10^{-3} \text{ m}$$

and

$$2\beta = 0.7^{\circ}$$

The actual length of the contact zone may be increased due to the influence of the scale layer. Therefore and to insure a conservative consideration of the problem, the contact-zone angle was chosen to be $2\beta = 2^{\circ}$, yielding a contact time of about 0.23 seconds.

Calculation of the HTC in zone 1:
Assuming the thickness of the scale layer between strand and roll to be about s = 0.1 mm, the HTC h_1 is calculated from the formula:

$$\dot{Q} = A \frac{k_s}{\ell} \; \Delta T = A \; h_1 \; \Delta T$$

$$h_1 = \frac{k_s}{\ell} = \frac{1.63}{0.1 \times 10^{-3}} = 16\;300 \text{ W/m}^2/\text{K}$$

$$h_1 = 16300 \text{ W/m}^2/\text{K}$$

k_s .... conductivity of the scale layer
$k_s \approx 1.63$ W/m/K [2]

The boundary temperature of the roller surface in zone 1 corresponds to the surface temperature of the strand, which amounts to about 840°C.

Zone 2,2a. Surface to surface radiation from the hot strand to the roller: To enable the application of the iterative algorithms for solving a system of equations described in chapter 3 it is necessary to linearize the radiation equation as follows:

$$\dot{Q}_r = \varepsilon \; C_s \; A \left[(\frac{T_1}{100})^4 - (\frac{T_2}{100})^4 \right] = h_r \; A \; (T_1 - T_2) \quad (6)$$

and:

$$h_r = \frac{C'}{(100)^4} \; (T_1^2 + T_2^2)(T_1 + T_2) \qquad (7)$$

where: $\varepsilon_1, \varepsilon_2$ emissivity of the surfaces
                  T_1, T_2 absolut surface temperatures K°
                  $C_s = 5.77$ W/m^2/K^4 :.... radiation constant
                  $C' = \dfrac{C_s}{1/_{\varepsilon_1} + 1/_{\varepsilon_2} - 1}$

using the actual values:

$$\varepsilon_1 = \varepsilon_{strand} \approx 0.9$$
$$\varepsilon_2 = \varepsilon_{roll} \approx 0.6$$

$$T_1 = T_{strand} \approx 1113°K \ (840°C)$$
$$T_2 = T_{roll} \approx 343°K \ (70°C)$$

$$h_r = 65 \ W/m^2/K$$

leads to

$$h_{2,2a} \approx 65 \ W/m^2/K$$

The corresponding boundary temperature for zone 2 is the surface temperature of the strand (T = 840°C).

Zone 3, 3a. Waterfilm cooling. Difficulties arise in the evaluation of the HTC for zone 3:

- The exact distribution of the water flow is not known; a part of the water sprayed on the roller surface by the nozzles is sprinkling backwards and does not reach zone 3.

- Rare information is to be found in the literature describing the heat transfer mechanism of water films on rotating surfaces.

- Generally there is a strong dependence of the HTC of water-cooling on the surface temperature of the cooled surface. In the considered temperature range the heat transfer coefficient varies between 1000 $W/m^2/K$ and 25000 $W/m^2/K$ at the so called burn-out-point, which in case of soft water cooling amounts to 130°C [1].

In the example under consideration, the surface temperature in zone 3, withdrawing from the hot contact zone 1 is estimated to range from 80 – 200°C; hence a HTC of 4000 is chosen, whereas the HTC in the cool zone 3a (assumed surface temperature ≈ 40°C) is set at a value of 1000 $W/m^2/K$. The water film temperature measured was 35°C.

Zone 4, 4a. Water spray (nozzle) cooling. The HTC in zone 4,4a can be calculated according to [1] using the formula:

$$h_4 = 82 \ V_S^{0.75} \ w^{0.4} \qquad\qquad (8)$$

V_S .... specific water flux $\ell/m^2/s$

w .... water velocity at the nozzle m/s

in our example:

$V \approx 12 \ \ell/min/nozzle$

$V_S \approx 90 \ \ell/m^2/sec$

$w \approx 10m/s$

$h_4 \approx 6000 \ W/m^2/K$

Due to the fact, that the HTC is again dependent on the surface temperature of the roller (see Ref.[1],[3]) the HTC for zone 4 (withdrawing from the
hot contact zone 1) is chosen to be 7000 W/m²/K and 5000 W/m²/K for zone
4a.

Zone 5. Heat transfer to the surroundings by convection and radiation. Due
to the low temperature and the small speed of the surface, the HTC amounts only
to about 15 W/m²/K. It is the sum of the convective-HTC ($\approx$ 10 W/m²/K) and the
linearized radiation-HTC ($\approx$ 5 W/m²/K) equ. (7). The boundary temperature of
zone 5 is about 35°C.

4.2 Thermal Model of the Roller

According to chapter 2 a one-dimensional model of a 0.5° sector of the
roller was developed (Figure 4). Since it was expected (and confirmed by the

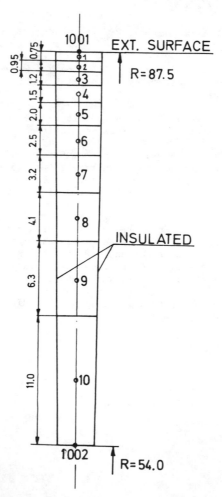

Fig. 4. One dimensional thermal model of
a 0.5° sector of the roller

result of the calculations) that considerable heat flow will only occur in a relatively small layer from the surface of the roller towards the inner parts, the modeling of the considered sector by cells takes account of this fact, i.e. small cells in the near of the surface (Figure 4). The thickness of cell 1 corresponds to a layer of coating material, which is applied to the surface in order to reduce the wear of the roller.

4.3 Procedure and Results of the Calculation

To achieve the required "quasi-steady state solution", i.e. to find the temperature distribution of the roller after a long operation period under the same boundary conditions, an iterative calculation method was used:

- Estimate the initial temperature distribution at $\varphi = 0^\circ$.

- Calculate the temperature field after one rotation.

- Compare the newly calculated temperature field with the initial one. If the quasi-steady state solution is achieved, the two temperature fields at $\varphi = 0^\circ$ and $\varphi = 360^\circ$ must be sufficiently identical.

- If there is no sufficient agreement between these temperature fields, calculate a new estimation for the steady state temperature distribution with the method described below and repeat the whole procedure until the condition mentioned above is met.

Now let us consider a practical hint, which allows to calculate an approximation of the steady state solution and thus considerably saves computing time. It has been found that the temperatures calculated at $\varphi = 0^\circ$, 360°, 720° and so on lie on an exponential curve which approaches the quasi-steady state relatively slowly (Figure 5).

The following formula can be obtained for the function T (t) in Figure 5.

$$T(t) = T_S - (T_S - T_0) \; \bar{e}^{\,at} \qquad\qquad (9)$$

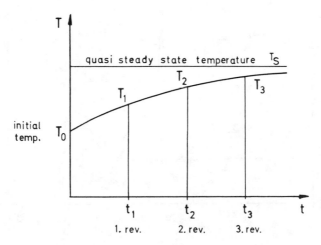

Fig. 5. Temperatures at $\varphi = 0^\circ$, 360°, 720° ...

$$(T_S - T_0) \; e^{-at_1} = T_S - T_1 \;\Big|\; \ln \qquad (10)$$

$$\ln(T_S - T_0) - at_1 = \ln(T_S - T_1) \;\Big|\; x - \frac{t_2}{t_1} \qquad (11)$$

$$\ln(T_S - T_0) - at_2 = \ln(T_S - T_2) \qquad (12)$$

$$\ln(T_S - T_0)\left[1 - \frac{t_2}{t_1}\right] - \ln(T_S - T_2) + \frac{t_2}{t_1} \; \ln(T_S - T_1) = \emptyset \qquad (13)$$

$$(T_S - T_0)^{\left[1 - \frac{t_2}{t_1}\right]} (T_S - T_1)^{\frac{t_2}{t_1}} = T_S - T_2 \qquad (14)$$

Using equation (14), T_0, T_1, T_2 and $\frac{t_2}{t_1}$ are known, the unknown steady state temperatures T_S at $\varphi = 0° = 360°$ can be calculated.

4.4 Results of the Calculation

The results of the calculation of the example under consideration are shown on two computer plots, Figure 6 and Figure 7.

Figure 6 is a plot which shows the temperature versus time for the five nodes 1 (nearest to the surface) 2, 4, 9 and 1∅.

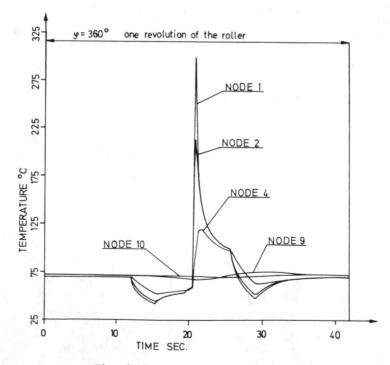

Fig. 6. Roller temperature versus time

Since one rotation of the roller lasts 41.23 sec, this time corresponds to an angle of 360°. The maximum value of the temperature curve of node 1 is 295°C, the minimum value is 43° C. The maximum temperature difference at node 1 during one rotation is 252° C, whereas it is negligible for the inner nodes 9 and 10.

Figure 7. This plot shows the radial temperature profile - from the inner node 1002 to the surface node 1001 - at various positions during one rotation of the roller. In spite of the relatively low surface speeds, the region where a temperature change worth mentioning takes place reaches only about 15 mm from the surface towards the inner parts of the roller.

4.5 Finite Element Thermal Stress Analysis

Using the results of this thermal analysis, a linear thermo-elastic finite element calculation using a two-dimensional model of the roller is executed, taking the temperature dependence of Youngs modulus into account.

Procedure of the FE calculation: First a global investigation, modeling the whole cross section of the roller by a coarse mesh of 8-node plane stress elements, is carried out (Figure 8).

The temperatures at the nodes are taken from the thermal analysis and interpolated using quadradic shape functions.

The displacements at the sectional area A-A between the lower and the upper half of the roller are taken as boundary conditions for a refined model of the lower half of the cross section (Fig.9) and used in a local investigation of the contact area.

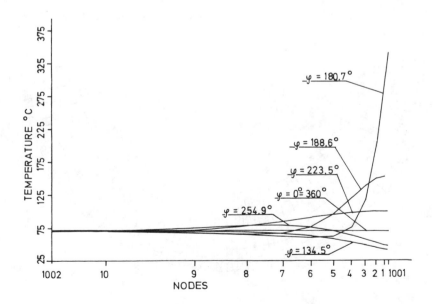

Fig. 7. Radial temperature distribution of the roller

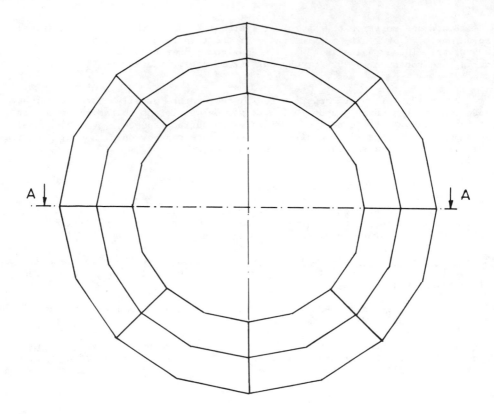

Fig. 8. Coarse FE mesh of the whole roller

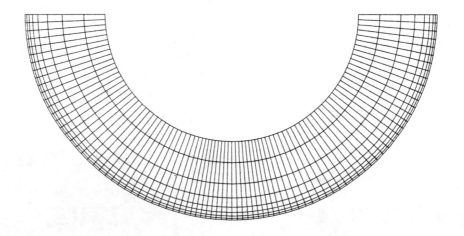

Fig. 9. Refined mesh of one half of the roller

Results of the FE calculation. The maximal stress occurs at the contact zone between roller and strand:

maximal principal stress: σ_{max} = 34.7 N/mm^2

minimal principal stress: σ_{min} = -378.3 N/mm^2

max. effective stress (Mises): σ_e = 380.3 N/mm^2

The maximal effective stress is calculated from the v. Mises formula valid for plane stress:

$$\sigma_e = \sqrt{\sigma_{xx}^2 + \sigma_{yy}^2 - \sigma_{xx}\sigma_{yy} + 3\sigma_{xy}^2} \qquad (15)$$

σ_{ij} .... stresses in rectangular coordinates

Figures 10 shows the most interesting area of the cross section of the roller. The stress field is demonstrated by lines of constant effective stress. It can be concluded that a very narrow domain is highly stressed (the contact zone). But due to the roation of the roller, each point at the surface undergoes this high loading in an alternating manner which has considerable consequences for the roller's wear and life time.

5. MATERIAL PROPERTIES

Base Material
Tensile strength: σ_T = 590-740 N/mm^2

Yield stress: σ_Y = 350 N/mm^2

Property	Density ρ	E-Modulus E	Coeff. of exp. mean value	spec. Heat c	Conductivity k
Unit	10^3 kg/m^3	kN/mm^2	10^6K^{-1}	10^3 J/kg/K	W/m/K
20°C	7.85	212	12.3	0.44	32
100°	7.83	207	12.5	0.48	34
200°	7.81	200	13.2	0.52	36
300°	7.79	193	13.7	0.56	35
400°	7.74	184	14.2	0.60	34
500°	7.68	175	14.6	0.66	33
600°	7.65	164	14.9	0.75	32

Coating Material
Tensile strength   : σ_T = 450-520 N/mm^2

Density   : ρ = 7.6x10^3 kg/m^3

Young's modulus   : E = 1.97 x 10^5 N/mm^2

Coefficient of thermal exp.: α = 13.3 x 10^{6}K^{-1}

specific heat   : c = 460 J/kg/K

Conductivity   : k = 58.5 W/m/K

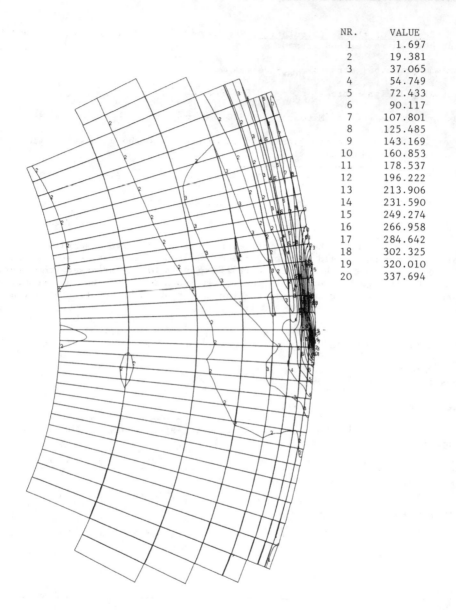

NR.	VALUE
1	1.697
2	19.381
3	37.065
4	54.749
5	72.433
6	90.117
7	107.801
8	125.485
9	143.169
10	160.853
11	178.537
12	196.222
13	213.906
14	231.590
15	249.274
16	266.958
17	284.642
18	302.325
19	320.010
20	337.694

Fig. 10. Lines of constant effective stress

6. SUMMARY

A one-dimensional thermal model was developed that allows to calculate the temperature distribution of a water-spray-cooled roller during one rotation and in the quasi-steady state, using an exponential formula to minimize the computing expense.

The calculation yields a rapid decrease of the radial temperature gradient from the roller surface towards the inner parts of the cross section. In the example under consideration, a water-spray-cooled supporting roller in a continuous casting plant - the maximum temperature difference at the roller surface during one rotation is 250° C, producing high thermal stresses which may lead to surface cracking.

The calculation model developed offers the advantage of enabling extensive parametric studies which are easily performed.

REFERENCES

1. Diener, A. 1976. Der Wärmeübergang beim Spritzkühlen heißer Stahlober-flächen. Stahl und Eisen 96, Nr. 4.

2. Pawelski, O. 1969. Berechnung der Wärmedurchgangszahl für das Warmwalzen und Schmieden. Archiv für Eisenhüttenwesen, Heft 10.

3. Etienne, A. und Mairy, B. 1979. Heat transfer in continuously cast strands Centre de Recherches Metallurgiques.

EXPERIMENTAL
TECHNIQUES

Design of Wide-Bandwidth Analogue Circuits for Heat Transfer Instrumentation in Transient Tunnels

M. L. G. OLDFIELD, H. J. BURD, and N. G. DOE
Department of Engineering Science
University of Oxford
Parks Road
Oxford, OXI 3PJ, England

ABSTRACT

The use of thin film surface thermometers to measure heat transfer rates is well established in the Oxford transient Isentropic Light Piston Tunnel, and has been used in many other facilities. In this technique a resistance-capacitance transmission line, or electrical analogue, is used to convert the surface temperature signal into a signal proportional to the heat transfer rate to the surface. This paper describes a new method for designing analogues with a logarithmic scaling of the elements. This uses far fewer elements for a given bandwidth than previous designs and can retain a 1% accuracy even when constructed from 20% tolerance capacitors, thus eliminating the need for expensive 1% tolerance components. A theoretical analysis is given which leads to computer programs to predict the performance of any analogue, both in the frequency domain and the time domain. A design example of a low noise analogue circuit using only 9 sections for a frequency range of 0.1 Hz to 100 kHz is given, together with experimental results showing high frequency heat transfer rate measurements on a turbine blade in the Oxford cascade tunnel.

NOMENCLATURE

C_i	ith capacitor in analogue	a	Arithmetic analogue constant
$C_T = \Sigma C_i$	Total capacitance	c	Specific heat
$F = \dfrac{\omega_2}{\omega_1} = \dfrac{\tau_2}{\tau_1}$	Working range of analogue	c	Capacitance per unit length
K	Electrical length of analogue	f_{min}, f_{max}	Minimum and maximum frequencies
N	Number of analogue sections	i	Current
R_i	ith resistor	k	Thermal conductivity
R_g	Thin film gauge resistance	ℓ	Length
T	Temperature	$\dot{q}$	Heat transfer rate per unit area
V_o	Set voltage across thin film gauge	$\underset{\sim}{q}_i$	Eigenvector
$Z = \dfrac{v}{i}$	Analogue impedance	r	Resistance per unit length
		s	Laplace transform variable

233

t	Time	λ_i	Eigenvalue
v	Varying voltage	ρ	Density
x	Distance	τ_1, τ_2	Analogue time limits
α	Thermal coefficient of resistivity	ω	Angular frequency
β	Resistor dividing ratio	ω_1, ω_2	Analogue frequency limits
γ	Logarithmic analogue constant	$-$	Laplace transforms

1. INTRODUCTION

The use of thin film heat transfer gauges and the principle of their operation has been extensively dealt with in the literature. A detailed treatment is included in [1] where a unified theory of thin and thick film gauges is given.

Originally developed for use in shock tubes and other very short duration facilities, thin film gauges are now being used for heat transfer measurements in turbine cascade tunnels with up to 0.5 seconds flow duration [3,4,5] and for measuring heat transfer rates on rotating turbine blades [6].

The technique depends on the measurement of the variation of surface temperature with time on an insulating substrate, (e.g. quartz, pyrex or Macor machinable glass) forming part or all of the test model. A thin film of platinum is deposited on the surface and used as a resistance thermometer through which a small constant current is passed. The changes of voltage across this thermometer are proportional to the surface temperature, T, and the heat transfer rate, $\dot{q}$, to the surface can be extracted by using the appropriate mathematical transformation. While it is possible to compute $\dot{q}$ from a digitally recorded T signal, it can be shown [5] that the quantization error when digitising the film voltage waveform gives rise to spurious noise on the reconstructed $\dot{q}$ signal. Consequently it is more usual to use a resistance-capacitance transmission line [1,7,8] as an electrical analogue of the substrate, to convert the film voltage into a current proportional to $\dot{q}$, and to record this current.

The mathematical transformation from $\dot{q}$ to T is not a noisy one [5] and so the number of instrumentation channels can be reduced by only recording the analogue current.

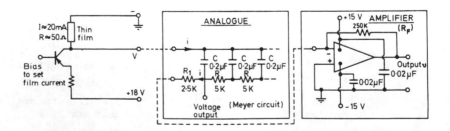

Figure 1   Typical analogue circuit used with thin film heat transfer gauges, from [5].

Figure 1, from [5] shows a typical analogue and associated circuitry. This analogue is made up of discrete lumped elements [7,8] and uses 19 sections to achieve a useful frequency range of 0.07 Hz to 6 kHz, by increasing the size of the elements arithmetically after the first three sections.

Boundary layer fluctuations caused by the passing of turbine blades behind stator blades give rise to high frequency fluctuations in the heat transfer rate to the blade surface, and new low noise analogues with a bandwidth of 0.1 Hz to 100 kHz are required.

The following sections describe a new method of analogue circuit design which can achieve this bandwidth with only 9 sections, using low tolerance capacitors, and give methods of predicting their performance.

2.   ANALOGUE THEORY

The equivalence between the one dimensional flow of heat through a medium and the flow of current through a continuous RC transmission line can be seen with the aid of Figure 2, following the treatment in [1].

For the thermal case the governing equations are:

$$\frac{\partial \dot{q}}{\partial x} = -\rho c \, \frac{\partial T}{\partial t} \tag{1}$$

$$\dot{q} = -k \, \frac{\partial T}{\partial x} \tag{2}$$

which gives the Diffusion Equation

$$\frac{\partial^2 T}{\partial x^2} = \frac{\rho c}{k} \, \frac{\partial T}{\partial t} \tag{3}$$

These may be solved for the heat transfer rate $\dot{q}$ at the surface of a semi-infinite medium in terms of the surface temperature T to give

$$\bar{\dot{q}} = \sqrt{\rho c k} \, \sqrt{s} \, \bar{T} \tag{4}$$

where the symbols with overlaps are Laplace transformed variables.

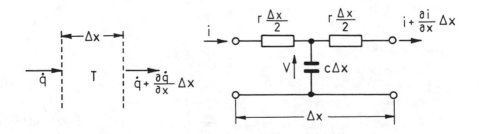

Figure 2   Derivation of the electrical analogue of heat conduction (after [1])

For the simple case where $\dot{q}$ is constant for $t > 0$, the surface temperature is a parabolic function of time:

$$T = \frac{2}{\sqrt{\rho c k}} \dot{q}\sqrt{t} \tag{5}$$

For the semi-infinite electrical analogue the equivalent equations to (1) - (4) are:

$$\frac{\partial i}{\partial x} = -c \frac{\partial v}{\partial t} , \tag{6}$$

$$i = -\frac{1}{r} \frac{\partial v}{\partial x} , \tag{7}$$

$$\frac{\partial^2 v}{\partial x^2} = rc \frac{\partial v}{\partial t} , \tag{8}$$

and $\bar{i} = \sqrt{\frac{c}{r}} \sqrt{s} \ \bar{v}$ $\qquad\qquad$ (9)

Thus it can be seen that i is analogous to $\dot{q}$ and v to T.

For the simple case where

$$v = \sqrt{t}$$

then $\quad \bar{v} = \frac{\sqrt{\pi}}{2} s^{-3/2}$

and $\quad \bar{i} = \frac{\sqrt{\pi}}{2} \sqrt{\frac{c}{r}} \frac{1}{s}$

which gives a constant current

$$i = \frac{\sqrt{\pi}}{2} \sqrt{\frac{c}{r}} \tag{10}$$

The characteristic impedance of the continuous electrical analogue is

$$\bar{Z} = \frac{\bar{v}}{\bar{i}} = \frac{1}{s} \sqrt{\frac{r}{c}} \tag{11}$$

or in the frequency domain

$$|Z| = \frac{1}{\sqrt{\omega}} \sqrt{\frac{r}{c}} \tag{12}$$

v is related to T, the change in surface temperature, by the thermal coefficient of resistivity of the thin film thermometer:

$$R_g = R_{go}(1 + \alpha T) \tag{13}$$

If the current source in Figure 1 is adjusted to give a voltage V_0 across the film for $T = 0$, then the change in voltage across the film due to a change in surface temperature T is

$$v = V_0 \alpha T \tag{14}$$

By combining (4), (9) and (14) it can be seen that the analogue current i is directly proportional to $\dot{q}$:

$$i = \frac{V_0 \alpha}{\sqrt{\frac{r}{c}} \sqrt{\rho c k}} \dot{q} \tag{15}$$

In practice the continuous semi-infinite analogue line is replaced by a finite length line of lumped "T" sections made from discrete components, and this is designed to give

$$|z| = \frac{1}{\sqrt{\omega}} \sqrt{\frac{r'}{c}}$$

over the required range of frequencies (equation (12)).

3. GENERAL DISCRETE FINITE ANALOGUE

Figure 3 shows a general finite length discrete element analogue, made up of "T" sections as shown in Figure 4.

The two resistive arms of the ith "T" section are distributed, with β_i, the resistor matching coefficient, to be determined, about C_i so as to preserve the designed $\frac{r}{c}$ ratio:

$$
\begin{aligned}
r_{i1} &= \beta_i \left(\tfrac{r}{c}\right) c_i \\
r_{i2} &= (1 - \beta_i) \left(\tfrac{r}{c}\right) c_i \\
r_{i1} + r_{i2} &= \left(\tfrac{r}{c}\right) c_i
\end{aligned}
\tag{16}
$$

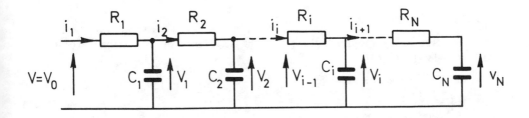

Figure 3   Finite length discrete element analogue

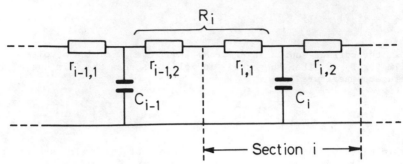

Figure 4   Analogue "T" sections

Then $R_i = r_{i-1,2} + r_{i,1}$

$$= (\frac{r}{c})((1 - \beta_i)C_{i-1} + \beta_i C_i), \; i > 1 \tag{17}$$

and $R_1 = \beta_1 C_1 (\frac{r}{c})$

The total capacitance of the line is given by

$$C_T = \sum_i C_i \tag{18}$$

There is, at this stage, no restriction on the way that C_i can vary along the analogue.

The <u>electrical length</u> of the analogue is defined as

$$K = C_T/C_i \tag{19}$$

In studying the performance of different types of analogue it is convenient to use <u>normalized analogues</u> for which

$$\frac{r}{c} = 1 \; \text{and} \; C_1 = 1 \tag{20}$$

4.   FREQUENCY RESPONSE

On a log z:log(ω) plot (Figure 5) the impedance of an analogue follows three asymptotic lines:

(i)   in the working range,

$$|z| \sim \sqrt{\frac{r}{c}} \frac{1}{\sqrt{\omega}} \tag{21}$$

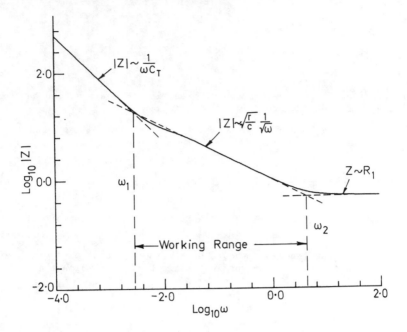

Figure 5   Impedance v. angular frequency plot of 20 constant-section normalized analogue

(ii)   at very high frequencies, all the capacitors are effectively short-circuited and

$$Z \sim R_1 = \beta_1 \frac{r}{c} C_1 \tag{22}$$

(iii)   at very low frequencies the impedance of the capacitances dominates,

and $$|Z| \sim \frac{1}{\omega C_T} = \frac{1}{\omega K C_1} \tag{23}$$

The <u>lower frequency limit</u> ω_1 is defined by the intersection of equations (21) and (23):

$$\omega_1 = \frac{c}{r} \frac{1}{C_T^2} = \frac{1}{K^2 C_1^2} \frac{c}{r} \tag{24}$$

The <u>upper frequency limit</u> is given, from equations (21) and (22) by

$$\omega_2 = \frac{r}{c} \frac{1}{R_1^2} = \frac{1}{\beta^2 C_1^2} \frac{c}{r} = \frac{1}{\beta R_1 C_1} \tag{25}$$

The working range is

$$F = \frac{\omega_2}{\omega_1} = \frac{K^2}{\beta^2} \tag{26}$$

The detailed frequency response of a particular analogue can be computed by using the iteration (Figure 3)

$$Z_N = R_N + \frac{1}{j\omega C_N}$$

$$Z_i = R_i + \frac{Z_{i+1}}{1 + j\omega C_i Z_{i+1}} , \quad i = N-1, \ldots 2,1 \tag{27}$$

The computer program to do this can be checked against two analytic solutions:

(i)   Near ω_1 the impedance should be the same as that of a continuous analogue of total capacitance C_T, of length $\ell = C_T/c$, for which the exact solution of equations (6), (7) and (8) with i = 0 at x = ℓ is

$$Z = \frac{1}{\sqrt{j\omega}} \frac{r}{c} \coth(\sqrt{j\omega} \sqrt{\frac{r}{c} C_T}) \tag{28}$$

(ii)   Near ω_2 the impedance should be the same as that of a semi-infinite discrete constant section line, with $C_i = C$, $R_i = R = \frac{r}{c} C$, $R_1 = {}^R/2$.   For this line, adding an extra section does not change the impedance, and so equation (27) becomes

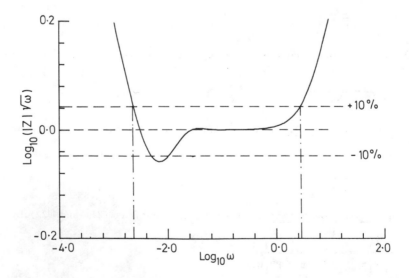

Figure 6   Frequency response of a 20 section constant-section normalized analogue

$$Z_2 = R + \frac{Z_2}{1 + j\omega C Z_2}$$

and $Z = Z_1 = \frac{R}{2} + \frac{Z_2}{1 + j\omega C Z_2} = Z_2 - \frac{R}{2}$

which gives $Z = \frac{1}{\sqrt{j\omega}} \sqrt{\frac{r}{c}} (1 + \frac{j\omega RC}{4})^{\frac{1}{2}}$ (29)

It is convenient to plot $\log (|Z|\sqrt{\omega})$ v. $\log \omega$ (Figure 6) so that analogue frequency responses can be compared with the ideal $|Z|\sqrt{\omega} = \sqrt{\frac{r}{c}}$

5. TIME RESPONSE

Equation (10) shows that the appropriate input voltage waveform for testing the time response of an analogue is $v = \sqrt{t}$, and that the response of an ideal analogue to this input is a step in i of height

$$i = \frac{\sqrt{\pi}}{2} \sqrt{\frac{c}{r}}$$

An estimate of the range of times over which the response of a real discrete finite analogue will be close to the ideal can be made from Figure 5:

The <u>lower time limit</u>

$$\tau_1 = \frac{1}{\omega_2} = \beta^2 C_1^2 \frac{r}{c}$$ (30)

The <u>upper time limit</u>

$$\tau_2 = \frac{1}{\omega_1} = K^2 C_1^2 \frac{r}{c}$$ (31)

The <u>working range</u>

$$F = \frac{\tau_2}{\tau_1} = \frac{\omega_2}{\omega_1} = \frac{K^2}{\beta^2}$$ (26)

as before.

The detailed time response of a particular analogue can be computed using the analysis given in Appendix 1. The computer program to do this can be checked near τ_1 against the analytic solution for a semi-infinite discrete constant section line: Using Laplace transforms, equation (29) can be written

$$\bar{Z}(s) = \frac{1}{\sqrt{s}} \sqrt{\frac{r}{c}} (1 + \frac{sRC}{4})^{\frac{1}{2}}.$$

with $\bar{v}(s) = \frac{\sqrt{\pi}}{2} s^{-3/2}$,

$$\bar{i}\ (s) = \sqrt{\frac{c}{r}} \sqrt{\frac{\pi}{RC}}\ \frac{1}{s(s + \frac{4}{RC})^{\frac{1}{2}}}$$

and the solution is

$$i\ (t) = \frac{\sqrt{\pi}}{2} \sqrt{\frac{c}{r}}\ erf(2\sqrt{t}) \tag{32}$$

This approaches the ideal for large t.

6.   CONSTANT SECTION ANALOGUE

A constant section analogue has all "T" sections identical and $\beta = 0.5$, so that

$$C_i = C,\ R_i = R = \frac{r}{c}\ C,\ R_1 = \frac{R}{2}.$$

If there are N sections,

$C_T = NC$, $k = N$ and the range is, from equation (26),

$F = 4N^2$

The number of sections required for a given range is

$N = 0.5\sqrt{F}$

The frequency response of a constant section normalized analogue with $N = 20$, computed using equation (27) is shown in Figures 5 and 6. The inter-sections of the asymptotes in Figure 5 give frequency limits which agree with the theoretical limits (equations (24) and (25)) of

$$\omega_1 = 2.5 \times 10^{-3},\ \omega_2 = 4$$

The (arbitrarily chosen) ± 10% lines on Figure 6 give frequency limits of $1.0\ \omega_1$ and $0.7\ \omega_2$.

At the low frequency end ($\log_{10}\omega < -1.0$) the curve agrees with the response of a finite length continuous analogue, equation (28). Interestingly, this shows that the dip in $\log_{10}\ (|Z|\sqrt{\omega})$ at $\log_{10}\ \omega \sim -2.0$ is caused by the finite length of the analogue, not by the use of discrete lumped elements.

At the high frequency end ($\log_{10}\omega > -1.0$) Figure 6 agrees with the response of the limiting case in equation (29).

The upper and lower time for this 20 section analogue are

$$\tau_1 = 0.25,\ \tau_2 = 400.$$

The time response of the line, computed using Appendix 1 is plotted in

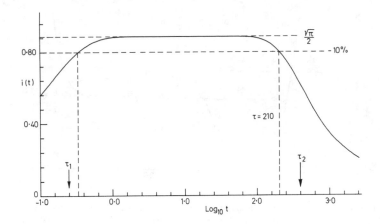

Figure 7   Time response of a 20 section constant—section normalized analogue

Figure 7. It will be seen that i(t) rises smoothly to the ideal value of $\frac{\sqrt{\pi}}{2}$ and, at long times, droops without any overshoot which might have been expected from the dip in Figure 6: i(t) rises to within 10% of the ideal after 0.34 secs (= 1.35 τ_1) and droops below the -10% level after 210 secs (= 0.53 τ_2).

Thus, for a constant section line, the analysis and the computer programs for anlysing the frequency and time response would seem to be vaild.

However, the design requirement is for an analogue useable from 0.1 Hz to 100 kHz, for which F = 10^6 would require,

$$N = 0.5\sqrt{10^6} = 500,$$

which is far too many sections for a practical analogue.

7.   ARITHMETIC ANALOGUES (MEYER TYPE)

To reduce the number of sections required, Meyer [7,8] designed analogues in which the capacitors increased in arithmetic progression:

$$C_i = C_1(1 + a(i-1)), \ i = 1, \ldots, N \tag{33}$$

Then $C_T = \Sigma C_i = \frac{C_1 N}{2}(2 + (N-1)a) = KC_1$

and with $\beta = 0.5$, from equation (26)

$$F = N^2(2 + a(N-1))^2$$

$$\doteq a^2 N^4 \text{ for } aN \gg 1 \tag{34}$$

The number of sections for a given F is then

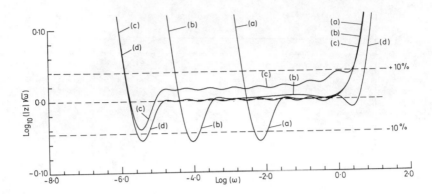

Figure 8   Frequency response of different normalized analogues: (a) 20 section constant-section, (b) 20 section Meyer arithmetic, a=1, β=0.5, (c) 9 section logarithmic, γ=10$^{1/3}$, β=0.5, (d) 9 section logarithmic, with logarithmic matching, γ=10$^{1/3}$, β=0.405

$$N \doteq \frac{F^{\frac{1}{4}}}{\sqrt{a}}$$

For $F = 10^6$ if we choose $a = 1$, then $N = 31.6$, and a 32 section analogue would be required, a vastly smaller number than the 400 that would be required for a constant section analogue.

Meyer used $a = 1$, and added 2 extra constant sections to the front.   The frequency and time responses of a normalised 20 section Meyer line are shown by curves (b) in Figures 8 and 9 from which it can be seen that the Meyer design is successful in considerably extending the frequency range.   However, it does have a slight hump in the mid range frequency response, with a corresponding slight dip in the time response, due to the addition of the constant sections at the

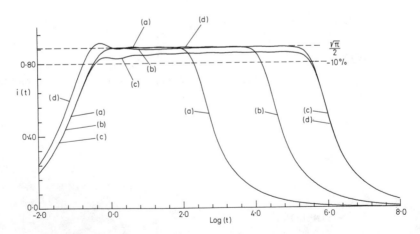

Figure 9   Time response of different normalized analogues: (a) 20 section constant -section, (b) 20 section Meyer arithmetic, a=1, β=0.5, (c) 9 section logarithmic, γ=10$^{1/3}$, β=0.5, (d) 9 section logarithmic, with logarithmic matching, γ=10$^{1/3}$, β=0.4

beginning of the analogue. The high frequency, short time, characteristics are the same as that of the constant section line (curves (a)), and the low frequency, long time, characteristics are simply displaced horizontally.

The Meyer line has been successfully used at Oxford over the last decade.

Arithmetic lines of this type, however, do require a large number of close tolerance components, and, as will be shown, it is possible to further reduce both the number and the tolerance of components required by using logarithmic analogues.

8. LOGARITHMIC ANALOGUES

A logarithmic analogue is one in which the values of C_i are evenly spaced on a logarithmic scale, and so form a geometric progression:

$$C_i = \gamma^{i-1} C_1, \quad i = 1, \ldots, N \tag{35}$$

Then $C_T = \Sigma C_i = C_1 \dfrac{(\gamma^N - 1)}{(\gamma - 1)} = K C_i$ $\tag{36}$

and, from equation (26),

$$F = \left(\frac{(\gamma^N - 1)}{\beta(\gamma - 1)} \right)^2 \tag{37}$$

The number of sections required for a given F is

$$N = \frac{\log(1 + \beta(\gamma - 1)\sqrt{F})}{\log\gamma} \tag{38}$$

N decreases as γ increases, and γ can be chosen to be as large as possible as long as the computed time and the frequency responses are acceptable. Setting $\gamma = 10^{1/3}$ has proved acceptable, for the capacitors form a 1, 2.154, 4.642, 10, 21.54 ... sequence which is close to the 1, 2.2, 4.7, 10, 22, ... sequence of commonly available capacitors.

For $F = 10^6$, if $\beta = 0.5$ again, and $\gamma = 10^{1/3}$, $N = 8.29$.

Thus a 9 section logarithmic analogue covers the same frequency range as a 23 section arithmetic analogue or a 500 section constant-section analogue.

Curves (c) on Figures 8 and 9 show the frequency and time responses of a 9 section normalised logarithmic analogue with $\gamma = 10^{1/3}$ and $\beta = 0.5$. The following features are apparent:

(i)   The usable time and frequency ranges are as predicted and the shapes of the ends of the curves are similar to those of the constant section analogue, curves (a).

(ii)   There is a slight (< 1%) acceptable ripple in both time and frequency due to the coarse spacing of the lumped elements.

(iii)   In the centre of the curves, $|Z|\sqrt{\omega} = 1.04$ instead of the theoretical 1.0, and $i(t) = 0.85$ instead of the expected $\sqrt{\pi}/2 = 0.886$.

(iv) The frequency response climbs above the expected value near the upper limit and the time response is low near t = 1.

While (i) and (ii) are satisfactory, (iii) and (iv) indicate that the modelling of the logarithmic analogue is not quite correct.

Running computer predictions of analogue responses with varying values of R_1 made it clear that although the capacitors were logarithmically spaced, the resistors of the fundamental "T" sections (Figure 4), with $\beta = \frac{1}{2}$, were still equally distributed on either side of each capacitor, and that this was not a satisfactory model. To cure this shortcoming the concept of logarithmic matching was developed.

9. LOGARITHMIC MATCHING

In a logarithmic line the capacitors C_i in Figure 4 increase in a geometric series and it is reasonable to assume that the resistors $r_{i-1,1}$, $r_{i-1,2}$, $r_{i,1}$, $r_{i,2}$ should do likewise. As there are twice as many resistors as capacitors the rate of increase should be

$$\frac{r_{i-1,2}}{r_{i-1,1}} = \frac{r_{i,1}}{r_{i-1,2}} = \frac{r_{i,2}}{r_{i,1}} = \sqrt{\gamma}$$

Using equation (16) this implies that

$$\frac{1 - \beta}{\beta} = \sqrt{\gamma}$$

or $$\beta = \frac{1}{1 + \sqrt{\gamma}} \tag{39}$$

For $\gamma = 10^{1/3}$, $\beta = 0.4052$ which is less than the value of 0.5 previously assumed.

Then from equation (17), noting that

$$C_{i-1} = \frac{1}{\gamma} C_i,$$

$$R_i = \frac{r}{c} \left(\left(1 - \frac{1}{(1 + \sqrt{\gamma})}\right) \frac{C_i}{\gamma} + \frac{C_i}{(1 + \sqrt{\gamma})} \right)$$

which simplifies to

$$R_i = \frac{r}{c} \frac{C_i}{\sqrt{\gamma}}, \quad i = 2, \ldots, N \tag{40}$$

and $$R_1 = \beta C_1 \frac{r}{c} = \frac{1}{(1 + \sqrt{\gamma})} \frac{r}{c} C_1 \tag{41}$$

Equations (35) – (38) still define the performance of the line, but as $\beta^2 = 0.1642$ for the logarithmically matched line instead of 0.25, there is a 52% increase in the working range F of the line brought about by decreasing R_1 from

$0.5 \frac{r}{c} C_1$ to $0.4052 \frac{r}{C} C_1$.

The frequency and time responses of a normalized 9 section logarithmic analogue with logarithmic matching, and $\gamma = 10^{1/3}$, is shown by curves (d) in Figures 8 and 9. The levels in the mid-range are now as expected and the high frequency, short time, response is now satisfactory. There is an acceptable slight 2.5% overshoot in around $\log t = 0$ in the time response, but the curve then follows the ideal line to within 1%. The predicted increase of high frequency response is also apparent.

For this analogue the frequency and time limits can be computed from equations (24), (25), (36) and (37):

$$\frac{r}{c} = 1, \qquad C_1 = 1, \qquad \gamma = 10^{1/3}, \qquad \beta = 0.4052.$$

$$k = 8.65 \times 10^2, \qquad F = 4.56 \times 10^6,$$

$$\omega_1 = 1.34 \times 10^{-6}, \qquad \omega_2 = 6.09 \text{ and}$$

$$\tau_1 = 0.164, \qquad \tau_2 = 7.49 \times 10^5.$$

From Figure 8 the +10% limits on $|Z|\sqrt{\omega}$ are $\omega = 1.3 \times 10^{-6} \doteq \omega_1$ and $\omega = 5.8 = 0.95 \, \omega_2$.

From Figure 9, the −10% time response limits are $t = 0.18$ sec $= 1.1 \, \tau_1$ and $t = 3.7 \times 10^5 = 0.5 \, \tau_2$.

The component values for this normalised line are given in Table 1:

Table 1 Components of normalised 9 section logarithmic analogue with logarithmic matching and $\gamma = 10^{1/3}$

i	C_i	R_i
1	1.000	0.405
2	2.154	1.468
3	4.642	3.163
4	10.00	6.813
5	21.54	14.68
6	46.42	31.62
7	100.0	68.13
8	215.4	146.8
9	464.2	316.2

Thus the logarithmically matched logarithmic analogue gives an excellent performance with relatively few sections. However, high tolerance (< 1%) capacitors would still be required unless a means could be found to deal with variations in capacitance.

10.   COMPENSATING CAPACITOR VARIATIONS

Close tolerance capacitors are difficult to obtain to specified values, whereas resistors can easily be made up to any value to 1% tolerance by adding lower values in series. Consequently, if capacitor variations can be compensated

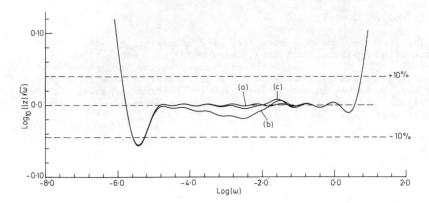

Figure 10 Affect of +20% error in C_4 on frequency response of analogue: (a)
   Accurate analogue as in Figure 8 curve (d), C_4 = 10.0. (b) C_4 = 12.0 (+20%).
   (c) C_4 = 12.0 (+20%) with R_4 and R_5 adjusted.

when designing an analogue, the cost of the circuit will be greatly reduced.

Figures 10 and 11, curves (b) show the effect of increasing the value of C_4
in Table 1 by 20% from 10 to 12 Farads. The time response is some 4% high near
$\log(t)$ = 2.5.

The analogue could be considered to be a logarithmic line in which the
capacitor ratio γ varies from section to section:

$$\gamma_i = \frac{C_i}{C_{i-1}} \; , \; i = 2, \ldots, N \tag{42}$$

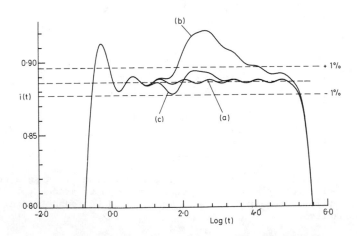

Figure 11 Affect of +20% error in C_4 on time response of analogue (note expanded
   i(t) scale): (a) Accurate analogue as in Figure 8 curve (d), C_4 = 10.0. (b)
   C_4 = 12.0 (+20%). (c) C_4 = 12.0 (+20%) with R_4 and R_5 adjusted.

Then, from equation (40) the resistors would be

$$R_i = \frac{r}{c} \frac{C_i}{\sqrt{\gamma_i}} \tag{43}$$

and

$$R_1 = \frac{r}{c} \frac{1}{1 + \sqrt{\gamma_2}} C_1 \tag{44}$$

as C_o does not exist.

Using equations (42) and (43), in the case where $C_4 = 12.0$, $\gamma_4 = 2.585$ and $\gamma_5 = 1.795$, which gives $R_4 = 7.464$ instead of 6.813 and $R_5 = 16.08$ instead of 14.68.

Curves (c) on Figures 10 and 11 show the results of adjusting R_4 and R_5 in this manner. In both the frequency and the time domains the response of the analogue has been restored to close to the ideal value, and in particular the time response is again within 1% of the ideal value.

Further studies with independent random variations of the values of all the capacitors in an analogue have further demonstrated the validity of this principle of resistor matching.

It is now clear that a practical logarithmic analogue with a time response within 1% of the ideal can be designed using capacitors of only approximately the ideal values, and that this will cost far less than an equivalent Meyer analogue using 1% components.

11. DESIGNING A PRACTICAL ANALOGUE

The process of designing a practical logarithmic analogue is now described, using a design example.

(i) Choose upper, f_{max}, and lower f_{min}, frequency limits required (100 KHz - 0.1 Hz) and the line constant $\gamma = 10^{1/3} = 2.154$.

Compute $F = f_{max}/f_{min} = 10^6$

$$\beta = \frac{1}{1 + \sqrt{\gamma}} = 0.4052 \tag{39}$$

and

$$N = \frac{\log(1 + \beta(\gamma-1)\sqrt{F})}{\log \gamma} = 8.01 \tag{38}$$

Choose N = 9 and recalculate F:

$$F = \left(\frac{(\gamma^N - 1)}{\beta(\gamma-1)}\right)^2 = 4.56 \times 10^6 \tag{37}$$

(ii) Find required values of ω_1 and ω_2: For 9 section $\gamma = 10^{1/3}$ analogue the +10% limits for $|z|/\sqrt{\omega}$ are ω_1 and $0.95 \, \omega_2$. Usually ω_2 is fixed and $\omega_1 = \omega_2/F$:

$$\omega_2 = \frac{2\pi f_{max}}{0.95} = 6.61 \times 10^5$$

$$\omega_1 = \frac{\omega_2}{F} = 0.145, \quad f_{min} = 0.023 \text{ Hz}$$

Calculate the lower and upper -10% time response limits:

$1.1/\omega_2 = 1.6$ µsec, $0.5/\omega_1 = 3.4$ sec.

(iii) Choose a nominal value for R_1. If Johnson circuit noise is to be minimised, R_1 should not be much greater than the resistances of the thin film thermometers $R_F(20 - 50\Omega)$ used. In this case the first resistor of the analogue will have to be adjusted to $(R_1 - R_F)$. If noise is not a problem, choose $R_1 \gg R_F$. For the design example low noise is required, so choose $R_1 = 200 \ \Omega$.

Estimate C_1 from

$$C_1 = \frac{1}{\beta R_1 \omega_2} = 18.7 \text{ nF} \tag{25}$$

Change C_1 to the nearest convenient standard value of 22nF and choose capacitor sequence 22, 47, 100, 220 ... Calculate $\frac{r}{c}$ and nominal R_1

$$\frac{r}{c} = \frac{1}{\omega_2 \beta^2 C_1^2} = 1.90 \times 10^{10} \tag{25}$$

$$R_1 = \frac{1}{\omega_2 \beta C_1} = 170 \ \Omega$$

(iv) Select set $C_1 - C_N$ from 5% tolerance polycarbonate or polyester film capacitors. Measure each capacitor and calculate $R_1 - R_N$:

$$\gamma_i = C_i/C_{i-1}, \quad i = 2, \ \dots \ N \tag{42}$$

$$R_i = \frac{r}{c} \frac{C_i}{\sqrt{\gamma_i}} = \frac{r}{c} \sqrt{C_i C_{i-1}}, \quad i = 2 \ \dots \ N \tag{43}$$

$$R_1 = \frac{r}{c} \frac{C_1}{(1 + \sqrt{\gamma_2})} \tag{44}$$

see Table 2.

(v) Use computer programs to calculate frequency response from equations (27) and time response from Appendix 1.

Figure 12 shows the predicted frequency response and Figure 13 the time response for the analogue in Table 2. The shapes of the curves are remarkably similar to the ideal curves (d) in Figures 8 and 9, and the levels, frequency limits and time limits are as predicted from the design equations. The compensation for capacitor values differing from the ideal is successful.

(vi) Since the resistors in Table 2 are not standard preferred values, it is necessary to use two series resistors in each section to obtain the desired 1% accuracy. Select the first resistor from the 2% tolerance E24 series with a value as close as possible below that required. Measure this resistor to an

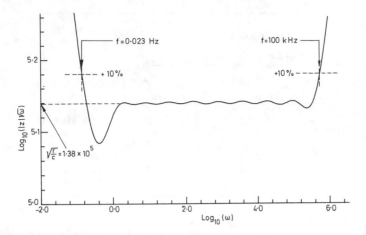

Figure 12 Frequency response of 0.02 Hz – 100 kHz analogue with component values
        given in Table 2.

accuracy of better than 0.5% and select another, much smaller, resistor to bring
the combination up to the required value. The combination will then be within
1% of the required value. It is suggested that Metal Oxide 2% E24 or equivalent
low noise, low temperature coefficient, resistors are used.

The measured frequency response of analogues using the above design has
agreed with the predicted curve to within the measurement accuracy.

The complete circuit of the Table 2 analogue, together with the associated
electronics, is given in Figure 14. The design has been optimised for low noise

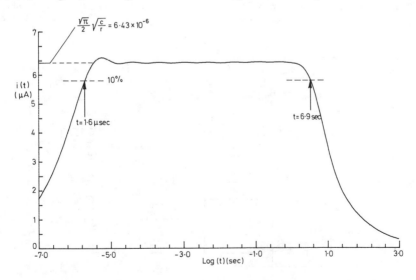

Figure 13 Time response of 0.02 Hz – 100 kHz analogue with component values
        given in Table 2.

Table 2 Design Example ($\frac{r}{c} = 1.90 \times 10^{10}$)

i	Nominal C_i		Measured C_i		R_i	
1	22.0	nF	23.10	nF	180.2	Ω
2	47.0	nF	47.60	nF	630.0	Ω
3	100.0	nF	100.4	nF	1.313	kΩ
4	220.0	nF	216.0	nF	2.798	kΩ
5	470.0	nF	463.0	nF	6.009	kΩ
6	1.0	μF	1.028	μF	13.11	kΩ
7	2.2	μF	2.20	μF	28.57	kΩ
8	4.7	μF	4.60	μF	60.44	kΩ
9	10.0	μF	10.11	μF	129.6	kΩ

[9], and consequently uses discrete transistors for the constant current supply and the current-to-voltage converter. A 100 kHz second-order Butterworth low-pass filter is used to reduce high frequency noise. The capacitors C_{14} and C_{15} can be increased to lower the bandwidth. In a multi-channel system separate isolated power supplies for each channel will reduce mains frequency interference due to earth loops.

Figure 15 shows 2 ms of a heat transfer record from a thin film gauge on the surface of a turbine cascade blade in a simulated upstream wake passing experiment conducted in the Oxford cascade tunnel [3,4,5]. The high frequency response of the 0.02 Hz -- 100 kHz analogue is clearly demonstrated.

12.   CONCLUSIONS

The analysis of heat transfer analogue circuits has led to methods of predicting the frequency and time response of any analogue. These methods have

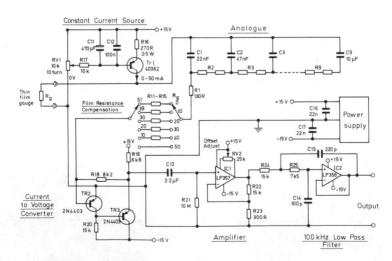

Figure 14   Complete circuit of 0.02 Hz - 100 kHz logarithmic analogue with associated electronics.

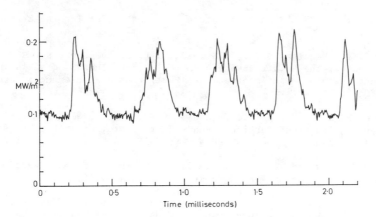

Figure 15   Heat transfer record showing fast response of 0.02 Hz – 100 kHz
            logarithmic analogue.

been used to design and develop the logarithmic analogue with logarithmic
resistor matching. This enables wide range analogues to be constructed from a
small number of sections, without the expense of close tolerance capacitors.

    A design example has been presented of a 9 section logarithmic analogue
having a frequency range of 0.02 Hz to 100 kHz, and this analogue has been used
to obtain heat transfer records containing high frequency fluctuations.

    The simplicity and low cost of these new types of analogue will add to the
advantages of using transient techniques in short duration test facilities for
obtaining both mean and fluctuating heat transfer rates. In particular, the use
of surface heat transfer as a boundary layer diagnostic can now be extended to
100 kHz without compromising the low frequency measurements.

ACKNOWLEDGEMENTS

    The inspiration for this work on analogues arose from a number of cascade
heat-transfer studies funded by the Science and Engineering Research Council and
Rolls-Royce Ltd. whose support and advice is gratefully acknowledged. Thanks
are also due to members of the Oxford University Cascade Group for their
invaluable help and advice. We thank Mr. D. J. Doorly for permission to include
Figure 15.

REFERENCES

1.   Schultz, D. L. and Jones, T. V.   1973.   Heat Transfer Measurements in Short
     Duration Hypersonic Facilities.   A.G.A.R.D.   AG-165.

2.   Wittliff, C. E.   1977.   Space Shuttle Leading Edge Heat Transfer
     Investigation in the CALSPAN 96-in Hypersonic Shock Tunnel.   Int. Cong. on
     Instrumentation in Aerospace Simulation, 77 CHI 251-8AES.

3.   Schultz, D. L., Jones, T. V., Oldfield, M. L. G. and Daniels, L. C.   1979.
     A New Transient Facility for the Measurement of Heat Transfer Rates.

A.G.A.R.D., CP-229, 'High Temperature Problems in Gas Turbine Engines', Ankara.

4. Jones, T. V., Schultz, D. L., Oldfield, M. L. G. and Daniels, L. C. 1978. Measurement of the Heat Transfer Rate to Turbine Blades and NGV's in a Transient Cascade. 6th International Heat Transfer Conference, Toronto, Paper EC-12.

5. Oldfield, M. L. G., Jones, T. V. and Schultz, D. L. 1978. On-line Computer for Transient Turbine Cascade Instrumentation. IEEE Transactions on Aerospace and Electronic Systems. Vol. AES-14, No. 5, pp. 738-749.

6. Dunn, M. G. and Stoddard, F. J. 1978. Measurement of Heat-Transfer Rate to a Gas Turbine Stator. ASME Paper No. 78-GT-119. 1978.

7. Meyer, R. F. 1960. A Heat Flux Meter for use with Thin Film Surface Thermometers, NRC Canada Aero. Rep. LR-279.

8. Meyer, R. F. 1963. Further Comments on Analogue Networks, to Obtain Heat Flux from Surface Temperature Measurements, NRC Canada Aero. Rep. LR-375.

9. Foord, A. 1981. Introduction to Low-Noise Amplifier Design. Wireless World, 71.

10. Abramowitz, M. and Stegun, J. A. 1972. Handbook of Mathematical Functions, Dover.

11. Cody, W. J., Paciorek, K. A. and Thatcher, H. C. Jnr. 1970. Chebyshev Approximations for Dawson's Integral. Math. Comp. 24, 171-178.

APPENDIX - THEORY OF TIME RESPONSE OF AN ANALOGUE

It has already been shown that the response of an ideal continuous analogue to a voltage input of $v = \sqrt{t}$ is a step function of height $i(t) \frac{\sqrt{\pi}}{2}\sqrt{\frac{c}{r}}$.

In the following analysis an expression for the response of any lumped section finite analogue is derived.

From Figure 3 it can be seen that

$$i_i = \frac{v_{i-1} - v_i}{R_i}$$

and $$i_i - i_{i+1} = C_i \frac{dv_i}{dt}$$

from which

$$\frac{dv_i}{dt} = \frac{v_{i-1}}{R_i C_i} - \frac{v_i}{C_i}\left[\frac{1}{R_i} + \frac{1}{R_{i+1}}\right] + \frac{v_{i+1}}{C_i R_{i+1}} \text{ for } i = 1, 2, \ldots N \qquad (45)$$

Note that $R_{N+1} = \infty$.

Equation (45) can be written in matrix form:-

$$\frac{d\underset{\sim}{y}}{dt} = \underset{\sim\sim}{A}\underset{\sim}{v} + \frac{v_o(t)}{R_1 C_1}\underset{\sim}{k}$$ (46)

where

$$\underset{\sim}{v} = \begin{bmatrix} v_1 \\ v_2 \\ \vdots \\ v_N \end{bmatrix}, \qquad \underset{\sim}{k} = \begin{bmatrix} 1 \\ 0 \\ \vdots \\ 0 \end{bmatrix}$$

and

$$\underset{\sim}{A} = \begin{bmatrix} -\dfrac{1}{C_1}\left(\dfrac{1}{R_1} + \dfrac{1}{R_2}\right), & \dfrac{1}{C_1 R_2} & , & 0 \cdots \cdots 0 \\ \dfrac{1}{C_2 R_2}, & -\dfrac{1}{C_2}\left(\dfrac{1}{R_2} + \dfrac{1}{R_3}\right), & \dfrac{1}{C_2 R_3}, & \\ 0 , & & & \\ \vdots & & & \\ 0 & & & -\dfrac{1}{R_N C_N} \end{bmatrix}$$

$\underset{\sim}{A}$ is a non symmetrical tri-diagonal matrix which can be changed into a symmetric tri-diagonal matrix $\underset{\sim}{B}$ by the substitution

$$v_i = \frac{u_i}{\sqrt{C_i}}$$

which changes equation (46) to

$$\frac{d\underset{\sim}{u}}{dt} = \underset{\sim}{B}\,\underset{\sim}{u} + \frac{v_o(t)}{R_1 \sqrt{C_1}}\underset{\sim}{k}$$ (47)

where

$$b_{ii} = -\frac{1}{C_i}\left(\frac{1}{R_i} + \frac{1}{R_{i+1}}\right), \quad (R_{N+1} = \infty)$$

$$b_{i,i+1} = b_{i+1,i} = \frac{1}{R_i \sqrt{C_i C_{i+1}}}$$

and $b_{ij} = 0$ for $|i - j| > 1$

Define a new set of coordinates $\underset{\sim}{y}$ such that

$$\underset{\sim}{u} = \underset{\sim\sim}{Q}\underset{\sim}{y}$$

where $\underset{\sim}{Q}$ is the matrix of the normalised eigenvectors $\underset{\sim}{q}_i$ of $\underset{\sim}{B}$:-

$$Q = [\underset{\sim}{q}_1, \underset{\sim}{q}_2, \cdots \underset{\sim}{q}_N] = [q_{ij}]$$

Substituting into equation (47) and premultiplying by $\underset{\sim}{Q}^{-1}$ gives

$$\underset{\sim}{Q}^{-1}\underset{\sim}{Q} \frac{d\underset{\sim}{y}}{dt} = \underset{\sim}{Q}^{-1}\underset{\sim}{B}\underset{\sim}{Q}\underset{\sim}{y} + \frac{\underset{\sim}{Q}^{-1}\underset{\sim}{k}}{R\sqrt{C}_1} v_o(t) \tag{48}$$

Since $\underset{\sim}{B}$ is a symmetric matrix, its eigenvalues are orthogonal, and

$$\underset{\sim}{Q}^{-1} = \underset{\sim}{Q}^T$$

Hence $\underset{\sim}{Q}^{-1}\underset{\sim}{Q} = \underset{\sim}{I}$ the unit matrix, and

$$\underset{\sim}{Q}^{-1}\underset{\sim}{B}\underset{\sim}{Q} = \underset{\sim}{\Lambda} = \begin{bmatrix} \lambda_1 & 0 & 0 & \cdot & \cdot & \\ 0 & \lambda_2 & 0 & & & \\ 0 & & \lambda_3 & & & \\ \cdot & & & & & \\ \cdot & \cdot & \cdot & \cdot & \cdot & \lambda_N \end{bmatrix}$$

where λ_i are the eigenvalues of $\underset{\sim}{B}$. It turns out that all $\lambda_i < 0$ for real analogues. Also,

$$\underset{\sim}{Q}^{-1}\underset{\sim}{k} = \underset{\sim}{Q}^T\underset{\sim}{k} = \begin{bmatrix} q_{11} \\ q_{12} \\ q_{13} \\ \vdots \\ q_{1N} \end{bmatrix}$$

where q_{1i} is the first element of $\underset{\sim}{q}_i$.

Equation (48) then simplifies to the set of decoupled equations

$$\frac{dy_i}{dt} = \lambda_i y_i + q_{1i} \frac{v_o(t)}{R_1\sqrt{C}_1} , \quad i = 1, \ldots, N \tag{49}$$

Taking Laplace transforms

$$\overline{y}_i(s) = \frac{q_{1i}\overline{v}_o(s)}{(s-\lambda_i)R_1\sqrt{C}_1} \tag{50}$$

In the particular case where the input voltage $v_o(t) = \sqrt{t}$,

$$\overline{v}_o(s) = \frac{\sqrt{\pi}}{2} s^{-\frac{3}{2}}$$

and so

$$\bar{y}_i(s) = \frac{q_{1i}\sqrt{\pi}}{2R_1\sqrt{C_1}} \frac{1}{s^{3/2}(s - \lambda_i)}$$

$$= \frac{q_{1i}\sqrt{\pi}}{2\lambda_i R_1\sqrt{C_1}} \left[\frac{1}{s^{\frac{1}{2}}(s - \lambda_i)} - \frac{1}{s^{3/2}} \right]$$

Inverting the Laplace Transform [10],

$$y_i(t) = \frac{q_{1i}}{\lambda_i R_1\sqrt{C_1}} \left[\frac{\text{Daw}(\sqrt{-\lambda_i}t)}{\sqrt{-\lambda_i}} - \sqrt{t} \right] \tag{51}$$

Where Dawson's Integral [10,11]

$$\text{Daw}(x) = e^{-x^2} \int_0^x e^{t^2} dt \tag{52}$$

Transforming back to the original voltages, the current i, through R_1 can be obtained:-

$$u_i = \sum_j q_{ij} y_j$$

$$v_i = \frac{u_i}{\sqrt{C_i}}$$

and hence $i_1 = \dfrac{v_o(t) - v_1}{R_1}$

or

$$i_1 = \frac{1}{R_1} \left[\sqrt{t}(1 + \frac{1}{R_1 C_1} \sum_j (\frac{q_{1j}^2}{\lambda_j})) - \frac{1}{R_1 C_1} \sum_j (\frac{q_{1j}^2}{\lambda_j} \frac{\text{Daw}\sqrt{-\lambda_j}t}{\sqrt{-\lambda_j}}) \right] \tag{53}$$

Equation (53) can be further simplified by showing that

$$1 + \frac{1}{R_1 C_1} \sum_j (\frac{q_{1j}^2}{\lambda_j}) = 0 :-$$

From the basic eigenvalue equation,

$$\underset{\sim}{B} \underset{\sim}{q}_i = \lambda_i \underset{\sim}{q}_i$$

or $\quad B \dfrac{\underset{\sim}{q}_i}{\lambda_i} = \underset{\sim}{q}_i, \quad i = 1, \ldots N.$

Define a new matrix

$$\underset{\sim}{Z} = \left[\frac{q_{ij}}{\lambda_j} \right]$$

Then $\underset{\sim}{B} \, \underset{\sim}{Z} = \underset{\sim}{Q}$

and $\underset{\sim}{B} \, \underset{\sim}{Z} \, \underset{\sim}{Q}^T = \underset{\sim}{Q} \, \underset{\sim}{Q}^T = \underset{\sim}{I}$.

Note that $(\underset{\sim}{Z} \, \underset{\sim}{Q}^T)_{11} = \sum_j \frac{q_{1j}^2}{\lambda_j}$

This term is extracted by pre-multiplying by

$$\underset{\sim}{F} = \begin{bmatrix} 1, & \sqrt{\dfrac{C_2}{C_1}}, & \sqrt{\dfrac{C_3}{C_1}}, & \cdots \sqrt{\dfrac{C_N}{C_1}} \\[2ex] 0, & 1, & \sqrt{\dfrac{C_3}{C_2}}, & \cdots \sqrt{\dfrac{C_N}{C_2}} \\[2ex] 0, & \ddots & \cdots & \end{bmatrix}$$

Then

$$\underset{\sim}{F} \, \underset{\sim}{B}(\underset{\sim}{Z} \, \underset{\sim}{Q}^T) = \underset{\sim}{F} \, \underset{\sim}{I} = \underset{\sim}{F} \tag{54}$$

But the first row of $\underset{\sim}{FB}$ is $(-\frac{1}{R_1C_1}, 0,0,0,0)$, and so the top left hand element of equation (54) is

$$-\frac{1}{R_1C_1} \sum_j \frac{q_{1j}^2}{\lambda_j} = 1$$

Substituting this into equation (53), the response of an analogue to a voltage input $v(t) = \sqrt{t}$ is

$$i(t) = \frac{1}{C_1 R_1^2} \sum_{j=1}^{N} \frac{q_{1j}^2}{(-\lambda_j)} \frac{\text{Daw}(\sqrt{-\lambda_j} \, t)}{\sqrt{-\lambda_j}} \tag{55}$$

Standard computing library routines are used to find the eigenvectors and eigenvalues of the symmetrical tri-diagonal matrix $\underset{\sim}{B}$, and Dawson's integral, equation (52), is calculated by using a Chebyshev series approximation [11].

A Composite Constant Heat Flux Test Surface

G. J. DiELSI and R. E. MAYLE
Department of Mechanical Engineering
Rensselaer Polytechnic Institute
Troy, N.Y., USA

ABSTRACT

A newly developed method for the construction of a constant heat flux test surface is described. The surface is a stainless steel, fiberglass-epoxy composite which is electrically heated. Error estimates are provided for a typical surface and the results of a demonstration test presented.

1. INTRODUCTION

The widespread use of finite-difference boundary-layer calculation programs has greatly reduced the need to conduct heat transfer tests at correct wall-to-mainstream temperature ratios. The main reason is that most programs automatically take into account fluid property variations with temperature. Hence, in many situations, the experimenter may conduct his test at reduced temperatures and temperature differences, compare his data with the calculations at the reduced levels, modify the program's physics in accordance with experiment, and extrapolate to the elevated temperatures using the newly modified program. In this way, for example, the effects of curvature, pressure gradient, separation and reattachment and free-stream turbulence on flow over a gas turbine airfoil may be investigated near room temperature and extrapolated to engine conditions.

Low-temperature, small-temperature-difference test surfaces for measuring convective heat transfer to a fluid vary widely in design and construction, e.g. [1] to [5], but, in general, all fall into one of two categories - either a constant temperature surface or a constant heat flux surface. As a consequence, for steady state experiments, the convective heat transfer coefficient is determined by measuring either the distribution of power from the surface in the former situation or the surface's temperature distribution in the latter. In either case, the convective heat transfer coefficient is determined from

$$h \equiv \frac{q}{T_o - T_\infty}$$

G. J. DiElsi is presently at Chevron, San Francisco.

where q is the heat flux (power per unit area) from the surface to the fluid, and T_0 and T_∞ are the surface and free-stream fluid temperatures, respectively.

In this paper, the construction of a nominally constant heat flux surface is described which uses foil strips for heating and thermocouples to measure the surface temperatures. Although foil heaters are commonly employed in heat transfer tests, e.g. [1] and [5], and the measuring technique is commonplace, the fabrication is unique and produces a test surface having some advantages over others of its type; the most notable being durability and ease of construction (once the technique is learned). Following the description of the surface and fabrication technique, a demonstration test is described and its results compared to theory.

2. TEST SURFACE DESCRIPTION

A section of the new heat transfer surface is shown in Fig.1. It consists of a number of thin stainless steel heater strips placed side by side and sandwiched between two layers of fiberglass and epoxy resin bonded on a balsa wood substrate. The balsa substrate not only forms the structural portion of the surface, but also reduces the heat loss by conduction through the surface. The individual metal strips are connected in series by heavy copper bus bars buried in the substrate, and the whole surface heated by passing a known electric current through the circuit. Small thermocouples, bonded between the strips and balsa substrate, are used to measure the surface temperature distribution. Since they are bonded at a constant and measureable distance from the strips and surface, corrections to ascertain the surface temperature are easily made. Their leads run in opposite directions along the surface for a short distance to minimize heat loss through them, and then through holes drilled in the substrate. Thermocouples are also attached to the back of the balsa at various locations in order to determine that portion of heat dissipated in the strips which is conducted through the substrate.

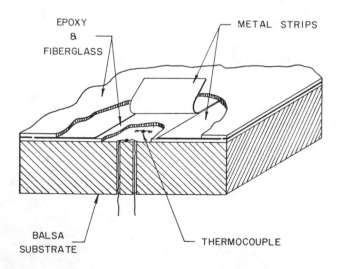

EPOXY
&
FIBERGLASS   METAL STRIPS

BALSA
SUBSTRATE   THERMOCOUPLE

Figure 1   Heat Transfer Test Surface

For the demonstration model, balsa wood sheets were glued to shaped wooden ribs, in model airplane fashion, to provide a 12.5 mm thick substrate. Once this was completed, 0.08 mm bead thermocouples were carefully attached using Superglue and the surface sealed with a light coat of lacquer in preparation for bonding on the heater sheet. The metal strips were 0.025 mm thick by 5.1 cm wide and were placed about 0.5 mm apart. They had a resistance of about 0.54 ohms/m and a temperature coefficient of resistivity of $8(10)^{-4}$ ohms/$^\circ$C. As a result of the latter, the variation in the strip resistance with temperature, and hence the generated heat flux variation, was small. For the demonstration test this amounted to a fraction of a percent, while in other situations a variation of 2 percent has been realized. Also, care was taken to use strips from the same roll and from the same area within the roll in order to minimize thickness variations which again could affect the uniformity of the generated heat flux. Micrometer measurements along a 30 m length showed no detectable thickness variation. The whole epoxy-fiberglass and metal strip layer was about 0.30 mm thick and could withstand temperatures up to about 70°C. Its thickness was found to be uniform and always equal to the sum of the fiberglass sheets and metal strip thicknesses. This feature facilitated the surface temperature calculation. The heated surface area consisted of thirteen strips and was about 67 cm wide by 90 cm long. Surfaces up to 2 m long have been constructed. Since a smooth test surface is naturally produced by the glueing process, only a finish coat of black paint with a known emissivity was applied to complete the model.

3. FABRICATION PROCEDURE

The basic manufacturing process is to first glue the metallic strips to a fiberglass layer using an epoxy resin, and then bond the resulting sheet to the prepared substrate using fiberglass and epoxy. In each step, a bag is formed around the items to be glued and a vacuum drawn within the bag through an absorbent cloth. This removes entrapped air and excess resin and produces layers of uniform thickness without either wrinkles or voids.

The materials used in fabricating the surface are listed in the accompanying table. These were selected as the best materials for the purpose only after much experimentation. The two most critical materials in the process are the glass fabric and the epoxy resin. The chosen fabric is epoxy compatible and reasonably thick to prevent both excessive stretching when wet with epoxy and tearing between the steel strips when dry. The Volan A treatment allows the glass to stretch over compound curves. The two-part epoxy has a low viscosity to facilitate flow during the evacuation stages of the process, producing as mentioned before, an even layer. This epoxy also retains its adhesive properties at temperatures higher than most other epoxies tested, allowing higher surface temperatures to be achieved.

Two materials, the bleeder cloth and porous release cloth, are necessary for extracting the excess epoxy and air from the fiberglass matrix. The release cloth is placed between the bleeder cloth and the layer to be bonded. The vacuum bag material was chosen because its elasticity and strength allowed it to stretch and fit smoothly around compound curves without either wrinkling or breaking. Lastly, it should be noted that the metal strips were ordered such that the burr from the cutting process ended up on the same side for each edge and arranged in the fabrication process such that the smooth surface faced outward.

TABLE OF MATERIALS FOR SURFACE MANUFACTURE

Materials	Designation
Heater Strips	Type 304 Annealed Stainless Steel .001" × 2" wide
Epoxy	Eccobond 24, Parts A and B
Fiberglass	Style 112 Volan A (or I-617) Treatment, 38" width
Porous Release	Style 100-1/5-104 Porous Teflon-Coated Fiberglass
Bleeder	Nexus Industrial Style 400 03048
Rigid Foam	Polyvinyl Chloride (PVC), 1/8" thick
Vacuum Bag	Polyvinyl Alcohol (PVA) Vacuum Film, .004" thick × 45" wide
Vacuum Bag Sealant	5126 Vacuum Bag Sealant, 1/2" × 1/8" × 30' rolls
Glueing Surface	High Density Polyethylene, 1/2" × 4" × 8'

In preparation for the first step of the glueing process, making the heater sheet, the metal strips are degreased by washing them with trichloroethylene followed by a distilled water rinse. After they are dry, pieces of masking tape are placed on their ends where the bus bar connections are to be made. Once the bonding process is completed, the tape is removed leaving the bare metal area for electrical contact. Next, a single piece of glass is cut to the correct length but about 5 cm wider than the heated area requirements. The teflon release cloth is cut much larger than the heated area and two layers of bleeder cloth are cut slightly smaller than the teflon in both dimensions. The teflon and bleeder both extended at least 10 cm beyond the strips in length; a vacuum is drawn through the bleeder from these ends later in the process. The heater sheet is constructed on a thick, high-density polyethylene sheet which not only provides a smooth flat glueing surface, but allows for release of the assembly after fabrication. This sheet is thoroughly cleaned with acetone, and an area approximately 5 cm larger than the bleeder layer in both dimensions is outlined on the table with the vacuum bag sealant. The paper cover on the sealant is not yet removed. A sheet of PVA vacuum film is cut to cover the outlined area.

With all the preparations completed, the heater sheet is now laid up. The epoxy is weighed and mixed. This and all succeeding steps are to be performed using vinyl gloves and a protective breathing apparatus. The glue is liberally brushed over the polyethylene surface covering an area equal to that of the glass fabric. The glass is carefully placed over the epoxy and the wrinkles worked out by hand. More glue is brushed over the glass to ensure complete

coverage and then rolled with a brayer to an even thickness. The strips are
then placed on the epoxy-saturated glass individually with their taped ends
extending beyond the glass and rolled along their length from the middle to
eliminate air pockets and excess epoxy. Each strip is laid parallel to and
about 0.5 mm from its neighbor.

Once all the strips are placed on the glass and correctly located, the
teflon sheet is spread over them and the bleeder cloth placed over the teflon.
A 10 cm square of PVC porous foam is placed on each end of the bleeder without
overlapping the strips and the lower half of the vacuum connectors, shown in
Fig.2, are positioned on each square. The paper is then removed from the
sealant and the vacuum bag formed by pressing the vacuum film to the sealant.
Using a razor knife, a small hole is cut in the film over the threaded hole in
each lower connector, and the upper portion of the connector screwed down
through the hole to seal against the material. A vacuum pump is connected to
each connector and the pump started. The pump is run for about three hours,
after which it is shut off and the vacuum bag removed to allow air to circulate
around the sheet.

The finished heater sheet is not removed for approximately forty-eight
hours to ensure complete curing. After removing the lower vacuum connectors,
the bleeder cloth is peeled off and discarded. The teflon is then slowly
removed from the back of the sheet; the teflon is reusable and is therefore
saved. Grasping the full width of one edge, the completed heater sheet is then
peeled from the polyethylene surface.

The remainder of the fabrication process involved bonding the heater sheet
to the model. This is done as just described except that the polyethylene
sheet is not used and all work is performed directly on the instrumented model.
Care must be taken at this time to remove all dirt and dust from the model,
otherwise surface bumps will arise and heater sheet separation can occur.
Again, a glass fabric sheet is cut to cover the heated area and larger teflon
and bleeder cloths are cut. In this case, however, it was found convenient to
tape all of the layers including the heater sheet along a common edge, one on
top of the other and in order, on the model to ensure a proper alignment during

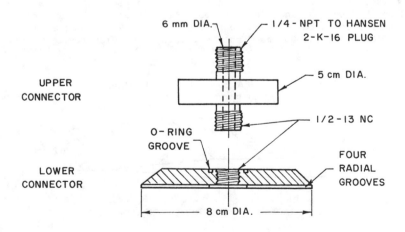

Figure 2   Vacuum Connector

the laying-up process. If the heated area is smaller than the surface, the
vacuum sealant may be attached directly to the surface; otherwise a vacuum bag
must be formed to enclose the model. In the latter case, the model obviously
must be constructed so as not to implode when the vacuum is applied.

As before, the vacuum is maintained for three hours after which the bag is
removed and the surface left to cure. The teflon and bleeder layers are then
peeled off leaving the finished surface. If too little resin was used in the
sheet construction, small holes will be visible in the final surface when
viewed under a microscope. These may be filled with epoxy by using a squeegee
and sanded smooth with crocous cloth before applying the finish coat of high
emissivity paint.

4. DEMONSTRATION TEST

The model used in the demonstration test is shown in Fig.3. It was 76 cm
wide, about 2.2 m long, and had a cylindrical leading edge with a 7.62 cm
radius. The top and bottom surfaces were parallel to one another for the first
1.1 meters and then tapered to form a trailing edge. It was made mostly of
wood. The heated area extended from the lower edge of the cylindrical leading
edge, around the front, and back about 60 cm along the top surface. The heater
strips ran lengthwise within this area as shown. Thermocouples were bonded
beneath the central strip along its length and across the model's span under
both the same strip and its neighbors at several locations. For this test,
measurements were taken only on the leading edge portion of the model. In the
heated region, thermocouples were also attached to the back of the balsa wood
substrate, as mentioned before, and commercial fiberglass insulation filled the
internal cavity of the model. In addition, pressure taps were located between
the two outermost heater strips on each side of the heated surface at various
distances from the leading edge. All of the thermocouple leads, pressure tap
tubing and power leads were led out a hole in the model's side and through the
sidewall of the wind tunnel.

The model was centrally located in a 46 cm by 76 cm wind tunnel such that
its top and bottom surfaces were parallel to the flow. Pressure taps on the
top and bottom surfaces were used to align the model so that stagnation
occurred on the leading edge along the model's midplane. The incident velocity
was uniform and could be adjusted to provide nominal Reynolds numbers of
80,000, 100,000 and 120,000 based on the leading edge diameter. The maximum

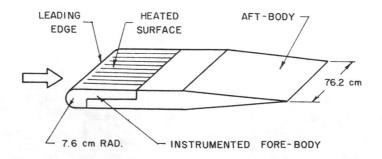

Figure 3 Model for Leading Edge Experiment

incident velocity was about 11.5 m/s. The free-stream turbulence measured at
the model's leading edge position but without the model in place was about
0.3 percent with all frequencies below 50 Hz filtered out. Throughout the test,
the free-stream temperature was maintained at about $16°C$. The generated heat
flux was about 670 w/m^2 which produced surface temperatures ranging from 27 to
$32°C$.

The test procedure was rather simple, in that once the flow was established, the power was turned on to the heater strips and a number of thermocouples monitored to assure that the surface did not overheat. After about
eight hours, once steady state was reached, the surface temperatures (refer-
enced to the free-stream temperature), free-stream temperature and power were
recorded.

The steady-state heat transfer boundary condition on the test model was
nominally that of constant heat flux. A slight modification of this condition,
however, was caused by surface radiation, heat conduction through the substrate, and conduction in the metal strips parallel to the surface. For this
test, these modifications amounted to less than one percent, however, variations in the convective heat flux of 10 - 15 percent can be realized in other
situations. In addition, the thermocouple junction and leads act as a heat
sink which resulted in measuring a lower temperature than would exist without
the thermocouple. These biased effects, along with local variations in strip
resistance resulting from local metal temperature differences, were included in
the calculation of heat transfer coefficients using

$$q = q_{gen} - q_{rad} - q_{cond}$$

and

$$h = q/(T_m - T_\infty - \Delta T) .$$

The approximate magnitudes of the correction are as follows: radiation,
thirteen percent of the generated heat flux; heat loss through the substrate,
one and one-half percent; heat conduction within the heater strips, less than
one-tenth of a percent; thermocouple temperature correction ΔT, about one-half
percent of the indicated wall-free-stream temperature difference. An uncertainty analysis following Kline and McClintock [6] was carried out after
accounting for the above biased effects. Uncertainty intervals for each of the
variables affecting the heat transfer coefficient were based on 20 to 1 odds
that the true value lay within the interval. This analysis showed the reported
heat transfer coefficients are certain to within 2.5 percent based on 20 to 1
odds.

5. RESULTS

The results of the test are shown in Fig.4 where the average Nusselt
number, scaled by the square root of the Reynolds number, for the three mainstream velocities is plotted as a function of the angular distance in radians
from stagnation. Although the Nusselt number was found to increase with
increasing Reynolds number, this only amounted to a total variation of about
four percent. At zero and x/R = 1 the average of the results from five spanwise
measurements were used to determine the Nusselt number at those locations. In
each case none of the spanwise results differed by more than one percent. Also
presented in the figure are Froessling's [7] solution for a constant temperature surface and his solution modified for a constant heat flux surface using

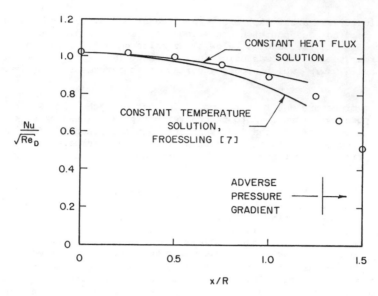

Figure 4   Leading Edge Results and Comparison to Theory

Lighthill's [8] solution for step changes in surface temperature together with
Tribus and Klein's [9] superposition technique. According to Spalding and
Pun [10], this should be accurate to an x/R of about 0.9. The free-stream
velocity distribution used in these calculations was obtained from the static
pressure measurements and is well represented in the accelerated region by

$$\frac{U}{U_\infty} = 2.10 \left(\frac{x}{R}\right) - 0.337 \left(\frac{x}{R}\right)^3 - 0.0228 \left(\frac{x}{R}\right)^5$$

where U_∞ is the incident velocity.

Within the limit of the theory, agreement is seen to be excellent. Beyond
this point, and as should be expected, the data falls consistently lower than
the theory as the mainstream velocity approaches its maximum at x/R = 1.3 and
the flow begins to decelerate towards separation. For this test body, separation occurs at an x/R between 1.5 and 1.6.

6. CONCLUDING REMARKS

A procedure has been developed for constructing a composite, constant
heat-flux test surface which is believed to be more durable than the usual
glued on metal strip surfaces. So far, even after numerous test hours and
much handling in the course of moving the model in and out of the tunnel
several times, no damage or delamination of the surface has occurred and no
thermocouple failures have taken place.

Although thermocouples were used on the present model, the surface temperatures could be measured by either infrared or liquid crystal techniques.
Each makes the fabrication procedure much simpler and less time consuming since

the major effort is usually expended in attaching and wiring the thermocouples. In this respect, the combination of the present fabrication technique with the recent developments made in producing encapsulated liquid crystal sheets is quite appealing. Further information on the liquid crystal technique may be found in Hippensteele, Russel and Stepka's [11] report.

NOMENCLATURE

h = convective heat transfer coefficient

Nu = Nusselt number based on the leading edge diameter

q = convective heat flux

q_{cond} = heat loss by conduction per unit area per unit time

q_{gen} = heat generated by metal strips per unit area per unit time

q_{rad} = heat loss by radiation per unit area per unit time

R = leading edge radius

Re_D = Reynolds number based on the incident velocity and leading edge diameter

T_o = surface temperature

T_∞ = free-stream temperature

T_m = measured surface temperature

U = local free-stream velocity

U_∞ = incident velocity

x = coordinate along the surface measured from stagnation

REFERENCES

1. Seban, R.A. and Doughty, D.L., "Heat Transfer to Turbulent Boundary Layers with Variable Free-Stream Velocity," ASME Trans., 78, 217 (1956).

2. Reynolds, W.C., Kays, W.M. and Kline, S.J., NASA Memo 12-1-58W, Washington, DC, 1958.

3. Hartnett, J.P., Birkebak, R.C. and Eckert, E.R.G., "Velocity Distributions, Temperature Distributions, Effectiveness and Heat Transfer for Air Injected Through a Tangential Slot into a Turbulent Boundary Layer," Journal of Heat Transfer, 83, 293-306 (1961).

4. Mayle, R.E., Blair, M.F. and Kopper, F.C., "Turbulent Boundary Layer Heat Transfer on Curved Surfaces," Journal of Heat Transfer, 101, 521-525 (1979).

5. Graziani, R.A., Blair, M.F., Taylor, J.R. and Mayle, R.E., "An Experimental Study of Endwall and Airfoil Surface Heat Transfer in a Large Scale Turbine Blade Cascade," Jour. of Engrg. for Power, 102, 257-267 (1980).

6. Kline, S.J. and McClintock, F.A., "Describing Uncertainties in Single Sample Experiments," Mechanical Engineering, 3-8 (January 1953).

7. Froessling, N., "Evaporation, Heat Transfer and Velocity Distribution in Two-Dimensional and Rotationally Symmetrical Laminar Boundary Layer Flow," NACA TM 1432.

8. Lighthill, M.J., "Contributions to the Theory of Heat Transfer Through a Laminar Boundary Layer," Proc. Royal Soc., A202, 369-377 (1950).

9. Tribus, M. and Klein, J., "Forced Convection from Nonisothermal Surfaces," Heat Transfer Symposium, Engineering Research Institute, University of Michigan, 1952, pp.211-235.

10. Spalding, D.B. and Pun, W.M., "A Review of Methods for Predicting Heat-Transfer Coefficients for Laminar Uniform-Property Boundary Layer Flows," Int. Jour. of Heat and Mass Transfer, 5, 239-250 (1962).

11. Hippensteele, S.A., Russell, L.M. and Stepka, F.A., "Evaluation of a Method for Heat Transfer Measurements and Thermal Visualization Using a Composite of a Heater Element and Liquid Crystals," NASA TM 81639, 1981.

General Analysis of the Operation of a Sampling Probe for Velocity and Concentration Measurements in Two-Phase Flows

GIANNI BENVENUTO and **MICHELE TROILO**
Facoltà di Ingegneria
Istituto di Macchine
Università di Genova, Italy

ABSTRACT

A potential flow model has been used for the theoretical prediction of the gas velocity field around a sampling probe. Results given by a proper numerical solution have been compared in a limit condition with those obtained from experimental measurements, showing a good agreement, in view of the number of simplifying hypotheses assumed and of the uncertainties due to the small scale experimental situation.

The isokinetic calibration of a specimen probe, here reported, puts into evidence the practical difficulties to achieve isokinetic operation; then calibration curves for concentration ratios in anisokinetic operation have to be used, given here as obtained from particle trajectory calculation.

Finally, the capability of anisokinetic operation to give more informations, if the particle velocity near the probe intake is also measured, is emphasized, on the base of the particle approach velocity analysis.

1. INTRODUCTION

The experimental investigation of turbomachinery operating in two-phase flows is concerned, among other things, with the dispersed phase concentration measurement. In this report attention is limited to those flows in which the dispersed phase is either solid or liquid, appearing as particles, which are currently supposed to be spherical; typical examples are the last stages of steam turbines operating in wetness condition, and turbocompressors operating in dusty environment.

A classical experimental tool for such flows is a sampling probe, connected to a concentration measuring device; the problem in this case is to obtain a sample representative of the main flow. This can be accomplished both by a proper choice of the sampling point inside the flow, where the particle distribution were uniform, and by a probe design and operation that cause a flow disturbance as low as possible. Due to the fact that, around an obstacle, particles do not follow the gas flow except for very low inertia parameter, the sampling probe should operate in such a way to keep the incoming flow undisturbed, allowing the same gas velocity to take place as in the free flow without the probe. This condition is referred to as "isokinetic" one.

269

In practical applications, two major difficulties may be encountered:
i) if the flow pressure level is well under the atmospheric one (e.g. in low pressure steam turbines), isokinetic conditions are possible at the expense of a large vacuum system;
ii) in every flow condition, and especially for low Mach numbers and for small probes, isokinetic operation control requires a careful reading of very low differential pressures, and a fine adjustement of the suction system according to an already available probe calibration map.

Anisokinetic operation of sampling probes seems to be of practical interest, according to the aforementioned arguments, provided that the particle concentration of the sample could be related to that of the main flow for a given operating condition of the probe. The problem has been already studied in the past from a theoretical point of view (see e.g. [1],[2]) and more recently in experimental way ([3]); in the present work the errors due to anisokinetic operation of a sampling probe are theoretically calculated with a different numerical method, and an experimental confirmation has been found by means of laser anemometry measurements near the probe mouth; the analysis has put into evidence other capabilities of the sampling technique as regards the particle size evaluation.

2. BASIC NOTATIONS

Consider in fig. 1 the typical case of an under-aspirated sampling probe, i.e. with the suction velocity V_{asp} lower than the free-stream velocity V_{∞} of the uniform flow. Let the "limiting streamline" be a streamline pertaining to the stream tube whose the probe inner wall is the extension; the limiting streamline will be unique for axisymmetric probes. Let the "limiting trajectory" be a trajectory of a particle that hits the probe entrance lip. In axisymmetric flow condition, the limiting trajectory defines a "particle stream tube", of those particles of equal size entering the probe. Consider the free-stream location of the limiting streamline (fig. 1) and let the gas and particle streams

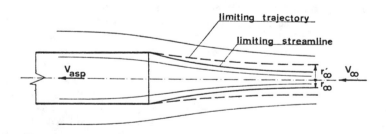

Fig. 1 - Sampling probe in typical under-aspirated condition.

have the same velocity in the undisturbed flow; then the capture efficiency is:

$$E = (r'_{oo}/r_S)^2 \tag{1}$$

The concentration ratio between the concentration c_S in the sample and the true concentration c_{oo} in the flow is then:

$$c_S/c_{oo} = E\, V_{oo}/V_{asp} \tag{2}$$

As may be shown from similarity laws, the quantities defined in (1) and (2), for incompressible flows, are depending upon two dimensionless characteristic numbers, i.e. the Stokes number

$$St = \rho_p d_p^2 V_{oo}/18\mu_g d_S \tag{3}$$

and the particle Reynolds number:

$$Re = \rho_g d_p V_{oo}/\mu_g \tag{4}$$

The Stokes number may be also regarded as the ratio between the particle relaxation time:

$$\tau_p = \rho_p d_p^2/18\mu_g \tag{5}$$

and the stream characteristic time near the obstacle:

$$\tau_S = V_{oo}/d_S \tag{6}$$

3. THEORETICAL ANALYSIS

3.1. Gas Flow Field Calculation around the Probe

The incompressible, inviscid, axisymmetric flow field around a cylindrical, circular cross-section sampling probe has been calculated by a finite-difference numerical technique ([1]), applied to the stream function equation:

$$\frac{\partial^2 \psi}{\partial z^2} - \frac{1}{r}\frac{\partial \psi}{\partial r} + \frac{\partial^2 \psi}{\partial r^2} = 0 \tag{7}$$

The probe wall is supposed to be infinitely thin, so it can be regarded as a streamline; furthermore the inlet stagnation point is supposed to be always located on the inlet probe contour. The boundary conditions may then be written as follows (fig. 2, [1]):

$\psi = 0$ on DA $\qquad\qquad \psi = -(V_{-oo}(r^2-r_S^2) + V_{asp} r_S^2)/2$ on CE

$\psi = -V_{oo} r^2/2$ on AB $\qquad \psi = -V_{asp} r^2/2$ on ED

$\psi = -V_{oo} r_B^2/2$ on BC $\qquad \psi = -V_{asp} r_S^2/2$ on EF

This mathematical model is likely inadequate to describe correctly those situations where the viscous effects are important, say the near-probe field in very low aspiration conditions; otherwise the hypothesis of infinitely thin wall, and the actual presence of boundary layers and separations make the model anyway approximated.

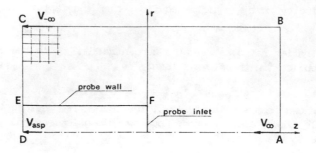

Fig. 2 - Gas flow field computational mesh.

Results obtained are presented, for two suction rates V_{asp}/V_{oo} in figs.3,4, where the solid lines are the gas flow streamlines. Results are in agreement with those obtained in [1], with a different numerical tchnique.

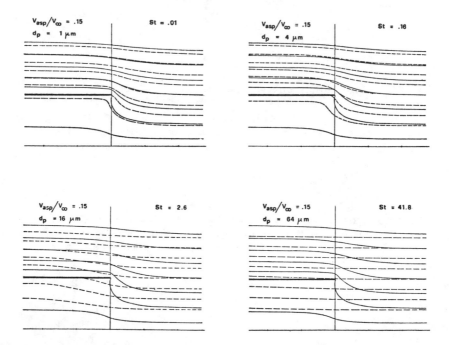

Fig. 3 - Gas streamlines and particle trajectories for under-aspirated probe.

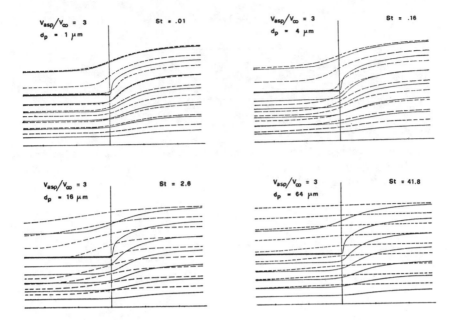

Fig. 4 - Gas streamlines and particle trajectories for over-aspirated probe.

3.2. Particle Trajectory Calculation

Particle trajectories have been calculated,following the "single particle" scheme -i.e. neglecting the gas flow changes due to the particles- and retaining as relevant the aerodynamic force only ([5]). The motion equations in the actual axisymmetric case are written as:

$$\frac{dV_{pz}}{dt} = Q \sqrt{(V_z - V_{pz})^2 + (V_r - V_{pr})^2} \; (V_z - V_{pz})$$

$$\frac{dV_{pr}}{dt} = Q \sqrt{(V_z - V_{pz})^2 + (V_r - V_{pr})^2} \; (V_r - V_{pr})$$

$$(8)$$

being

$$Q = \frac{3 \, \rho_g \, c_D}{4 \, \rho_p \, d_p}$$

$$(9)$$

Substituting for the Stokes number (3) and Reynolds number (4) one has:

$$Q = \frac{c_D \, Re}{24 \, d_S \, St}$$

$$(10)$$

Introducing the non-dimensional time, velocity and lengths:

$t' = t \, V_{oo}/d_S$; $V' = V/V_{oo}$; $z' = z/d_S$; $r' = r/d_S$; the motion equations can be written in non-dimensional form:

$$\frac{dV'_{pz}}{dt'} = \frac{c_D Re}{24 \, St} \sqrt{(V'_z - V'_{pz})^2 + (V'_r - V'_{pr})^2} \; (V'_z - V'_{pz})$$

$$\frac{dV'_{pr}}{dt'} = \frac{c_D Re}{24 \, St} \sqrt{(V'_z - V'_{pz})^2 + (V'_r - V'_{pr})^2} \; (V'_r - V'_{pr})$$

(11)

These equations confirm the St and Re dependence of the trajectories, and hence of the capture efficiency and concentration ratio, as stated in chapter 2. If the simplified drag correlation given by Stokes law is used:

$$c_D = \frac{24 \, \mu_g}{\rho_g d_p |\bar{V} - \bar{V}_p|}$$

(12)

then the system (11) sounds as:

$$\frac{dV'_{pz}}{dt'} = \frac{1}{St} \, (V'_z - V'_{pz})$$

$$\frac{dV'_{pr}}{dt'} = \frac{1}{St} \, (V'_r - V'_{pr})$$

(13)

So, if (12) were verified, particle trajectories should depend only on St number and V_{asp}/V_{oo}, which determines the gas flow field. Taking into account the number of simplifying hypotheses already assumed, the small error introduced by (12) can be accepted, as compared with the presumable model unadequacy.

Some trajectories, calculated by means of the aforementioned equations, nemerically integrated, are plotted in figs. 3,4 as dotted lines. They are relative to particles of selected diameters, and of ρ_p = 1000 kg/m^3 (water), suspended in a stream of air at standard conditions as far as viscosity is concerned (μ_g = 1.814·10^{-5} Pa·s); the pertinent Stokes number is indicated on the figs.

3.3. Capture Efficiency and Concentration Ratio Calculation

On the basis of the notations already introduced, the capture efficiency and concentration ratio have been calculated, searching for the limiting trajectory by an iterative procedure, starting far upstream of the probe inlet, as permitted by the mesh used in solving the gas flow field. Then the capture efficiency E has been evaluated by (1) and concentration ratio by (2), and the results are plotted in figs. 5,6. As expected, the capture efficiency tends towards the unit value as long as St increases, and towards V_{asp}/V_{oo} for decreasing St. Practically, for .15 < V_{asp}/V_{oo} < 1.5, and St > 10, one has E = 1. ± .05; on the other hand, the limiting value for low St, can be reached within a small error only for St values of poor practical interest. For a given Stokes number, the almost linear dependence (fig. 5b) of E on V_{asp}/V_{oo} is an interesting feature for calibration purposes, since it allows the curve (E, V_{asp}/V_{oo}) to be drawn with

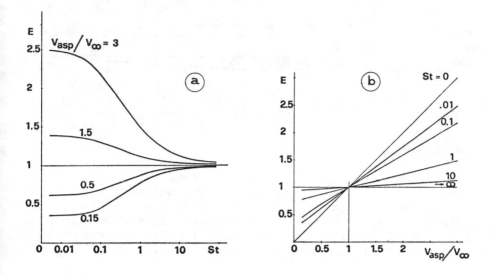

Fig. 5 - Capture efficiency in function of Stokes number and suction rate.

only one experimental point. Another interesting feature appears in fig. 5a, for
$.1 < St < 1.$ and $V_{asp}/V_{oo} \neq 1$: in this range the strong dependence of E from St
allows St - and hence the particle diameter - to be determined by an experimental
measurement of E and V_{asp}/V_{oo}. Remembering (2), the above remarks can be extended with proper changes to the concentration dependence on St and V_{asp}/V_{oo}.
In particular, as shown in fig. 6b, it is interesting to note that an under-aspirated sampling probe gives flow samples where the particle concentration is magni-
fied, and it could be useful for measurements in very low concentration flows.

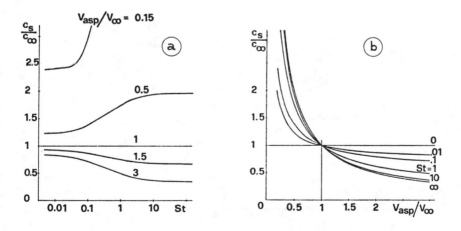

Fig. 6 - Concentration ratio in function of Stokes number and suction rate.

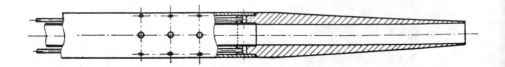

Fig. 7 - The sampling probe head.

4. EXPERIMENTAL ANALYSIS

The experimental analysis had two main purposes:
i) to verify the practical possibilities of an isokinetic operation
ii) to obtain at least a partial confirmation of the theoretical model.
The probe used for the investigation was basically an NPL standard pitot-static tube with tapered nose ([6]), modified with additional pressure taps to detect the internal static pressure, as shown in fig. 7.

4.1. Isokinetic Calibration

The probe, connected to a vacuum pump through a flow rate meter, was positioned and aligned in the flow direction at the outlet of a calibration wind tunnel. For a set of free-stream velocities, the aspirated flow rate and internal external pressure differences $-\Delta p_{ie}$ - have been measured. Results are plotted in fig. 8, as suction velocity V_{asp} in function of Δp_{ie}, for a given flow total pressure. On the same plot, the dotted line represents the isokinetic condition,

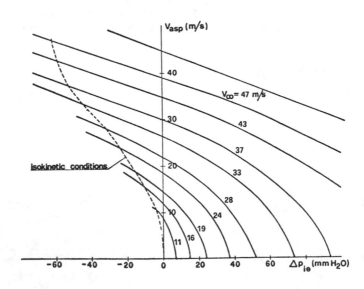

Fig. 8 - Suction velocity calibration curves of the tested probe.

for it crosses the constant velocity lines in points where one has $V_{asp} = V_{oo}$, i.e. an aspirated mass flow rate equal to that could enter the probe if the flow were undisturbed. The calibration curve thus obtained may be used for the experimental measurement of V_{asp}, which will be useful in anisokinetic operation, as shown later on. Except for a specific design, isokinetic operation cannot be guaranteed, as in the present case, by vanishing Δp_{ie}; as previously stated, a calibration curve would then be accounted for during a probe exploitation, to obtain sample concentration equal to that in the flow. Moreover, for the lowest free-stream velocities, small errors in Δp_{ie} readings, would give rise to large errors in the sampled mass evaluation.

4.2. Flow Field Investigation

By means of an L2F laser anemometer ([7]), expressly set up for the purpose, the velocity has been measured upstream of the probe along its axis, for a set of free-stream and suction velocities, using natural laboratory dust particles as tracers. Comparison has been made with the values predicted by the analytical model previously outlined, for the case $V_{asp} = 0$, i.e. with the probe shut; these results are presented in fig. 9. A very high spatial resolution can be achieved in this case by the use of laser anemometer, because its non-intrusive character becomes crucial in the presence of large velocity gradients.

Minor deviations between observed and predicted values can be ascribed to both small air stream fluctuations and velocity lag between dust particles and gas flow. In author's opinion, the agreement is good enough to validate the theoretical model, taking into account that the stagnation condition is the worst one both from the theoretical and the experimental point of view.

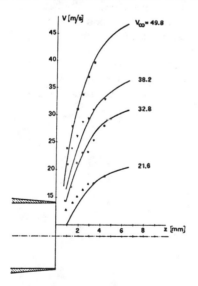

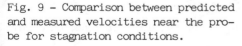

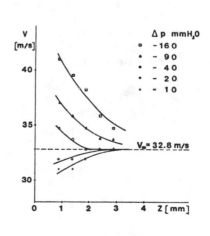

Fig. 9 – Comparison between predicted and measured velocities near the probe for stagnation conditions.

Fig. 10 – Particle approach velocities for different suction rates.

The velocities measured along the probe axis at various suction rates are presented in figs. 10,11 for a selected value of V_{oo}. These results confirm in a different way the practical difficulties to achieve isokinetic operation. Indeed, even small dust particles such that examined here, maintain the free-stream velocity up to a short distance from the probe inlet, in a too wide range of suction rates, for it could be a practical isokinetic control means. On the contrary, a particle velocity measurement near the probe inlet, can give useful informations if sufficient anisokinetic operation is provided for, as explained in the following paragraph.

5. ANISOKINETIC OPERATION OF SAMPLING PROBES

Turning back to the figs. 5,6, it appears that their use for the purpose to correct data obtained in anisokinetic operation of the probe, is possible only if the particle size is known. On the basis of the experimental observations here reported, a way to overcome the lack of knowledge about particle size, could be the establishment of a relation between the capture efficiency and the particle velocity V_{ps} on the probe axis, near the inlet. Neglecting the radial component of the particle velocity, one has:

$$V_{ps} d_S^2/4 = V_{oo} r_{oo}'^2 \tag{14}$$

that gives:

$$E = V_{ps}/V_{oo} \tag{15}$$

It could be reasonably expected that the true dependence of E on V_{ps}/V_{oo} were not far from (15), except for very low Stokes numbers.

For the purpose of testing this hypothesis, calculations have been made of the particle velocities V_{ps} at $z = 0.1 d_S$ upstream of the probe inlet, with the equations already described, for selected suction velocities and particle sizes. The corresponding E values have been plotted (fig. 12) in function of V_{ps}/V_{oo} at a given V_{asp}/V_{oo}. As can be seen, the determination of V_{ps}/V_{oo} could permit a reliable evaluation of E, when $V_{asp}/V_{oo} \neq 1$ (only under-aspirated con-

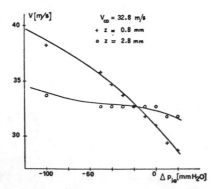

Fig.11 – Particle velocities in points near the probe inlet, in function of the suction rate.

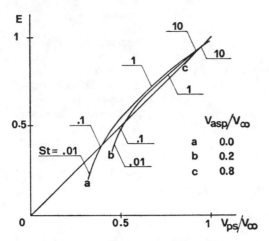

Fig. 12 - Capture efficiency in function of the particle velocity at the probe inlet.

ditions have been considered here); advantage could be taken of that, in view of an experimental evaluation of E.

As already suggested, the knowledge of E permits the particle size evaluation, by means of the plot in fig. 5a, likewise particle diameters may be shown on the curves in fig. 12, as the corresponding St numbers.

6. CONCLUSIONS

A theoretical analysis of a sampling probe for gas-particle flow concentration measurements, confirmed by experimental results, led to some observations concerning the anisokinetic operation:
i) under-aspirated anisokinetic operation could be in practice more handy then isokinetic one, since reliable corrections can be applied by means of calculated characteristics;
ii) the capture efficiency knowledge permits, in a wide range of Stokes numbers, the evaluation of the particle size;
iii) low concentration flows can take advantage of the fast rise in sample concentration in under-aspirated conditions;
iv) capture efficiency is well related with the particle approach velocity near the probe inlet, and then it could be actually measured, as better as lower is the suction rate.

All the above conclusions are readily effective for monodispersed flows, where the concentration does not exceed a level compatible with the "single particle" theoretical model. The same conclusions seem to be valid, after a proper mathematical arrangement, for polydispersed flows, provided that some informations on the size distribution are available.

SYMBOLS

		Subscripts	
c	coefficient	asp	aspirated
d	diameter	D	drag
E	capture efficiency	g	gas
p	pressure	ie	internal-external
r	radius	p	particle
Re	Reynolds number	r	radial component
St	Stokes number	S	probe
t	time	z	axial component
V	velocity	oo	free-stream (upstream)
z	axial coordinate	-oo	free-stream (downstream)
μ	dynamic viscosity		
ρ	density		
τ	characteristic time		
ψ	stream function		

REFERENCES

1. V.Vitols: "Determination of theoretical collection efficiencies of aspirated particulate matter sampling probes under anisokinetic flow" - Ph.D. Thesis, University of Michigan, 1964.

2. M.Bohnet: "Staubgehaltsbestimmung in strömenden Gases mit Absang-Sonden" - Chemie Ing.Techn. 39 Jahrg. 1967, Heft 16.

3. F.H.Smith: "The effects of sampling probe design and sampling techniques on aerosol measurements" AEDC-TR-74-119, 1975.

4. T.A.Temple: "Field computation in Engineering and Physics" - Van Nostrand N.Y. 1961.

5. G.Rudinger: "Fundamentals and applications of gas-particle flow" AGARD-AG 222, 1976.

6. E.Ower, R.C.Pankhurst: "The measurement of air flow" - Pergamon Press, 1966.

7. R.Schödl: "The laser dual focus flow velocimeter" - AGARD-CP 193, 1976.

oOo

Laser-Doppler- and Laser-Dual-Focus Measurements in Laminar and Fully Turbulent Boundary Layers

S. ERIKSEN, KH. SAKBANI, and S. L. K. WITTIG
Lehrstuhl und Institut für Thermische Strömungsmaschinen
Universität Karlsruhe, FRG

ABSTRACT

Optical techniques were applied to boundary layer measurements on a flat plate with natural transition. At free stream velocities of approximately 41 m/s ($Re_L \sim 10^6$) and turbulence intensity of 1,6%, the laminar boundary layer was investigated using a Laser-Dual-Focus Velocimeter, whereas the characteristics of the turbulent boundary layer were determined with a Laser-Doppler-Anemometer. Off-line calculations of the flow and integral parameters of the laminar and turbulent boundary layers were carried out on a minicomputer. Utilizing the stored data of the boundary layer velocity profiles, the shear stress, boundary layer thickness and turbulence intensity distribution are calculated. The comparison with Pitot-tube measurements, hot wire measurements and available analytical solutions shows excellent agreement.

1. INTRODUCTION

In recent years, attention has been directed towards a significant rise of turbine inlet temperatures of gas turbines for aircraft propulsion as well as stationary power plant application. Blade cooling techniques are, therefore, presently under intensive study. The knowledge of the local heat transfer coefficient is of vital importance for the optimization of the cooling system. As of yet, codes for the calculation of the local heat transfer coefficient have not led to fully satisfactory results.

The present investigations are part of a research program at the Institute for Thermal Turbomachinery, University of Karlsruhe, to develop experimentally verified mathematical models for heat transfer calculations.

As measurements in high-temperature gases are necessary, optical techniques,i.e. Laser-Doppler Anemometry (LDA) and Laser-Dual-Focus Velocimetry (L2F) are used to determine the velocity- and turbulence profiles of the boundary layers as well as the free stream parameters, such as intensity of turbulence and velocity.

2. EXPERIMENTAL ARRANGEMENT

In the initial phase of the program, measurements have been

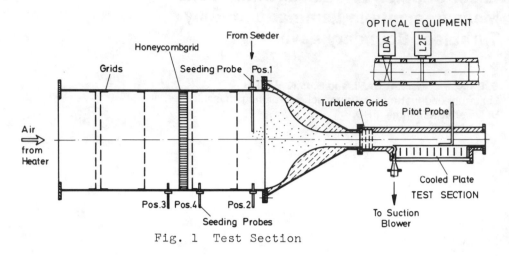

Fig. 1   Test Section

made on a flat plate with natural transition. These experiments were
intended primarily for comparison with available data and extension
to high temperature flow.

2.1 Test Section

The hot gas tunnel is shown schematically in Fig. 1. Briefly,
heated air is supplied to a settling chamber equipped with several
grids and honeycomb flowstraighteners followed by a]2,7 : 1 contraction nozzle. The test section itself is 760 mm long and has
a 460 mm long cooled flat plate with boundary layer suction. The
test conditions chosen were at approximately 325 K and atmospheric
pressure, resulting in Reynolds number based on the length of the
flat plate of up to 10^6 ($U_\infty \approx 41$ m/s). Typically, these conditions
yield a 150 mm laminar boundary layer followed by a 250 mm long
transition region and a 60 mm turbulent boundary layer. Conventional
instrumentation serves for comparison with optoelectronic data
acquisition.

2.2 Optical Techniques

The Laser-Doppler Anemometer which can be operated in forward
as well as in back-scattering mode is a dual beam one-component
optical arrangement equipped with a 4 W Argon/Ion Laser (see Fig. 2).
The probe volume is 0,22 mm in diameter, but is reduced by means of
a mask at the photomultiplier to 0,1 mm. Through rotation of the
two beams about the optical axis, the magnitude and direction of
the local velocity in two dimensional flows can be determined. The
Doppler frequencies are determined using a counter. The data reduction system is based on a PDP-11/34, 128 K-minicomputer which
is connected on-line to the data interface of the counter. The
maximum data rate which is set by the computer, is about 1000 per
second. The data reduction system is shown on Fig. 2, where also a
second system is illustrated. Here, the Doppler frequencies are
detected by a fast (500 MHz) wave form digitizer controlled by a
16 K-microcomputer, which is connected to the PDP-11/34 for further data storage and reduction.

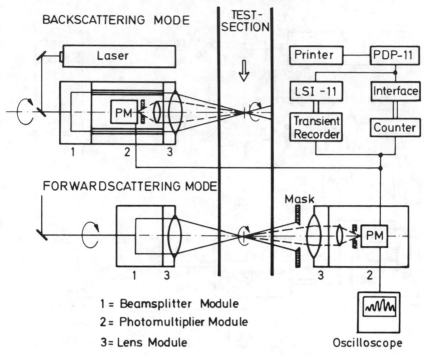

BACKSCATTERING MODE

FORWARDSCATTERING MODE

1 = Beamsplitter Module
2 = Photomultiplier Module
3 = Lens Module

Fig. 2   Laser-Doppler Anemometer

For the measurement of the free-stream velocity, and especially for boundary layer profile analysis, a Laser-Dual-Focus Velocimeter was used as shown on Fig. 3. It works in the back-scattering mode only. The distance between the two foci amounts to 0.58 mm with a 0.02 mm beam diameter. The time of flight, detected by two photomultipliers is recorded with a Time-Amplitude-Converter. The measurements are controlled by means of a microcomputer and the data reduction is achieved with the minicomputer. The probability histogram of the measurements can be visualized on a Multi-Channel-Analyzer.

3. MEASUREMENT PROCEDURE

The optical equipment is supported by a x-y-z coordinate platform. The required accurate positioning of the probe volume to within 0.01 mm close to the wall, which is especially important for the boundary layer measurements, is illustrated in Fig. 4. The plane of the two laser beams is inclined to the wall at an angle of 0.15°. By observing the probe volume by means of a microscope through a window in the top wall of the test section, it is adjusted towards the plate until it just disappears. By retracting the probe volume by half of its diameter, the origin of the measurement is obtained.

One major problem in applying optical techniques in hot gas flow - the final goal of the present study - is seeding. In opti-

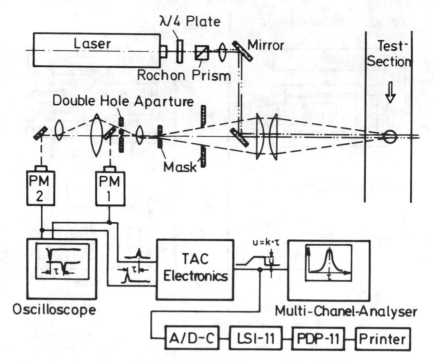

Fig. 3 Laser-Dual-Focus Velocimeter

mizing all parameters of importance, it was found that SiO_2-parti-
cles lead to the best results. The scattering particles are produced in a fluidized bed with individual particle diameters of
0,02 μm (See Fig. 5). The fluidized bed is generated by a small
compressor which simultaneously drives two cyclones to separate the
large agglomerated particles. The particles are led through a spiral
for further deagglomeration before they are seeded into the flow.
The size distribution of the agglomerated particles was measured
with a Cascade Impactor and the particle density was determined by
means of an optical Particle Sizer. Solving the Basset equation for
the motion of discrete particles in a turbulent fluid [1] it is
shown that the particles are following turbulent motions without
slip up to a frequency of at least 10 kHz.

Considerable effort has been directed to avoid disturbances
of the flow caused by seeding. The various positions chosen in
finding an optimum location for the seeding probe are shown in
Fig. 1. Position No. 4 provides the best results.

4. DATA ANALYSIS

4.1 Laser-Doppler Signals

When the frequency of the Doppler signal is known, the
velocity can be calculated from

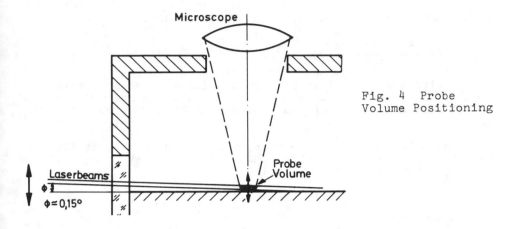

Fig. 4 Probe
Volume Positioning

$$u = f \cdot \frac{\lambda}{2\sin(\beta/2)}$$

The Doppler frequencies are obtained using the information from the counter which determines the number of fringes crossed during the time t_B.

The mean velocity is calculated from a large number of measurements using the time t_B as a weighting factor:

$$\bar{u} = \frac{\Sigma\, u \cdot t_B}{\Sigma\, t_B}$$

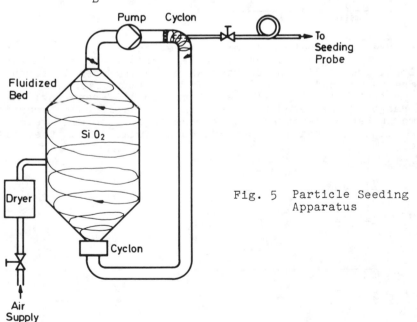

Fig. 5 Particle Seeding
Apparatus

and the mean square of the turbulent fluctuations:

$$\overline{u'^2} = \frac{\Sigma\ (u-\bar{u})^2 \cdot t_B}{\Sigma\ t_B}$$

By using the time t_B as a weighting factor, the proper mean values are obtained without velocity bias, which might exist in individual realization of LDA measurements of highly turbulent flows [7].

A second component of the mean velocity and turbulent fluctuations can be determined by obtaining these values at three different angles and by means of the following equations [2]:

$$\bar{u}_\alpha = \bar{u}_x \cos\alpha + \bar{u}_y \sin\alpha$$

$$\overline{u'^2_\alpha} = \overline{u'^2_x} \cos^2\alpha + 2\ \overline{u'_x u'_y}\ \cos\alpha\sin\alpha + \overline{u'^2_y} \sin^2\alpha$$

4.2 Laser-Dual-Focus Signals

The flow velocity is determined from the time of flight for a particle crossing the two foci.

$$u = a/\tau$$

From a large number of measurements, the probability histogram of the times τ_i is obtained and the mean velocity can be calculated from

$$\bar{u} = \frac{\Sigma\ u_i \cdot N_i}{\Sigma\ N_i} = \frac{\Sigma\ \dfrac{a}{\tau_i} \cdot N_i}{\Sigma\ N_i}$$

Because of the extremely small acceptance angle of the probe volume of the L-2-F Velocimeter, it is necessary to obtain two-dimensional probability histograms when measurements in turbulent flows are to be conducted [3]. This can be accomplished by setting the plane containing the two laser beams to different angles within the range of the velocity direction fluctuations and counting the same number of measurements at each angle. The mean velocities and turbulence intensities then can be calculated from the following equations:

$$\bar{u}_x = \frac{\displaystyle\sum_{i=1}^{M}\ \sum_{j=1}^{N} u_j\ (\alpha_i)\ \cos\alpha_i\ N_j(\alpha_i)}{\displaystyle\sum_{i=1}^{M}\ \sum_{j=1}^{N}\ N_j\ (\alpha_i)}$$

$$\bar{u}_y = \frac{\displaystyle\sum_{i=1}^{M}\ \sum_{j=1}^{N} u_j\ (\alpha_i)\ \sin\alpha_i\ N_j(\alpha_i)}{\displaystyle\sum_{i=1}^{M}\ \sum_{j=1}^{N}\ N_j\ (\alpha_i)}$$

$$\overline{u_y'^2} = \frac{\sum\limits_{i=1}^{M} \sum\limits_{j=1}^{N} [u_j (\alpha_i) \cos\alpha_i - u_x]^2 \cdot N_j (\alpha_i)}{\sum\limits_{i=1}^{M} \sum\limits_{j=1}^{N} N_j (\alpha_i)}$$

$$\overline{u_y'^2} = \frac{\sum\limits_{i=1}^{M} \sum\limits_{j=1}^{N} [u_j (\alpha_i) \sin\alpha_i - u_y]^2 \cdot N_j (\alpha_i)}{\sum\limits_{i=1}^{M} \sum\limits_{j=1}^{N} N_j (\alpha_i)}$$

The velocity bias is eliminated using the reciprocal of the velocity as a weighting factor [7]:

$$\overline{u} = \left(\frac{1}{N} \sum\limits_{1}^{N} u_i^{-1} \right)^{-1}$$

5. RESULTS AND DISCUSSION

Boundary layer measurements were carried out on a flat plate with natural transition. The measurements obtained with the optical equipment are compared with Pitot-tube measurements in the same test section and under the same test conditions.

The shear stress, boundary layer thickness, displacement thickness, momentum thickness, form parameter and turbulence intensity distribution of the boundary layer are calculated from the stored data of the boundary layer velocity profiles.

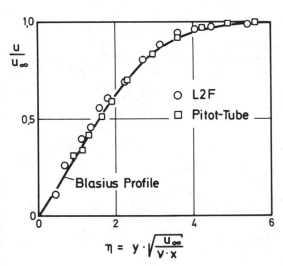

Fig. 6   Laminar Boundary Layer Profile at x = 110 mm

Fig. 6 shows a sample of the measured velocity profiles, here for the laminar boundary layer at x = 110 mm. The comparison with the Blasius profile as well as the Pitot-tube measurements reflects very good agreement.

5.1 Wall Shear Stress

For laminar boundary layers, the computer program contains three different methods. The wall shear stress can be computed from the velocity gradient $\partial u/\partial y$ directly from the Blasius solution or indirectly from the drag coefficient through the dimensionless momentum equation.

$$c_f = \frac{\delta_2}{x}$$

In the case of turbulent boundary layers, the wall shear stress is obtained from the law of the wall, which is gained from the measurements by means of a least-square fit assuming a value of 0.41 for κ.

5.2 Determination of Displacement- and Momentum Thickness

The integral parameters are calculated by least-square profile fit as will be shown in Table 1.

The laminar boundary layer is approximated with the Pohlhausen profile [4]:

$$\frac{u}{u_\delta} = 1 - (1 - \eta)^3 \cdot (1 + \alpha_1 \cdot \eta)$$

For the turbulent boundary layer, the wake is approximated with the Bull method [5], i.e.:

$$u^+ = \frac{1}{\kappa} \ln y^+ + \frac{\Pi}{\kappa} \cdot \omega \left(\frac{y}{\delta}\right)$$

$$\omega\left(\frac{y}{\delta}\right) = 1 - \cos \left(\frac{\pi \cdot \left(\frac{y}{\delta} - A\right)}{1 - A}\right)$$

where

$$A = 0,08;$$

whereas the boundary layer is extended to the wall by means of the Walz-Neubert method [6]:

$$u^+ = \left[(1-a_2 c - \frac{1}{\kappa}) y^+ + k \cdot y^{+2} - c\right] \cdot e^{-a_2 y^+} + c + \frac{1}{\kappa} \ln (1+y^+)$$

$a_2 = 0,3$ and $k = 0$ for a flat plate.

The laser-Dual-Focus Velocimeter was used for the investigation of the laminar boundary layer. Fig. 7 shows an example of the meas-

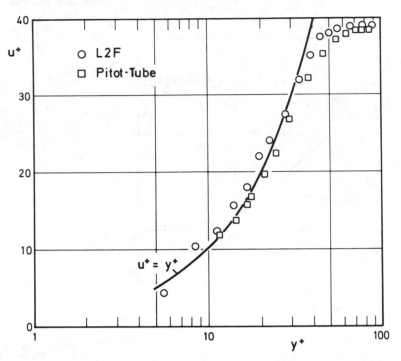

Fig. 7   Laminar Boundary Layer Profile at x = 110 mm

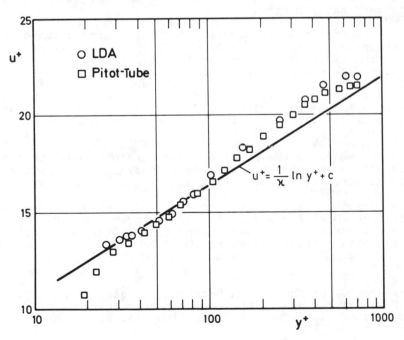

Fig. 8   Turbulent Boundary Layer Profile at x = 410 mm

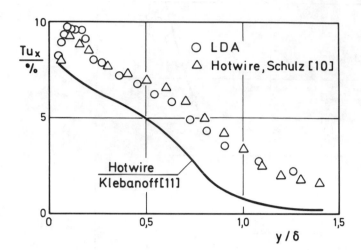

Fig. 9 Turbulence
Intensity Distribu-
tion of the Turbulent
Boundary Layer at
x = 410 mm

urements at x = 110 mm plotted in universal coordinates. The com-
parison with Pitot-tube measurements shows excellent agreement.
Close to the wall, the measurements are very similar to the
$u^+ = y^+$. The point nearst to the wall is at y = 0.1 mm.

    Further downstream, the characteristics of the fully turbulent
boundary layer were determined using the Laser-Doppler Anemometer.
Plotted in universal coordinates, Fig. 8 shows a velocity profile
obtained at x = 410 mm. Also, here the agreement with the Pitot-tube
results is very satisfactory.

    Fig. 9 presents the turbulence intensity distribution of the
turbulent boundary layer at x = 410 mm. The present measurements obtained with the Laser-Doppler Anemometer at a free-stream turbulence
level of 1.6 % are compared with the hot wire measurements of
Schulz [10] for a flat plate at a free-stream turbulence intensity
of 1.5 %. In order to clarify the effect of free-stream turbulence
on the measurements, the well known data of Klebanoff [9] obtained
at an extremely low turbulence level (0.04 %) are also presented in
Fig. 9.

    Table 1 summarizes the profile parameters obtained for the two
previously mentioned examples. The discrepancy between optical and
Pitot-tube measurements is less than 10 %.

    In summary, the experiments show that Laser-Doppler Anemometer
as well as Laser-Dual-Focus Velocimeter measurements can be applied
to laminar and turbulent boundary layer flows. Their application
to high-temperature flows exceeding 700 K is under way.

ACKNOWLEDGEMENTS

    Thank are due to the Danish Research Council for partial
support of S. Eriksen and to the Forschungsvereinigung fur Verbrennungskraftmaschinen for providing financial assistance.

	Laminar		Turbulent	
	LDF	Pitot	LDA	Pitot
U_∞ $|m/s|$	40,240	40,246	40,953	41,003
δ $|mm|$	1,344	1,195	5,534	5,493
δ_1 $|mm|$	0,369	0,389	0,687	0,695
δ_2 $|mm|$	0,139	0,155	0,481	0,491
H_{12}	2,649	2,514	1,429	1,414
C	—	—	5,236	5,205
Π	—	—	0,301	0,229
$\tau_w |N/m^2|$	1,115	1,180	3,788	3,944
Re_{δ_2}	307	318	1080	1054

Table 1   Comparison between optical- and Pitot-
tube boundary-layer measurements

NOMENCLATURE

a	–	distance between the two parallel beams of the L2F
a_1	–	Pohlhausen profile parameter
C	–	constant in the law of the wall
c_f	–	drag coefficient
f	–	frequency
H_{12}	–	boundary layer form parameter
N	–	number of measurements
u	–	velocity
$\bar{u}$	–	mean velocity
u_∞	–	free stream velocity
u_δ	–	$u_\delta = 0,995 \cdot u_\infty$
u'	–	velocity fluctuation
u^+	–	dimensionless velocity
x	–	distance along the wall
y	–	distance normal to the wall
y^+	–	dimensionless normal distance
α	–	angle between mean flow direction and plane of probe volume
β	–	angle between the two laser beams of the LDA
δ	–	boundary layer thickness

δ_1 –   displacement thickness

δ_2 –   momentum thickness

η –   dimensionless distance normal to the wall $\eta = y/\delta$

κ –   constant in the law of the wall

λ –   laser beam wave length

μ –   viscocity

Π –   wake-parameter

τ –   time of flight between the two laser beams of the LDF

τ_w –   wall shear stress

REFERENCES

1.   Hjelmfelt, A.T.;Mockros, L.F.
      Appl. Sci. Ros. Vol. 16, p. 149

2.   Stevenson, W.H.;Thompson, H.D.;Bremmer, R.; Roesler, T. 1980.
      Technical Report AFAPL-TR-79-2009, Part II,
      School of Mechanical Engineering, Purdue University,
      Indiana U.S.A.

3.   Schodl, R. 1977.
      Dissertation TH Aachen, W.Germany.

4.   Walz, A. 1966.
      Strömungs- und Temperaturgrenzschichten
      Verlag Braun, Karlsruhe, W.Germany.

5.   Murlis, J. 1975.
      Department of Aeronautics, Imperial College of Science
      and Technology, London.

6.   Festschrift zum 60. Geburtstag von Prof. Dr.-Ing.
      Erich Truckenbrodt 1977.
      TU-München.

7.   McLaughlin, D.;Tiederman, W. 1973.
      The Physics of Fluids, Vol. 16, No. 12, p. 2082.

8.   Schlichting, H. 1968.
      Boundary Layer Theory
      Mc. Graw Hill, New York.

9.   Klebanoff, P.S. 1955.
      NACA-Report 1247.

10.   Schulz, K.-J. 1976.
      Dissertation TU Kaiserslautern, W.Germany.

Analysis of the Uncertainties in Velocity Measurements with Triple Hot-Wire Probes

MAURICIO N. FROTA and ROBERT J. MOFFAT
Department of Mechanical Enginerring
Stanford University
Stanford, California 94305, USA

ABSTRACT

This paper presents a purely analytical investigation of the uncertainties in a triple hot-wire system and shows them to be a natural consequence of probe-manufacturing tolerances. The size of the probe relative to the velocity gradient is shown to be an important source of spurious signals.

Exact equations have been derived for the sensitivity coefficients of the three instantaneous velocity components with respect to each the variables involved. These equations were derived using the MACSYMA Multiaccess Computer System (the MC-Machine at the Laboratory for Computer Sciences, M.I.T.). Absolute and relative uncertainties for the instantaneous triple hot-wire outputs have been calculated, based upon a constant probability combination of the uncertainties in the inputs. These uncertainties have been calculated as functions of roll and pitch angles.

Results show that the uncertainty band associated with measurements of the streamwise velocity component is on the order of 2% regardless of roll and pitch. Uncertainties associated with the V and W channels are sensitive to both roll and pitch angles and are on the order of 2% and 4%, respectively, when the probe is aligned with the flow. These uncertainties reach ± 5% at 20° misalignment, and are not symmetrical with respect to pitch.

In addition to evaluating the stochastic component of the uncertainties, this analysis has also allowed prediction of the deterministic components (fixed errors) present in some of the outputs. Comparison between computations and data taken in a channel flow and in a thin boundary layer flow shows that the fixed errors (spurious signals) present in the V and W channels of the triple wire output are due to the non-zero measuring volume of the triple wire probe.

The technique of uncertainty analysis has proved to be a useful tool in the design stage of a new instrument or experiment and may save considerable time and cost. This analysis can be used as a design criterion, since it shows the tolerances necessary to achieve the overall accuracy desired.

Mauricio N. Frota is on leave from Pontifícia Universidade Católica, R. J., Brazil.

1.0 INTRODUCTION

Identification of the different sources of errors and experimental uncertainties associated with measurements of different functions of engineering importance usually is not a trivial task. Calibration of a triple hot-wire anemometer system for measurements of velocity fluctuations is an example. The best practice consists of calibrating the probe against a flow of known shear stresses as well as mean velocity (Frota and Moffat, 1982a). Accuracy of the calibration will depend upon imperfections of the reference flow, uncertainties associated with parameters on Bernoulli's equation, uncertainties associated with curve fitting, temperature drift instrumentation setting, probe misalignment, wire geometry, etc. Overall analysis of these uncertainties seems to be ambitious, and very little work has been reported in this area. Zank (1981) studied the errors caused by non-ideal wire geometry; Andreopoulos (1981) estimated the magnitude of the errors associated with triaxial hot-wire probes resulting from high angular excursions of the velocity vector in turbulent flows. His work is a general extension of the work of Tutu and Chevray (1977), who investigated the errors resulting from truncation of higher-order terms, for single and cross-wire probes.

The present analysis consists of a purely analytical study of the propagation of uncertainties through the triple hot-wire equations. Exact equations for the sensitivity coefficients provide for the relative importance of each individual variable. The overall uncertainties associated with measurements of the indicated velocity components are discussed.

Measurements of turbulence structure using the triple hot-wire probe have already been reported (Frota and Moffat, 1982b).

2.0 THE HOT-WIRE RESPONSE EQUATION

The concept of "effective velocity" greatly simplifies the problem of calibrating a hot-wire for the effects of a misaligned flow. The "effective velocity" may be defined as the velocity which, if perpendicular to the wire and in the plane formed by the axis of the prongs, will produce the same bridge output as does the unknown velocity; i.e., a velocity perpendicular to the wire which would cause the same "cooling effect" in the wire.

Several expressions for the effective cooling velocity, U_{eff}, have been suggested as the basis for empirical correlation of the directional sensitivity of hot wires (Champagne and Sleicher, 1967; Friehe and Schwarz, 1968; Hinze, 1975; Jørgensen, 1971). Comprehensive hot-wire response equations have been suggested in the literature (Butler and Wagner, 1982; Andreopoulos, 1982b; Lakshminarayana, 1982) accounting for complex nonlinear depencence of yaw/pitch, flow angle and velocity magnitude on the hot-wire directional sensitivities. Since this nonlinearity introduces difficulties in signal processing, we shall adhere to Jørgensen's decomposition (Eq. (1)). The present work has given analytical support to previous findings that the principal cause of error is in geometry defects and in failure to match the electronics, not in the form of the decomposition equation.

$$U_{eff}^2 = X^2 + k_1^2 Y^2 + k_2^2 Z^2 \tag{1}$$

In this equation, k_1 and k_2 are the directional sensitivity coefficients of the wire, which, to some extent, depend upon the hot-wire probe design, the probe identity, and the flow Reynolds number. The components X, Y, and Z are instantaneous velocity components in the wire coordinate system x', y', z', as seen in Fig. 1, which illustrates a single sensor wire (the wire may be normal or

slanted with respect to its prongs). U, V, and W are the corresponding velocity components in laboratory coordinates x, y, z. The slant angle φ and the
roll angle α are also shown.

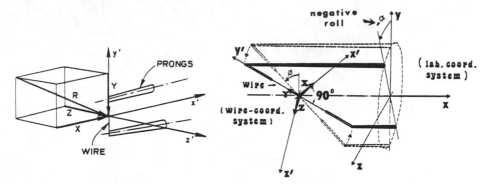

Fig. 1-a,b. Wire (x,'y',z') and laboratory (x,y,z) coordinate systems.

3.0 THE TRIPLE HOT-WIRE APPROACH

Assume that three slanted wires (like the wire shown in Fig. 1) are simultaneously placed in the flow, at an arbitrary roll angle α. Figure 2 illustrates the new wire configuration and defines the corresponding wire and laboratory coordinate systems. In this case, x', y', and z' have been arbitrarily
chosen along wires #1, #2, and #3, respectively.

Jørgensen's decomposition may thus be used to describe the effective velocity of each one of the three wires, yielding a set of three equations and nine
unknowns (three velocity components: X_j, Y_j, and Z_j, for each wire), as follows:

$$U^2_{eff_j} = X^2_j + k^2_{1j}Y^2_j + k^2_{2j}Z^2_j \quad (j = 1,2,3) \tag{2}$$

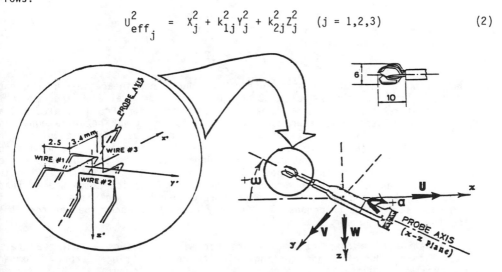

Fig. 2. The orthogonal triple hot-wire probe. Wire (x',y',z') and
         laboratory (x,y,z) coordinate systems.

Note that the directional sensitivities k_{1j} are always applied to the Y velocity component along wire #j $(j = 1,2,3)$, whereas the k_{2j} sensitivities correct the binormal velocity components Z of each wire. The effective velocities, U_{eff}, are known from the calibration of the wires, the probe geometry, and known values of the k_{ij} $(i = 1,2; j = 1,2,3)$ sensitivity coefficients, based upon previous calibration of the hot wires.

Alternatively, Eq. (2) may be written as:

$$
U^2_{eff_j} = \begin{bmatrix} X_j & Y_j & Z_j \end{bmatrix} \underbrace{\begin{bmatrix} 1 & 0 & 0 \\ 0 & k^2_{1j} & 0 \\ 0 & 0 & k^2_{2j} \end{bmatrix}}_{\triangleq \; \kappa_j} \underbrace{\begin{bmatrix} X_j \\ Y_j \\ Z_j \end{bmatrix}}_{\triangleq \; \chi_j} = \chi^T_j \cdot \kappa_j \cdot \chi_j \tag{3}
$$

However, if the wires are mutually orthogonal, these three wire coordinate systems collapse onto a single frame such that the velocity component X for one wire is the Y component and Z component for the others, i.e.,

$$
\begin{aligned}
X_3 &= Y_1 = Z_2 \triangleq X \\
X_1 &= Y_2 = Z_3 \triangleq Y \\
X_2 &= Y_3 = Z_1 \triangleq Z
\end{aligned} \tag{4}
$$

Therefore, Eq. (3), subjected to the above wire orthogonality condition, yields a compatible linear set of three equations and three unknowns:

$$
\begin{aligned}
U^2_{eff_1} &= k^2_{11}X^2 + Y^2 + k^2_{21}Z^2 \\
U^2_{eff_2} &= k^2_{22}X^2 + k^2_{12}Y^2 + Z^2 \\
U^2_{eff_3} &= X^2 + k^2_{23}Y^2 + k^2_{13}Z^2
\end{aligned} \tag{5}
$$

Non-orthogonality of the three wires introduces nonlinearities in the hot-wire equations; e.g., the velocity components cannot be evaluated independently of the cross-terms XY, XZ, YZ. It has already been shown (Frota et al., 1981a) that expansion of this nonlinear equation leads to the slant hot-wire equations which introduce the inconvenience of the so-called time-averaging ambiguities, as described by Moffat et al., 1978. Although different approximate techniques for handling the nonlinear equations have been suggested, it is the authors' belief that the level of sophistication should be kept as simple as possible. Extensive experience at Stanford in probe design and probe construction demonstrates that, with moderate skill and by means of an optical comparator, the wires of the triaxial probe can be made orthogonal within very tight tolerances (better than 1°), leaving the problem linear.

Calibration of a constant-temperature hot wire, provides a nonlinear relationship between the effective cooling velocity and the bridge output. Commercially available "linearizers" produce a linear relationship which speeds up digital processing and greatly simplifies analog processing (Bradshaw, 1975).

Replacing the effective velocities U_{eff_i} by a linear function of the linearizer's output, Eq. (5) can be solved for the three unknowns as follows:

$$\begin{bmatrix} X^2 \\ Y^2 \\ Z^2 \end{bmatrix} = \begin{bmatrix} K_{ij} \end{bmatrix}^{-1} \begin{bmatrix} (A_{eff_1} + B_{eff_1}E)^2 \\ (A_{eff_2} + B_{eff_2}E)^2 \\ (A_{eff_3} + B_{eff_3}E)^2 \end{bmatrix} \tag{6}$$

where

$$\begin{bmatrix} K_{ij} \end{bmatrix}^{-1} = \frac{1}{\Delta} \begin{bmatrix} k_{12}^2 k_{13}^2 - k_{23}^2 & k_{21}^2 k_{23}^2 - k_{13}^2 & 1 - k_{12}^2 k_{21}^2 \\ 1 - k_{13}^2 k_{22}^2 & k_{11}^2 k_{13}^2 - k_{21}^2 & k_{21}^2 k_{22}^2 - k_{11}^2 \\ k_{22}^2 k_{23}^2 - k_{12}^2 & 1 - k_{11}^2 k_{23}^2 & k_{11}^2 k_{12}^2 - k_{22}^2 \end{bmatrix} \tag{7}$$

In this equation, Δ is the determinant of the Jørgensen $[K_{ij}]$ matrix:

$$\Delta = 1 + (k_{11}^2 k_{12}^2 k_{13}^2) + (k_{21}^2 k_{22}^2 k_{23}^2) - (k_{21}^2 k_{12}^2) - (k_{23}^2 k_{11}^2) - (k_{13}^2 k_{22}^2) \tag{8}$$

The instantanous velocity components (U,V,W) in the laboratory coordinate system can be found by a transformation of coordinates:

$$\begin{bmatrix} U \\ V \\ W \end{bmatrix} = \begin{bmatrix} \cos\omega & 0 & -\sin\omega \\ 0 & 1 & 0 \\ \sin\omega & 0 & \cos\omega \end{bmatrix} \begin{bmatrix} 1 & 0 & 0 \\ 0 & \cos\alpha & \sin\alpha \\ 0 & -\sin\alpha & \cos\alpha \end{bmatrix} \begin{bmatrix} \sqrt{3}/3 & \sqrt{3}/3 & \sqrt{3}/3 \\ -\sqrt{6}/6 & -\sqrt{6}/6 & \sqrt{6}/3 \\ -\sqrt{2}/2 & \sqrt{2}/2 & 0 \end{bmatrix} \begin{bmatrix} X \\ Y \\ Z \end{bmatrix} \tag{9}$$

In this equation, ω and α are the pitch and roll angles as defined in Fig. 2. The pitch angle ω is measured in the x-z plane and is taken to be zero when the axis of the probe is in the x-direction, in laboratory coordinates, with the probe facing upstream. The angle of roll, α, is here consistent with the definition given in Fig. 1 for the slant wire. However, the position of zero roll for the triple wire is defined to be the position in which wire #3 is in the x-y plane with the longer prong of wire #3 at the smaller y value.

The third elemental matrix in Eq. (9) results from the <u>orthogonal</u> wire geometry referred to the <u>orthogonal</u> laboratory coordinate system.

Expansion of Eq. (9) may be instructive. It can be shown that the ω-pitch dependence drops out of the equation for the V-component of the velocity. It can also be shown that, regardless of the values of k_1 and k_2, the equations for V and W always vanish if the three linearizers' output are matched. If one of the linearizer's output is disturbed by 1%, however, errors in $\overline{V}$ and $\overline{W}$, expressed by the ratios $\overline{V}/\overline{U}$ and $\overline{W}/\overline{U}$ an order of $\sim 0.5°$ will be detected, i.e., $\tan^{-1}(\overline{V}/\overline{U}) = 0.5°$.

4.0 SENSITIVITY ANALYSIS

Even assuming that the wires of the triple wire are mutually orthogonal (to leave the problem linear) and also assuming that one can accurately account for temperature effects on the hot-wire signals (Frota and Moffat, 1981b), the triple hot-wire equations still depend upon several variables. The instantaneous velocity components F (F = U,V,W, the outputs of the triple hot-wire processor) may be written as a function of the x_i variables:

$$F = F(\alpha, \omega, k_{ij}, E_{cta,i}, A'_i, B'_i, n_i) \qquad (10)$$

α and ω are the already defined roll and pitch angles describing the relative position of the wires with respect to the approaching flow; $E_{cta,i}$ (i = 1,2,3) are the nonlinearized anemometer outputs, and A'_i, B'_i, and n_i (i = 1,2,3) are the parameters in King's Law, usually determined by a calibrating procedure.

The use of linearizers reduces the number of variables involved in Eq. (10):

$$F = F(\alpha, \omega, k_{ij}, E_i, A_i, B_i) \qquad (11)$$

In this equation, E_i are the linearizers' output and A_i and B_i are the modified linearized calibrating constants.

Calculation of the partial derivative of each one of the triple hot-wire outputs F (F = U, V, or W) with respect to each one of the x_i variables in Eq. (11) is desirable. Physically, these partial derivatives represent the main effect of each factor individually. Dependence of these sensitivity coefficients on each one of the x_i variables describes the "interaction" among variables.

Knowledge of these sensitivity coefficients would, for instance, make evident the relative importance of small misalignment in roll α or pitch ω in measurements of the velocity components. Such a sensitivity analysis would even guide the experimenter to strategically allocate (within the overall setup) pieces of hardware with the purpose of increasing the accuracy on measurements of specific components. It will be shown, for instance (Table I) that, regardless of the pitch angle ω and for roll angle α = 90°, measurements of the instantaneous transverse velocity component V is strongly and equally sensitive to the output of wires #1 and #2, but weakly dependent on the output of wire #3. Therefore, an unstable anemometer which eventually drifts may not much affect the measurements of $\overline{V}$, v'^2, though it would compromise the measurements of the W component.

The equations for the velocity components are algebraically complex, and derivation of their partial derivatives is difficult, tedious, and time-consuming. This difficulty has been overcome by the development of sophisticated computers devoted to the manipulation of algebraic expressions. By means of a special computer, operating in symbolics,* it was possible to evaluate exact solutions for the partial derivatives $\partial F/\partial x_j$. Availability of these sensitivites makes evident the relative importance of each variable individually. Absolute and relative uncertainties on the instantaneous triple hot-wire equations can be evaluated based upon the root-sum-square approach. By this means, one can simulate the behavior of a very complex scientific instrument. This analysis

---

*MACSYMA multiaccess computer system, which is a collection of implemented programs running on a PDP-10, known as the MC-Machine, at the Laboratory for Computer Science at M.I.T.

allows for estimation of the absolute and relative stochastic component of the uncertainties of each one of the system outputs based upon uncertainties in each of the different inputs.

The following section briefly describes the framework upon which this uncertainty analysis on the triple hot-wire equation has been derived.

5.0 THE MATHEMATICAL FORMULATION

As suggested by Kline and McClintock (1953) and more recently emphasized by Moffat (1980), it will be assumed that each independent variable x_i can be described by either one of the following equations:

$$x_i = \underbrace{\overline{x_i}}_{\substack{\text{Best} \\ \text{Estimate}}} \pm \underbrace{(\delta x_i)}_{\substack{\text{Absolute Uncertainty} \\ \text{Interval}}} \qquad (\text{@ odds})$$

or   (12)

$$x_i = \underbrace{\overline{x_i}}_{\substack{\text{Best} \\ \text{Estimate}}} \pm \underbrace{(\delta x_i / x_i)}_{\substack{\text{Relative Uncertainty} \\ \text{Interval}}} \qquad (\text{@ odds})$$

The variables x_i, usually taken as each data bit, can be thought of as the different input variables of the triple hot-wire processor. In these equations, δx_i and $\delta x_i / x_i$ are the absolute and relative uncertainty intervals, respectively. As prescribed in probability theory, one can suitably associate to the above equations some odds or confidence level. (In this sense 20:1 odds correspond to confidence intervals at $P = 0.952$.)

It has been shown that, if each of the x_i comes from a Gaussian distribution and if all the x_i are independent variables, one can assess the uncertainty with good accuracy for most functions of engineering importance by means of the root-sum-square approach, as follows:

$$\delta F = \left\{ \sum_{i=1}^{N} \left(\frac{\partial F}{\partial x_i} \, \delta x_i \right)^2 \right\}^{1/2} \qquad (13)$$

where F can be either U, V, or W.

The individual terms in Eq. (7) can be identified as:

$(\partial F / \partial x_i) \triangleq x_i$-sensitivity coefficients,

$(\partial F / \partial x_i) \delta x_i \triangleq x_i$-contribution,

δF = absolute uncertainty.

It is understood that the calculated value of δF will be appropriate at the same odds as those stated for the individual x_i-statements. The odds must be the same for all x_i.

The computerized general expressions for the sensitivity coefficients and for the absolute and relative stochastic components of the uncertainties associated with the triple hot-wire equations are algebraically complex and have not been included, due to space limitation. This information, however, is available elsewhere (Frota et al., 1981a).

5.1 Calculation of the Stochastic Components of the Uncertainties

Tables I and II (in appendix) summarize these results. Table I shows the effects of the pitch angle ω (for $\alpha = 90°$) on these sensitivity coefficients $\partial F/\partial x_j$ ($F = U$, V, or W; $x_j = \alpha$, ω, k_1, k_2, E_j, etc.) for two sets of k_1, k_2 values. Table II also contains the absolute and relative uncertainties evaluated based upon the root-sum-square approach, as prescribed by Eq. (13).

The data in Tables I and II represent a pure analytical calculation of the uncertainties of the three instantaneous velocity comonents U, V, and W, calculated by means of the triple hot-wire equations. All the sensitivity coefficients have been calculated based upon the theoretically correct values of the linearized output E_i ($i = 1,2,3$), using the triple hot-wire equations for an ideal probe. Since there is no experimental uncertainty involved in the present analysis, it can be used as a criterion to describe the best that can be expected from the triple hot-wire system. The results are summarized in Fig. 3, which compares the relative uncertainty $\delta U/U$, $\delta V/U$, and $\delta W/U$ as a function of the pitch angle for two roll angles ($\alpha = 0°$ and $\alpha = 90°$). Note that the uncertainty associated with measurements of the U components are independent of the pitch angle = 2% (for $-35° < \omega < 35°$, the instantaneous approach velocity vector may lie outside the cone formed by the directions of the wires).

From Fig. 3, one concludes:

- For zero roll ($\alpha = 0°$ and any pitch angle ω, the uncertainty associated with measurements of W is always larger than the uncertainty in measurements of V. These uncertainties are minimum at $\omega = 0°$; i.e., probe aligned with the flow (2% for V and 4% for W) and increases (symmetrically) for a positive or negative pitch misalignment.

- For $\alpha = 90°$, however, the response is quite asymmetric. For positive pitch, the uncertainties in measurements of V are extremely dependent on the pitch angle ω. For negative pitch, this trend is reversed. Due to symmetry, one should expect similar behavior at different roll angles, and it seems safe to conclude that there is no such roll angle α that would lead to a flat response simultaneously for V and W, in the same range.

- The uncertainty associated with measurements of U is on the order of 2.2% and does not depend on pitch or roll angles (remember that three wires suffice for measurements of U; therefore, the U-output of a triple wire is an average of all three wires).

---

[*]Note that the uncertainty on V and W have been scaled on U rather than on V and W, respectively. U is a better choice because V and/or W could be zero, which is the case when the probe stem is aligned with the flow and because the spurious V and W signals arise from improper handling of the U-component. It is a form of common-mode interference.

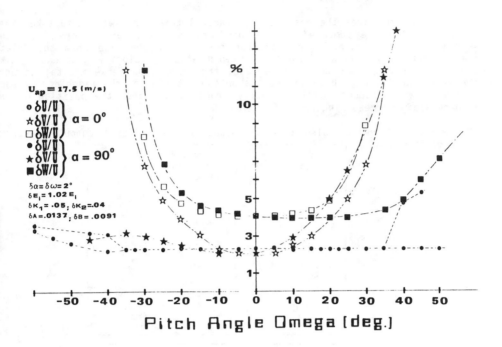

Fig. 3. Prediction of the deterministic components of the uncertainty.

Tables I and II are the basis for comparison of the relative order of magnitude of the uncertainties on U, V, and W. Note, for instance, that at $\omega = 0$ $\delta V < \delta U$, although this is not caused by the fact that $\partial V/\partial \omega = 0$, because $\partial U/\partial \omega$ is also very small when compared to other sensitivity coefficients. From Table I, it is clear that at $\omega = 0°$, $\partial U/\partial k_1 \gg \partial V/\partial k_1$ and also $\partial U/\partial k_2 \gg \partial V/\partial k_2$, and these are the dominant terms in Eq. (13) for the U-component. The same argument could be used to explain the "large" uncertainty associated with the W-component. (Note that $\partial W/\partial \omega \gg \partial U/\partial \omega$; $\partial W/\partial E_i \gg \partial U/\partial E_i$ (i = 1,2,3), Table I.) In this sense, the value of the x_i-sensitivity coefficient proves to be a very useful descriptor of the importance of each variable, independently.

5.2 Prediction of the Deterministic Components of the Uncertainty

As a result of special techniques accounting for drift in flow temperature level and for small differences in response between channels, it has been possible to reduce substantially the level of the spurious signals generated in the V and W channels. It has been proved that the remaining spurious signals are caused by the fact that each wire of the triple-wire probe "sees" a different velocity vector, not the true velocity, at the assigned measuring point. With calibrated sensor wires, a knowledge of the velocity gradient, and accurate knowledge of the position of each wire, it has been possible to predict the spurious signals completely. Therefore, the remaining artificial velocity components present in the near-wall velocity gradient are seen to be real, predictable, and completely understood. It is therefore feasible to correct for them during data reduction. The velocity gradient can easily be evaluated at each probe location by the individual output of two wires or, alternatively by the measurement of $\overline{U}$ at two neighbor locations.

Figure 4 summarizes the results where actual triple hot-wire data taken in a 2-D channel flow are compared to the predictions which take into account the "shear effect". Figure 5 shows similar results for the case of a thin flat-plate boundary layer flow, displaying a much steeper velocity gradient. Like the fully developed 2-D channel flow case, the comparison between data and predictions is excellent. Both figures show the position of the midpoints of each sensor wire with respect to the velocity gradient and also the position of the imaginary assigned measuring point (labeled ζ_L).

The computation was also done by means of the MACSYMA computer system, already mentioned.

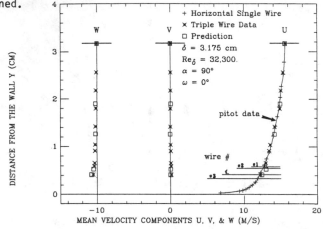

Fig. 4. Comparison between data and computation (channel flow).

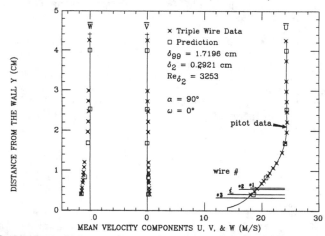

Fig. 5. Comparison between data and computation (boundary layer flow).

---

*"Shear effect" in the actual sense of the word would be a second-order effect, as opposed to what could be called a "velocity effect". In the present work, the expression "shear effect" is used to emphasize that the probe is exposed to a velocity gradient region; therefore each wire "sees" a different value of velocity (regardless of the local shear value). However, it is a common practice in the literature to refer to this effect as a "shear effect" when applied to pressure probes, and we continue the habit.

6.0   CONCLUSIONS

A detailed computerized sensitivity analysis has been completed on the triple hot-wire equations to delineate the uncertainties associated with measurements of the velocity components. Numerical simulation of the overall system revealed that the small inherent difficulties associated with the triple hot-wire data do not reflect artifacts introduced by the data processing. The spurious signals (fixed errors) generated in the V and W channels are entirely predictable, and the behavior of the overall system is believed to be fully understood.

ACKNOWLEDGMENTS

This research was supported by the Air Force Office of Scientific Research, grant AFOSR-0010, and the National Aeronautics and Space Administration, contract NAG 3/3.

The first author is grateful to the National Research Council of Brazil, CNPq, and to Pontificia Universidade Catolica do Rio de Janeiro for sponsoring his doctoral studies at Stanford University.

APPENDIX

TABLE I. The Computerized x_i-Sensitivity Coefficients

#	F	ω	K_1	K_2	$\partial F/K_1$	$\partial F/\partial K_2$	$\partial F/\partial \alpha$	$\partial F/\partial \omega$	$\partial F/\partial A$	$\partial F/\partial B$	$\partial F/\partial E_1$	$\partial F/\partial E_2$	$\partial F/\partial E_3$
	m/s	deg	-	-	m/s	m/s	(m/s)/rad	(m/s)/rad	-	V	(m/s)/V	(m/s)/V	(m/s)/V
1	U	0.	0.100	0.980	0.91E 00	0.89E 01	0.30E-03	0.50E-03	0.12E 01	0.80E 01	0.8338	0.8444	0.5274
2	U	10.	0.100	0.980	0.91E 00	0.89E 01	0.30E-03	0.67E-03	0.12E 01	0.80E 01	0.7932	0.7983	0.6481
3	U	20.	0.100	0.980	0.91E 00	0.89E 01	0.30E-03	0.50E-03	0.12E 01	0.80E 01	0.8338	0.8444	0.5274
4	U	30.	0.100	0.980	0.91E 00	0.89E 01	0.20E-03	0.60E-03	0.12E 01	0.81E 01	0.8683	0.8224	0.3924
5	U	-10.	0.100	0.980	0.91E 00	0.89E 01	0.34E-04	0.13E-02	0.12E 01	0.80E 01	0.7095	0.7041	0.8291
6	U	-20.	0.100	0.980	0.91E 00	0.89E 01	0.40E-03	0.10E-03	0,12E 01	0.80E 01	0.6776	0.6677	0.8840
7	U	-30.	0.100	0.980	0.91E 00	0.89E 01	0.30E-03	0.54E-02	0.12E 01	0.80E 01	0.6620	0.6495	0.9090
8	U	0.	0.150	1.020	0.13E 01	0.87E 01	0.00E 00	0.00E 00	0.12E 01	0.78E 01	0.7328	0.7328	0.7328
9	U	10.	0.150	1.020	0.13E 01	0.87E 01	0.30E-03	0.10E-03	0.12E 01	0.78E 01	0.7797	0.7741	0.6352
10	U	20.	0.150	1.020	0.13E 01	0.87E 01	0.40E-03	0.19E-04	0.12E 01	0.78E 01	0.8234	0.8138	0.5201
11	U	-10.	0.150	1.020	0.13E 01	0.87E 01	0.20E-03	0.18E-02	0.12E 01	0.78E 01	0.6890	0.6941	0.8089
12	U	-30.	0.150	1.020	0.13E 01	0.87E 01	0.50E-03	0.52E-02	0.12E 01	0.78E 01	0.6367	0.6487	0.8858
13	V	0.	0.100	0.980	0.22E-07	0.17E-07	0.44E-07	0.00E 00	0.55E-08	0.44E-07	1.9006	1.8246	0.0760
14	V	5.	0.100	0.980	0.12E-01	0.31E 01	0.16E 01	0.00E 00	0.20E-03	0.80E-03	2.0923	2.0160	0.0760
15	V	10.	0.100	0.980	0.27E-01	0.67E 01	0.31E 01	0.00E 00	0.90E-03	0.23E-02	2.3320	2.2544	0.0760
16	V	15.	0.100	0.980	0.45E-01	0.11E 02	0.46E 01	0.00E 00	0.23E-02	0.44E-02	2.6356	2.5554	0.0761
17	V	20.	0.100	0.980	0.68E-01	0.17E 02	0.61E 01	0.00E 00	0.46E-02	0.64E-02	3.0297	2.9445	0.0768
18	V	25.	0.100	0.980	0.95E-01	0.23E 02	0.76E 01	0.00E 00	0.85E-02	0.14E-01	3.5555	3.4634	0.0767
19	V	30.	0.100	0.980	0.13E 00	0.33E 02	0.90E 01	0.00E 00	0.14E-01	0.22E-01	4.2948	4.1909	0.0775
20	V	-5.	0.100	0.980	0.11E-01	0.25E 01	0.16E 01	0.00E 00	0.10E-03	0.70E-03	1.7467	1.6705	0.0759
21	V	-10.	0.100	0.980	0.19E-01	0.46E 01	0.31E 01	0.00E 00	0.50E-03	0.70E-03	1.6235	1.5466	0.0759
22	V	-20.	0.100	0.980	0.31E-01	0.76E 01	0.61E 01	0.00E 00	0.16E-02	0.17E-02	1.4540	1.3751	0.0760
23	V	-25.	0.100	0.980	0.34E-01	0.85E 01	0.76E 01	0.00E 00	0.20E-02	0.26E-02	1.4026	1.3230	0.0759
24	V	-30.	0.100	0.980	0.37E-01	0.90E 01	0.90E 01	0.00E 00	0.23E-02	0.31E-02	1.3716	1.2915	0.0759
25 →	V	0.	0.150	1.020	0.19E-08	0.13E-06	0.00E 00	0.00E 00	0.52E-08	0.00E 00	1.1350	1.0050	2.1400 ← α = 0°
26	V	0.	0.150	1.020	0.00E 00	0.00E 00	0.00E 00	0.00E 00	0.52E-08	0.83E-07	1.8158	1.8909	0.0759
27	V	10.	0.150	1.020	0.40E-01	0.64E 01	0.30E 01	0.00E 00	0.80E-03	0.20E-03	2.2408	2.3180	0.0757
28	V	20.	0.150	1.020	0.95E-01	0.16E 02	0.60E 01	0.00E 00	0.44E-02	0.72E-02	2.9247	3.0089	0.0761
29	V	-10.	0.150	1.020	0.26E-01	0.44E 01	0.30E 01	0.00E 00	0.50E-03	0.20E-03	1.5409	1.6168	0.0749
30	V	-20.	0.150	1.020	0.43E-01	0.72E 01	0.60E 01	0.00E 00	0.15E-02	0.17E-02	1.3717	1.4493	0.0748
32	W	5.	0.100	0.980	0.47E 00	0.72E-01	0.90E-03	0.18E 02	0.10E 00	0.16E 00	0.8606	0.9939	2.0434
33	W	10.	0.100	0.980	0.94E 00	0.14E 00	0.18E-02	0.18E 02	0.20E 00	0.30E 00	0.7349	0.8727	1.9746
34	W	15.	0.100	0.980	0.14E 01	0.22E 00	0.17E-02	0.18E 02	0.30E 00	0.44E 00	0.6235	0.7688	1.9321
35	W	20.	0.100	0.980	0.20E 01	0.31E 00	0.80E-03	0.18E 02	0.39E 00	0.59E 00	0.5215	0.6768	1.9083
36	W	25.	0.100	0.980	0.27E 01	0.42E 00	0.14E-02	0.18E 02	0.48E 00	0.72E 00	0.4231	0.5965	1.8991
37	W	30.	0.100	0.980	0.36E 01	0.55E 00	0.30E-03	0.18E 02	0.58E 00	0.87E 00	0.3269	0.5248	1.9036
38	W	-5.	0.100	0.980	0.50E 00	0.80E-01	0.80E-03	0.18E 02	0.11E 00	0.17E 00	1.1955	1.3292	2.3172
39	W	-10.	0.100	0.980	0.11E 01	0.17E 00	0.20E-03	0.18E 02	0.25E 00	0.37E 00	1.4445	1.5844	2.5807
40	W	-15.	0.100	0.980	0.18E 01	0.28E 00	0.20E-03	0.18E 02	0.41E 00	0.62E 00	1.8064	1.9601	3.0137
41	W	-20.	0.100	0.980	0.29E 01	0.44E 00	0.11E-02	0.18E 02	0.65E 00	0.98E 00	2.3984	2.5804	3.7878
42	W	-25.	0.100	0.980	0.47E 01	0.73E 00	0.70E-03	0.18E 02	0.11E 01	0.16E 01	3.5619	3.8091	5.4034
43	W	-30.	0.100	0.980	0.98E 01	0.15E 01	0.60E-03	0.18E 02	0.22E 01	0.34E 01	6.9546	7.4093	10.2899
44	W	0.	0.150	1.020	0.00E 00	0.00E 00	0.00E 00	0.18E 02	0.10E-07	0.21E-07	1.1350	1.0050	2.1400
45	W	10.	0.150	1.020	0.13E 01	0.28E 00	0.18E-02	0.18E 02	0.20E 00	0.30E 00	0.8712	0.7364	1.9668
46	W	20.	0.150	1.020	0.29E 01	0.60E 00	0.10E-02	0.18E 02	0.38E 00	0.57E 00	0.6830	0.5289	1.9091
47	W	-10.	0.150	1.020	0.15E 01	0.32E 00	0.10E-02	0.18E 02	0.24E 00	0.36E 00	1.5730	1.4346	2.5696
48	W	-20.	0.150	1.020	0.41E 01	0.86E 00	0.10E-02	0.18E 02	0.64E 00	0.96E 00	2.5612	2.3801	3.7767

TABLE II. The x_i-Contributions to the Overall Uncertainties

#	F	deg	k_1	k_2	$\frac{\partial F}{\partial k_1}\delta k_1$	$\frac{\partial F}{\partial k_2}\delta k_2$	$\frac{\partial F}{\partial \alpha}\delta\alpha$	$\frac{\partial F}{\partial \omega}\delta\omega$	$\frac{\partial F}{\partial A}\delta A$	$\frac{\partial F}{\partial B}\delta B$	$\frac{\partial F}{\partial E_1}\delta E_1$	$\frac{\partial F}{\partial E_2}\delta E_2$	$\frac{\partial F}{\partial E_3}\delta E_3$	δF	$\frac{\delta F}{U}$ %
1	U	0.	0.100	0.980	0.0455	0.3570	0.0000	0.0000	0.0166	0.0731	0.1080	0.1093	0.0683	0.4765	2.72
2	U	10.	0.100	0.980	0.0455	0.3571	0.0000	0.0000	0.0168	0.0728	0.1027	0.1034	0.0839	0.4902	2.79
3	U	20.	0.100	0.980	0.0455	0.3570	0.0000	0.0000	0.0166	0.0731	0.1093	0.1093	0.0683	0.4765	2.72
4	U	30.	0.100	0.980	0.0455	0.3571	0.0000	0.0000	0.0161	0.0735	0.1124	0.1065	0.0508	0.4583	2.61
5	U	-10.	0.100	0.980	0.0455	0.3570	0.0000	0.0000	0.0168	0.0728	0.0919	0.0912	0.1074	0.5092	2.90
6	U	-20.	0.100	0.980	0.0455	0.3571	0.0000	0.0000	0.0167	0.0729	0.0877	0.0865	0.1145	0.5146	2.93
7	U	-30.	0.100	0.980	0.0457	0.3571	0.0000	0.0000	0.0165	0.0729	0.0857	0.0841	0.1177	0.5170	2.95
8	U	0.	0.150	1.020	0.0638	0.3469	0.0000	0.0000	0.0164	0.0711	0.0949	0.1002	0.0949	0.4926	2.81
9	U	10.	0.150	1.020	0.0638	0.3469	0.0000	0.0000	0.0162	0.0711	0.1010	0.1054	0.0823	0.4819	2.75
10	U	20.	0.150	1.020	0.0638	0.3469	0.0000	0.0000	0.0165	0.0714	0.1066	0.0899	0.0674	0.4686	2.67
11	U	-10.	0.150	1.020	0.0638	0.3469	0.0001	0.0000	0.0163	0.0711	0.0892	0.0840	0.1048	0.5005	2.85
12	U	-20.	0.150	1.020	0.0640	0.3469	0.0002	0.0000	0.0162	0.0713	0.0825	0.0840	0.1147	0.5083	2.90
13	V	0.	0.100	0.980	0.0006	0.0000	0.0000	0.0000	0.0000	0.0000	0.2461	0.2363	0.0098	0.3553	2.03
14	V	5.	0.100	0.980	0.0013	0.1226	0.0546	0.0000	0.0000	0.0000	0.2710	0.2611	0.0098	0.4116	2.35
15	V	10.	0.100	0.980	0.0022	0.2697	0.1088	0.0000	0.0001	0.0000	0.3020	0.2919	0.0098	0.5204	2.97
16	V	15.	0.100	0.980	0.0034	0.4474	0.1621	0.0000	0.0001	0.0001	0.3413	0.3309	0.0099	0.6799	3.88
17	V	20.	0.100	0.980	0.0048	0.6656	0.2143	0.0000	0.0002	0.0001	0.3923	0.3813	0.0099	0.8934	5.09
18	V	25.	0.100	0.980	0.0066	0.9400	0.2647	0.0000	0.0000	0.0002	0.4604	0.4405	0.0099	1.1733	6.69
19	V	30.	0.100	0.980	0.0005	1.3024	0.3132	0.0000	0.0000	0.0000	0.5562	0.5427	0.0100	1.5519	8.85
20	V	-5.	0.100	0.980	0.0009	0.1018	0.0546	0.0000	0.0000	0.0000	0.2262	0.2163	0.0098	0.3480	1.98
21	V	-10.	0.100	0.980	0.0015	0.1853	0.1088	0.0000	0.0000	0.0000	0.2102	0.2003	0.0098	0.3746	2.14
22	V	-20.	0.100	0.980	0.0017	0.3034	0.2142	0.0000	0.0000	0.0000	0.1883	0.1781	0.0098	0.4636	2.64
23	V	-25.	0.100	0.980	0.0018	0.3398	0.2647	0.0000	0.0002	0.0001	0.1816	0.1713	0.0098	0.5076	2.89
24	V	-30.	0.100	0.980	0.0000	0.3617	0.3133	0.0000	0.0000	0.0000	0.1776	0.1672	0.0098	0.5463	3.11
25	V	0.	0.150	1.020	0.0000	0.0000	0.0000	0.0000	0.0000	0.0000	0.1470	0.1301	0.2771	0.5618	3.20
26	V	10.	0.150	1.020	0.0020	0.2561	0.1063	0.0000	0.0000	0.0000	0.2351	0.2449	0.0098	0.3537	2.02
27	V	20.	0.150	1.020	0.0048	0.6321	0.2094	0.0000	0.0001	0.0000	0.2902	0.3002	0.0098	0.5109	2.91
28	V	-10.	0.150	1.020	0.0013	0.1761	0.1063	0.0000	0.0000	0.0001	0.3787	0.3897	0.0099	0.8652	4.93
29	V	-20.	0.150	1.020	0.0022	0.2882	0.2094	0.0000	0.0000	0.0000	0.1995	0.2094	0.0097	0.3683	2.10
30	V	-30.	0.150	1.020	0.0236	0.3034	0.0000	0.0000	0.0000	0.0000	0.1776	0.1877	0.0097	0.4510	2.57
31	H	0.	0.100	0.980	0.0472	0.0029	0.0000	0.6264	0.0000	0.0000	0.1307	0.1478	0.2785	0.8425	4.80
32	H	5.	0.100	0.980	0.0724	0.0058	0.0000	0.6264	0.0014	0.0014	0.1114	0.1287	0.2646	0.8286	4.72
33	H	10.	0.100	0.980	0.1008	0.0089	0.0000	0.6264	0.0028	0.0028	0.0952	0.1130	0.2557	0.8198	4.67
34	H	15.	0.100	0.980	0.1347	0.0125	0.0000	0.6264	0.0041	0.0040	0.0807	0.0996	0.2502	0.8151	4.65
35	H	20.	0.100	0.980	0.1784	0.0168	0.0000	0.6263	0.0053	0.0053	0.0675	0.0876	0.2471	0.8137	4.64
36	H	25.	0.100	0.980	0.0251	0.0221	0.0000	0.6263	0.0066	0.0066	0.0548	0.0772	0.2459	0.8159	4.65
37	H	30.	0.100	0.980	0.0540	0.0032	0.0000	0.6263	0.0079	0.0079	0.0423	0.0680	0.2465	0.8232	4.69
38	H	-5.	0.100	0.980	0.0906	0.0067	0.0000	0.6264	0.0016	0.0015	0.1548	0.1721	0.3001	0.8641	4.93
39	H	-10.	0.100	0.980	0.0176	0.0112	0.0000	0.6263	0.0034	0.0034	0.1871	0.2052	0.3342	0.8981	5.12
40	H	-15.	0.100	0.980	0.1434	0.0176	0.0000	0.6264	0.0057	0.0056	0.2339	0.2538	0.3903	0.9540	5.44
41	H	-20.	0.100	0.980	0.2373	0.0292	0.0000	0.6264	0.0089	0.0089	0.3106	0.3342	0.4905	1.0545	6.01
42	H	-25.	0.100	0.980	0.4922	0.0606	0.0000	0.6264	0.0148	0.0148	0.4613	0.4933	0.6997	1.2672	7.22
43	H	-30.	0.100	0.980	0.0000	0.0000	0.0000	0.6264	0.0306	0.0306	0.9006	0.9595	1.3325	1.9247	10.97
44	H	0.	0.150	1.020	0.0323	0.0029	0.0000	0.6122	0.0000	0.0000	0.1470	0.1301	0.2771	0.8309	4.74
45	H	10.	0.150	1.020	0.0669	0.0114	0.0000	0.6122	0.0027	0.0027	0.1128	0.0954	0.2547	0.8099	4.62
46	H	20.	0.150	1.020	0.1423	0.0241	0.0001	0.6122	0.0052	0.0052	0.0884	0.0685	0.2472	0.8096	4.62
47	H	-10.	0.150	1.020	0.0766	0.0129	0.0000	0.6122	0.0033	0.0033	0.2037	0.1858	0.3328	0.8886	5.07
48	H	-20.	0.150	1.020	0.2033	0.0342	0.0000	0.6122	0.0087	0.0087	0.3317	0.3082	0.4891	1.0543	6.01

$\rightarrow \alpha = 0^\circ$ (row 25)

REFERENCES

Andreopoulos, J., 1982a, "Statistical Errors Associated with Probe Geometry and Turbulence Intensity in Triple Hot-Wire Anemometry," paper to appear in J. Phys. E: Sci. Instrument.

Bradshaw, P., 1975, An Introduction to Turbulence and Its Measurement, Pergamon International Library of Science, Technology, Engineering, and Social Studies, Pergamon Press, pp. 167-168.

Butler, T. L., and Wagner, J. H., 1982, "An Improved Method for Calibration and Use of a Three-Sensor Hot-Wire Probe in Turbomachinery Flows," paper No. 82-0195, AIAA 20th Aerospace Sciences meeting, Orlando, Florida, Jan. 1-14.

Champagne, F. H., and Sleicher, C. A., 1967, "Turbulence Measurements with Inclined Hot Wires, Part 2: Hot-Wire Response Equations," J. Fluid Mech., 28, Part 1, pp. 177-182.

Friehe, C. A., and Schwarz, W. H., 1968, "Deviation form the Cosine Law for Yawed Cylindrical Anemometer Sensors," J. Appl. Mech., 35E, Trans. ASME, Vol. 90, Series E, p. 622.

Frota, M. N., Moffat, R. J., and Ferziger, J. H., 1981-a, "The Triple Hot-Wire Equations and Related Sensitivity Coefficients," Thermosci. Div., Dept. of Mech. Engrg., Stanford University, HTTM Report IL-37.

Frota, M. N., and Moffat, R. J., 1981-b, "Temperature Compensation for Hot-Wire Anemometry," Thermosci. Div., Dept. of Mech. Engrg., Stanford University, HTTM Report IL-47.

Frota, M. N., and Moffat, R. J., 1982a, "effect of Combined Roll and Pitch Angles on Triple Hot-Wire Measurements of Mean and Turbulence Structure," to appear in DISA Bulletin, DISA Information.

Frota, M. N., and Moffat, R. J., 1982b, "Triple Hot-Wire Technique for Measurements of Turbulence in Heated Flows," submitted to the 7th Int'l. Heat Transfer Conference, Munich, West Germany, Sept. 6-12.

Hinze, J. O., 1975, Turbulence, McGraw-Hill, pp. 100-105.

Jørgensen, F. E., 1971, "Directional Sensitivity of Wire and Fiber Film Probes: An Experimental Study," DISA Information No. 15, pp. 31-37.

Kline, S. J., and McClintock, F. F., Jan. 1953, "Describing Uncertainties in Single-Sample Experiments," Mechanical Engineering.

Lakshminarayana, B., Jan.-March 1981, "Three Sensor Hot-Wire/Film Technique for Three-Dimensional Mean and Turbulence Flow Field Measurement, TSI Quarterly.

Moffat, R. J., Yavuzkurt, S., and Crawford, M. E., Sept. 1978, "Real-Time Measurements of Turbulence Quantities with a Triple Hot-Wire System," Proc. of the Dynamic Flow Conference: Dynamic Measurements in Unsteady Flows, Johns Hopkins University, Baltimore, MD, USA, pp. 1013-1035.

Moffat, R. J., 1980, "Contributions to the Theory of Uncertainty Analysis for Single-Sample Experiments, Thermosci. Div., Mech. Engrg. Dept., Stanford University. A position paper for the 1980-81 AFOSR-HTTM-Stanford Conference on Complex Turbulent Flows.

Tutu, H. K., and Chevray, R., 1975, "Cross-Wire Anemometry in High-Intensity Turbulence," J. Fluid Mech., 71, Part 4, pp. 785-800.

Zank, I., 1981, "Sources of Errors and Running Calibration of Three-Dimensional Hot-Film Anemometers, Especially near the Sea Surface," DISA Information No. 26, pp. 11-18, Feb.

The Simulation and Design of Heat-Transfer Experiments on a Rotating Cylindrical Cavity Rig

C. A. LONG and J. M. OWEN
School of Engineering and Applied Sciences
University of Sussex
Falmer, Brighton, Sussex BN1 9QT, England

ABSTRACT

Using a transient analysis technique, temperature measurements on the heated disc of a rotating cavity rig have been simulated. The computed times to reach steady-state temperatures were consistent with experimental observations, and these computed temperatures were in good agreement with those measured. The technique has been used to design a heater, comprising five independently-controlled annular radiant elements, that can generate prescribed temperature profiles on a rotating disc. The simulation and design exercises have demonstrated that it is feasible to determine Nusselt numbers from rotating disc rigs operating under simulated turbine transients.

NOMENCLATURE

a	Inner radius of disc
a'	Inner radius of finite-difference grid
b	Outer radius of disc
$Bi \equiv h\ell/k_d$	Biot number
c	Disc thickness
$C_D \equiv \tau/\frac{1}{2}\rho_a\Omega^2 b^2$	Surface drag coefficient
C_p	Specific heat
$C_w \equiv \dot{m}/\mu b$	Dimensionless mass flow rate
F	View factor
$G \equiv s/b$	Gap ratio
h	Heat transfer coefficient
k	Thermal conductivity
ℓ	Half-thickness of disc
$\dot{m}$	Coolant mass flow rate
n	Surface-temperature power-law index
N	Number of heater circuits
$Nu \equiv hr/k_a$	Nusselt number
$Pr \equiv \mu\, C_{p_a}/k_a$	Prandtl number
Q	Local radiant power input to disc
Q'	Heater radiant power output
Q_t	Total heater radiant power output
r	Radial coordinate from centre-line of disc
$Re_\theta \equiv \Omega b^2/\nu$	Rotational Reynolds number

$Re_z \equiv 2m/\pi\mu$	Axial Reynolds number
s	Axial clearance between discs in cavity
t	Time
T	Temperature
T_s	Disc surface temperature
z	Axial distance from back face of disc
$\alpha \equiv k_d/\rho_d c_{p_d}$	Thermal diffusivity
$\Delta F_o \equiv \alpha\Delta t/\ell^2$	Fourier step-length
Δr	Radial step-length
Δt	Time step-length
Δz	Axial step-length
μ	Dynamic viscosity
ν	Kinematic viscosity
ρ	Density
τ	Fluid surface shear stress
ϕ	Power distribution factor
Ω	Rotational speed of disc

Subscripts

a	Pertaining to air
b	Pertaining to back face of disc ($z = 0$)
d	Pertaining to the disc
f	Pertaining to the front face of disc ($z = c$)
i	Pertaining to the ith heater
j	Finite difference point in radial direction
s	Pertaining to disc surface

1. INTRODUCTION

In order to calculate the stresses in a turbine or compressor disc, the designer first needs to know its temperature distribution. As fatigue life depends on the variation of stress with time, it is necessary to calculate this temperature under both transient and steady-state conditions. This is particularly important for aero-engines where there is a significant change in the radial temperature profile during take-off and land conditions: in the former case, the disc tip is usually hotter than the centre; in the latter case, the converse is true.

At the present time, published heat transfer measurements in rotating disc systems (rotating cavities and rotor-stator systems) have been mainly limited to steady-state conditions. Fig. 1 shows a schematic diagram of the rig used by Owen and Bilimoria (1) and Owen and Onur (2) to measure heat transfer in a rotating cavity. This work concentrated on the determination of heat transfer from the downstream disc of the cavity, which was heated by an external radiant heater. The measurements were confined to the steady-state case, and the radial temperature profile generated by the heater was arbitrary (that is, only the level of power, not its radial distribution, was controllable).Using the measured temperatures on the heated disc as boundary conditions, Laplace's equation was solved numerically and the surface heat fluxes (and hence the Nusselt numbers) were determined by numerical differentiation of the computed temperatures.

Recently, Owen (3) showed that it should be possible to calculate heat

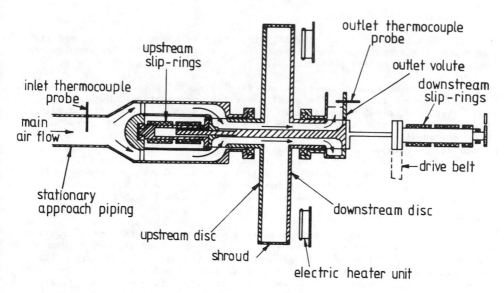

Fig.1 A schematic diagram of the rotating cylindrical cavity assembly

transfer coefficients, on a rotating disc operating under transient conditions,
from the numerical solution of the two-dimensional axisymmetric form of
Fourier's equation. In fact, this technique has since been used to analyse
the data obtained from transient tests of gas turbine rotors. Its application
to rig testing offers two important advantages: (i) an engine transient can be
simulated; (ii) the experimental time can (under many conditions) be significantly reduced.

As stated above, the radial temperature distribution in a turbine disc can
vary significantly with time. The simulation of an engine transient on an
experimental rig requires the correct form of the disc temperature profile to
be generated. To do this, it is necessary to design, and provide control over,
an appropriate heater unit.

The purpose of this paper is to demonstrate that, by solving Fourier's
equation, it is possible to: (i) simulate experimentally-measured temperature
profiles; and (ii) design a heater unit that can produce prescribed temperature profiles on the heated disc. In the future, it is intended to apply this
analysis technique to the computation of Nusselt numbers in rotating disc
systems operating under transient conditions.

Section 2 briefly describes the experimental apparatus, and Section 3
discusses the numerical methods. In Section 4, the computed results are
compared with available temperature measurements. Section 5 outlines the
design of the heater system and compares simulated temperature profiles with
those prescribed.

2 THE EXPERIMENTAL APPARATUS

Full details of the experimental rig, which is shown in fig. 1, and its associated instrumentation can be found in refs (1,2 and 4). For completeness, the salient features are presented below.

The cavity comprised two stainless steel discs of outer radius b = 381 mm and thickness c = 12.7 mm. The axial separation was variable, but, for the tests discussed below, the gap ratio (G $\equiv$ s/b , where s is the axial clearance between the two discs) was G = 0.133 and 0.267. Cooling air (at approximately 20°C) was admitted through a central hole, of radius a = 38.1 mm, in the up-stream disc. For the axial throughflow tests, the air left through a similar central hole in the downstream disc; for the radial outflow tests, it left through a series of holes in the peripheral shroud. The shroud was made from paxolin, 1.5 mm thickness (thermal conductivity, k, of 0.21 W/m K), sealed to the outside of each disc; for the axial throughflow tests, there were no holes in the shroud.

The downstream disc was heated up to a maximum temperature of approximately 100°C, by a bank of thirty cylindrical electrical 'fire-bar elements', each with a maximum output of 750 W. Each element was positioned with its axis on a radial line at an axial distance,from the surface of element to the surface of the disc, of 30 mm. The elements extended radially from $0.6 \leq r/b \leq 1$. Each disc face was instrumented with eleven thermocouples, positioned at equal radial intervals, from $0.2 \leq r/b \leq 1$. For the computation, the inner radius, a', corresponded to the location of the innermost thermocouple (that is, a'/b = 0.2).

3 NUMERICAL AND COMPUTATIONAL METHODS

3.1 An outline of the techniques used

The mathematical and numerical techniques employed for both the simulation exercise and the design study were essentially similar. The simulation work, where the radial temperature profiles were allowed to 'float' according to the imposed boundary conditions, required the solution of Fourier's equation in the 'direct' mode. (In the 'direct' mode, the heat transfer coefficients are specified and the temperatures are calculated; in the 'inverse' mode, the converse applies). The aim of both the simulation work and the design study was to compute a simulated transient temperature response to a given set of boundary conditions. Sections 3.2 and 3.3 outline the methods used in the design study to obtain these boundary conditions; Section 3.4, describing the solution of Fourier's equation, bears equal relevance to both investigations.

For the design study, the complete solution procedure (from the specifica-tion of the prescribed temperature profile to the transient response of the disc for a particular heater configuration) was computed in three distinct stages.

(i)   Laplace's conduction equation was solved to obtain a set of radiant heat fluxes corresponding to the prescribed steady-state radial temperature distribution on the back (heated) face of the disc.

(ii)   The fluxes calculated from (i) above were matched to a set of fluxes derived from an assumed heater configuration (where the axial location of each heater, relative to the disc surface was specified). In this way, the steady-state heater powers were calculated.

(iii) Fourier's (time-dependent) conduction equation was then solved using the flux distribution, calculated from (ii) above, as a boundary condition for the back face of the disc. The computation was carried through to the steady-state enabling a comparison to be made with the prescribed temperature profile.

For the simulation work, the steady-state temperatures obtained from (iii) above were compared with the experimental data of Onur (4).

Numerical methods for the finite-difference solution of both Laplace's equation and Fourier's equation are extensively documented in the literature (see, for example, Gladwell and Wait (5)), and a detailed description of the solution method will, for the sake of briefness, be omitted. For the computation, the finite-difference grid comprised 21 radial and 6 axial nodes (that is, δr = 15.24 mm and δz = 2.54 mm, where δr and δz are the radial and axial step-lengths, respectively). The boundary conditions used in the solution of both equations are described in the respective sub-sections below.

3.2 Laplace's equation

The solution of the steady-state temperature field was obtained subject to the following boundary conditions:-

(i)   On the heated back face of the disc, z = 0, the surface temperature at each grid point was specified.

(ii) On the cooled front face of the disc, z = c, the local Nusselt numbers (representing either 'free-disc' conditions or typical cavity data) were specified. For the free-disc, the correlations given by Dorfman (6) (for a power-law temperature profile where $T_s \propto r^n$) were used. In laminar flow $Re_\theta < 3 \times 10^5$:

$$Nu = 0.308 \ (n + 2)^{0.5} \ Re_\theta^{0.5} \ Pr^{0.6}; \qquad\qquad (3.1)$$

and for turbulent flow ($Re_\theta > 3 \times 10^5$):

$$Nu = 0.0197 \ (n + 2.6)^{0.2} \ Re_\theta^{0.8} \ Pr^{0.6}; \qquad\qquad (3.2)$$

where Nu is the local Nusselt number, Re_θ the local rotational Reynolds number ($\Omega r^2 / \nu$), and Pr is the Prandtl number.

(iii) At the inner radius, r = a', a 'one-dimensional' conduction boundary condition (see refs.1, 2 and 4) was taken such that

$$\left\{ \frac{1}{r} \frac{\partial}{\partial r} \left(r \frac{\partial T}{\partial r} \right) \right\}_{r=a'} = 0 \qquad\qquad (3.3)$$

In practice, however, little difference was found to exist between temperatures computed using eqn.(3.3) and those computed using an adiabatic inner radius boundary condition.

(iv) For the outer radius, r = b, three separate boundary conditions were examined:

(a) the adiabatic outer radius, with $\left(\frac{\partial T}{\partial r} \right)_{r=b} = 0$;   (3.4)

(b) the 'one-dimensional' outer radius with

$$\left\{\frac{1}{r} \frac{\partial}{\partial r} \left(r \frac{\partial T}{\partial r}\right)\right\}_{r=b} = 0; \tag{3.5}$$

(c) the 'rotating cylinder' outer radius. For this latter case, the heat transfer coefficient for a rotating cylinder was specified. The correlation of Kays and Bjorklund (7) was used, where

$$Nu = \frac{Re_\theta Pr \ (C_D/2)^{0.5}}{5Pr + 5\ln (3Pr + 1) + (2/C_D)^{0.5} - 12} \tag{3.6}$$

The drag coefficient, C_D, being obtained from either

$$Re_\theta/B = -1.828 + 1.77 \ln B, \tag{3.7}$$

for $B > 950$ ($B \equiv Re_\theta \ C_D^{0.5}$), or

$$Re_\theta/B = -3.68 + 2.04 \ln B, \tag{3.8}$$

for $B < 950$.

The solution of Laplace's equation was used for the design of the new heater system, where the incident fluxes were a function of some prescribed surface temperature profile, $(T_s \propto r^n)$. Having calculated the internal temperature field, the heat flux at each surface grid point on the heated back face of the disc was evaluated by a second-order backward-difference numerical differentiation of the axial temperature field. Knowing the heat fluxes at these points (and assuming a free-disc convective heat transfer coefficient, given by equation (3.1) or (3.2), for the back face) enabled the local 'ideal' radiant fluxes to be calculated. The individual heater powers and axial location of each heater circuit were then determined as described below.

3.3 The radiant heater solution.

The heater was assumed to comprise a number(five was chosen) of independently-controlled annular elements. The total heater power output is controlled by: (i) the number of independent elements; (ii) the physical dimensions and axial location of each element; and (iii) the partitioning of the required total power to each element. The heat input to any annular ring on the disc surface is therefore a function of these variables.

The various geometric parameters may be expressed in terms of a local view factor, F_{ij}, where i and j refer to the ith heater circuit and the grid point j on the disc, respectively. Wong (8) provides an expression for the view factor for two opposing, annular, concentric rings, and this was evaluated to specify a radiant flux boundary condition. This boundary condition was used for both the simulation work (for the existing 'fixed-profile' heater system used by Onur (4)) and for the design of a variable-temperature-profile heater system.

The local radiant heat input, Q_j, at a grid point of location j on the disc may be written as the summation of the contributions from each heater circuit. That is, for N heaters

$$Q_j = \sum_{i=1}^{N} F_{i,j} Q_i' \tag{3.9}$$

where Q_i' is the radiant power output from the ith heater and $F_{i,j}$ the local
view factor (which is a function of the geometric variables of each heater
element). For the purpose of an engineering analysis, the number of elements
and the dimensions (ring width and diameter) of each heater were assumed to be
fixed. The remaining variables (individual heater power and axial location of
each heater) were used to match the 'derived' fluxes (from equation (3.9)) to
the 'ideal' flux distribution (found as described in Section 3.2).

     In practice, the matching was achieved quite simply by the solution of a
set of simultaneous equations for Q_i', formulated from equation (3.9), where
the surface grid point under consideration was that directly opposite the ith
heater. The solution of such a system ensured correspondence between the
'derived' and 'ideal' fluxes at each of these grid points; however, there was
some degree of mismatch of fluxes between these points. This discrepancy was
'minimized' by varying the axial location of independent heaters relative to
the disc surface.

     Specification of both axial position and power output from each heater en-
abled a set of power 'distribution factors', ϕ_j, to be calculated. That is,

$$Q_j = Q_t \phi_j = \sum_{i=1}^{N} F_{i,j} Q_i' \tag{3.10}$$

where Q_t is the total radiant power output of the entire heater system. In
addition to the radiant heat, Q_j, the free disc convective heat transfer coefficient was used to provide the boundary condition for the heated face of the
disc in the solution of Fourier's equation, which is discussed below.

3.4 Fourier's equation.

     The solution of the time-dependent equation was obtained using a Crank-
Nicholson (semi-implicit) scheme. The boundary conditions were the same as
those specified in Section 3.1, with the exception of the heated back face,
$z = 0$. For this face, Q_t and ϕ_j, calculated as described in Section 3.3, were
used in conjunction with the appropriate free-disc correlation; this involved
the superposition of the local radiant and convective fluxes at each grid point.

     The computation was carried through to the steady-state using a time-step,
Δt, calculated from the criterion of Owen (3). This criterion, which was based
on the truncation error for the so-called one-dimensional 'quenching problem',
is

$$\Delta F_o = 0.01 \, Bi^{-1} \tag{3.11}$$

where $\Delta F_o \equiv \alpha \Delta t / \ell^2$ and $Bi \equiv h\ell/k_d$; $\alpha, \Delta t, k_d, \ell$ and h being the thermal diffusivity, time-step, thermal conductivity of the disc, half-thickness of the disc,
and heat transfer coefficient, respectively. For the stainless-steel disc,
$\alpha = 5.1 \times 10^6 \, m^2/s$, $k_d = 16.8 \, W/mK$, and $\ell = \frac{1}{2}c = 6.35$ mm.

     It should be noted that, for the 'quenching problem', the surface temperature
is within 0.1% of the steady-state value after 640 Δt, where the latter is calculated from eqn (3.11). The transient computations discussed in Section 4 were
stopped after 640 time-steps, and the resulting temperature is referred to as

the 'steady-state'. These computations were started, at time t = 0, by assuming
that the disc temperature was uniform, and equal to the coolant inlet temperature, and a step-change in the radiant flux was introduced.

4 SIMULATION OF EXPERIMENTALLY-MEASURED TEMPERATURES

In this section, the experimental data of Onur (4), who used the rig described in Section 2, are used to test the transient analysis technique described
in Section 3. For simplicity of analysis, the radiant heater was assumed to be
an annulus of inner and outer radii 228 mm and 381 mm, respectively, located at
an axial distance of 30 mm from the back face of the downstream disc, which was
of radius 381 mm.

For the inner and outer radii of the disc, eqns (3.3) and (3.6)(the agreement between the computed and measured temperatures was better when eqn (3.6)
was used than when either (3.4) or (3.5) was used) were used as boundary conditions; and for the front (cavity-side) face of the disc, Onur's Nusselt
numbers were specified. For the back (free-disc) face, a combined radiant flux
(based on the above heater geometry and the view factors of Wong (8))and a
convective heat transfer coefficient (based on eqns (3.1) and (3.2) with Pr =
0.72 and n = 2) were used as the boundary conditions. In all cases, the
radiant power was chosen to produce agreement between Onur's maximum measured
temperature (which was approximately $100^{\circ}C$ and occurred at $r/b \simeq 0.8$) and the
computed temperature at this radial location. Apart from this one point, agreement between measured and computed temperatures depends on the validity of the
numerical method and the associated boundary conditions.

Typical comparisons between measured and computed front-face temperature
profiles, $T_{s,f}$, for both the radial outflow and axial throughflow cases, are
shown in figs 2 to 5; for each figure, there are ten computed curves. Each
curve corresponds to times of the following multiples of the time-step,Δt: 1,2,
4,10,20,40,80,160,320 and 640. The value of Δt for all tests was chosen to be
2,4,8 or 16 s, according to the value calculated from eqn (3.11) (where h was
calculated from Onur's mean Nusselt numbers). It should also be pointed out
that, in figs 2 to 5, eqn (3.6) was used for the boundary conditions at the
outer radius; in the main, as stated above, tests using eqns (3.4) and (3.5)
produced less satisfactory results.

Fig. 2 shows the results for the radial outflow case with G = 0.133,
C_w = 3300 ($C_w \equiv \dot{m}/\mu b$, where $\dot{m}$ is the coolant flow rate and μ the absolute
viscosity) and Re_θ = 1.9 x 10^6 ($Re_\theta \equiv \mu b^2/\nu$). The overall agreement between the
measured steady-state temperatures (on the front face of the heated disc) and
those computed at t = 1280 s (160 Δt) is reasonable. It should be noted that the
results for 320 Δt and 640 Δt are superimposed on those for 160 Δt. The agreement for $r/b \gtrsim 0.8$ is excellent, but at the smaller radii there is a significant
difference between measured and computed values. This is attributed, in part
at least, to uncertainty in the boundary conditions near the centre of the
cavity. The assumed Nusselt numbers (Onur's measured values on the front face
and laminar free disc values on the back face) are likely to be inaccurate at
these smaller radii.

For fig.3, at the same gap ratio but with C_w = 1.4 x 10^4 and Re_θ = 2.5 x 10^5,
the agreement is better, particularly at the smaller radii. It can also be
seen, by comparing figs 2 and 3, that there is a significant difference(which
is caused by the convective boundary conditions) between the shapes of the two

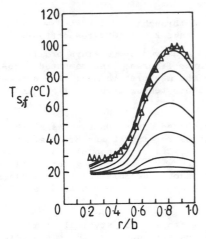

Fig. 2. Radial temperature distribution for the radial outflow case.
$G = 0.133$, $C_w = 3300$, $Re_\theta = 1.9 \times 10^6$
△ Measured temperature
— Computed temperature($\Delta t = 8s$)

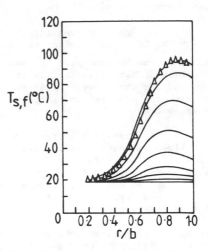

Fig 3 Radial temperature distribution for the radial outflow case
$G = 0.133$, $C_w = 14230$, $Re_\theta = 2.5 \times 10^5$
△ Measured temperature
— Computed temperature ($\Delta t = 4$ s)

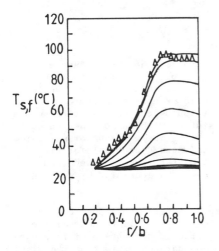

Fig 4 Radial temperature distribution for axial throughflow case
$G = 0.267$, $Re_z = 0.9 \times 10^5$, $Re_\theta = 0.9 \times 10^5$
△ Measured temperature
— Computed temperature($\Delta t = 8s$)

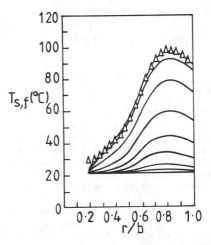

Fig 5 Radial temperature distribution for axial throughflow case
$G = 0.267$, $Re_z = 0.9 \times 10^5$, $Re_\theta = 2.6 \times 10^5$
△ Measured temperature
— Computed temperature($\Delta t = 16s$)

profiles for r/b $\gtrsim$ 0.8.

Figs 4 and 5 show the results for the axial throughflow case at G = 0.267 Re_z= 0.9 x 10^5($Re_z \equiv 2\dot{m}/\pi\mu a$) for Re_θ = 0.9 x 10^5 and 2.6 x 10^5, respectively. In fig. 4, it can be seen that the shape of the measured temperature profile is significantly different to that in the other three figures; the computed temperature profiles show very similar features to the measured values. The agreement between the measured and computed steady-state temperatures in fig.5 is extremely good.

Although no transient temperatures were recorded during Onur's tests, the computed times to reach a steady state (43 minutes for figs 2 and 3, 85 minutes for fig.4, 171 minutes for fig.5) were consistent with experimental observations. It can be appreciated that steady-state measurements require a long period of settling time. By contrast, transient analysis should be able (apart from flows in which the heat transfer coefficients are time-dependent) to reduce experimental time significantly.

It should be pointed out that the above results are not atypical: some experimental conditions produce slightly better comparisons and others significantly worse. In particular, radial outflow experiments with low Re_θ and high values of C_w (the so-called regime I of Onur) tended to produce data that were impossible to simulate; the reason for this is not understood.

The above tests have demonstrated that the analysis techniques can, under most conditions, simulate existing steady-state data with reasonable accuracy. It is now appropriate to consider the design of a heater system, for transient experiments, where control is required over the form of the radial temperature profile.

5. THE 'DESIGN AND TESTING' OF A DISC HEATER

During acceleration of a gas turbine, the tips of the turbine and compressor discs tend to heat up faster than the centres; during deceleration, the converse occurs. Also, the speed transient is usually much shorter than the thermal transient. If an experimental rig were to be used to measure the Nusselt numbers under transient conditions, it would be necessary to generate temperature profiles, on the heated disc, consistent with those experienced in the engine. Power law profiles of the form $T_s \propto r^n$ (where n > 0 for simulated engine accelerations, and n < 0 for decelerations), are convenient for this purpose. Experiments could be carried out at the steady-state speed (and coolant flow rate), and the thermal transient could be generated by putting a step-change into the heater.

The rig described in Section 2 has a simple heater: it is possible to control the level of the radiant flux but it is not possible to control the radial distribution of flux. As a consequence, the temperature profile generated by this heater depends on the convective boundary conditions on the disc; the disc temperature cannot, therefore, be prescribed.

In order to generate power-law profiles, it is necessary to use a compound heater comprising a number of independently-controlled annular elements. For the design discussed below, five concentric annular elements were chosen to extend from 0.4 $\leq$ r/b $\leq$ 1. The fluxes for each of the five elements were calculated to produce an acceptable approximation to the required temperature profile.

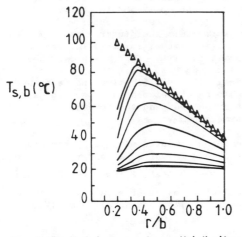

Fig. 6 Radial temperature distribution
for the free disc case: $Re_\theta = 2 \times 10^6$
$\triangle$ Prescribed temperature($T_{s,b} \propto r^{-1}$)
— Computed temperature ($\Delta t = 4s$)

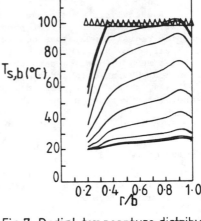

Fig. 7 Radial temperature distribution
for the free disc case: $Re_\theta = 2 \times 10^6$
$\triangle$ Prescribed temperature($T_{s,b} \propto r^0$)
— Computed temperature($\Delta t = 4s$)

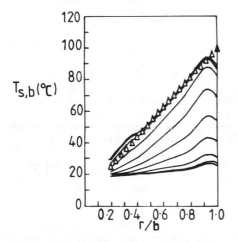

Fig. 8 Radial temperature distribution
for the free disc case: $Re_\theta = 2 \times 10^6$
$\triangle$ Prescribed temperature($T_{s,b} \propto r^1$)
— Computed temperature ($\Delta t = 4s$)

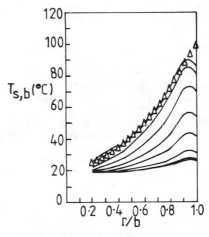

Fig. 9 Radial temperature profile for the
free disc case: $Re_\theta = 2 \times 10^6$
$\triangle$ Prescribed temperature($T_{s,b} \propto r^2$)
— Computed temperature($\Delta t = 4s$)

The technique was 'tested' on a free disc, in which the Nusselt numbers on both back and front faces were calculated from eqns (3.1) and (3.2) (with Pr = 0.72). Eqns (3.3) and (3.6) were used for the inner and outer radii, respectively. The analysis was conducted in two parts: the 'inverse problem' was solved first, and then the 'direct problem' was solved. For the 'inverse problem', the radial temperature on the back face of the disc was prescribed, and the 'ideal' flux distribution on this face was computed. The 'derived' heater fluxes were then matched to the 'ideal' flux distribution, as described in Section 3.3. Having obtained an acceptable match between the 'derived' and 'ideal' fluxes, the 'direct problem' was then solved, and the computed back-face temperatures were compared with the prescribed values.

Figs 6 to 9 show the comparison between the computed and the prescribed back-face temperature profiles, $T_{s,b}$, on a free disc, with $Re_\theta = 2 \times 10^6$. The four figures correspond to $T_{s,b} \propto r^n$ with n = -1,0,1 and 2, respectively. Each curve corresponds to times of the following multiples of Δt: 1,2,4,10,20,40,80, 160, 320 and 640. In accordance with eqn (3.11), Δt was chosen to be 4 s for all these computations.

Fig.6, for n = -1, is appropriate to a thermal transient following an enging deceleration. The agreement between the steady-state prescribed and computed temperatures is good for $r/b \geq 0.4$. For smaller radii, where the heater power is limited, the agreement is less satisfactory. However, at these radii, the temperature distribution would be expected to exercise only a small influence on heat transfer. The results for figs 7,8 and 9 also show good agreement, apart from the region $0.2 \lesssim r/b \lesssim 0.4$, between prescribed and computed temperatures.

Computations have also demonstrated that it is, in principle, possible to generate power-law profiles for a rotating cavity. The technique shows great potential for determining the Nusselt numbers from experimental rigs operating under simulated engine conditions.

6. CONCLUSIONS

A numerical method has been developed to: (i) simulate the results of existing heat transfer experiments on rotating discs; and (ii) design a radiant heater that will generate prescribed temperature profiles on a heated disc.

For the simulation, steady-state temperature measurements, obtained from the heated disc of a rotating cavity rig, were reproduced using transient analysis. For this analysis, Fourier's equation was solved by modelling the heater as an annular radiant source and by using appropriate convective boundary conditions for the disc surface. The computed steady-state surface temperatures were, in most cases, in good agreement with measured values, and the computed time to reach the steady-state was consistent with experimental observations.

For the design, the heater was considered to comprise five concentric independently-controlled annular radiant elements. By prescribing a power-law temperature profile (with the form $T_s \propto r^n$, on the heated face of the rotating disc), the required flux for each element could be computed from the 'inverse' solution of Laplace's equation. Using these heater fluxes, Fourier's equation was solved in the 'direct' mode to simulate a transient experiment on a rotating disc, and the surface temperatures were calculated. For the free disc test cases considered (with n = -1, 0, 1, 2), the simulated temperature profiles were in good agreement with the prescribed values, apart from $0.2 < r/b < 0.4$, where insufficient heater power was available.

The simulation and design exercises have demonstrated that it is possible to determine Nusselt numbers from rotating disc rigs operating under conditions that simulate gas turbine transients.

ACKNOWLEDGEMENTS

The authors wish to thank Motoren-und Turbinen-Union, Rolls Royce Limited and the Science and Engineering Research Council for supporting the above research.

REFERENCES

1. Owen, J.M. and Bilimoria, E.D. 1977. Heat transfer in rotating cylindrical cavities. J.Mech.Engng Sci., 19, p.175.

2. Owen, J.M. and Onur, H.S. 1982. Convective heat transfer in a rotating cylindrical cavity. ASME 27th Intl.Gas Turbine Conference,Paper No.82-GT-145.

3. Owen,J.M. 1979. On the computation of heat-transfer coefficients from imperfect temperature measurements. J.Mech.Engng Sci. 21, p.323.

4. Onur, H.S. 1980. Convective heat transfer in rotating cavities. D.Phil Thesis, University of Sussex.

5. Gladwell,I. and Wait, R. 1979. A survey of numerical methods for partial differential equations. Oxford University Press.

6. Dorfman, L.A. 1963. Hydrodynamic resistance and heat loss of rotating solids. Oliver and Boyd, Edinburgh.

7. Kays, W.M. and Bjorklund, I.S. 1958. Heat transfer from a rotating cylinder with and without cross flow. Trans. ASME, 80, p.70.

8. Wong, H.Y. 1977. Handbook of essential formulae and data on heat transfer for engineers. Longman, London.

Steam Condensate Droplet Evolution: Experimental Results

ROBERT A. KANTOLA
Turbine Business Group
General Electric Company
Lynn, Mass., USA

PAUL V. HEBERLING
Corporate Research and Development
Schenectady, N.Y., USA

ABSTRACT

This paper discusses the results of experiments on the non-equilibrium condensation of steam. The thermodynamic conditions studied here replicate those seen in actual steam turbines. The appearance of moisture in turbine is well-known to cause both thermodynamic and mechanical losses and, in some cases, erosion of turbine blading. To simulate the expansion of steam in a turbine, a non-steady expansion wave technique is used. This method allows selection of steam expansion rate and starting conditions. An optical droplet measurement technique using the attenuation of laser light at two wave lengths allows determination of the onset of nucleation and the droplet size and number density histories.

By assuming an isentropic expansion, the measured steam pressure can be used along with the droplet size and number density to predict the ratio of condensate mass to the moisture available from an equilibrium basis. From this comparison a considerable lag in the approach to equilibrium wetness is seen. Some of the prior investigators have assumed that the amount of condensate quickly approached equilibrium and then used this assumption to calculate droplet size from single color attenuation measurements. In this paper it is shown that this "return to equilibrium" assumption can produce a serious underestimation of droplet size.

The basic issue of what mechanism controls the droplet growth rate is raised. It is extremely important to know whether isolated droplet growth via impinging steam molecules (the assumption on which the classical nucleation theory stands) is dominant or whether agglomeration plays a role. The knowledge of when these different mechanisms are controlling can be of vital importance in trying to reduce the moisture-induced losses in steam turbines.

1.0 INTRODUCTION

The first serious studies of condensation in steam turbines were published by Stodala (Ref. 1) in 1927. This pioneering work has been followed by research and development on the effects of moisture in turbines for the past fifty-odd years. In fossil-fired turbines, the moisture doesn't usually appear until the low pressure stages. Even though the effects of moisture are limited to the last stages, they impose a lower limit on the expansion of the steam. For turbines in nuclear power plants, due to the low temperature

Robert A. Kantola was formerly with General Electric Corporate Research and Development.

available, moisture can occur throughout the steam path and water separators are a necessary part of the design.

Reducing the moisture penalties requires knowledge of the point of nucleation onset, and the droplet size growth and number densities must be known over a range of pressures and expansion rates. The key to achieving higher efficiency in the wet stages of steam turbines is to keep the condensed steam droplet size very small throughout the turbine. It is the goal of the work reported herein to provide experimental data on nonequilibrium steam condensation in its simplest form - that is, without the effects of turbulence, sheared flows, temperature gradients and interactions between the condensation front and shock waves. The data will concentrate on the size growth of the droplets and the droplet number density from onset to the later stages of condensation.

The technique for investigating steam condensation that has been developed overcomes the above objections to the prior "state-of-the-art" data. This subsonic technique uses a nonsteady expansion wave to provide a controlled, but rapid, expansion of slightly superheated steam. An optical attenuation technique (with two wavelengths, 457.9 nm and 632.8 nm) is used and is able to separately determine both droplet size and number density. The experimental approach and measurement techniques used in this study are described in a companion paper, Reference 2, written by one of the authors. A brief description of the experimental technique is given in the following section.

2.0 EXPERIMENTAL APPROACH

The expansion wave technique that is used in these experiments provides a controlled, but rapid expansion of slightly superheated steam. Conceptually, the apparatus called the nucleation tube consists of an externally heated tube which is permanently closed at one end and sealed with a set of rupture diaphragms at the other. The nucleation tube is filled with a slightly superheated (20°F) steam upstream from the rupture diaphragms. When the diaphragms are ruptured, a non-steady expansion wave centered at the orifice/diaphragm station propagates through the steam. The expansion wave reflects from the closed end of the tube and processes the steam a second time as it travels toward the orifice. To vary the expansion rate the tube length is changed, keeping the test section near the closed end. The pressure decay in this wave is very rapid and after arrival of the wave onset usually occurs within a 1 ms time interval. After onset the droplet growth transient lasts up to 1.5 ms. To capture the optical and pressure information, high speed transient recorders are used. Time delays are set in such a manner that the data sampling occurs only in the time frame of interest.

3.0 EXPERIMENTAL RESULTS

3.1 Scope of Testing

The experimental parameters of interest are the pressure at onset and the expansion rate at onset. To insure that the nucleation tube is "dry" prior to the test the steam is superheated about 20°F above saturation conditions. The initial steam pressure is varied to change the onset pressure and a range of 23 psia to 750 psia has been tested. This provides onset pressures ranging from about 8 psia to about 450 psia. To vary the expansion rate two nominal tube lengths have been used, 42 and 18 inches, these provide expansion rates of about 375 S^{-1} to 470 S^{-1} and 460 S^{-1} to 1100 S^{-1}, respectively. For a given mechanical configuration the expansion rate increases slightly with initial pressure. In a companion paper (Ref. 2) the sensitivity of the nucleation process to low impurity levels was demonstrated. To reduce the effect of impurities all the

testing reported herein was done with demineralized, triple-distilled feedwater. To quantify the steam purity, steam was allowed to flow through the
heated nucleation tube, and then was condensed in a heat exchanger. In-line
analyzers measured both conductivity and the sodium ion concentration of this
condensate with typical values ranging from 0.2 to 2.0 megaohm-cm and 0.3 to
0.6 ppb Na, respectively. To reiterate, to define this process three basic
measurements are required, onset of nucleation, droplet size and droplet number
density histories. Discussion will start with the onset data.

3.2   Onset Data

An example of the measured signals is shown on Figure 1. Onset is easily
located as the sudden change in the attenuation signals. Since a common time
base is used for pressure as well as attenuation the pressure at onset can be
accurately measured. By assuming an isentropic expansion, up to onset, a
Mollier chart can be used to locate the equilibrium wetness at onset (Wilson
line), as shown in Figure 2. To locate the onset point the supersaturated vapor is assumed to expand as a perfect gas with the properties of the initially
superheated steam. For the same pressure this supersaturated line is located
at higher entropies than the equilibrium condition and therefore shifts the
onset point (about 1/4 to 1/2 percent) to a drier equivalent equilibrium value.
This procedure has been used previously by Yellot(Ref.3 and 4). The starting
point and the expansion rate at onset are also indicated in Figure 2. Although
the expansion rates are given to 4 significant figures this is only to identify different test points, the expansion rate accuracy is no better than 2 significant figures. In some cases the optical signal exhibited a slow change
prior to the typical sudden change, the typical signal is shown in Figure 1.
The start of this slow change has been labelled as "early" onset. These instances of early onset occur at the higher pressures(above 75 psia) and could
possibly be due to hetergenous nucleation on vaporized impurities. At the
lower onset pressures the number of nucleation sites is greatly reduced due to
the lower steam density and the lower solubility of the impurities. The resulting lower signal could cause any hetergenous nucleation to go undetected.
In Figure 1, it can be seen that the equilibrium moisture at onset is only
slightly increased with onset pressure and very little affected by the expansion rate.

To compare these data to prior work, the data of Yellot (Ref. 4) seems to
be the most appropriate. Yellot used a convergent-divergent nozzle with a high
expansion rate (around $5 \times 10^4 S^{-1}$). In a manner somewhat similar to that used
here, Yellot detected onset by observing light scattering from the droplet fog.
The envelope of Yellot's data is presented in Figure 2. In general Yellot's
data lies between 4 and 5 percent moisture with a slight trend to higher moisture as onset pressure increases. Yellot also pointed out that there was a
close correlation between the optically determined onset and a sharp break in
the pressure curve. From our testing, conducted with the nucleation tube, the
change in the pressure decay curve, caused by the release of latent heat,
doesn't occur until well after onset when significant condensation has taken
place. For this reason it would be expected that Yellot's onset data should
be at somewhat lower pressure (higher moisture) than the data reported herein.
From Figure 2 it can be seen that is what generally happens. At the high
pressures the onset data are in good agreement with the earlier data of Gyarmathy, et al, (Ref. 5). Although onset can be located accurately in a given
test, repeat tests can, at times, give substantilly different onset conditions.
This has occurred with both mechanical configurations and over a wide range of
onset pressures. This variation causes the Wilson Line to be between the 3 to
5 percent wetness lines. It is suspected that hetergenous nucleation plays

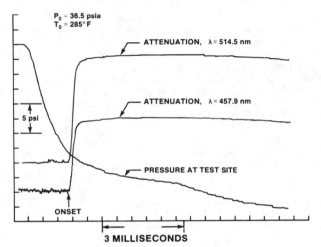

FIGURE 1. TYPICAL PRESSURE AND ATTENUATION
TRACES

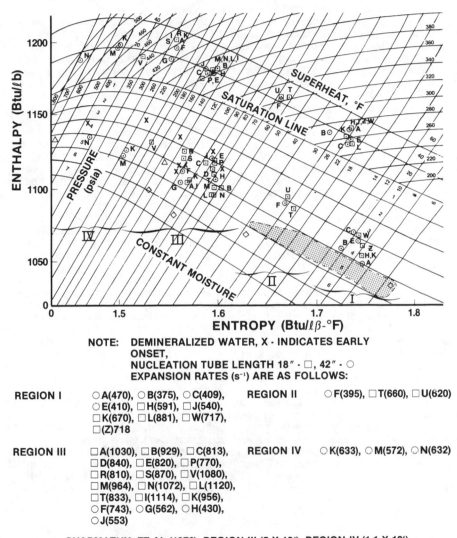

NOTE: DEMINERALIZED WATER, X - INDICATES EARLY ONSET,
NUCLEATION TUBE LENGTH 18″ - □, 42″ - ○
EXPANSION RATES (s⁻¹) ARE AS FOLLOWS:

REGION I	○A(470), ○B(375), ○C(409), ○E(410), □H(591), □J(540), □K(670), □L(881), □W(717), □(Z)718	**REGION II** ○F(395), □T(660), □U(620)
REGION III	□A(1030), □B(929), □C(813), □D(840), □E(820), □P(770), □R(810), □S(870), □V(1080), □M(964), □N(1072), □L(1120), □T(833), □I(1114), □K(956), ○F(743), □G(562), ○H(430), ○J(553)	**REGION IV** ○K(633), ○M(572), ○N(632)

△ GYARMATHY, ET AL (1973), REGION III (5 X 10⁴), REGION IV (1.1 X 10⁵)

◇ SALTANOV, ET AL (1973) 3 X 10⁶ TO 6 X 10⁶

CROSS-HATCHED AREA - YELLOT, ET AL (1937)

FIGURE 2. EQUILIBRIUM WETNESS AT NUCLEATION ONSET (WILSON LINE)

some part in these variations.

3.3   Droplet Growth

Reliable measurements of droplet sizes occur 20 to 30 microseconds
after onset. At times closer to onset the signal to noise ratios are not adequate for accurate droplet sizing. High speed sampling with digital transient
recorders allows sample intervals as low as 0.2 microseconds. With up to
4096 samples, averaging over many samples is used to reduce the effects of
random background noise and "smooth" the droplet time histories. The procedure for determining the droplet size is disucssed in Reference 2.

An example of a typical droplet growth history is shown in Figure 3.
With the high sampling rates used here a detailed history of the droplet growth
can be determined. Also shown in this figure are the steam pressure and the
corresponding available equilibrium wetness based on an isentropic expansion
from the initial conditions. To characterize the droplet growth and study the
effects of onset pressure and expansion rate, the final or plateau diameters
are used. In some cases the droplet diameter did not level out during the test
interval and that data could not be used in this comparison. Using this basis
does, however, lead to an interesting result, as shown in Figure 4. Here the
final droplet sizes are plotted against the steam pressure, at onset, for both
the long and short nucleation tube configurations. The onset expansion rate,
$\dot{p}/p$, is shown as a parameter in this figure. To a large degree the higher expansion rate data (18" tube) exhibit droplet sizes nearly one-half of those
measured in the long tube (42"). Increasing steam pressure, at onset, over a
ten to one range causes the droplets to approximately double in size. Both of
the effects are in general agreement with the arguments presented by Gyarmathy
(Ref. 5) which are based on classical nucleation theory. For comparison with
these measurements, data on droplet sizes are available from prior publications,
References 5,6,7,8 and 9. All the prior data with the exception of Ref. 9 were
obtained from nozzle tests. In Ref. 9 a small turbine, with an expansion channel, was used. Absolute comparisons are very difficult to make, because in the
prior data the evolutionary stage of the condensate cloud is hard to determine.
This is due to the small number of downstream sample points usually taken in
the typical nozzle or turbine test. For the comparison, the final droplet sizes,
quoted in these prior investigations, are included in Figure 5. Only the data
of Dibelius (Ref. 9) spans the same thermodynamic conditions as the data reported herein, and there the agreement is reasonably good. The trends with expansion rate as seen in the prior data, are consistent with the data reported
herein, although the conditions do not match. It should be noted that in our
total test program the maximum droplet diameter measured was 1.4μm, this value
is within the upper bound (1.5μm) of other reported data (Smith Ref. 10).
Further discussion of this subject will be taken up in a later section.

3.4   Number Density

With the droplet diameter determined the attenuation measurements can be
used to establish droplet number density, assuming a monodisperse droplet
cloud. To gain a better understanding of the process, both the droplet growth
and number density histories are shown together in Figures 6,7 and 8, which
also serve to illustrate typical features seen in the data. A most interesting difference is seen in the number density trends for some of the short
tube tests versus the long tube results. The short(18") tube result, as seen
in Figure 7, shows a rising number density with time, then sometimes, as in the
case here, a slight peak occurs. On the other hand, the long(42")tube data (Figure 6) generally shows an initial drop in number density and then a levelling
off. A decreasing number density with time can be explained by an agglomeration

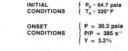

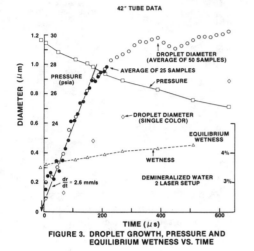

FIGURE 3. DROPLET GROWTH, PRESSURE AND
EQUILIBRIUM WETNESS VS. TIME

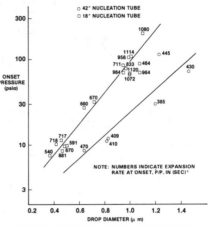

FIGURE 4. DROPLET DIAMETER VS. ONSET PRESSURE

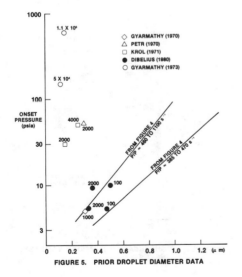

FIGURE 5. PRIOR DROPLET DIAMETER DATA

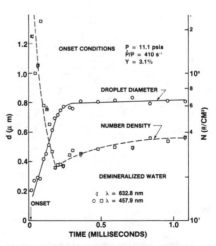

FIGURE 6. DROPLET DIAMETER AND DENSITY HISTORY,
P_0 = 23.5 psia, T_0 = 250°F, 42″ TUBE

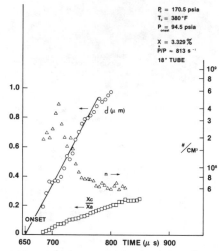

FIGURE 7.   DROPLET CLOUD HISTORY, AGGLOMERATION
            CONDITION

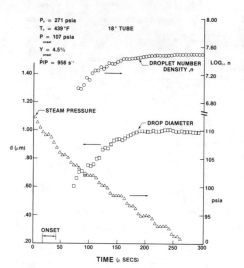

FIGURE 8. DROPLET CLOUD HISTORY,
           HIGH EXPANSION RATE

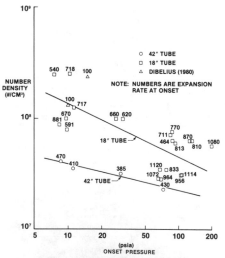

FIGURE 9.   DROPLET NUMBER DENSITY VS. ONSET
            PRESSURE

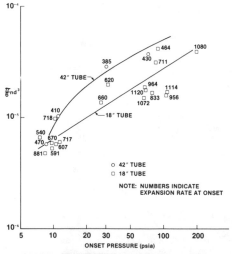

FIGURE 10.   CONDENSATE VOLUME/UNIT VOLUME VS.
             ONSET PRESSURE

process. For the lower expansion rate cases the initial increase and peak are not seen, only the decrease is noted, however, the peak may occur between on-set and the first measureable diameter. It should be noted that determining the correct droplet sizes is crucial. If the size is underestimated, for the small sizes, the number density will be overestimated as the attenuation measurement for small drop size varies with the sixth power of diameter. In the long tube results every test run exhibited this initial drop in number density. Likewise for the data case shown in Figure 7, three other tests at these same starting conditions were conducted and exhibited a similar peak in the number density hisotry. At other conditions, for the short tube, Figure 8 illustrates the typical data. Here a sharp rise in number density is followed by a flat plateau. This latter rising trend is in agreement with the nozzle experiments.

To determine the effects of onset pressure and expansion rate on the final number densities the data from both the long and short tube are plotted versus onset pressure with expansion as a parameter. Again as in the droplet growth data (Figure 4) the data splits into two groups as shown in Figure 9. The short tube tests with their higher expansion rates have about three to four times more drops per unit volume than the long tube results. This again agrees with the arguments of Gyarmathy (Ref. 5) which are based on classical nucleation theory. Again from Figure 9, the effect of onset pressure is to lower the droplet number density, despite the slight increase in expansion rate.

Earlier it was mentioned that a monodisperse droplet cloud was assumed. To qualify that assumption and assess the polydispersity of the droplet fog, the attenuation levels from both wavelengths can be used. From the work of Dobbins and Jizmagian (Ref. 11), it is known that when the scattering coeffi-cient is integrated over a polydisperse droplet distribution the "effective" coefficient will be changed. Since the original monodisperse droplet scat-tering coefficient was used to determine the functional relation between drop-let size and the ratio of the attenuation levels, then a necessary, but not sufficient condition, is that both wavelengths should predict the same droplet number density if the droplet fog is truly monodisperse. If the droplet fog is polydisperse, the two calculations of the number density could be different. In Figure 6, typical results of such a calculation can be seen. The blue line (λ = 457.9 nm) predicts a slightly higher droplet number density. This sligh-tly higher droplet number density is consistent with a droplet size distribu-tion that is slightly biased to the smaller sizes. The difference here is quite small, about 3% to 4%. Although it is not conclusive, a nearly monodis-perse droplet cloud is indicated. Now with measures of both the droplet size and the number density, based on a monodisperse droplet cloud, the condensed moisture can be calculated.

3.5   Condensate Accumulation

The density of water does not change greatly over the range of test tem-perature and pressure. Consequently, a simple product of number density and diameter to the third power will be proportional to the condensate volume, and will also be a good estimate of the condensed moisture mass. To obtain an overall comparison the plateau values of droplet diameter and number density are used to calculate the volume of the condensate and the results are shown in Figure 10. The effect of increasing the expansion rate, shown as a para-meter on Figure 10, is not as definitive as seen for droplet size and number density, since the increase in number density is counterbalanced by a reduction in droplet diameter. As the onset pressure is increased the increase in con-densate volume lags behind indicating that the ratio of actual condensate, Y, to that available at equilibrium, Ye, is decreased at the higher onset pressures. This effect is illustrated in Figures 11 and 12. The available moisture is found by assuming that an isentropic expansion from the initial starting conditions is occurring, so that only the pressures need to be known to define all the properties. This calculation provides an additional check on the quality of data as the condensate mass should approach the equilibrium values but not exceed them. In reviewing the data, it was seen that the ratio of condensed-to-available moisture does indeed approach, but does not exceed unity. What was particularly evident though was that for a long time the ratio is much less than unity, therefore a commonly used assumption concerning a complete "return to equilibrium" may not be valid. Moreover, due to the subcooling, even the equilibrium moisture is not available at onset. To understand this comment, a brief discussion follows.

3.5   Condensate Cloud Thermodynamics

        In this section a brief look will be taken at the thermodynamics involved with the condensate cloud and the surrounding steam. The principle variable is the subcooling of the surrounding steam with respect to the droplets. At equilibrium both the droplets and the steam are at the same temperature, so without subcooling no net condensation is taking place. In the case under study here, it is necessary to define a quantity called the subcooled available moisture, Y_s, to differentiate it from the equilibrium available moisture, Y_e. This quantity, Y_s, can be determined as follows. Consider a mixture of subcooled steam at temperature T_s, pressure P, and droplets at temperature T_d. At this pressure, the equilibrium temperature would be T_e, the saturation temperature at pressure P. The enthalpy, h_m, of the mixture then is given by

$$h_m = Y_s h_f(T_d) + (1 - Y_s) h_g(T_s),$$

where subscripts f and g refer to liquid and gaseous phases, respectively.

The enthalpy, h_e, at equilibrium is given by

$$h_e = Y_e h_f(T_e) + (1 - Y_e) h_g(T_e)$$

Assuming that this is an adiabatic system, we can equate h_e and h_m, and through the use of the additional assumption that the drops are at the equilibrium temperature, T_e, we can obtain the following relation

$$(Y_e - Y_s) \quad h_{fg}(T_e) = (1 - Y_s) C_{pg} \Delta T_s$$

where

$$C_{pg}(T_e - T_s) = C_{pg} \Delta T_s = h_g(T_e) - h_g(T_s)$$

and h_{fg} is the latent heat of vaporization.

Solving for Y_s yields,

$$Y_s = \frac{Y_e h_{fg}(T_e) - C_{pg} \Delta T_s}{h_{fg}(T_e) - C_{pg} \Delta T_s} .$$

A very good approximation of the above is given by the following,

$$Y_s \approx Y_e - \frac{C_{pg} \Delta T_s}{h_{fg}(T_e)} .$$

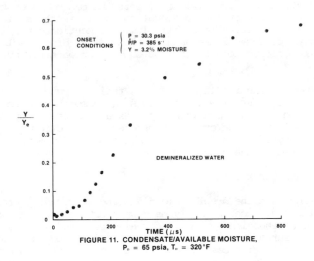

FIGURE 11. CONDENSATE/AVAILABLE MOISTURE,
P_0 = 65 psia, T_0 = 320°F

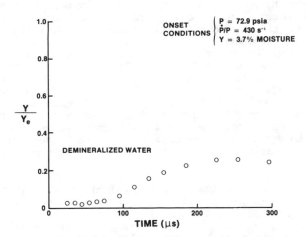

FIGURE 12. CONDENSATE/AVAILABLE MOISTURE,
P_0 = 155 psia, T_0 = 380°F

It can now be seen that the available subcooled moisture during a non-equilibrium condensation process is considerably less than the equilibrium moisture.
The sub-cooled available moisture, Y_S, represents the maximum amount of the
condensate that is available during condensation and Y_e is the maximum amount
available as the mixture returns to equilibrium. As mentioned earlier, the
other measure of the accuracy of the data is that as the droplet cloud returns
to equilibrium, the measured condensate mass should approach but not exceed the
equilibrium value.

As can be seen in Figures 11 and 12, the ratio of condensed-to-available
moisture saturates at levels below unity. This is reasonable on the grounds
that the deviation from the assumed isentropic expansion caused by the supersaturation effects will move the expansion process into a drier portion of the
Mollier diagram and therefore the final ratio of condensed-to-available moisture cannot be expected to reach unity.

3.7   Comparison with Other Data

The assumption that the droplet cloud and the surrounding steam have "returned to equilibrium" has been commonly used when droplet fogs are measured
with a single color attenuation system (References 5, 6 and 7). It allows
separation of the measured attenuation parameter, Knd^2, into the individual
terms. K, the total scattering coefficient, is a known function of droplet diameter and n and d can be defined through the use of nd^3, which is proprotional
to the condensate mass. However, when the "return to equilibrium" assumption
is made, and the equilibrium condensate is equated to the actual condensate,
serious errors in the estimate of droplet size and number density can result.
The direction of this error is to underestimate the droplet size, for droplet
diameters of less than 2.2λ or 1.2 μm (for $\lambda = 457.9$ nm).

To illustrate this, an example will be taken from the data used in Figure
3. The droplet diameter, as measured with the two color attenuation system, is
0.4 μm. By assuming that the process had returned to equilibrium and using a
single attenuation measurement, the droplet diameter would be estimated to be
only 0.13 μm. This difference in the two estimates tends to disappear as equilibrium is approached as seen in Figure 3. In the example given here only
4.2% of the available equilibrium moisture has condensed. Returning to Figure
5 and the discussion of comparison to the data obtained by prior investigators,
it is seen that the use of the "return to equilibrium" assumption could help to
explain the very small droplet sizes that these investigators reported. The
extent to which the droplet sizes were underestimated depends on the ratio of
actual condensate to equilibrium moisture, which in most cases could not be
determined. Other comparisons were not limited by this assumption and are discussed next.

The turbine data of Dibelius (Ref. 9) is in the same range of onset pressure and expansion rate (a side scattering technique was used) as the data reported herein. The agreement with Dibelius' (Ref. 9) results is very good
(see Figure 5). In general the prior data shown on Figure 5 and that reported
here (Figure 4) follow the general trends of decreasing droplet size with increasing expansion rate and increasing droplet size with increased onset pressure. These trends are in qualitative agreement with classical nucleation
theory. Let us now turn to the number density comparisons.

It is difficult to draw accurate comparisons to prior data on number density since the use of the "return to equilibrium assumption" has a very drastic effect on number density. For small droplets the calculated number density can be inversely proportional to a high power of the drop diameter. Again, the data of Dibelius (Ref.9) agrees in average sense with the data reported herein as seen in Figure 9. Dibelius' data to not, however, follow the trend of decreasing number density with onset pressure as classical theory predicts.

Throughout this discussion the droplet fog has been assumed to be monodisperse. How valid is this assumption? In Dibelius' (Ref.9) very careful experiments droplet size distributions from 0.2μm to 1.0μm were found. A direct comparison is not possible, however, since his experiments were conducted in a turbine or immediately downstream of a turbine. Where nucleation occurred in a three dimensional flow pattern. In such a circumstance the droplets would experience different pressure histories and therefore different growth rates, leading to an eventual polydisperse droplet size distribution. Other evidence on this subject is given in Reference 12, a turbine was used to vary the amount of superheat at the entry to a downstream expansion nozzle. An eighteen color attenuation system, which used wavelengths from 310 nm to 800 nm was employed to determine the mean droplet size and the size distribution. By comparing the shape of the attenuation coefficient versus wavelength with that predicted from Mie's theory (Ref. 13), it was found that in the early stages of condensation, the droplets were very monodisperse with a droplet diameter of 0.46μm. The expansion rate was about 1000 s^{-1}. Unfortunately, the onset pressure was not reported, so that a droplet size comparison cannot be made. Although the evidence is not overwhelming, it does suggest that nucleating steam produces a nearly monodisperse droplet cloud.

3.8 Droplet Cloud Evolution

In this section certain aspects of the classical nucleation theory and how it relates to the experiments will be discussed.

According to nucleation theory, nucleation onset is defined as the point at which the subcooling of the steam allows stable clusters of water molecules to form and grow. During the initial phases of nucleation, the droplet sizes are so extremely small that very little condensate is generated and the subcooling continues. Since both the rate of droplet generation and growth rate of existing droplets increase rapidly with subcooling, a polydisperse droplet cloud is predicted. Once the amount of condensed droplets reach a level that the transfer of the latent heat of condensation from the droplets warms the surrounding steam, the subcooling will peak and then decrease. This peak in the subcooling defines the end of the nucleation phase and while the steam temperature approaches the droplet temperature (equilibrates), condensation on the existing droplets continues.

The classical nonequilibrium theory also considers the droplets to be stationary in space and growth occurs as a result of steam molecules impinging on the droplet surface. This assumption has been questioned by later investigators who have argued that the droplets should be allowed to move. This alternation of classical theory has a dramatic effect on the nucleation rate, increasing it by many orders of magnitude, as discussed in Reference 14. Another consequence of droplet motion would be collisions with other droplets. These collisions could result in agglomeration and or also shattering of the droplets. Agglomeration would result in a lower number of droplets and an increase in droplet size growth rates.

With regard to our experiments, direct evidence of possible agglomeration in the form of a decreasing number density with time was exhibited by all the data taken with the long (42") nucleation tube. For these tests, the onset expansion rate varied from 385 to 470 S^{-1}, with an onset pressure variation from 8 psia to 120 psia. These falling number densities occurred very early in the condensation transient. Generally, the number density was reduced from above 1×10^8 drops/cm^3 to a lower value that ranged from 4×10^6 to 2×10^7 drops/cm^3. A typical history is shown in Figure 6. The tests with the higher expansion rate tube (18") gave a mixed result, only one particular condition showed signs of agglomeration. At an onset pressure of about 90 psia and an expansion rate of about 850 S^{-1} a peaking of the number density was seen shortly after onset. Four tests were conducted near this condition and all have very similar transients, Figure 7 is a typical example.

In the other cases of high expansion rate test conditions the droplet number density did not exhibit falling trends and rose steadily to a plateau value (see Figure 8). This trend agrees qualitatively with the mechanisms of classical nucleation theory. However, the quantitative predictions of the classical theory are very wide of the mark even for the data such as shown in Figure 8. In particular, predicted number densities are several orders of magnitude higher than the measured values reported herein. The droplet diameters likewise are smaller than reported here. In summary a large portion of the measured data is in qualitative agreement with trends of classical nucleation theory, while a smaller portion of the data indicates that agglomeration may be going on. Under the agglomeration conditions the condensation process deviates drastically from the classical nucleation model of condensation on a non-interacting droplet. It is the current thinking to expand the steam rapidly and produce a large number of small droplets. This is what the classical theories predict will happen at high expansion rates. If, however, this dense cloud of fine droplets undergoes agglomeration, as (under certain conditions) the data in this report indicates, the end result would be droplet sizes and number densities that vary much less with expansion rate than predicted by classical nucleation theory. The resolution of this point could have a profound impact on the way in which steam turbine designers would try to reduce moisture induced losses.

5.0 CONCLUSIONS

A technique using a nonsteady expansion wave to provide a controlled, but rapid, expansion of slightly superheated steam has been developed. The onset of droplet formation and its subsequent growth are determined by measuring the attenuation of laser light. The location of nucleation onset is identified as the point at which the attenuation signals move away from their baselines. The sudden change in these signals provides a very accurate location of the point of onset. The Wilson zone was found to lie between 3 and 5 percent moisture. Hetergenous nucleation on vaporized impurities is believed to be a major contributor to the width of the Wilson zone. The test conditions used in this project span an onset pressure range from 8 psia to over 400 psia and expansion rates from 370 to 1100 S^{-1}.

Since a two-color attenuation technique is used, both the droplet diameter and the droplet number density have been determined. The final condensate droplet sizes were seen to increase with onset pressure and decrease with the expansion rate at onset. The final droplet number densities decreased slightly with onset pressure and increased with onset expansion rate. These observed trends are in qualitative agreement with classical nucleation theory. With the droplet size and number density information, the rate at which the condensate mass grows has been measured and compared to the condensate available at equilibrium. There is a considerable lag in the approach of the condensate mass to the equilibrium value. Prior investigators using single-color attenuation methods have assumed that the amount of condensate quickly approached the equilibrium and used this assumption to then calculate the droplet size. As a result of the experiments described herein, it has been shown that this "return to equilibrium" assumption can produce a serious underestimation of droplet sizes. Measured final droplet sizes from these tests are in the range of 0.4 to 1.4 microns and these results have been compared with the results reported in prior references. To summarize this comparison, it can be said that the apparent differences in droplet sizes as measured in prior experiments and as found here are due to different experimental conditions or use of the "return to equilibrium" assumption. In those cases where conditions were reasonably well matched and the "return to equilibrium" assumption was not used, very good agreement on droplet sizes was found.

The basic issue of what mechanism controls the droplet growth rate has been raised. It is extremely important to know whether the isolated droplet growth via impinging steam molecules (the assumption on which the classical theory stands) is dominant or whether agglomeration plays a role. The knowledge of when these different mechanisms are controlling can be of vital importance in trying to reduce the moisture-induced losses in steam turbines.

ACKNOWLEDGEMENTS

The research contained in this paper was supported by the Electric Power Research Institute under Contract RP 735-1.

REFERENCES

1. Stodala, A., "Steam and Gas Turbines," McGraw-Hill Book Co., 1927.

2. Kantola, R.A., "Steam Condensate Droplet Evolution: Experimental Technique" XIV ICHMT Symposium, 1982.

3. Yellot, J.I.,"Supersaturated Steam," Transactions of the American Society of Mechanical Engineers, Vol. 56, 1934, p. 411.

4. Yellot, J.I. and Holland, C.K.,"The Condensation of Flowing Steam, Part I: Condensation in Divergent Nozzles," Transactions of the American Society of Mechanical Engineers, Vol. 59, 1937,p. 171.

5. Gyarmathy, G., Burkhard, H.P., Lesch, F., and Siegenthaler, A. "Spontaneous Condensation of Steam at High Pressure: First Experimental Results," Proceedings of the Institution of Mechanical Engineers, Vol. 187, 1973, p. 192.

6. Gyarmathy, G. and Lesch, F., "Fog Droplet Observations in Laval Nozzles and in an Experimental Turbine," Proc. Inst. Mech. Eng. (London), Vol. 184, pt. 3G(III), pp. 29-36, 1969-1970.

7. Petr, V., "Measurement of an Average Size and Number of Droplets During Spontaneous Condensation of Supersaturated Steam," Proc. Inst. Mech.Eng. (London), Vol. 184, pt. 3G(III), pp. 22-28, 1969-1970.

8. Krol, T.,"Results of Optical Measurements of Diameters of Drops Formed Due to Condensation of Steam in a Laval Nozzle (in Polish), Trans. Inst. Fluid Flow Mech. (Poland), No. 57, pp. 19-30, 1971.

9. Dibelius, G. and Mertens, K.,"Investigation of Drop Formation and Their Growth During Expansion in a Turbine," Institute for Steam and Gas Turbines of the RWTH Report 80-5455, 1980, Aachen.

10.   Smith, A., "Wilson Point Experiments in Steam Turbines," Third Conference on Steam Turbines of Great Output, Sept. 24-27, 1974, at Gdansk, Poland.

11.   Dobbins, R.A. and Jizmagian, G.S., J. Opt. Soc. Amer., Vol. 56, p.1345, 1966.

12.   Walters, P.T., "Practical Applications of Inverting Special Turbidity Data to Provide Aerosol Size Distributions", Applied Optics, Vol. 19, No. 14, July 15, 1980.

13.   Mie. G. (1900), Ann. Physik, 25, 377.

14.   Springer, G.S., "Homogeneous Nucleation," Advances in Heat Transfer, Vol. 14, Academic Press, pp. 281-346, 1978.

Temperature Measurement Techniques and Their Application on Gas Turbine Rotor Heat Transfer Research

W. KÜHL and U. STÖCKER
Institut für Strahlantriebe
RWTH Aachen, FRG

ABSTRACT

Detailed knowledge about local surface temperature and material temperature gradients of highly stressed turbine components is required for the determination of heat transfer and material stress in order to perform lifetime prediction and control. The rotor of the engine poses some particularly difficult measurement problems.

In this paper the temperature measuring technique applied to the rotor system of a test gas turbine of the Institut für Strahlantriebe, RWTH Aachen, is described. Three methods are used for measuring blade surface temperatures: thermocouple technique, temperature indicating paints, infrared pyrometry. Examinations have been made with reference to the accuracy of measurement. Some results obtained to date by application of the different techniques are presented and discussed with respect to the determination of local gas to blade heat transfer.

1. INTRODUCTION

The knowledge of the variation of heat transfer over the surface of gas turbine blades is presently one of the most important requirements for blade design. For the determination of local heat transfer, experimental methods |1,2,3,4| utilising temperature measurements on stationary or rotating cascades are known. The boundary layer heat transfer calculation along the profile |5| as well as the FEM calculation of the temperature field within the blade |1;6| needs the distribution of the surface temperature. The quality of heat transfer results obtained both by measurement and calculation depends on a high degree of accuracy in measuring surface temperatures. In this connection the rotating blade of a gas turbine poses a particularly difficult problem of temperature measurement |7|.

At the Institut für Strahlantriebe of the Technical University Aachen (RWTH Aachen) a research program for measuring temperatures of gas turbine blades during operation has been conducted during the last years |2;8;9|. The measured temperature data were used as boundary values for the determination of local heat transfer coefficients over the blade surface.

337

The following three methods of measuring blade surface temperatures were applied:
(1) use of thermocouples,
(2) use of thermal paints for the indication of temperature,
(3) use of an infrared pyrometer.

In this paper these measurement techniques will be described as applied to a test turbine rotor. The important aspect of accuracy will be discussed in detail.

2. METHODS OF MEASURING BLADE SURFACE TEMPERATURES

2.1 Blade temperature measurements using thermocouples

The application of the thermocouple technique on turbine rotor blades requires a great deal of preparation and preliminary tests because of the unique conditions to be considered at each measuring point. In particular the problems to be solved are the installation of the sensor and the transmission of the measured data from the rotating blade to the stationary instruments.

For our turbine experiments we used an instrumentation consisting of the following components:
(1) 48 sheathed chromel/alumel thermocouple assemblies, one millimeter in diameter, mounted and brazed on the blades in grooves (fig. 1), distributed over the profile sections at the blade hub, the blade pitch, and the blade tip,
(2) one slipring rotor system (fig. 2) for eight thermocouple signals to be transmitted from the rotating blades to the stationary registration instruments,
(3) one co-rotating selector switch (fig. 2) which connects in each of the ten possible positions eight thermocouple assemblies to the sliprings.

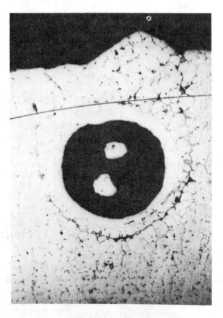

Fig. 1. Micrograph of a sheathed thermocouple of 1 mm diameter brazed into the blade near the gas side surface

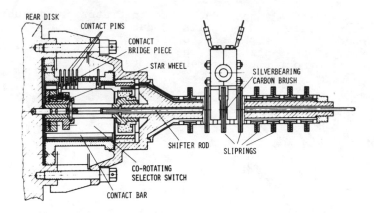

Fig. 2. Sectional view of the slipring rotor system and the
co-rotating selector switch

A high accuracy of temperature measurement could be obtained by the application of calibrated thermocouples, by extensive testing of the slipring rotor system with respect to
the suitability of sliding carbon brushes, and by the investigation of the thermocouple installation error.

Fig. 3 shows the electrical circuit of two thermocouple
assemblies. All contact parts of this arrangement are made of
thermocouple material, so that from the hot junction to the
sliprings no error can be caused by a foreign material. The
carbon brushes represent the only foreign body of the circuit.
From tests we know, that in an unfavourable case the used carbon causes a voltage deviation of -0.3 mV, being independent
of the rotor speed.

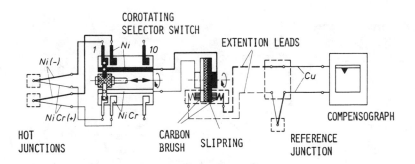

Fig. 3. Electrical circuit arrangement of the thermocouple
assemblies.

In fig. 4 an analogy test result demonstrates the temperature field disturbance (undashed lines) due to the installation
of the thermocouple in comparison to the undisturbed field
(dashed lines). After testing all the thermocouples in this way
we were able to correct the measured results.

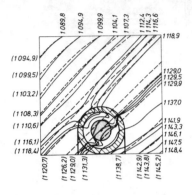

Fig. 4. Disturbance of the
temperature field
by thermocouple in-
stallation

By such investigations it is possible to minimize the
measurement errors which in our case led to a relative error
of less than 0.5 %. This accuracy can be realized only under
test conditions, nevertheless it is a fundamental requirement
for the determination of the local heat transfer. An accepted
error of ± 5 K in measuring blade temperatures of about 1000 K
would result in an error of ± 2 % in heat transfer rate.

The disadvantage of measuring blade temperatures using
thermocouples is the impossibility to incorporate a large number of measuring points needed for a well-defined temperature
field, as an excessive number of such points reduce the resistance of the blade material and affect its homogeneity. Moreover
there are such blade regions as the trailing edge where the installation of a thermocouple is not possible.

2.2 Blade temperature measurements using temperature indicating
     paints

The application of temperature indicating paints for measure-
ments on rotor blades |10| includes the analysis of the problems
of

(a) time dependency of the colour change and temperature,
(b) adhesion of the paints on the blade surface,
(c) influence of exhaust gas components on the colour change,
(d) influence of coating thickness on the temperature field,
(e) high temperature resistivity of the paints,
(f) erosion of the paint due to high speed flow particles.

We have analysed these problems by investigating the application of thermal paints not only to the rotor blades of our test
turbine |11| but also to the turbine blades of a jet engine and to
the guide vanes of a cascade section |12|. Furthermore we had the
great advantage to compare the results of the indicating paints
with those of the thermocouple measurements. The paints used for
the experiments were of the colour changing type in comparison

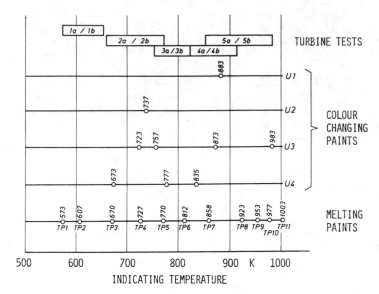

Fig. 5. Thermal paints applicated to turbine tests

to those of the melting type. In fig. 5 the usable paints are
shown in relation to the temperature range of the turbine tests
up to 1000 K. The temperature values noted there are taken from
the producers' specifications for an exposure time of 1800 seconds.

In all our tests on turbine blades the exposure time was
limited to 600 seconds.

The conclusions derived from our experiments are as follows:
(1) In order to get accurate temperature indication, the paints
have to be calibrated under working conditions. In particular
this procedure must include the time dependency of the colour
change and temperature.

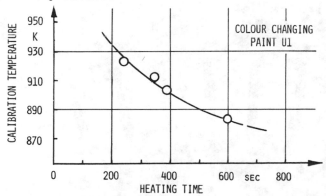

Fig. 6. Influence of heating time on the colour change of the
temperature indicating paint U1 during calibration under
test conditions

(2) Our calibration tests led to more or less different results compared with the specifications of the producers. A typical relationship between colour change calibration temperature and heating time obtained from our tests is illustrated in fig. 6.
(3) The adhesion of most of the paints is good at the suction side of the blade. At the pressure side the paint will always be worn away due to erosion processes. The use of binding materials proposed by the producers don't improve remarkably the adhesivity.
(4) Exhaust gas components resulting from the combustion like natural gas, diesel oil, or kerosene, don't influence remarkably the temperature indication of the paints used.
(5) Due to its resistance against heat conduction the coating thickness of the paint on the blade causes a decrease of the temperature level. By measuring the coating thickness, the temperature results can be corrected if the heat conductivity of the blade material and of the paint is known. By spraying the paint on the blade, a thin film can be surfaced so that its influence becomes negligible.
(6) The resistivity of the paints against burning especially at temperatures above 1000 K is not very high.

The application of temperature indicating paints lead to problems of determining the line of temperature change. Colour changes of surfaces with low temperature gradients show a transition region where an exact change line can not be found. For the interpretation one needs a great deal of experience, so that under the best conditions the temperature results can have an accuracy of about ± 10 K with respect to the calibration temperature. For qualitative measurements or for measurements to get information of exceeded temperature limits the use of thermal paints can be recommended.

Fig. 7 illustrates a part of the turbine blading after the test as seen from the rear side. The thermal paint has changed its colour from red at the hub to white. The transition line at all blades has nearly the same pattern, so that the temperature field can be assumed to be equal from blade to blade.

Fig. 7. Temperature indication on turbine blades by the application of the colour changing paint U1

2.3 Application of an infrared optical pyrometer

As a third method we applied the infrared pyrometry to measure turbine blade temperatures |23|. The special advantage of this method is the possibility of non-contact measurements in those regions of the blade where a detailed information of the temperature field is needed, like those at the leading edge or where the installation of sensors is not possible like those at the trailing edge.

In principle the optical pyrometer measures the radiance of a spot on the blade surface which is viewed by the optical system of it. The radiant energy is picked up by a detector the output of which is primarily a function of the surface spot temperature. The temperature measurement range and the sensitivity of the system is established by the selection of the detector |14;15;16;17|.

In fig. 8 a block diagram of the infrared pyrometer set-up is shown as applied to our turbine tests. The radiance of the blade surface spot, 1.3 mm in diameter, is transmitted by the optical system to the detector. The detector output signal has to be amplified and can be viewed on the oscillograph. The choice of a particular blade is done by a trigger system using impuls transmitters and a counter comparator.

For measuring blade surface temperatures of the leading edge region a water cooled probe is positioned in the axial space between the vanes and the rotor blades of the turbine stage. The probe head can be shifted in the radial direction and can also be turned to observe either the pressure side or the suction side of the blade. The optical system incorporated within the probe head

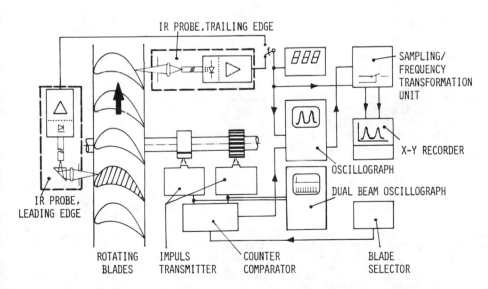

Fig. 8. Block diagram of the infrared temperature measurement system in application to turbine blades

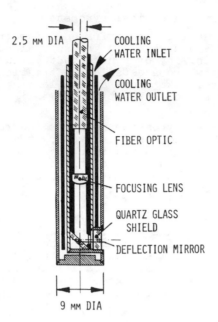

2.5 MM DIA — COOLING WATER INLET

COOLING WATER OUTLET

FIBER OPTIC

FOCUSING LENS

QUARTZ GLASS SHIELD

DEFLECTION MIRROR

9 MM DIA

Fig. 9. Head of the infrared probe with the optical system

(fig. 9) consists of a quartz glass shield, a deflection mirror, and a system of one or more lenses that is focused on the desired target of the blade. The signal is passed through a fiber optic bundle to a relatively cool location of the detector, remote from the hot turbine casing.

The optical system must be designed in such a way that the radiant energy is transferred under a minimum of transmission losses |19|. Watercooling of the probe head must be provided to avoid overheating of the head and to protect the detector from radiation influences of the probe head itself.

The detector selected for measuring blade temperatures during turbine operation has to fulfil the following requirements:
(1) a fast time response,
(2) a high cutoff frequency,
(3) a sensitivity down to blade temperatures of 600 K,
(4) no sensitivity to the absorption lines of water vapour, CO_2, NO_x, and other gas components.

Finally it was decided to use a planar silicon photo cell of p-i-n semiconductor configuration. It is of high detectivity combined with a favourable ratio of noise to signal voltage, having the maximum relative spectral sensitivity at a wave length of 950 nm. The response time of the detector is lower than 10 nsec. The temperature drift caused by the heating of the photo cell is compensated by a reference cell arranged in the detector circuit.

The calibration of the pyrometer system was conducted under test conditions viewing a surface of well-known emission characteristic and measuring its temperature accurately by thermo-

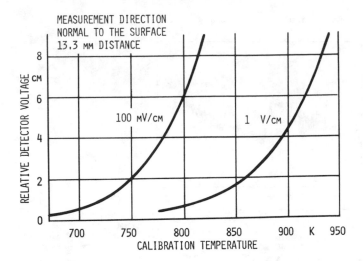

Fig. 10. Plot of relative detector response vs. calibration
temperature

couples. One of the results of this calibration is shown in fig.
10 where the relative detector voltage is plotted against the
calibration temperature. For lower temperatures the detector
output can be scattered by using a higher amplification factor.
At temperatures of about 620 K the limit of measurement is
reached where a filtering of the noise from the signal is just
possible. The relations, shown in fig. 10, are valid for measurements normal to the calibration surface with a distance of 13.3 mm
between the probe head and the spot of 1.3 mm diameter. A
deviation up to 45 degrees from the normal direction causes
only negligible changes of the signal value.

The following impairments on the accuracy of the infrared
temperature measurement are to be mentioned:
(1) the difficulty of complete compensation of the temperature
drift in the detector circuit,
(2) radiation from hot components in the surroundings of the
blade,
(3) effects of emissivity differences between the calibration
target surface and the blades.

The temperature drift can be determined by using a shutter
during the measurements.

Surface temperature measurements of turbine blades which
were partly coated with a paint of known emissivity resulted in
negligible differences compared with those of uncoated blades.
By these experiments the assumption is confirmed that the
emissivity variation has only a small influence on the accuracy
of the measurement because the output of the silicon cell is
only proportional to the surface emissivity, but is a 20th power

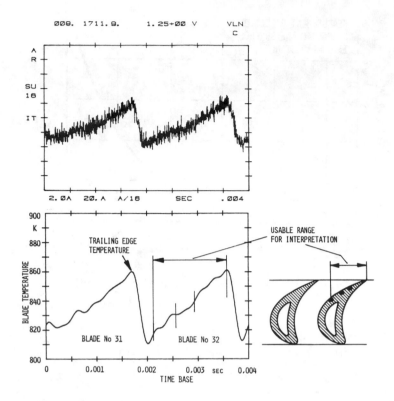

Fig. 11. Evaluation of the infrared probe signal for measuring
the rotor blade temperature of the trailing edge
region

function of the surface temperature. For the temperature range
of the investigation an uncertainty of 5 % in surface effective
emissivity would result in a temperature change of 0.3 %.

Fig. 11 illustrates the evaluation of an infrared pyrometer
signal delivered from the trailing edge. The upper graph shows
the signal displayed on the oscillograph. The temperature distribution on the lower graph is determined by using the calibration
curve and the Fast-Fourier-Transformation to filter the signal.
The limit of usable range of interpretation is reached when the
angle between the view direction of the probe optic and the
normal to the blade surface exceedes a maximum of 45 degrees.

3.   COMPARISON OF THE TEMPERATURE MEASUREMENT METHODS USED

The three methods of temperature measurement can be compared as used for measuring surface temperatures of turbine
blades.

Although the temperature indicating paints are easy to

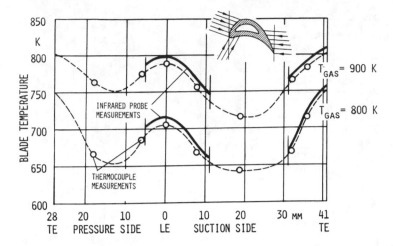

Fig. 12. Comparison of infrared probe temperature and thermocouple measurements at midspan profile surface

applicate, the accuracy of their results is not adequate for
heat transfer determination.

Results obtained by the use of thermocouples are compared
with those of infrared probe measurements in fig. 12 where the
blade surface temperature is plotted over the profile contour
for gas temperatures of 800 K and 900 K. The dashed line represents a connection of the distinct points of thermocouple
measurement, the undashed line the result of infrared probe
measurement. Due to installation the thermocouples deliver
too low temperature values as against the infrared probe output
which must be assumed to be higher than the correct ones, as
the influence of radiation reflected from hotter turbine blade
parts can not be excluded.

4.   INFLUENCE OF TEMPERATURE MEASUREMENTS ON THE DETERMINATION
OF HEAT TRANSFER

Based on measurements of blade temperatures on a test turbine
the following three methods of determining the heat transfer
along the blade surface are used:
(1) the twodimensional finite element method (FEM) |6;18|,
(2) the threedimensional analogy method |2;3;8|, and
(3) the twodimensional boundary layer calculation method (program
STAN 5) |5|.
While the FEM and the analogy method both need the measured
temperature distribution directly to evaluate the temperature
field within the blade, the temperature gradients at the blade
surface, and thus determine the local heat transfer, the STAN 5-
program calculates the flow conditions and the boundary layer at
the gas side of the blade using the surface temperature distribution indirectly, taking into account effects of pressure gradients, turbulence level, laminar-turbulent transition, etc.

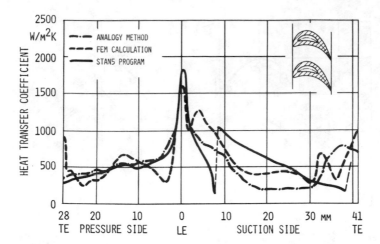

Fig. 13. Comparison of heat transfer coefficient determined
         by three different methods using the same temperature
         distribution

    Fig. 13 shows distributions of the heat transfer coeffi-
cient obtained by the three methods. Differences can be observed
at the profile suction side. The STAN 5 calculation leads to a
sudden rise of heat transfer due to a laminar-turbulent boundary
layer transition. To indicate this transition by the other two
methods, the measurements in this blade region must deliver a
more accurate temperature distribution needing measuring points
of high closeness.
    A study of this transition problem showed that the sudden
rise of heat transfer results in only a slight increase of the
surface temperature due to the transient phenomena within the
temperature field. The temperature curve of this region may be
expected to have a more or less remarkable inflection point.

5.   CONCLUSIONS

    This paper deals with three methods of measurement to ob-
tain the surface temperature field of a turbine blade during
operation.

    Within the range of blade temperatures between 600 K and
1200 K a high accuracy of ± 5 K is postulated for the determination of local heat transfer rates of the blade surface.

    Tests were conducted on a hot gas turbine equipped with
an advanced thermocouple measurement technique. Results of surface temperature measurements using an infrared pyrometer and
temperature indicating paints were compared with those of the
thermocouple measurements.

    The results can be summarized as follows:
(1) The thermocouple measurement technique can be aided or even
replaced by the non-contact infrared pyrometry.

(2) The infrared pyrometry is as equally accurate as the thermocouple technique for measuring turbine blade temperatures.
(3) The main advantage of the infrared pyrometry is the ability
to measure at every accessible location of the temperature field.
(4) Owing to their inaccuracy temperature indicating paints can
only be used for qualitative measurements.
(5) The requirements of high accuracy for the determination of
heat transfer can only be reached by infrared pyrometry and
thermocouple technique, both after skilled calibration.

6.   REFERENCES

1. Turner, A.B.: Local Heat Transfer Measurements on a Gas
   Turbine Blade. J. Mech. Eng. Science, Vol. 13 (1971), No. 11,
   p. 1-12.

2. Kühl, W.: Investigation on the Local Heat Transfer Coefficient
   of a Convection Cooled Rotor Blade. AGARD CP Nr. 229, High
   Temperature Problems in Gas Turbine Engines, 1977.

3. Kühl, W.: Untersuchungen zur Bestimmung örtlicher Wärmeüber-
   ganskoeffizienten an der luftgekühlten Laufschaufel einer
   Gasturbine. VDI-Berichte Nr. 264 (1976), p. 41-49.

4. Consigny, H. and B.E. Richards: Short Duration Measurements
   of Heat Transfer Rate to a Gas Turbine Rotor Blade. ASME-Paper
   Nr. 81-GT-146, 1981.

5. Crawford, M.E. and W.M. Kays: STAN 5 - A Program for Numerical
   Computation of Two-Dimensional Internal and External Boundary
   Layer Flows. NASA-CR 2742 (1976).

6. Lötzerich, M.: FEM-Rechenprogramm zur Berechnung zweidimen-
   sionaler, rotationssymmetrischer und quasi-dreidimensionaler
   Temperaturfelder in gekühlten Turbinenschaufeln unter Verwen-
   dung eines automatischen Maschengenerators. Institut für Strahlantriebe und Turboarbeitsmaschinen, RWTH Aachen, 1980, unpublished.

7. Alwang, W.G.: Problems in the Measurement of Metal Temperature,
   Gas Temperature, Heat Flux and Strain in Combustors and Tur-
   bines. AGARD-CP-281, Testing and Measurement Technique in
   Heat Transfer and Combustion, 1980.

8. Kühl, W.: Experimental Investigation on a Single-Stage Air-
   Cooled Gas Turbine. AGARD-CP-73-71, High Temperature Turbines,
   1971.

9. Kühl, W. and U. Stöcker: Wärmeübergangsprobleme an Turbinenschaufeln unter besonderer Berücksichtigung instationärer
   Effekte. Mitteilung Nr. 79-02 of the Institut für Strahlantriebe, RWTH Aachen, 1979.

10. Tyte, L.C.: Temperature Indicating Paints. Inst. Mech. Eng.
    (1943), p. 226-231.

11. Koschel, W. and W. Kühl: Untersuchungen von Temperaturmeß-
    farben beim Einsatz in einer Versuchsgasturbine. Institut
    für Strahlantriebe und Turboarbeitsmaschinen, RWTH Aachen,
    1975, unpublished.

12. Koschel, W. and W. Kühl: Vergleichsmessungen mit Thermo-
    elementen und Temperaturmeßfarben in einem Turbinengitter-
    prüfstand und Erprobung der Temperaturmeßfarben in der Turbine
    eines Strahltriebwerks. Institut für Strahlantriebe und Turbo-
    arbeitsmaschinen, RWTH Aachen, 1975, unpublished.

13. Stöcker, U. and H. Britten and W. Kühl: Ein Infrarot-Meß-
    system zur Bestimmung örtlicher Temperaturen an Turbinen-
    laufschaufeln. Mitteilung Nr. 81-02 of the Institut für
    Strahlantriebe, RWTH Aachen, 1981.

14. Wiederhold, P.R.: Infrared Pyrometer for Temperature Moni-
    toring of Train Wheels and Jet Engine Rotors. Materials
    Evaluation, Vol. 32 (1974), No. 11, p. 239-48.

15. Charpenel, M. and J. Wilhelm: Pyrometer optique adapté à la
    mésure de la température de surface des aubes de turbine.
    ATMA, Paris, 1981.

16. Rohy, D.A., T.E. Duffy and W.A. Compton: Radiation Pyrometer
    for Gas Turbine Blades. SAE-Paper, Nr. 720159, 1972.

17. Barber, R.: A Radiation Pyrometer Designed for In-Flight
    Measurement of Turbine Blade Temperatures. SAE-Paper, No.
    690432, 1969.

18. Reid, J.K. and A.B. Turner: Fortran Subroutines for the
    Solution of Laplace's Equation over a General Region in Two
    Dimensions. A.E.R.E. TP 422 Harwell, 1970.

19. Salden, D.: Optimierung und Einsatz optischer Systeme in
    Infrarotsonden zur Messung von Materialtemperaturen am um-
    laufenden Turbinenrad. Institut für Strahlantriebe und
    Turboarbeitsmaschinen, RWTH Aachen, 1981, unpublished.

GAS TURBINES

Heat Transfer Problems in Aero-Engines

DIETMAR K. HENNECKE
MTU Motoren- und Turbinen-Union Munchen GmbH
Munich, FRG

ABSTRACT

A survey of the heat transfer problems to be solved when developing an aircraft gas turbine engine is presented looking at each engine component separately. It is shown that the responsibilities of the heat transfer engineer have greatly increased. In addition to the design of the cooling system of the "hot parts", such as the combustor and turbine, these include the evaluation and assessment of the thermal behavior of nearly all other parts such as the fan, compressor, labyrinth seals, bearings, etc. This is the result of the trend in aero-engines toward higher cycle pressures and temperatures, increased reliability with reduced engine weight, and ever tighter clearances to improve performance and, thus, achieve a further reduction in fuel consumption.

It is shown that many heat transfer phenomena in aero-engines are not yet understood well enough. Therefore, heat transfer areas that need continued or intensified research are highlighted, demonstrating that many challenging problems remain to be tackled.

1. INTRODUCTION

The purpose of this paper is to describe the major heat transfer areas that have to be dealt with in the development of a modern aero-engine. The focus will be on gas turbine engines; piston and rocket engines will be excluded. Although the focus will be on aero-engines, many statements will also be relevant for stationary gas turbines.

Gas turbines, being heat engines, have always received attention by heat transfer specialists. The present paper will show that, owing to the trends in aircraft gas turbine design, which will be described briefly, the scope of heat transfer problems has greatly increased. By looking at each component separately it will be demonstrated that many new challenging problems have arisen besides the hitherto classical areas of combustor and turbine cooling.

2. SOME TRENDS IN AERO-ENGINE DESIGN

Modern aero-engines may be characterized by their extremely high power concentration and low specific fuel consumption leading to relatively low weight, small size, high thrust-to-weight ratio and low mission fuel consumption. To illustrate this trend, two fighter engines with about the same thrust are shown in Fig. 1. The top one, the RB 199, was developed two decades later than the bottom one, the J79. The dramatic reduction in the engine size should be noted.

For civil engines the fuel consumption is of paramount importance. With an advanced technology engine such as the PW 2037, shown in Fig. 2, which is currently under development, the Boeing 757 will consume about 40% less fuel per passenger seat than the Boeing 727.

It is well known that two of the main factors contributing toward this achievement are the high compression ratio and the high turbine entry gas temperature, providing for high thermal efficiency of the thermodynamic cycle and high power density.

Fig. 3 depicts the development of the turbine entry temperature of actual aero-engines since 1950 (Ref. 1). Also shown is the increase in the maximum allowable blade material temperature. It may be seen that the turbine blades operate in an environment in which the gas temperature is well above the material temperature even for ceramics. And the tendency is increasing. Today's aero-engines have a turbine entry temperature of 1600 to 1700 K and above. It is obvious that intensive cooling is required.

The reduction in engine mass achieved to date is presented in Fig. 4 (Ref. 2). The thrust-to-mass ratio has increased from about 3:1 to 8:1 for military engines, with a similar gain for commercial engines.

The following chapter will show that the tendency to light weight parts, high temperatures, highly loaded compressors and turbines combined with the extremely severe requirement for reliability has greatly increased the importance of the heat transfer engineer in the design of aero-engines. In addition, a large number of challenging new heat transfer problems has emerged, calling for intensified research, which will be identified.

3. HEAT TRANSFER PROBLEMS IN THE MAJOR ENGINE COMPONENTS

Fig. 5 shows the cross-section of a modern aero-engine with the major components indicated. The heat transfer areas will be described for these components.

3.1 Inlet, Fan and Compressor

Inlet and fan. In certain altitude, flight and atmospheric conditions the engine inlet and fan may be subject to severe icing problems. The task of the heat transfer engineer is to identify and assess these conditions and to design appropriate

Fig. 1 Progress in aero-engine design
       Top: RB 199 engine; bottom: J79 engine

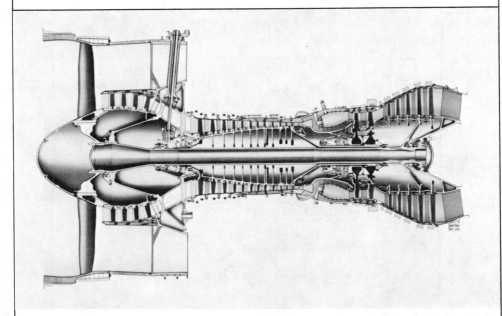

Fig. 2 Modern civil high bypass ratio engine: The PW 2037

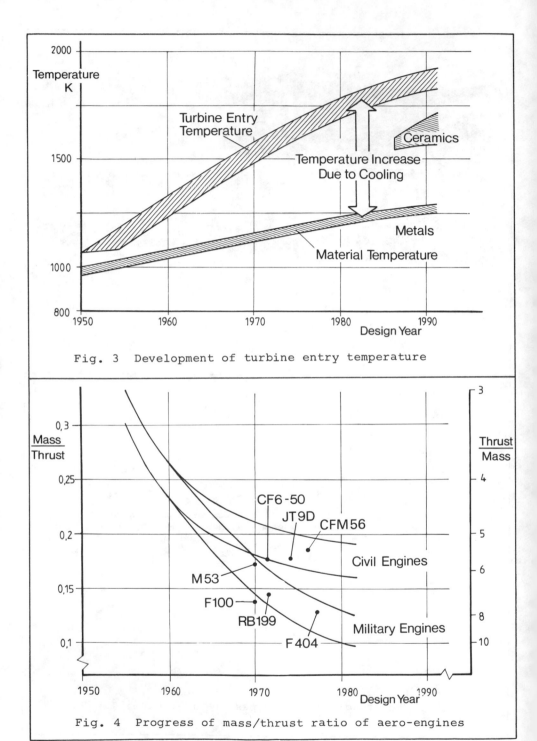

Fig. 3 Development of turbine entry temperature

Fig. 4 Progress of mass/thrust ratio of aero-engines

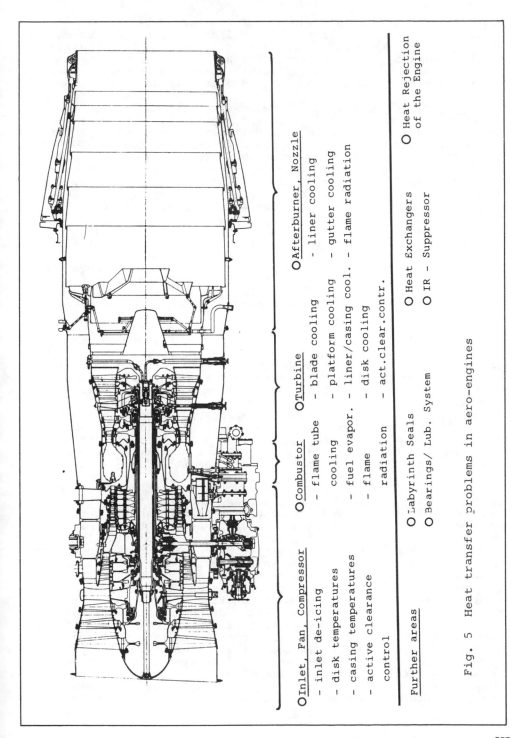

○ Inlet, Fan, Compressor
- inlet de-icing
- disk temperatures
- casing temperatures
- active clearance
  control

○ Combustor
- flame tube
  cooling
- fuel evapor.
- flame
  radiation

○ Turbine
- blade cooling
- platform cooling
- liner/casing cool.
- disk cooling
- act.clear.contr.

○ Afterburner, Nozzle
- liner cooling
- gutter cooling
- flame radiation

Further areas

○ Labyrinth Seals
○ Bearings/ Lub. System

○ Heat Exchangers
○ IR - Suppressor

○ Heat Rejection
  of the Engine

Fig. 5 Heat transfer problems in aero-engines

means to prevent the build-up of harmful ice. This may be
achieved by internally heating the bullet nose, the fan vanes and
blades, etc. by warm air from higher stages. Even film heating
may be employed in severe cases. Surface areas that are prone to
icing may also be covered with a thin electrically heated mat.
The problem of icing is gone into in detail in Ref. 3.

Compressor. A typical compressor is sketched in Fig. 6. The
trend in compressor design is toward higher pressure ratio (now
up to 30 - 40) resulting in high exit temperature (800 - 900 K),
increased rotor speed, lower weight, as well as higher aerodynamic efficiency and loading. This has called for a much more detailed analytical design, including thermal analysis, bringing
the heat transfer engineer into the picture. His task is to predict accurately the transient temperature distribution of the
compressor disk and casing for the whole flight cycle from start
to landing with two objectives:
- To facilitate computation of the life of the disks and
  casing and, on this basis, to design for the required life
  at minimum weight, and
- To predict the transient rotor and casing clearances and,
  again, to actively influence the design such that the
  transient thermal behavior of the parts is optimal, i.e.
  resulting in tight clearances throughout the flight cycle.

The life calculation is of particular importance for the
rotor. In addition to the stresses resulting from the centrifugal
forces there may be strong thermal stresses, in particular during
transient conditions. The disk rim, which is close to the main
stream, follows a change in power setting much faster than the
bulkier hub section which is removed from the main stream. An
example is shown in Fig. 7 for a typical compressor disk. The
large intermittent temperature difference, causing high thermal
stress and possibly severe consumption of disk life, should be
noted. The main stream temperature is also plotted. It changes
as rapidly as the rotational speed, i.e. typically within about
5 - 10 seconds from idle to full-power condition.

The resulting radial deflection of the rotor during transient
conditions is also relatively slow. In contrast, the radial movement of the casing is usually much faster, because the temperature of the thin-walled casing, which is exposed to the main
stream, follows a change in power setting very rapidly. The principal variation in the radial deflection of a typical compressor,
(shown in Fig. 6), is sketched in Fig. 8 for an acceleration from
steady-state idle to full power and a deceleration back to idle.
The deflection of casing and rotor is shown at the top of Fig. 8,
with the resulting clearance as a function of time shown at the
bottom. Initially, there is a short closing of the gap when the
rotor speeds up and expands because of the increased centrifugal
force. The transient thermal behavior then results in large gaps
during warm-up, leading to a loss of efficiency and possibly
even to surge. These gaps close to the desired small size under
steady-state full-power condition. During deceleration, however,
they increase briefly owing to the reduced centrifugal load on
the rotor and then decrease to negative values, i.e. the rotor

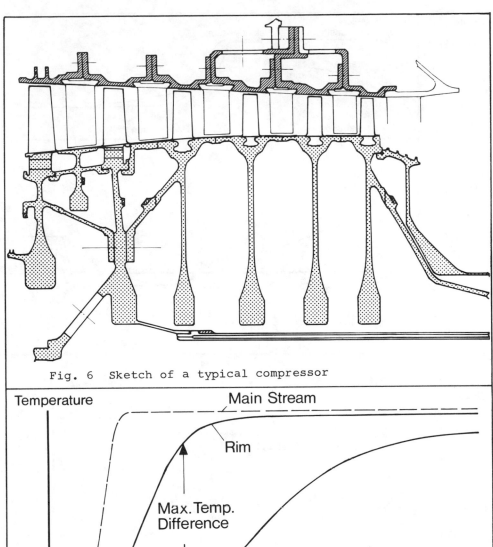

Fig. 6   Sketch of a typical compressor

Fig. 7   Temperature response of a typical
compressor disk during acceleration

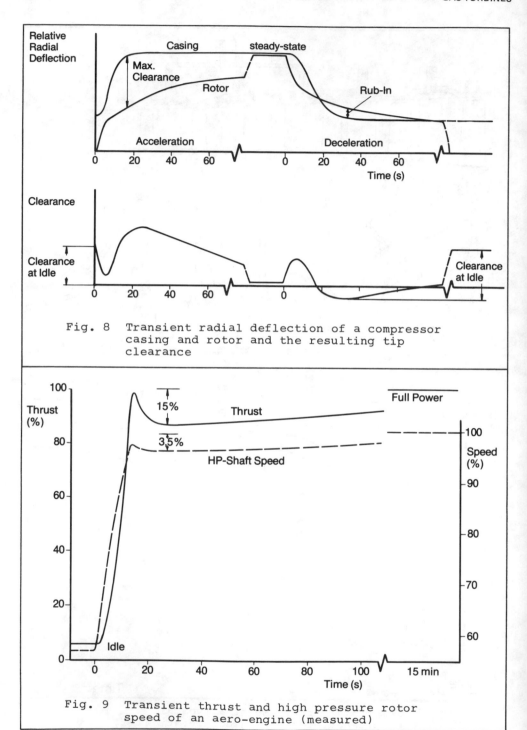

Fig. 8 Transient radial deflection of a compressor
        casing and rotor and the resulting tip
        clearance

Fig. 9 Transient thrust and high pressure rotor
        speed of an aero-engine (measured)

blades rub into the abradable coating on the casing. Then the clearances are enlarged by the amount of rub-in, even in the steady-state full-power condition, leading to a further deterioration in the efficiency of the compressor and in the surge margin.

To emphasize the importance of the transient variation of the gaps, the high-pressure spool speed and thrust of an actual aero-engine as a function of time are shown in Fig. 9. It should be noted that the thrust increases from its value at idle to full power rather rapidly, then decreases to a minimum and increases slowly, finally to reach its full-power value again after 10 - 15 minutes. The minimum of about 15% below the full-power value occurs after about 20 - 40 seconds, when the aircraft may have reached the end of the runway and is about to take off. This is precisely when maximum thrust is required.

Transient engine behavior, as shown in Fig. 9, is obviously unacceptable. Since it is to a large degree the result of the transient variation of the clearances, one tries to achieve in advanced engines a more favorable transient thermal behavior of the components by careful design of the compressor and turbine. With compressors the aim is to speed up the temperature response of the rotor and to slow down that of the casing. For the rotor this can be achieved, for instance, by venting the cavities between the disks by a small amount of air taken from the main stream (a review of the heat transfer in vented rotating cavities simulating those in aero-engine compressors is given in Ref. 4). To slow down the casing one may simply add mass which, of course, is not exactly desirable in aero-engines. A more effective method is to construct the casing of an outer supporting structure and to hook separate segments to its inner side to form the outer wall of the compressor annulus. In this way the radial thermal movement of the casing is controlled solely by the outer structure, which is not exposed to the main stream and, therefore, reacts slowly to changes in the power setting.

A thermally compensated casing of this type was designed, built, comprehensively instrumented and tested. Back-to-back engine tests, with the compressor having a standard casing and the new casing, were carried out for demonstrating the advances to be obtained from the casing alone.

Fig. 10 (inset) shows a sketch of the thermally compensated casing. An example of a comparison between calculated and measured temperatures as a function of time is also shown. Good agreement can be observed indicating that the heat transfer model used in the design is satisfactory. The improvements of the compressor using the thermally compensated casing compared to using the standard casing is demonstrated in Fig. 11. The relative gap (shown for stage 6) is reduced by about 0.6%, resulting in an increase in efficiency of about 1% and a gain of 6% in the surge margin.

This method of optimizing the gaps in turbomachinery may be called "passive clearance control", because the clearances react passively to changes in the power setting. Another approach is

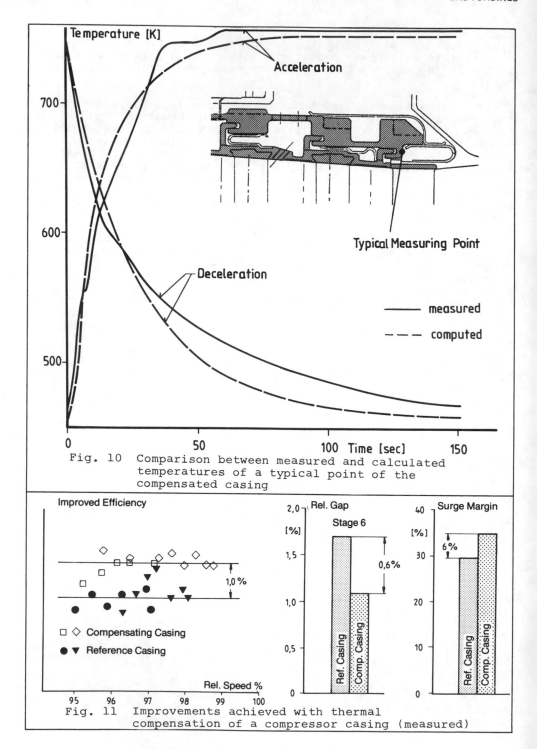

Fig. 10 Comparison between measured and calculated
temperatures of a typical point of the
compensated casing

Fig. 11 Improvements achieved with thermal
compensation of a compressor casing (measured)

referred to as "active clearance control", where the engine control system is used to change the clearances actively with the
engine running. This can be achieved, for instance, by placing
tubes around the casing, for blowing cold air taken from a lower
compressor stage through a row of small holes against the casing
to facilitate impingement cooling and shrinkage of casing. In
this way it is possible to have larger clearances during certain
flight conditions with the coolant flow turned off, avoiding
excessive abrasion that would occur as a result of high maneuver
loads, thermal deflections, etc. During other flight conditions,
such as cruise, when most of the fuel is consumed, the cooling
system may be turned on to actively close the clearances and
increase the efficiency of the compressor.

Finally, it should be noted that before one can design a
compressor with thermal behavior which will achieve the desired
life and optimum clearance variation throughout the flight cycle,
very advanced computational methods and an accurate knowledge
of the convective heat transfer are required. The following problem areas are of particular interest:
- Disk heat transfer: rotating disks near stationary walls
  and rotating cavities, both cases with axial throughflow
  and/or radial net inflow or outflow; disks with different
  temperatures and, thus, buoyancy-induced flow in a centrifugal force field, especially during the transient phase
  when the disks are heated or cooled;
- Heat transfer in the bladed annulus: heat transfer to the
  blades, platforms and blade attachments and into the disks;
  heat transfer into the casing including the effect of
  secondary and blade tip clearance flows; heat generation
  when the blades rub into the abradable coating;
- Casing heat transfer: heat transfer within the casing usually controlled by leakage flows through narrow channels
  and irregular passages with three dimensional flows, often
  affected by natural convection; influence of bleed flow;
  radiation between surfaces of intricate geometry; heat
  conduction through contact surfaces.

3.2 Combustor

The highest temperatures in aero-engines occur in the combustor, where the combustion gases may reach values of about
2400 K in the primary zone. Thus, it is obvious that extensive
cooling of the flame tube walls is required.

Modern combustors of aero-engines are usually annular and
either straight axial type (see Fig. 12, Ref. 5) or, especially
for smaller engines, of the reverse flow type. They are characterized by their short length and extremely small volume relative to the amount of heat released.

Cooling of the flame tube is typically effected by film
cooling, using a large portion of the compressor delivery air.
Today's engines use about 40% to 50% of this air flow. Future
design trends will make cooling more difficult, because

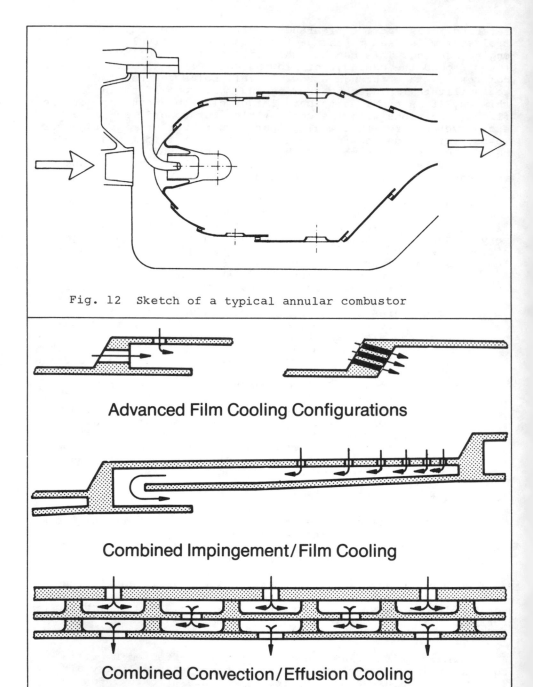

Fig. 12   Sketch of a typical annular combustor

Advanced Film Cooling Configurations

Combined Impingement/Film Cooling

Combined Convection/Effusion Cooling

Fig. 13   Conceptual designs of improved flame
tube cooling configurations

- The compression ratio will continue to increase, leading to higher compressor exit temperatures, (now about 800 - 900 K, as stated above), diminishing the cooling potential of the combustor cooling air,
- Higher turbine inlet temperatures (Fig. 3) will raise the gas temperatures in the dilution zone of the combustor,
- Increased turbine cooling flows, required because of the higher gas temperatures, bypass the combustor, reducing the amount of cooling air for the flame tube and raising the dilution zone temperature even further,
- Application of heat exchangers in a regenerative cycle (discussed in Chapter 3.5) will bring the air temperature at the combustor inlet close to the level that the flame tube material can withstand, with consequent reduction in the cooling potential, and
- Fuel shortages are likely to oblige Air Lines and Air Forces to use alternative fuels with a wider specification and a lower hydrogen content, resulting in a higher level of flame radiation (Ref. 6 surveys the impact of alternative fuels).

Therefore, an improvement in the effectiveness of flame tube cooling is called for. A considerable increase in the cooling air flow is usually not possible, because most of the compressor delivery air is required for the combustion as well as in the dilution zone to achieve the desired combustor exit temperature profile. Thus, improvements in the cooling configuration are required in order to attain greater cooling effectiveness for given or even reduced cooling flows. The trend has been from simple louvers, wiggle strips, etc. to highly intricate machined cooling rings. The aim is to introduce the film into the flame tube as uniformly as possible, with a low level of film turbulence and retaining a rugged configuration that can withstand high temperatures without warping. Fig. 13 (top) shows examples of these machined cooling rings.

Further improvement in the cooling effectiveness is possible if the cooling air is used for convective cooling before it is ejected as a film. An example of this combined convection and film cooling is shown in the middle of Fig. 13. The flame tube is double-walled. The outer wall is perforated, the cooling air passes through the holes, impinges against the inner wall, flows to one side and exits as a film. The perforations may be designed to cool especially the region where the film has lost its high effectiveness. This approach results in a very uniform wall temperature, which is one of the design goals. Effective impingement cooling, however, requires a relatively high pressure drop. Therefore, in some application, different designs of convective cooling may be preferable. These may comprise ribs, pimples or other means to enhance heat transfer in the double-wall structure (Ref. 7).

Still higher levels of cooling effectiveness can be achieved by combining convection and effusion cooling. An example is indicated at the bottom of Fig. 13. The cooling air is passed through several layers of perforated walls separated by ribs and/or pimples to promote turbulence and form large surface areas of the inner passages. The holes of the outer layer may be designed to meter the air flow, by which means the desired variation of the flow rate of the effused air at the inner surface of the flame tube may be achieved (Ref. 8).

Obviously, the design of complex configurations of this type calls for very sophisticated computational methods with good models for the flame radiation, the heat transfer coefficients at the inner and outer flame tube surfaces and the hot gas tempera- ture effective for the gas side convection. The latter is parti- cularly complex, because the flow inside the flame tube is usu- ally highly three-dimensional, recirculating and very turbulent with large temperature variations. And the flow is affected greatly by the chemical reactions taking place. In the future, careful designs will require reliable computer codes for solving the full Navier-Stokes equations.

Furthermore, an optimum design necessitates close coopera- tion between the heat transfer engineer and the materials and manufacturing specialists to ensure that the configuration re- mains cost-effective.

The use of thermal barrier coatings in ceramic materials, such as zirconium oxide, has become feasible since new spraying and bonding techniques have become available (Ref. 9). The effect of these coatings is twofold: they insulate, thanks to their low heat conductivity, which is particularly effective when additio- nal convective cooling is employed, and they absorb less radia- tive heat because of their high reflectance, thus reducing the heat flow into the flame tube wall.

Even if future combustors are all-ceramic, the heat transfer engineer will still be greatly involved in the design because detailed temperature calculations will be required if the desired life of the flame tube is to be achieved (Ref. 10).

Another heat transfer problem in combustors involves vapori- zation of the fuel droplets. The speed of this process is one of the factors controlling the subsequent combustion. Therefore, de- tailed knowledge of the heat and mass transfer during the vapori- zation of clouds of droplets in complex swirling flow fields is required.

Summing up, the design of durable and reliable combustors calls for detailed methods for predicting
- The effectiveness of film and convection cooling in complex configurations to achieve a uniform, acceptable wall tem- perature with a minimum of coolant flow and pressure drop,
- The flame radiation at all points of the flame tube,
- The gas-side heat transfer coefficients, and
- The droplet vaporization process.

In the light of today's highly-loaded combustors and in- creasing cycle pressures, the current methods are not sufficiently accurate, and a considerable amount of research effort is still required.

3.3 Turbine

Similar to the combustor, the first turbine stages of modern aero-engines operate in an environment where the gas temperatures are well above the allowable metal temperature (Fig. 3). There-fore, efficient cooling is one of the key requirements.

High temperature parts are prone to failure. Therefore, in aero-engines with their extremely severe requirement for relia-bility, turbine cooling is a particularly important subject. It is not surprising that the largest portion of the maintenance cost of an aero-engine originates from its hot end (Ref. 11). Hence, the design of the turbine cooling system requires parti-cular care. It calls for highly accurate computational methods, sophisticated experimental facilities for realistic testing, as well as advanced materials (Ref. 12) and manufacturing techniques.

Fig. 14 shows a typical axial turbine of an aero-engine, with the parts that require cooling indicated. These are the vanes and blades, platforms, casing liner or shroud, and the disks. Typical values of the cooling air mass flow rate as a percentage of the main stream mass flow are also given.

In general, cooling of a part means the removal of heat and for this a cooling medium is needed. A very efficient coolant would be a liquid, such as water or a liquid metal, because of their high heat transfer capability, large thermal capacity and the possibility to utilize their latent heat of vaporization. In aero-engines, however, air has been used exclusively so far. Although its thermal properties make it a good insulator and a very poor coolant, air still has decisive advantages for aero-engines:
- It is immediate available,
- No special storage or pumping devices are required, if compressor delivery air is used,
- Leakage is not disastrous,
- Air can be fed back into the main gas stream conveniently, so no collection and recirculation devices are needed,
- Air is light, rotating blades don't have to carry the centrifugal load of a heavy liquid,
- Start-up, off-design power and shut-down (all occuring frequently with aero-engines) do not pose great problems, in contrast to liquid coolants.
A survey is given in Ref. 13.

For stationary gas turbines the advantages of air over liquid as coolant are not so decisive. Therefore, liquid coolants are considered for actual applications (e.g. Ref. 14). In aero-engines, liquids may be useful for emergency cases. If, for instance, in a twin-engined helicopter, one engine fails the other can de-liver much more power for a short time if water injection into the turbine cooling air is used for increasing the turbine entry temperature.

Typical paths of the turbine cooling air are shown in Fig. 14. As can be seen, the air is taken from the compressor and

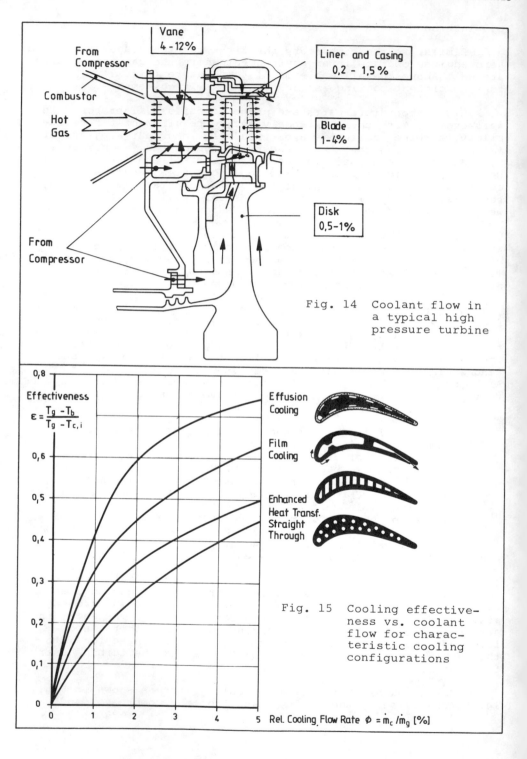

Fig. 14   Coolant flow in
          a typical high
          pressure turbine

Fig. 15   Cooling effective-
          ness vs. coolant
          flow for charac-
          teristic cooling
          configurations

ducted into the vanes from either side, into the blades via their roots, to the platforms, casing and liners, as well as to the disks. Afterwards, the coolant is fed back into the main gas stream to perform aerodynamic work in later turbine stages. It should be noted that the air to cool the blades passes through a preswirler on leaving the stationary parts. This minimizes the degree of heat-up resulting from dissipation. If it is an inter-mediate- or low-pressure turbine, the cooling air is taken from an intermediate compressor stage, where it is "thermodynamically cheaper" (i.e. not so much compression work has been put into it) and is cooler.

In some cases, the pressure of the coolant at the blade root is not sufficient to pass the required amount of cooling air through the blade. Special pressure boost systems then come into question.

Blades and vanes are the most critical parts of cooled tur-bines (see Ref. 1). Although the ways of arranging the cooling system are innumerable, classification into characteristic groups is possible. Examples are shown in Fig. 15.

The first group is "internal cooling" where the coolant cools by internal convection only and has no further cooling effect after leaving the blade. This group can be subdivided into
- Systems employing straight-through radial holes or channels that may be connected to form multipass systems in which the air passes the blade several times before leaving,
- Systems involving enhanced heat transfer, by means of ribs, pimples, pedestals or similar devices to enlarge the sur-face and promote turbulence, or impingement cooling,
- Systems employing an insert that provides a high degree of flexibility for distributing the air according to the cooling requirements.

The second group is "film cooling". Single, multiple or full coverage films are used. For structural reasons, the films usu-ally originate from rows of closely-spaced holes rather than slots.

The third group is "effusion cooling" where the air leaves the blade at every point through a porous surface.

For typical cooling configurations, the cooling effective-ness is also plotted as a function of the relative cooling flow rate. It can be seen that, for a given cooling flow, the effec-tiveness of internal cooling is lowest and that of effusion cool-ing highest with film cooling in between. Alternatively, a re-quired effectiveness can be achieved in the case of the effusion-cooled blade with less cooling air compared to the straight-through type. Hence, a compromise between the amount of cooling, i.e. cycle efficiency, and level of cooling technology, i.e. cost, is called for.

Blade cooling configurations may combine elements from each group. Modern aero-engines usually employ cooling systems from the first or second group. Effusion cooling, although the most

effective, has not found its way into production yet, mainly because of the mechanical problems, the manufacturing difficulties
and the detrimental effect on the aerodynamic efficiency of the
turbine.

The requirements made of the blade cooling system are
- Minimum amount of cooling air to minimize its negative
  effect on the engine performance, calling for high cooling
  effectiveness and efficiency,
- Uniform blade temperature distribution as far as possible,
  to reduce thermal stresses to achieve high reliability
  and long life,
- Low cost, requiring cheap materials and manufacturing
  techniques.

Since these requirements are contradictory, a careful compromise is necessary for each application. This means that all
of the cooling configurations shown in Fig. 15, although they
represent different levels of technology, will be used in the
future. For certain applications, a simple radial hole blade,
for instance, may be the optimum blade and be more suitable than
the more advanced film- or effusion-cooled blades. Therefore,
potential improvements should continue to be investigated and
developed for each of the characteristic blade-cooling configurations.

An example of the cooling configuration of a turbine vane
of a high bypass ratio engine is shown in Fig. 16. The inserts
facilitating impingement cooling and the leading edge with full
coverage film cooling should be noted. As a further example, a
cooled gas generator turbine of a helicopter engine (Ref. 15) is
shown in Fig. 17. It is interesting to observe that not just
one particular cooling configuration was chosen for the whole
turbine, but that blades and vanes have quite different systems,
which were individually optimized. The second stage cooling
system differs from the first stage system, with uncooled second
stage blades. All of the various features mentioned above are
found, except for effusion cooling.

The tendency to increase the turbine entry temperature more
quickly than the allowable material temperature, as shown in Fig.
3, generally means higher heat flow into the turbine blade. Since
this heat is energy taken out of the main gas stream, it is lost
for the performance of work at that particular turbine stage,
which may be expressed as a loss of turbine efficiency. The heat
flow into the blades of an advanced high temperature turbine may
be of the order of 1% to 2% of the turbine work. With today's
aerodynamic efficiency levels of around 90%, this is not negli-
ible and efforts to reduce the heat flow may be worthwhile, not
just from the cooling point of view.

The use of thermal barrier coatings (Ref. 9) presents one
possibility for reducing the heat flow into the blade and makes
cooling simpler. All-ceramic turbines are still in the research
stage. Their design requires very detailed three-dimensional
temperature and stress calculations.

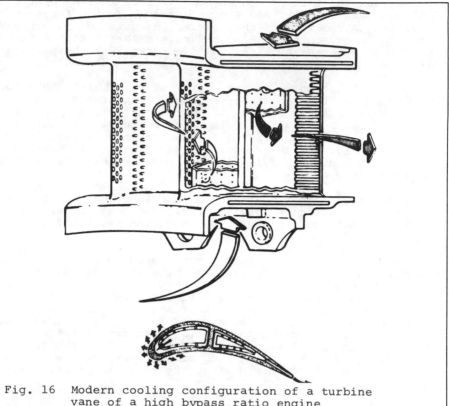

Fig. 16   Modern cooling configuration of a turbine
          vane of a high bypass ratio engine

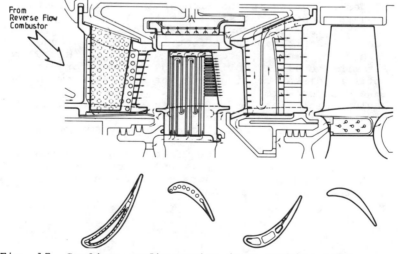

Fig. 17   Cooling configuration in a turbine of a
          helicopter engine

Besides the task of cooling the various turbine parts, such as blades, vanes, disks and liners, careful design for optimum clearances is required if the high efficiencies of modern turbines are to be achieved. A procedure similar to that with the compressor is necessary, i.e. detailed transient thermal analysis throughout the flight cycle for the disk and the casing.

As an illustration of the effect of the thermal behavior, the movement of the tip fin of a shrouded turbine through a flight cycle is shown in Fig. 18. Large deflections, both in the radial and in the axial direction, can be seen. This particular graph has been plotted from data obtained from x-rays. An example is given in Fig. 19. X-rays may be used by the heat transfer engineer, in addition to temperature measurements, to validate his theoretical models.

Since tight clearances are so important, new commercial engines, such as the PW 2037 (Fig. 2), are designed with an active clearance control system of the type described in Chapter 3.1.

One of the most important future research areas in turbine design is the effect of cooling on the aerodynamic efficiency. At today's level of cooling, the turbine efficiency may drop by around 5% to 10% when the turbine is operated in a hot environment, compared to cold aerodynamic tests (Refs. 16 and 17). The aerodynamicist can only hope to recover at least part of this loss if he works in close cooperation with the heat transfer engineer. Then the blade profile and the internal cooling system may be designed such that cooling films, for instance, are avoided at locations where they disturb the flow around the aerofoil and hence reduce the efficiency.

Summarizing the heat transfer problems in turbines, the following areas that need special attention may be mentioned:
- Blades and vanes: prediction of the local gas-side heat transfer coefficient still unsatisfactory; effect of secondary and tip flows on blade, casing and platform heat transfer; cooling effect in complex internal passages with fins, pimples, etc., including the effect of rotation with the resulting Coriolis and buoyancy forces;
- Rotating disks: cooling with the coolant flowing radially inward or outward;
- Casing: heat transfer in complex structures with leakage flows, contact resistances; and
- Interaction between cooling and aerodynamics.

An idea of the accuracies required can be obtained if one considers that today's turbine blades are operated at temperatures which are so high that a further increase by just 15 K would halve the life of the blade.

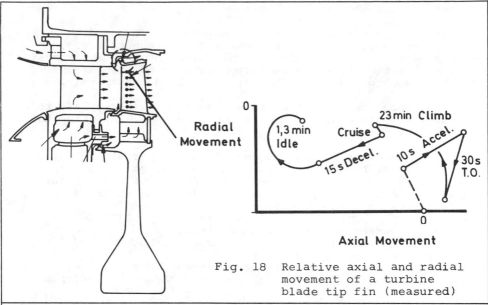

Fig. 18 Relative axial and radial movement of a turbine blade tip fin (measured)

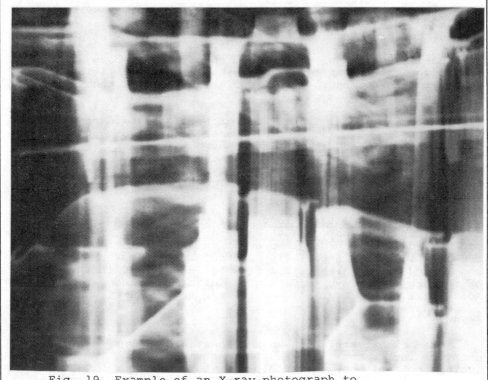

Fig. 19 Example of an X-ray photograph to measure clearances

3.4 Afterburner and Exhaust Nozzle

Afterburners and exhaust nozzles are also exposed to hot gas temperatures well above the allowable metal temperature. The corresponding cooling problems to be solved by the heat transfer engineer are similar to those that have already been discussed for the combustor. Again film cooling is used extensively.

The temperatures of the flameholders, generally, have to be analyzed, too, in order to ensure the structural integrity and reliability.

Variable-geometry exhaust nozzles need special attention, because unfavorable temperature distributions may cause warping of the flaps and failure of the hinges and actuating mechanism. The situation is complicated by leakage flows that are difficult to predict (Ref. 18).

Heat transfer problems of importance are
- Film cooling with high main stream turbulence and large gas-to-film temperature ratio,
- Film cooling in convergent/divergent nozzles, i.e. reaching from subsonic flow through the throat to supersonic flows,
- Flame radiation, and
- Droplet vaporization.

3.5 Further Heat Transfer Areas

Besides the major engine modules discussed above, there are many components in an aero-engine that require the attention of the heat transfer engineer. Lack of space permits them to be mentioned only.

Labyrinth seals. Since labyrinth seals have to be designed for very tight clearances, the thermal behavior of the static and rotating member has to be evaluated for the whole flight cycle, just as is the case with the compressor and turbine. Thermal analysis is still hampered by a lack of understanding of the heat transfer inside a modern high-performance step seal (see Ref. 19).

Bearings and lubrication. Detailed thermal analysis of the bearings is required, because certain temperature limits must not be exceeded in order to prevent the oil from cracking and possibly igniting. Especially critical situations also to be considered occur when the oil flow is interrupted in the event of failure, in zero-g-condition or during inverted flight. Additionally, the heat balances of the whole oil circulation system have to be evaluated. Papers on this subject are published in Ref. 20.

Heat exchangers. Aero-engines have a number of heat exchangers for various purposes, such as for cooling the oil (using the fuel, for instance), for cooling the cabin air, etc. Future applications may also include extracting heat from the exhaust stream and transferring it into the air stream between the compressor and combustor to reduce the fuel consumption. This seems particularly attractive for helicopter engines (Ref. 21). Another future application is likely to be the cooling of turbine cooling air. Since the temperature of the cooling air will increase and thus its cooling potential will decrease with increasing compression ratios, cooling of the cooling air using part of the cold by-pass flow, for instance, will be very effective. Of course, the extra complication, weight and safety risk have to be weighed against the advantages.

The heat exchangers for aero-engines may be characterized by their compactness, light weight, reliability, and large values of efficiency with low pressure drops. A great deal of research is still necessary for designing these heat exchangers, especially for future applications.

Infrared radiation suppressor. For many military applications, a reduction in infrared radiation is required to reduce the detectability of the aircraft by sensors. Most of the radiation is emitted by the hot turbine and nozzle surfaces and by the exhaust gas. This is effected by attaching the suppressor to the rear of the engine. The usual principle is to shield the hot parts from direct view and to cool the hot gas by admixing cold air (e.g. Ref. 22). Detailed information is scarce because of the sensitivity of the subject.

Heat rejection by the engine. The heat loss from the engine through its outer casing to the environment is also one of the problem areas of the heat transfer engineer. The amount of heat rejected has to be known for the design of the structure and the ventilation of the engine bay or nacelle. In addition, especially for smaller engines, the heat lost has to be accounted for in a proper heat balance when analyzing an engine's performance.

4. COMPUTATIONAL METHOD

Prediction of the transient temperature distribution in the various engine parts is made with the aid of an appropriate computer program system. The structure of such a system is illustrated in Fig. 20. The left-hand column shows the computation sequence, and the center column the data files and subroutine library. The first step is to take the relevant geometry data and generate a finite element grid. Then the thermal property values are evaluated. After that the boundary conditions (heat transfer coefficients, air temperatures, etc.) for each surface element are computed using cycle, air system and mission data and calling-up the relevant subroutines. These subroutines contain relationships describing the heat transfer process for typical situations, such as flat plates, channels of various geometries, rotating disks, etc. In some cases, especially for cooled turbine blades, detailed boundary layer calculations are performed to evaluate the gas side heat transfer coefficient. The next block, the heart of the program, solves the heat conduction equation for one time step. Then the procedure is repeated for all time steps throughout the flight mission. The temperatures are stored and, as indicated in the right-hand column in Fig. 20, are presented together with the geometry in

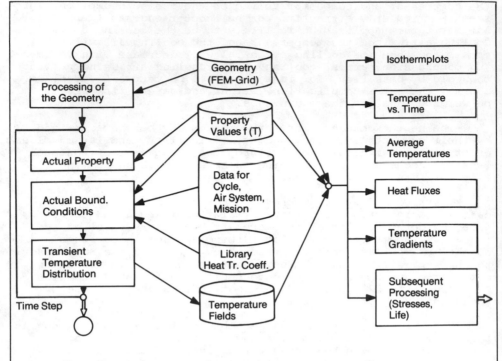

Fig. 20   Structure of computer program system
           for thermal analysis

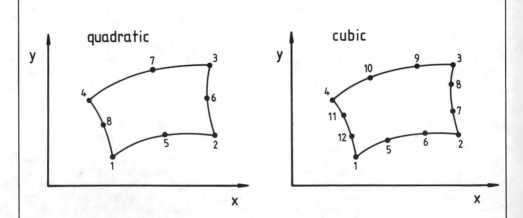

Fig. 21   Typical finite element shapes

isotherm plots, in graphs as function of time, and as mean values. In some cases, heat fluxes and temperature gradients are of interest. These results are used by the heat transfer engineer for making an initial evaluation as to whether the desired thermal behavior or cooling performance has been achieved. Then, the results are processed further in programs that determine the stress distribution and the life.

For describing the generally highly complex geometry with curved surfaces it is convenient to employ isoparametric quadratic or cubic elements as shown in Fig. 21 (Refs. 23 and 24). They can be used directly for stress calculations without the necessity for interpolation onto another finite element grid. In this way, computation is simplified and there is no unnecessary loss of accuracy.

5. CONCLUSIONS

It has been shown that the activity of the heat transfer engineer in the field of aero-engine development has greatly expanded and has gained importance in recent years. Traditionally, he concentrated mainly on the cooling of the hot parts., notably the combustor and turbine. Now, his responsibilities include the thermal behavior of nearly all other parts, including the "cold parts", such as the fan, compressor, labyrinth seals, bearings, etc. This is the result of the trends in aero-engines toward higher pressure ratios and turbine temperatures. Furthermore, costly redesigns can be avoided and engine reliability increased with simultaneous reduction in engine size and weight if thermal analyses are performed from the start of a new project. Finally, future engines will have to maintain tight clearances throughout the flight cycle. Only then will it be possible to achieve still higher aerodynamic efficiencies of the compressors and turbines and, thus, to attain the reduction in fuel consumption which is so much strived after.

A discussion of the heat transfer tasks involved in major engine components has demonstrated that nearly all aspects of heat transfer technology are met with aero-engines. Very detailed models, describing the heat transfer processes, will have to be developed in order that the accuracies necessary in engine design may be achieved. Since many phenomena are still not fully understood, continued, intensive research is called for. Particularly important areas have been highlighted in this report.

REFERENCES

1. Hennecke, D.K. 1982. Turbine blade cooling in aero engines. Von Karman Institute for Fluid Dynamics. LS 3 on Film cooling and turbine blade heat transfer.

2. Winterfeld, G. 1979. Prospects for propulsion and energetics. AGARD Highlights 79/1, pp. 12-25.

3. AGARD, 1978. Icing testing for aircraft engines. AGARD-CP-236.

4.   Owen, J.M. 1982. Rotating disks and enclosures. XIV ICHMT
      Symp. on Heat and mass transfer in rotating machinery,
      Dubrovnik, Yugoslavia.

5.   Krockow, W., Simon, B. and Parnell, E.C. 1979. A chemical
      reactor model and its application to a practical combustor.
      AGARD-CP-275, Paper 23.

6.   AGARD 1978. Aircraft engine future fuels and energy conser-
      vation. AGARD-LS-96.

7.   Singleton, R.E. 1982. Aeropropulsion research for the U.S.
      Army. ASME-Paper 82-GT-203.

8.   Essman, D.J., Vogel, R.E., Tomlinson, J.G. and Novick, A.S.
      1981. TF41/Lamilloy accelerated mission test. AIAA-81-1349.

9.   Stepka, F.S., Liebert, C.H. and Stecura, S. 1977. Summary of
      NASA research on thermal-barrier coatings. SAE, Int. Auto-
      motive Eng. Congr., Detroit.

10.   Kappler, G. 1979. Status of combustor development from
      ceramic materials. 6th Army Materials Techn. Conf. on
      Ceramics for High Performance Application - III. Relia-
      bility, Orcas Island, Wash.

11.   Swan, W.C., Bouwer, D.W. and Tolle, F.F. 1976. Life cycle
      cost impact on design considerations for civil transport
      aircraft propulsion systems, in Proceedings 3rd Int. Symp.
      Air Breathing Engines, D.K. Hennecke and G. Winterfeld, Eds.,
      DGLR-Fachbuch Nr. 6.

12.   Freche, J.C. and Ault, G.M. 1977. Progress in advanced high
      temperature turbine materials, coatings, and technology,
      AGARD-CP-229, Paper 3.

13.   Japikse, D. 1976. Alternative turbine cooling technology,
      von Karman Inst., LS 83

14.   Horner, M.W., Day, W.H., Smith, D.P. and Cohn, A. 1978.
      Development of a water-cooled gas turbine, ASME Paper
      78-GT-72.

15.   Hourmouziadis, J. and Kreiner, H.B. 1981. Advanced component
      design basis for next generation medium power helicopter
      engines, AGARD-CP-302.

16.   Barry, G. 1976. Effect of cooling on aerodynamic performance,
      von Karman Inst., LS 83.

17.   McDonel, J.D. and Eiswerth, J.E. 1977. Effects of film in-
      jection on performance of a cooled turbine, AGARD-CP-229,
      Paper 29.

18.   Grieb, H., Vedova, R., Enderle, H. and Nagel, H. 1981.
      Comparison of different nozzle concepts for a reheated
      turbofan. AGARD-CP-301.

19. AGARD 1979. Seal technology in gas turbine engines. AGARD-CP-237.

20. AGARD 1982. Problems in bearings and lubrication. To be published as AGARD Conference Proceedings.

21. Grieb, H. and Klussmann, W. 1981. Regenerative helicopter engines - Advances in performance and expected development problems. AGARD-CP-302.

22. Barlow, B. and Petach, A. 1977. Advanced design infrared suppressor for turboshaft engines. 33rd Annual Nat. Forum of the Am. Helicopter Soc., Wash., D.C.

23. Zienkiewicz, O.C. and Parakh, C.H. 1970. Transient field problems: two-dimensional and three-dimensional analysis by isoparametric finite elements, Int. J. Num. Meth. Engng. 2, p. 67 - 71.

24. Köhler, W. and Pittr, J. 1974. Calculation of transient temperature fields with finite elements in space and time dimensions, Int. J. Num. Meth. Engng. 8, p. 625 - 631.

Fundamental Heat Transfer Research for Gas Turbine Engines: NASA Workshop Overview

DARRYL E. METZGER
Mechanical and Aerospace Engineering
Arizona State University
Tempe, Ariz., USA

ABSTRACT

A review and discussion are presented of the objectives, organization, and results of a workshop sponsored by the Aerothermodynamics and Fuels Division, NASA Lewis Research Center, Cleveland, Ohio. The objective of the meeting was to assemble heat transfer specialists from USA industries, universities and government agencies to discuss approaches and priorities for the Center's future plans for research into gas turbine heat transfer fundamentals. Extensive input was solicited from and distributed to participants prior to the meeting. Those inputs, together with the subsequent meeting reports and recommendations, are reproduced and discussed in the present paper.

1. INTRODUCTION

In October of 1980, a workshop on Fundamental Heat Transfer Research for Gas Turbine Engines was held at the NASA Lewis Research Center, Cleveland Ohio. The two-day workshop was organized to accomplish several objectives. First, to provide a common meeting ground and forum for discussions among practitioners from all USA segments of gas turbine engine technology: industry, government, and universities. Second, to uncover industry - wide problem areas where heat transfer research can contribute to product improvement, and to discuss priorities and strategies for attacks on these problems. Third, to provide a forum for presentation of current and planned NASA work and description of NASA facilities; and fourth, to provide an opportunity for collective industry and university input to NASA planning.

In terms of the first two objectives, the workshop was the third in a series of recent gatherings of USA workers in gas turbine heat transfer. The first of these, sponsored by the Office of Naval Research, was held at the US Naval Postgraduate School, Monterey, California, in September 1978. The second, sponsored the American Society of Mechanical Engineers, was a session at the 1980 International Gas Turbine Conference in March 1980.

This active interest in meetings to discuss common problems and possible approaches in gas turbine heat transfer research underscores the important role heat transfer plays in the continued development of these engines, in terms of both improved performance and improved durability. It is also indicative of the overall complexity of fluid mechanic and convective heat transfer phenomena in turbomachinery; and the necessity, as continued

381

improvement is sought, to recognize and deal with more and more of these complexities.

This paper will present an overview of the workshop preparations, proceedings, and conclusions. Those desiring more information may wish to consult NASA Conference Publication 2178.

2. ORGANIZATION OF THE WORKSHOP

The meeting organizers (the author and Robert Simoneau of NASA Lewis) attempted to give the two-day gathering some initial momentum by requesting advance input from the participants in the form of written goals and identification of problem areas. This pre-workshop input was assembled and distributed to all participants in advance of the meeting.

The morning of the first workshop day was devoted to presentations of NASA Lewis current programs and plans by various NASA personnel. The first afternoon and second morning were mainly devoted to working group deliberations, with working group reports and open discussion occupying the second afternoon.

The working groups were organized along the lines of methods of attack, experimental and analytical, rather than around specific problem areas. Based partially on the pre-workshop input, and on the member of expected attendees, four workshop groups were planned: two in Near Engine Environment Experiments, and one each in Computational Analysis and Basic Experiments. Robert Fish of Pratt and Whitney Aircraft and David Nealy of Detroit Diesel Allison chaired the near engine experiments groups. Richard Pletcher of Iowa State University and Robert Mayle of Rensselaer Polytechnic Institute chaired the computation and basic experiments groups, respectively.

3. PRE-WORKSHOP INPUT

Participants were asked to provide the workshop organizers with written input in advance of the meeting. Both goals for their work over the next few years and suggested questions for the workshop to address were requested. Response to this request was extensive and the input was assembled and mailed to all participants approximately one week prior to the meeting.

The questions that were proposed for discussion generally fall into two categories: Those that relate to generic topics such as film cooling and rotational effects, and those that relate to analytical methods or experimental methods in general. These statements reflect individual and organizational concerns and priorities at the time of the workshop. As such, they provide in themselves a reasonably up to date view of heat transfer in relation to turbine engine design. The questions submitted are reproduced below essentially unedited. There is some duplication which presumably reflects common concerns among the participants.

3.1 Questions Related to Generic Topics

> Life and Durability. "What are the weakest links in existing life prediction methods?"

"Can we effectively increase component life by minimizing the effects of thermal stress cycling by finding ways to keep a component at a near-isothermal condition at any instant during a flight cycle?"

"Most thermal analyses of new gas turbine combustors are based on the assumption of steady flow. How can we identify those aspects of the unsteady flow which contribute to thermal cycle fatigue? What useful dimensionless groups can be identified for predicting failure times? Can design strategies be developed based on unsteady thermal analyses which allow thermal cycle fatigue damage to be minimized?"

"Based on the uncertainties in the analytical prediction of correlation of variables such as h_g, h_i, to at best ± 10% a NASA report shows that even then we will not be able to predict steady state metal temperatures with certainty better than ± 50°F. This temperature uncertainty can account for order of magnitude in life. Doesn't this indicate that experimentation with the designed hardware will always be necessary to reduce the uncertainty and doesn't this also raise questions about the need or practicality of complex analyses which may not improve the accuracy of the first design?"

Boundary and/or Environmental Conditions. "What features of the engine environment should be modeled; e.g., is it necessary to know gas path turbulence levels and scales? How would these be measured?"

"Should there be more emphasis on the complex gas temperature distribution phenomena through the turbine and in the airfoil passages due to stage energy distribution and secondary flows in our prediction technique's research and development programs?"

"The local gas environment, particularly local gas temperatures, are not well-known, predictable or presently controllable. Since these are directly related to predicting metal temperatures, what plans are underway to improve knowledge and prediction capability in this important variable? What are suggested research directions to improve the prediction capability for this important variable?"

Film Cooling. "There is a need for better correlations to unify the wide variety of data with its many parameters."

"Do we really understand the basic problem of mixing of two streams of different temperature and different velocity?"

"Film cooling effectiveness. Do we keep to the traditional approach to the concept of film cooling effectiveness, or adopt the Nusselt number approach? What about the new transient techniques?"

"The controversy of isothermal versus adiabatic film cooling effectiveness research has long been lingering. A concensus on the definition of the needs for both may place this question in its proper perspective."

"What is the effect of extreme variable property environments?"

"Where do we stand on the accurate prediction of effects of transpiration and discrete hole cooling on heat transfer under the extreme variable property environments characteristic of actual gas turbine operating conditions? Much of the experimental data which have guided the development of turbulence models for these effects has been taken under nearly constant property conditions. Have these models been adequately evaluated for the much higher temperatures and heat fluxes of actual engines?"

"How do you model a multislot film cooling configuration?"

"What is the future role of transpiration cooling?"

"We know that we must cool and protect some turbine hot section parts by film, transpiration and multihole cooling. What are the ways that we can manage the injection of this cooling air to minimize (and even reduce) main gas path aerodynamic losses?"

"Aerodynamic losses associated with film cooling--how do you analyze them?"

"What is the importance of shock wave interaction with film cooling?"

"Uncertainties regarding the separate influences of distributed surface injection (film cooling) on turbulence generation, near wall temperature dilution, and downstream thickening of the boundary layer."

"What will be the role of film cooling, if future fuel specifications for aircraft engines allow a particle laden hot gas flow?"

Turbulence and Unsteady Flows. "Many recent advances in the analytical or computational treatment of complex turbulent flows have been guided primarily by experiments in which temperatures or heat fluxes have not been measured. Can predictions of heat transfer be made for most of these flows using a form of the Reynolds analogy in the turbulence modeling for heat transfer parameters?"

"It appears that the total free stream turbulent disturbance as well as its intensity and length and time scales have a large effect on airfoil heat transfer which has not yet been correlated. What is the status of efforts on airfoils under realistic Mach No., Reynolds No. and wall cooling conditions?"

"Uncertainties regarding the influence of free-stream turbulence on local heat transfer rates in the laminar region as well as on initiation and extent of the transition region."

"Limited understanding of role of airfoil surface curvature on turbulence production/dissipation and boundary layer stability. A corrollary concern involves the role of curvature in the generation of Goertler vortices along the concave surface."

"What is the influence of free stream turbulence on laminar heat transfer in the stagnation region?"

"What are the periodic flow effects due to blade passing--especially in the thin boundary layer at the leading edge."

Transition and Separated Regions. "Uncertainties regarding the surface location at which transition is initiated as well as the surface extent of the transition zone."

"Length of transition zone along surface?"

"Location of transition with competing effects present: cooling of boundary layers; pressure gradient; curvature; formulation of transition model?"

"How are separation bubbles initiated--pressure gradient required? What determines size? Prediction of heat transfer inside a bubble?"

"What turbulence modeling do we use in conjunction with predicting separation and reattachment?"

Rotational Effects. "Uncertainties regarding the influence of Coriolis and buoyancy forces on thermal/momentum boundary layer development and stability for rotor blade surfaces with and without film cooling."

"How is the effect of rotation being included in the calculation of gas side heat transfer coefficients? Is there any experimental evidence to show such an effect?"

"What are the effects of rotation on external flow, boundary layer stability, film cooling flows, internal blade passage, and inside disk cavities?"

Curvature, 3-D, and Other Complex Effects. "Advances in computational fluid mechanics along with available experimental data permit the prediction of turbine airfoil heat transfer coefficients, including the effects of curvature, pressure gradients and turbulence intensity. However, questions on boundary layer transition, the effects of wakes, rotation and secondary flows still remain unanswered. Is it realistic to anticipate that these effects will be able to be predicted analytically in the foreseeable future?"

"Have the effects of surface curvature and free stream acceleration been adequately separated and understood for airfoil heat transfer?"

"Effects of streamline curvature on wall heat transfer."

"Where do we stand on endwall regions, corner regions, free stream convergence effects, secondary flows?"

Internal Flows. "Should there be additional work directed towards internal heat transfer coefficients (impingement, pins, roughened surfaces, entrance effects, etc.)? If so, in what areas should the emphasis be placed?"

"Passages of irregular cross-section inside blade need study."

Miscellaneous.   "How are heat and mass transfer mechanisms interrelated in a catalytic combustor? How does this interrelation affect performance?"

"Any significant heat loads due to thermal radiation?"

"What are the effects of combustor radiation?--with more soot due to alternate fuels?"

"How well do analytical descriptions of heat transfer in the presence of a thermal barrier coating agree with experience? Is the operative mechanism well understood?"

Priorities and Cooperative Arrangements.   "Which areas should be worked that would give the quickest/greatest pay-off?

"What do others in the technical community perceive as important areas of turbine design which are overly dependent on previous experience and rough correlations as opposed to basic analytical models."

"What form of results are most useful, i.e., computer programs, raw data, correlations, etc.?"

"What is the natural divide between research work industry can (and must) perform and that which a universiy can carry out?"

"What is the best way to transfer information between the engine companies and the university system?   Can this method be implemented?"

"To what extent would industry or the academic community be interested in a cooperative research project utilizing any existing NASA facilities--including the new HPT facility.   What are some possible combined research programs or in what areas are they possible?"

"To what extent do you believe there should be duplication of facilities by Lewis when such facilities exist in industry? What are the views on Lewis building a large low speed turbine facility when time to get it operational may be four to five years and there are such facilities plus experience already existing in industry?"

3.2 Questions Related to Analytical Approaches

"What are the limitations and level of capabilities of current computational techniques?"

"Wake passing frequency, turbulence, curvature, temperature ratio, and secondary flows, to name a few, all contribute to the external heat transfer problem and prediction techniques. Since present test procedures appear prohibitively expensive and complex to handle all of the existing environmental levels of effects and geometry variations, what is the present status and prognosis of validated computer programs that can or will in the future reasonably evaluate these heat transfer effects on a variety of turbine airfoils?"

"How complex can a computer code be before it becomes useless (i.e., is it more economical to build several engines and test them)?"

"With respect to calculations of turbine heat transfer, which approach should have top priority?

- Development of complex 3-D analyses, including high level turbulence models.
- Quasi 3-D analyses with generalized correlations to handle special regimes (i.e., free stream turbulence effects, boundary layer transition, separation, etc.)"

"What are the most urgent needs in the following three categories?

- Numerical prediction schemes
- Turbulence modeling
- Experiments to guide development and verification of prediction methods."

"What are the more promising approaches to be followed in developing a more complete, general, and reliable theory for the prediction of transition on turbine vanes and blades?"

"What are the more promising analytical approaches to be pursued in accounting for the effect of free stream turbulence level on turbine blade transition and heat transfer?"

"Lack of precision in the prediction of the inviscid flow field around the airfoil, particularly in the forward, highly accelerated stagnation region."

"What is status of analysis of 3-D boundary layer with heat transfer—especially with respect to endwall heat transfer?

3.3   Questions Related to Experimental Approaches

Data Base.   "Lack of a systematic data base relative to the influence of airfoil surface roughness on boundary layer turbulence."

"Is there a need for data with variable properties—data on cooled surfaces rather than heated surfaces?"

"In order to properly verify any analytical models for airfoil heat transfer, an appropriate data base of actual turbine measurements is needed.   What techniques should be used to get this data, and what is the status to date?"

"What should be the balance between benchmark tests, engine simulation tests and full scale engine tests; i.e., how should developing prediction methods be evaluated?"

Similitude and Scale.   "Heat transfer and flow phenomena in rotating cavities are very complex.   Presently, heat tranfer analyses in engine applications are based on data obtained for the specific configuration.   The question of the utility of experimental programs based on generic compressor and turbine rotor configurations deserves consideration as a source of data for verification of computational models and for specific design applications."

"How much confidence can be placed in very large scale experiments where only limited similitude is achieved?"

"Are tests conducted at scaled condition, (i.e., low T and atmospheric pressure) useful?"

"How far must we go for similitude in experiments? True similitude of all variables is impossible, short of an actual engine test. How much similitude is required before extrapolation of the data to engine conditions is reasonable?"

"What can be done in regard to measuring the actual engine conditions?"

"What is being done to determine the engine conditions such as turbine turbulence level, scale, etc.?"

Techniques and Instrumentation.   "To what extent do researchers feel comfortable conducting heat transfer tests with heat flow out of surfaces when actual heat flow is inward and with small T when actual is large?"

"Recent publications and work indicate increased interest in shock tunnels and isothermal light piston tunnels facilities in order to measure and evaluate airfoil surface heat transfer and film effectiveness distributions. Could this be considered as a cheaper and more viable approach to future test programs over conventional steady state cascade testing?"

"How can we translate cascade results to rotational components?"

"Since verification of any prediction method requires use of a near-engine environment, what instruments are operationally available and to what extent should instruments be developed to obtain the necessary measurements (e.g., accurate measurement of metal temperatures, gas side local temperature, pressure, heat flux, turbulence, and turbulent structure)?"

"What is the status of efforts to define the turbulence characteristics at the entrance to successive turbine blade rows? What instrumentation is required/preferred?"

"What is the feasibility of liquid crystal thermography to visualize surface characteristics of flows?"

3.4 Summary

The key words and phrases which appeared, many repeatedly, in the pre-workshop input are represented reasonably well by the following list:

- Film Cooling
- Turbulence
- Unsteady Flows
- Transition
- Separated Flow Regions
- Curvature
- Secondary Flows
- Internal Flows
- Optimization
- Effects of New Fuels
- Coatings

These topics are those for which additional heat transfer information is indicated as needed to help achieve design goals. The pre-workshop input on goals is not reproduced here; but these statements did indicate that durability is, currently, a foremost goal for much of the industry. The needed additional heat transfer understanding and information will be generated by both analytical and experimental research. Tied to the key topics listed above are questions related to the desired mix between analysis and experiment, and the degree of approach to the real engine environment needed in experiments.

By whatever means the needed information is acquired, it will ultimately feed through the various design steps including structural analysis and life prediction. However, uncertainties in both the acquired information and in the life prediction schemes necessarily reduce the utility of the research results in achieving design goals. In setting priorities for future research it seems prudent to consider not only the feasibility of the work but also the degree of real gain to be achieved considering these uncertainties.

4. NASA LEWIS WORK

The workshop included a comprehensive presentation of current and planned NASA-Lewis work, both in-house and outside contracted. The present overview of the workshop will not include the specifics of this presentation. For details, including a description of several pertinent NASA experimental facilities, the reader is referred to the workshop proceedings.

Richard A. Rudey, Chief of the Aerodynamics and Fuel Division, the sponsoring division of the workshop, presented an introduction to the NASA presentations in terms of trends in the related aeronautical research and technology and the factors which drive them. Four groups of factors appear to be the primary driving forces: (1) energy conservation pressures, (2) economic pressures, (3) environmental pressures, and (4) political/social pressures. The net result of all these influences, as perceived by NASA, will be continued emphasis on high pressure ratio and high temperature cycles. There is an anticipated emphasis on the use of alternatives to some current practices, particularly in the areas of materials and fuel use. Also anticipated is an increase of emphasis on durability,maintainability, and related ideas. From NASA's viewpoint, the fundamental knowledge of heat transfer phenomena in the gas turbine engine hot section will be the key to the required improvements in durability and maintainability.

5. REPORTS

At the conclusion of the workshop, the chairs of the four working groups each presented reports in the form of recommendations, as given below. The reports of the two working groups on near engine environment experiments were very similar and have been combined here for clarity.

5.1 Near Engine Environment Experiments - Recommendations

Engine or near engine environmental work is needed because:

- Temperature scaling is questioned
- Turbulence effects are important
- Rotation effects are important
- Geometry effects are important

Specifically engine or near engine environmental work is needed in the following areas, listed in rough priority order:

- External heat transfer
- Gas temperature distributions
- Rotor/Stator endwall region cooling
- Rotational effects on internal cooling and flow distribution
- Blade tip and tip shroud heat transfer
- Instrumentation development, especially for external heat transfer and gas temperature distributions
- Cavity and rim flow effects
- Combustor liner cooling
- Thermal and aero performance of thermal barrier coatings
- Film cooling
- Radiation heat load on vanes

The latter two topics, film cooling and radiation heat loads, would become relatively more important if TIT were to increase substantially. However, the current consensus is that commercial engine TIT's will not go much higher than 2800F or so.

External heat transfer is listed first in the above list, indicating that it is probably the most pressing current problem area. This area can be further broken down as follows:

- Transition behavior with pressure gradients and curvature (i.e., vane and blade geometry)
- Effects of turbulence
- Effects of roughness
- Effects of rotation and wakes

Most basic experiments to check transitional turbulence models must include pressure gradients and high turbulence level.

Approaches, especially where NASA could contribute, are suggested as follows:

- Instrumentation development for hostile environment, especially heat flux, turbulence, and temperature sensors.

- Full engine testing. This approach is expensive with major difficulties associated with getting instrumentation out.

- High spool engine testing. This approach gives up only pressure level for easier instrumentation. Pressure level is not important except for cyclic endurance testing.

- NASA Vane Cascade and High Pressure Turbine Facilty. These facilities provide good engine simulation of temperature level and turbulence. Work should continue on development of high temperature heat flux sensors to make direct heat flux measurements in facilities like these.

Finally, the overall need for coordination between the experimenter and the analyst is emphasized. Experiments must be well designed to give the analyst the right data to check out prediction systems properly; and analysts must develop prediction systems which reflect the needs of the real world. Such coordination is essential for success in improving engine durability.

5.2   Basic Experiments - Recommendations

First, the group defines basic experiments to be those which isolate and investigate an important aspect of a problem area. In addition, they must be well documented and provide the physical understanding required to allow an extrapolation beyond the data set.

As a general prioritization of needs in basic experiments, the consensus is that problems that need most immediate attention are primarily those associated with predicting external heat loads. Second in priority are problems associated with internal cooling.

The following list are those recommended basic experiments (in rough priority order) which satisfy an immediate need for present airfoil heat transfer prediction schemes. These are two dimensional or quasi-two dimensional experiments.

- Transition
- Turbulence/unsteady flows
- Separation/reattachment
- Curvature
- Large acceleration
- Film cooling
- Internal cooling with rotation
- Temperature ratio

A second list is recommended as basic experiments which will help to establish future gas turbine gas path heat transfer prediction schemes. These are three dimensional experiments.

- Intrarow gas temperature re-distribution to determine local driving potential for heat transfer
- Three dimensional turbulent boundary layer heat transfer
- Separation
- Unsteady flows
- Off-design flows

As a general recommendation, the group agrees that experiments must be designed to verify the design system at various levels of sophistication. Inherent in this recommendation is a belief that such verification is essential to achieve confidence in the design system as an accurate predictive tool for new designs.

The general area of basic experiments is one where contributions can be made jointly by NASA, industry, and the universities.

- Wind tunnel experiments are natural for universities
- Rotating, multi-row turbine experiments are natural for industry
- NASA can do either

Finally, some discussion concerned information flow between industry and the universities, and it is recommended that NASA increase its role as an interface between these two groups.

5.3 Analysis - Recommendations

- Establish a clearing house to collect and evaluate available data. After evalauation, needed experiments can be recommended for use in validation of prediction methods.

- Concentrate on development of program modules rather than general codes for design.

- Many questions remain on airfoil external heat transfer. Among these are effects of transition, turbulence, and thermal scaling. The data base is inadequate for the needs of analysts, and needs to be enlarged, particularly a data base for combined effects.

- Three dimensional inviscid analyses are needed which include adequate solutions to the energy equation. This capability is needed for blade rows to define the local driving potential for heat transfer.

- More attention is needed on turbulence modeling. This is the pacing item for prediction of very complex three dimension flows.

- Basic studies are needed on the modeling of Reynolds heat flux terms. The present Reynolds analogy approach is questionable for some flows.

- Work is needed in film cooling, in terms of better organization of existing data and more work on basic analyses. Prediction of film cooling at present is a weak point.

- More work is needed on unsteady flow effects on heat transfer; both additional data and analytical work are needed.

- Analytical methods and data are needed for rotating cavity heat transfer.

- More analytical effort is needed on endwall heat transfer both with and without film cooling.

- More information is needed on the importance of the horseshoe vortex to heat transfer.

6. DISCUSSION

This section is the author's attempt to convey both a summary of the working group recommendations, and also impressions of the meeting as a whole. These impressions are based on observations of all the working groups in session, study of the notes and transcripts from the deliberation and reporting sessions, and discussions with many of the participants.

6.1 Overall Impressions

The organizers attempted, through the mechanism of the pre-workshop input distributed in advance, to bring the most pressing problem areas into focus early. The desired result was that the workshop groups could devote the primary efforts at Cleveland to deciding which problems were of the highest priority, which methods of attack were the most promising, and which groups were the best suited to the various tasks.

The pre-workshop input did highlight many areas of common concern; but the working groups, once in session, displayed a strong reluctance to be focused on any limited number of problem areas. A wide diversity of opinions and perceptions were aired. All of the groups had great difficulty in prioritizing problem areas. The rough rankings given in the working group reports represent a considerable amount of compromise. Promising methods of attack and allocation of resources were discussed only in general terms.

Part of the working group difficulty in getting down to specifics is certainly a result of the sizes of the groups. Two groups each for analysis and basic experiments and only one group for near engine experiments would have been, in retrospect, a better arrangement.

Realistically, however, the difficulties experienced by the groups in formulating specific recommendations are probably just reflective of the underlying complexity of flow and heat tranfer in gas turbine engines. There are many phenomena occuring in these engines that are not understood. There are names to describe the phenomena; but all workers are not in agreement even on what the names imply. There is no sure knowledge of what effects or questions are the most important in terms of improving design systems. There was agreement that fundamental work is needed to build our understanding of the phenomena, especially when they occur in combination.

It may well be that the unexpected overpopulation of the analysis and basic experiments groups, and the lower than expected turnout for the near engine experiments group, reflects recognition on the part of the participants that improved understanding of the basic phenomena is essential; and that this is not likely to be achieved through near engine environment testing. Indeed, the consensus of all the groups appeared to be that testing at the very near engine conditions should concentrate on the task of better defining the engine environment.

In comparison with previous similar meetings in recent years, there was much more discussion emphasis on the overall flow and heat transfer problem in terms of combined effects of the many phenomena that exist in the real engine. There was also much more discussion of numerical computation of the overall problem. These two observations are of course not unrelated. The fast growing capability in computation is one of the major driving forces creating need for a better definition of the engine environment and need for a better understanding of the flow physics.

The traditional basic experiments that study an isolated single effect have obvious shortcomings. First, they are usually conducted under conditions far removed from the actual severe engine environment. Second, they leave open the question of possible important interaction between multiple effects. Third, they often do not document all of the information needed as boundary conditions in a numerical simulation of the experiment.

On the other hand, experiments conducted with more near prototype geometries and conditions have their own set of shortcomings. They are more often than not non-generic, and are conducted with questionable uncertainty levels. They usually compound the problem of acquiring full documentation of information needed for numerical simulation.

In the working group and reporting sessions, there was clearly not full agreement on what types of work have already been adequately done, on what had already been demonstrated, etc. There was also recognition that in many areas, film cooling for one, large volumes of data from diffferent sources exist, but are in need of a systematic evaluation and summarization. Many participants felt that this situation could be improved through the creation of some sort of clearinghouse or data repository, but this is obviously difficult to do in a way that would be truly useful to designers.

The people most involved in the actual design process also expressed concern about the direct usefulness of the rapidly developing computational capability as a design tool. They pointed out the time that has been traditionally required to incorporate a new technique into a particular design system. The consensus of industry seemed to be that these computational tools would probably be most readily used if they were made available in modules that could be tailored by each company for their particular use. Along these lines the point was made that industry could make good immediate use of modules that stop short of full 3-D viscous solution capability. One example mentioned several times was the need for a good 3-D inviscid code that could predict gas path temperatures downstream of the combustor exit profile.

6.2 General Recommendations

The recommendations from the working groups focus on problems associated with heat transfer from the external gas path; they do not focus on cooling techniques per se. This is partly the result of recognition throughout the industry that there are very significant gaps in our knowledge of the gas path heat transfer. The uncertainty in gas path heat loads prevents optimal decisions on the use of cooling air and truly hinders efforts to improve durability and performance.

This focus at the workshop was probably also strongly influenced by the fact that all of the discussions became very much entwined with the developments in numerical computation. These computational developments at present are concentrated on gas path analysis rather than internal cooling flow analysis. As a result, some important problem areas that were mentioned in the pre-workshop input received little attention. For example, the problem of cavity heat transfer and its implications on turbine durability and clearance control was probably discussed far less than deserved.

For the gas path heat transfer and all other problem areas, it appears that there are advocates and sound arguments for a wide range of future experimental work. The goals of this work are both to provide immediate help to the designers and to provide the necessary validation of computational techniques at several levels of flow complexity.

These experiments must span the spectrum from single effect modeling experiments to very near engine environment experiments. More emphasis than in the past must be placed on modeling experiments that treat several effects simultaneously; and more emphasis must be placed on developing near engine experiments that are as generic as possible. The consensus seems to be that experiments that are designed to run at or very close to actual engine conditions should, for the present, have as their principal objective the definition of the engine environment. Along with this, efforts should be increased to develop the instrumentation necessary for environment definition.

6.3 Recommendation for NASA

Since one of the main purposes of the workshop effort was to provide NASA with advice concerning its own future work in fundamental heat transfer, it is perhaps appropriate to conclude with a summary of expressed needs that NASA appears to be uniquely in a position to fill.

First is a recommendation that NASA increase its present activities as an interface between industry and the academic community. It should continue to identify and fund work at universities that are compatible with university time scales and resources. In this way it can encourage utilization of the academic community in gas turbine technology and encourage graduate study in areas relevant to the industry. As part of the interface activity is a recommendation that if data repositories and data evaluation functions are to be established, that NASA assume responsibility for initiating and coordinating such activities.

Second, NASA should continue and probably increase its efforts in the development of instrumentation that can be used to better determine the engine environment under actual operating conditions. A particularly critical need is the ability to measure turbulence in real environments. The actual engine testing should probably be done in industry so that information from a wide spectrum of different current technology engines can be accumulated and summarized.

Third, NASA should continue to develop an intense and ongoing dialogue with industry in regard to its own efforts in both the computational area and in the HOST program. In order to gain maximum usefulness from both these important activities, it is important for NASA personnel to be continually well informed about the character of the design system that the information will feed into.

The fourth, and perhaps most important recommendation concerns NASA's own in-house experimental work. NASA has resources that enable it to do a wide range of testing from very basic through near engine work. However, the best use of these resources appears to be in the realm of basic experiments in cases where conditions are beyond most university capabilities, and in full scale testing with engine geometries. NASA should seek to develop flexible facilities that are capable of interchanging hardware of different manufacturer's designs, so that the overall results are representative of a range of current design practice.

Finally, NASA should endeavor, in all ways possible, to facilitate communication, coordination, and support among all segments of U.S. gas turbine technology, industry, government, and universities.

Rotating Machinery Heat and Mass Transfer Research in the People's Republic of China

CHUNG-HUA WU, SHAO-YEN KO, LIU DENGYUN,
SHEN JIARUI, and XU JING-ZHONG
Institute of Engineering Thermophysics
Chinese Academy of Sciences
Beijing, People's Republic of China

ABSTRACT

A survey of research on rotating machinery heat and mass transfer in the People's Republic of China has been made. Since the later part of 1950s, considerable research and development work has been conducted in this field in China in order to improve the performance and prolong the life of rotating machinery. The emphasis of gas turbine heat transfer has been made in this survey. The water cooling of generator and the heat transfer of rotary piston engine are also included. Researches on the measuring technique of rotating machinery such as the temperature measurement, heat flux gauge, turbulence measurement, optical measurement and flow field visualization are discussed. The following topics of gas turbine heat and mass transfer are included: numerical analysis of air cooling of turbine blades, internal cooling passage heat transfer, impingement cooling, film cooling, transpiration cooling of turbine blades, cooling of blade root tenon, cooling of rotor disc, film cooling of flame tube and cooling of afterburner.

NOMEMCLATURE

I   momentum blowing rate, $\rho_s U_s^2 / \rho_g U_g^2$

M   mass blowing rate, $\rho_s U_s / \rho_g U_g$

Taw   adiabatic wall temperature, K

Tg   combustion gas temperature, K

Ts   cooling air temperature, K

Tw   equilibrium wall temperature, K

Ug   combustion gas velocity, m/s

U_s cooling air velocity at the exit of film cooling slots, m/s

x   downstream distance along the wall from the exit of
       cooling slot, m

ρ density, kg/m^3

η film cooling effectiveness, (T_g-T_{aw}) / (T_g-T_s)

ν kinematic viscosity

Nu   Nusselt number, hx/k

Re   Reynolds number, U_g x/ν

Pr   Prandtl number, uCp/k

I. HEAT TRANSFER OF GAS TURBINE

    In China, considerable efforts have been made in the research
of gas turbine heat transfer.

TURBINE BLADE COOLING

    Two dimensional steady and transient temperature distributions
of air cooled turbine blade have been calculated /1,2/. The coefficient of heat transfer in the internal channel of air cooled turbine blade is calculated by heat balance equation. Some results
are shown in Figs. 1 and 2. The finite element method had been
adopted in evaluating the local coefficients of heat transfer of
film cooled turbine blade /3/. In this treatment, the heat carried
away by the cooling air inside the film cooling holes was considr-
ed as a heat sink. Iteration method was adopted to determine the
temperature distributions at the tip, root and mid-span of a turbine blade.

    Thermal contact resistance has been considered in the analysis
of temperature distribution at the root and the mounting groove of
a turbine blade /4/.

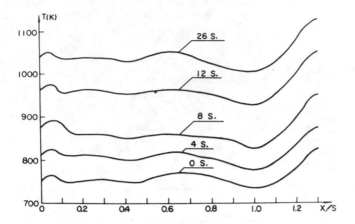

Fig. 1   The transient temperature distributions over the
suction surface of the air-cooled blade

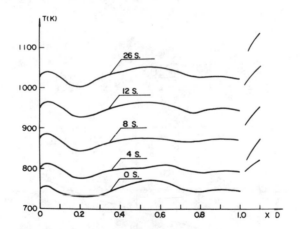

Fig. 2   The transient temperature distributions over the
pressure surface of the air-cooled blade

Two dimensional boundary layer computer program STAN5 has
been used in calculating the adiabatic effectiveness of film cooling process /5/. At lower blowing rate, the numerical results of
flat-plate adiabatic film cooling effectiveness agree well with
experimental data. But, as shown in Fig. 3, at higher blowing rate,
numerical results are higher than experimental data. The calculated
local coefficients of heat transfer of film cooling in the near slot
region agree well with author's experimental results as shown in

Fig. 4.

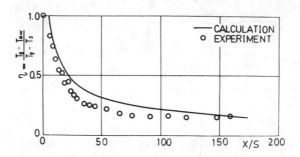

KASE	M	U$_s$	U$_g$	T$_s$	T$_g$	M	NO.
28	0.21	15.1	76.5	49.3	68.2	0.197	156

Fig. 3   Calculation and experimental results of film
         cooling effectiveness at lower M values

KASE	AF	M	m	U$_g$	U$_s$	T$_g$	T$_s$
39	321	0.44	0.39	75.8	29.9	57.7	25.0

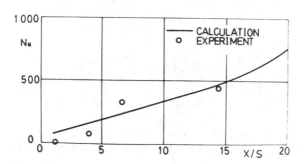

Fig. 4   Calculation and experimental results of film
         cooling effectiveness at high M values

In order to calculate the heat transfer characteristics of
film cooled turbine blade, computer program has been developed to
evaluate the three-dimensional flow field of film cooling over a
curved surface with discrete holes /6/. At lower blowing rate,

partial parabolic Navier-stokes equations and stagnation enthalpy equation have been solved to determine cooling effectiveness and local coefficient of heat transfer near the film cooling holes. The basic equation is:

$$\frac{\partial}{\partial x}(\rho u\phi) + \frac{\partial}{\partial y}\left[(1+ky)(\rho v\phi - \Gamma_\phi \frac{\partial\phi}{\partial y})\right] + \frac{\partial}{\partial z}\left[(1+ky)(\rho w\phi - \Gamma_\phi \frac{\partial\phi}{\partial z})\right] = S_\phi$$

Setting ϕ successively equal to 1, u, v, w, H, K and $\mathcal{E}$, continuity equation, three momentum equations, stagnation enthalpy equation and K-$\mathcal{E}$ equations are obtained. The unknown pressure gradient term is included in the source term S_ϕ. Pressure correction equation is derived to evaluate the pressure distribution at each point by iteration in order to satisfy the continuity equation. Some results of such computation are shown in Figs. 5 and 6.

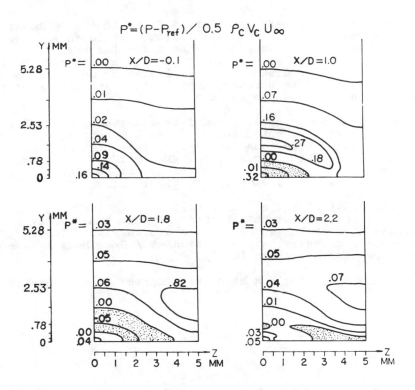

Fig. 5   The pressure contours in Y-Z plane

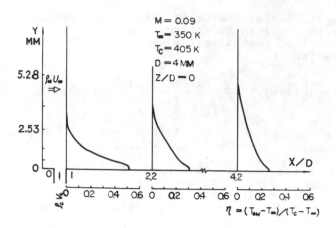

Fig. 6   Effectiveness distribution with normal ejection
          at low blowing rate, M= 0.09

Three-dimensional heat transfer of flame in the combustion
chamber have been calculated by using finite difference method and
Monte-Carlo method /7/.   Convection and diffusion terms are evaluated by finite difference method in the energy equation.   The thermal radiation in the source term is evaluated by multiple integration method.   The 3-D temperature distribution and wall heat flux
of combustion chamber can be calculated.

Simultaneous solution of fluid flow equations and heat conduction equation have been obtained /8/.   Two-Dimensional compressible
flow boundary layer equations and steady-state heat conduction equation are solved simultaneously by integral method.   As shown in
Fig. 7, the temperature distribution of a turbine blade can be evaluated for arbitrary wall temperature distribution.

Ceramic coating of the inner cooling channel of the air cooled
turbine blade and enhancement of internal heat transfer by rough
surface have been evaluated /9/.   As shown in Fig. 8, three kinds
of cooling-hole geometry have been studied /10/.   Some of the experimental results are shown in Fig. 9.

Impingement cooling has been investigated experimentally
/11,12/.

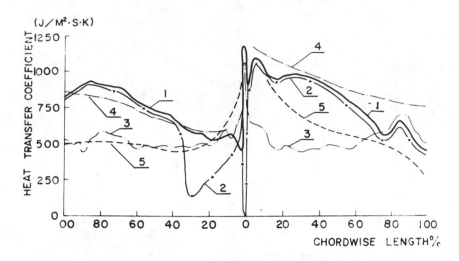

Fig. 7 Heat transfer coefficient distribution over a
blade surface
1. author's calculation
   results
2. calculation results from
   NASA TN D-5681, (1970)
3. A.B. Turner's experi-
   mental results
4. calculation results of
   turbulent flat plate
5. Spalding's theoretical
   results

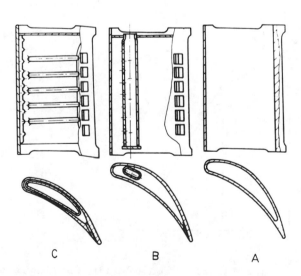

Fig. 8 Three cooling structures for guide vanes

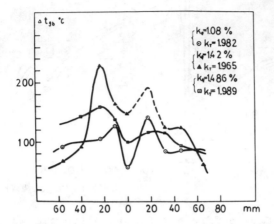

Fig. 9 Comparison of effectiveness of three cooling struc-
        tures for guide vanes at different gas and coolant
        temperature ratios

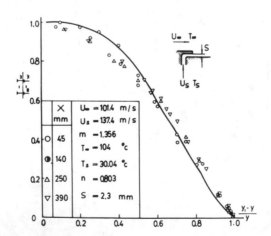

Fig. 10 Temperature distribution of film cooling over
         adiabatic flat plate

FILM COOLING

Extensive research has been carried out in film cooling. Since film cooling is the most effective cooling technique in high temperature gas turbine components, such as turbine blade, combustor and afterburner, the effectiveness of film cooling of a continuous slot over a flat plate has been measured /13,29/. The temperature distribution of the flow field and adiabatic wall effectiveness are shown in Figs. 10 and 11 and a set of film cooling effectiveness correlation equations derived from the experimental results is given below:

For $\quad M = \rho s U s / \rho_\infty U_\infty < 1$,

$$\eta = 2.73 \ M^{0.4} \ (x/s)^{-0.38}, \ x/s \leq 20$$

$$\eta = 5.44 \ M^{0.4} \ (x/s)^{-0.58}, \ 20 \leq x/s \leq 150$$

$$\eta = 2.04 \ (x/s)^{-0.38}$$

For $\quad 1 \leq M < 2$,

$$\eta = 1.96 \ M^{0.55} \ (x/s)^{-0.38}, \ x/s < 150$$

For $\quad 2 \leq M < 3.5$

$$\eta = 2.71 \ (x/s)^{-0.38}, \ x/s < 150.$$

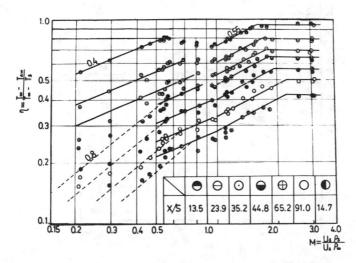

Fig. 11 Variations of effectiveness with respect to blowing rate M

Local coefficient of heat transfer near the film cooling slot region is different from that over a conventional flat plate. For conventional forced convection over a flat plate, the local coefficient of heat transfer is monotonously decreasing downstream from the slot, while for film cooling, the local coefficient of heat transfer varies in different flow regions and at different blowing rates. The following corelations are suggested /13/:

For   M < 1 and 8 ≤ x/s < 60,

$$Nu_g = 0.144 \ Re_g^{0.66} \ M^{-0.1}$$

For   1 < M < 2,   x/s < 10,

$$Nu_s = 0.057 \ Re_x^{0.7}$$

For   1 < M < 2,   10 ≤ x/s ≤ 35,

$$Nu_s = 6.39 \times 10^{-5} \ (Re_x \ / \ M)^{1.3}$$

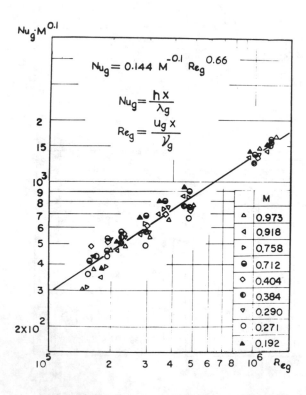

Fig. 12   Heat transfer coefficient of film cooled flat plate at 8 ≤ x/s ≤ 35 (M ≤ 1)

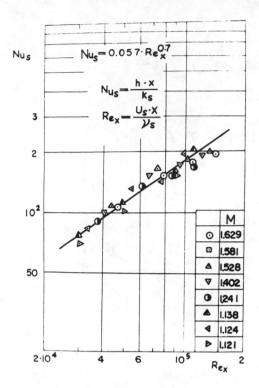

Fig. 13   Effectiveness of film cooled flat plate with
          ejection angle of 30°

The above equations were derived from the experimental data
shown in Figs. 12 and 13.

Near the inclined film slot region, there is a recirculating
zone; the effectiveness in this region has been investigated for
the slot inclination angle of 30° /14/. Experimental correlations
in terms of mass blowing rate M and momentum blowing rate I are
given below:

For $20 < x/s < 80$, $M < 1$,

$$\eta = 8.14 \, M^{1.2} \left(\frac{x}{s}\right)^{-0.65}$$

For $5 < x/s < 20$, $M < 0.6$,

$$\eta = 5.1 \, M^{1.1} \left(\frac{x}{s}\right)^{-0.5}$$

For $20 < x/s < 8$,

$$\eta = 9.96 \ (I \ Cos^2 \ )^{0.6} \ (\frac{x}{s})^{-0.65}$$

For $5 < x/s < 20$,

$$\eta = 5.76 \ (I \ Cos^2 \ )^{0.51} \ (\frac{x}{s})^{-0.5}$$

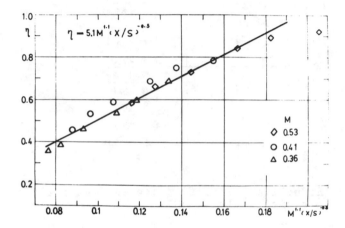

Fig. 14   Heat transfer coefficient of film cooled flat
          plate at $x/s < 10$ ($1 < M < 2$)

Fig. 14 shows some of the experimental results.

Film cooling effectiveness of discrete holes, compound-angle
holes and rectangular holes have been investigated /15/ by mass
transfer technique. Concentration of carbon dioxide was measured
to determine the film cooling effectiveness. The concentration
profiles at two different blowing rates are given in Figs. 15 and
16 in which the sparation of jet can be seen clearly. The effectiveness of rectangular slot with width/height ratio of 7.5 can
be expressed as

$$\eta = 1.89 \ M^{0.64} \ (x/D)^{-0.4}$$

which gives higher η value than that of round holes.

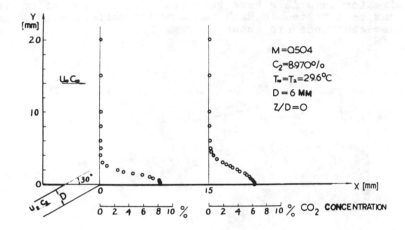

Fig. 15   Concentrations at low blowing rate , M= 0.504

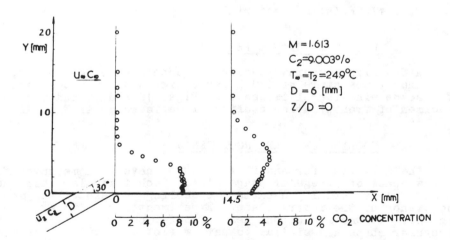

Fig. 16   Concentrations at high blowing rate , M= 1.613

Full coverage film cooling and transpiration cooling studies have been made /16/. Heat transfer and friction losses of different transpiration materials have been tested. Transpiration cooling blade has been tested in high temperature cascade wind tunnel. Some test specimens are shown in Fig. 17.

Fig. 17 Rotating test rig

HIGH TEMPERATURE TEST RIGS

A few high temperature cascade tunnels have been built, and a maximum temperature of 1300K can be reached. The test section of cascade wind tunnel is shown in Figs. 18 and 19 and the test specimen of transpiration cooling blade is shown in Fig. 20.

COOLING OF BLADE ROOT TENON

The heat transfer and flow friction have been measured for a narrow space of irregular shape /17/. Such irregular shape as shown in Fig. 21 is normally used in the cooling passage of the mounting root tenon of the rotary blade. Experimental results are given in Fig. 22. It was found that heat transfer characteristics of the irregular shape was similar to narrow slot except that the transition region was longer. In this region the heat transfer coefficient can be correlated as:

$$Nu = 0.00351 \ Re^{0.947}$$

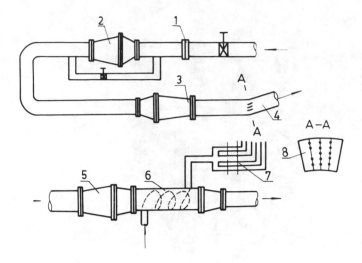

Fig. 18 Testing rig of turbine blade cascade wind tunnel

1. gas flow diaphram   2. pre-combustor
3. combustor   4. test cascade
5. heater   6. heat exchanger
7. coolant flow pipe   8. temperature probe

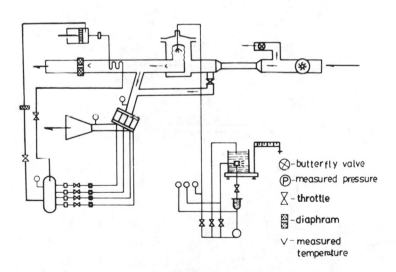

⊗ - butterfly valve
Ⓟ - measured pressure
⋈ - throttle
▥ - diaphram
∨ - measured temperature

Fig. 19 Experimental equipment for turbine blade
          transpiration cooling

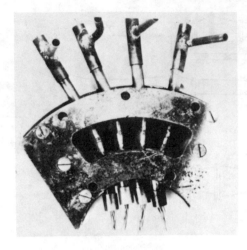

Fig. 20   Cascade of transpiration cooling blades

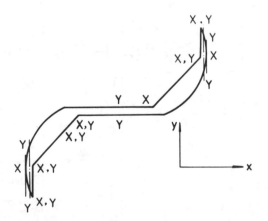

Fig. 21   Configurations of the assembly gap of
          blade-root tenon

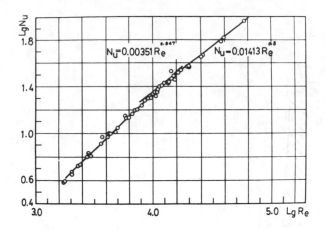

Fig. 22 Relation between Nu and Re

In the turbulent region, friction loss was higher than in circular tube, while the coefficient of heat transfer was lower. The following experimental equations are suggested:

$$\mathfrak{Z} = 0.332 \, / \, Re^{0.25}$$

$$Nu = 0.01413 \, Re^{\,0.80}$$

COOLING OF TURBINE ROTOR DISC

The temperature distribution of the turbine rotor disc has been analyzed and measured by conduction-paper analogy method /18/ and the radial cooling system of the rotor disc is shown in Fig. 23. Other electric analogy techniques such as electro-bath analogy and resistance-capacitance analogy were also used in the study of steady state temperature distribution of rotor disc. Some of the results are given in Figs. 24 and 25.

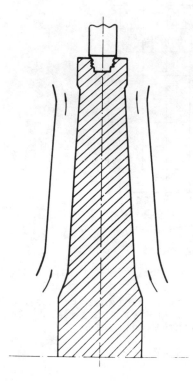

Fig. 23   Radial flow of
          cooling air over
          a disc surface

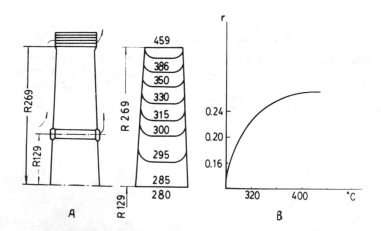

Fig. 24   The disc of gas turbine
          A Structure of the disc
          B Temperature field of the disc

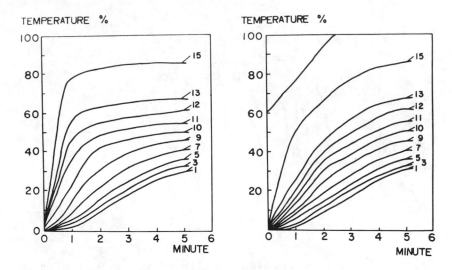

Fig. 25   Temperature-time curves at various radial
positions of turbine rotor disc

COMBUSTOR COOLING

Flame tube is one of the shortest life components of a gas
turbine. Film cooling of the inner wall and forced convection
cooling of the outer wall of a flame tube are common practice
of cooling design of flame tube. A set of heat balance equations
and charts for flame tube wall temperature calculation have been
suggested /19/. Experimental results of film cooling effectiveness
were introduced in the heat balance equations /13/. Some of the
typical calculated results are given in Fig. 26.

Computer programs are available for the evaluation of film-
cooled wall temperature and heat flux /20/. The influeuce of turbulent strength has been considered. Some parameters which can
be evaluated by the user from optional equations are considered
as input data. The program is appropriate for comparing various
designing schemes.

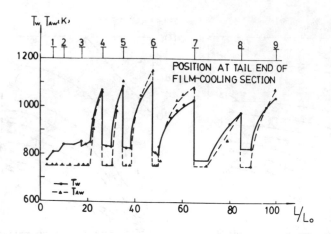

Fig. 26   Typical calculated results of the film-cooled
          flame tube wall temperature distributions

AFTERBURNER COOLING

The enhancement of cooling in the annular space of afterburner
by using short circular fins has been investigated experimentally
/21/. In this work, the heat transfer and friction coefficient
have been studied. Experimental results indicated that the
heat transfer coefficient for rough annuli were about 1.5- 2.1
times higher than that of smooth annulus.

Some of the results obtained are shown in Fig. 27. Heat transfer
coefficients can be corelated as:

$$Nu_f = 0.12 \ Pr_f^{0.52} \ Re_f^{0.8} \ (S/h)^{-0.45}$$

Simplified calculation equations have been suggested in eva-luating the wall temperature of air-cooled afterburner /22/.

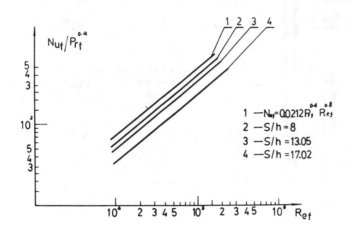

Fig. 27   Comparison of the coefficients of transfer for
          rough annuli with that of smooth annulus

II. HEAT AND MASS TRANSFER IN ELECTRIC GENERATORS

Extensive research and development work in the cooling of generators has been carried out /23/. Both the stator and rotor of the generator are cooled by water and the results are very satisfactory. The maximum capacity of water cooled generator in production is 300 MW.

Evaporative cooling of rotating electrical machinery has been studied. A test rig of two-phase flow is shown in Fig. 28. The friction loss and volume ratio of fluid and steam inside the rectangular channel in vertical and horizontal positions have been carefully investigated.

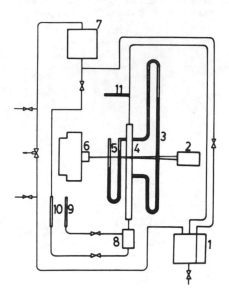

1. lower water tank
2. x-ray detecting element
3. pressure difference probe
4. test section
5. U-pressure gauge
6. x-ray generator
7. upper water tank
8. air-water mixer
9. gasometer
10. watermeter
11. semiconductor point thermometer

Fig. 28   Testing rig of the air-water cooling system

Fig. 29   The testing rig of the rotating condensation system

The forced convection heat transfer and boiling of Freon inside the copper tube and copper rectangular coil have been investigated. The location of starting point of boiling along the tube and the coil was correlated from experimental results. Friction losses were also correlated.

Condensation heat transfer of closed loop rotating system has been studied. The test rig of the rotating condensation system is shown in Fig. 29.

III. HEAT AND MASS TRANSFER OF OTHER ROTATING MACHINERY

STEAM TURBINE

Temperature distributions in the steam turbine blade and shaft have been studied in order to improve the efficiency and reliability of the steam power plant. Floating pumping and spray cooling system have been constructed and their features studied in order to achieve more effective cooling of the upper layer of water of the steam power plant cooling pond /24/. The floating pumping and spray cooling system are shown in Fig. 30.

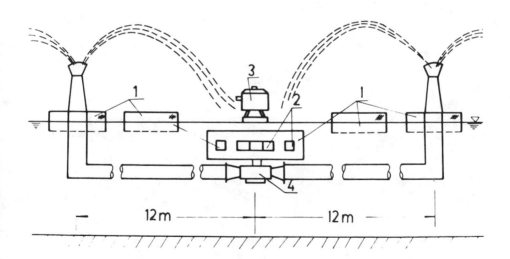

Fig. 30 The floating pumping and spray cooling system

    1. floating system   2. suction pipe
    3. motor   4. pump

WANKEL ROTARY ENGINE

During the early seventies, considerable research has been conducted in the heat transfer and combustion of Wankel rotary engine in China. Experimental results of temperature distribution of air cooled Wankel engine wall has been measured and compared with numerical results /25/. Some results are shown in Fig. 31. It was found that at the hot zone, the maximum heat flux could reach a value of 5×10^5 (kcal/hr m^2).In this region, the cooling effectiveness can no longer be improved by increasing the cooling air speed. Film thermocouples indicated that the wall temperature fluctuation of Wankel engine was larger than that of ordinary internal combustion engine.

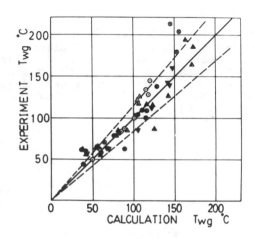

Fig. 31   Comparison between the experimental results and the calculated results of the wall temperature of an engine cylinder

IV. TECHNIQUE OF THERMOPHYSICAL MEASUREMENTS

TEMPERATURE MEASUREMENTS

    Dynamic correction and radiation correction have been suggested for the temperature measurement by thermocouples in the high speed, high temperature flow field /26/. Errors can be reduced to 2% for temperature range of 1300-2500°C and speed range of 100-200 m/s.

    Thin film thermocouple of 0.03 mm thickness has been developed for the purpose of study of the wall temperature of the gas turbine /18/. Its response speed was 5-10 times faster than for ordinary thermocouple. Four-channel temperature sensor transmitter has been developed for the purpose of measuring the rotor disc temperatures /27/.

Infrared temperature measurement technique has been widely adopted in China. An accuracy of $\pm$ 1.5% can be obtained.

Laser holograph has been used to determine the 3-D free convection temperature distribution of vertical plate with high accuracy /28/. Some of the results are given in Fig. 32.

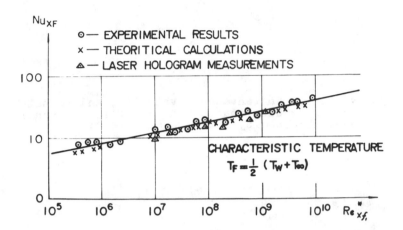

Fig. 32 Comparison of laser holograph measurements with experimental results and theoritical calculations

VELOCITY AND TURBULENCE MEASUREMENTS

Velocity and turbulence measurements have been made by the use of hot wire anemometer in the near slot region of film cooling /13/. Some of the results are shown in Fig. 33.

OPTICAL MEASUREMENTS

Flow visualization technique has been used in the study of velocity and temperature distributions of rotary machinery. Laser interferometer has been developed for the observation of shock formation. Clear pictures were obtained near the injection region of the discrete film cooling holes by using laser interferometer. Typical picture is given in Fig. 34.

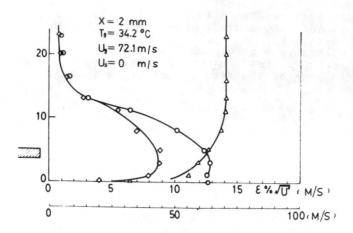

Fig. 33 **Velocity, instantaneous velocity, and turbulent
intensity distributions of the film cooling process obtained by hot wire anemometer**

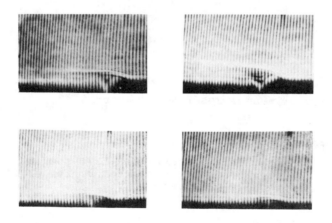

Fig. 34 Laser interferometer pictures

REFERENCES

1. Ge Xinshi, Guo Kuanliang, Sun Xiaolan: "Variational Principle
   on Heat Conduction in Solid", Journal of Chinese University
   of Science and Technology, Vol. 9, No. 2, 1979.

2. Guo Kuanliang, Ge Xinshi, Sun Xiaolan: "Application of the Finite Element Method to the Solution of Transient Two-Dimensional Temperature Field for Air-Cooled Turbine Blade", Journal of Engineering Thermophysics, Vol. 1, No. 2, May 1980.

3. Sheng Rongchang: "The Finite Element Analysis of Temperature Field for Film Cooled Turbine Blade and Calculation of Heat Transfer Coefficient", Chinese Socity of Engineering Thermophysics, Paper 81-3005, 1980.

4. Lin Hantao: "Calculation of Temperature Field of the Fir-Shape Tenon with Finite Element Method", Paper 80-2002 Presented at the Third Annual Meeting of Chinese Socity of Engineering Thermophysics. Paper 80-2002, 1980.

5. Shao-Yen Ko, Li Jin, Liu Dengyun: "Numerical Computation of Turbulent Flow over a Film Cooled Flat Plate", Technical Paper. Institute of Engineering Thermophysics, Chinese Academy of Sciences, 1980.

6. Xu Jingzhong, Shao-Yen Ko: "Numerical Calculation for Flow over a Curved Surface with Injection Through Discrete Holes", Technical Report, Institute of Engineering Thermophysics, Chinese Academy of Sciences, 1982.

7. Xu Xuchang: "Mathematical Simulation for Three-Dimensional Heat Transfer Process of Flame in Combustion Chamber", Technical Report, Department of Heat Energy Engineering, Qinhua University, 1979.

8. Jin Denian, Lin Junxiao: "Numerical Computation of local Coefficients of Heat Transfer over the Surface of Air-Cooled Turbine Blade", Technical Report QH 81031 (No. 103), Qinhua University, 1981.

9. Gu Weizao, "Thermal Performance of Turbine Vane with Ceramic Coatings and Enhanced Cooling", Journal of Engineering Thermophysics, Vol. 3, No. 2, 1982.

10. Gas Turbine Research Institute: "Experimental Report on the Cooling Effect of the First Stage Guide Vanes of Turbojet Engine 6B", Technical Report, 1982.

11. Zheng Jirui: "Experimental Investigation of Heat Transfer by a Single - and a Triple - Row Round Jets Impinging on Semi-Cylindrical Concave Surfaces", National Conference on Heat and Mass Transfer and Combustion, Huangshan, Anhui Province, 1981.

12. Qiu Xuguang: "Experimental Investigations of Impingement Flow in Parabola-Shape Target Cavity", Technical Report BH-B762, Beijing Aeronautical Institute, 1981.

13. Shao-Yen Ko, Liu Dengyun: "Experimental Investigations on Effectiveness, Heat Transfer Coefficient, and Turbulence of Film Cooling", AIAA Journal, Vol. 18, No. 8, August, 1980.

14. Din Binyuan: "Experimental Investigation and Numerical Compu-

tation for Flow over a Two-Dimensional Film Cooled Flat Plate",
Master's Thesis, Institute of Engineering Thermophysics,
Chinese Academy of Sciences, 1981.

15. Shao-Yen Ko, Liu Dengyun, Yao Yongqing, Li Jing, F.K. Tsou,
    "Film Cooling Effectiveness of Discrete Holes Measured by
    Mass Transfer and Laser Interferometer", Paper No. CP21, 7th.
    International Heat Transfer Conference, Munchen, Federal Re-
    public of Germany, September 6-10, 1982.

16. Huang Guorui, "Studies of Transpiration Cooled Turbine Blade
    and Vane in Dongfang Steam Turbine Factory", National Conferen-
    ce on Heat and Mass Transfer and Combustion, Huangshan, Anhui
    Province, Paper 81-3057. Oct. 16-21, 1981.

17. Min Guirong: "Experimental Study on Flow Resistance and Cool-
    ing Effect in Assembly Gap of Blade-Root Tenon", Technical Re-
    port, Institute of Mechanics, Chinese Academy of Sciences, 1967.

18. Ge Yongle, Lu Jiancheng, "Monograph on the Temperature Field
    of Turbine High-Temperature Parts", Publishing House of Na-
    tional Defence Industy, 1978.

19. Wang Hengyue, Ge Shaoyan(Shao-Yen Ko), Liu Dengyun, Yang Yaxian:
    "Graphic Solution and Charts for the Evaluation of Wall Tempera-
    ture and Heat Flux of Film-Cooled Wall", Journal of Mechanical
    Engineering. Vol. 15, March 1979. English translation: Enginee-
    ring Thermophysics in China, Vol. 1, No. 1, pp. 95-110, Jan.1980

20. Shao-Yen Ko, "Computer Program of Temperature and Heat Flux
    on the Wall of Film Cooled Combustion Chamber", Technical
    Report, Institute of Engineering Thermophysics, Chinese Aca-
    demy of Sciences, 1976.

21. Ma Tongze, Zhang Baodong, An Zhiqing, Yang Yaxian, Qiao Yunshan:
    "Heat Transfer and Friction of Turbulent Flow in Annuli with
    Repeated-Rib Roughness", Journal of Engineering Thermophysics
    Vol. 2, No. 3, Aug, 1981.

22. Zhu Changqing: "Calculation of Temperature of Convection and
    Film Cooled Afterburner Wall", National Conference of Heat and
    Mass Transfer and Combustion, Huangshan, Anhui Province, Oct.
    16-21, 1981.

23. Technical Report, Electric Engineering Institute, Chinese
    Academy of Sciences, 1982.

24. Zhao Gu: "Cooling Experiments for Floating Ejection", Electric
    Power Technology, Vol. 8, 1981.

25. Ma Chongfang, et al: "Combustion and Heat Transfer in Rotary
    Engine", People's Communication Publishing House, 1981.

26. Dynamic Response of Temperature Measurement Group, Qinhua Uni-
    versity: "Transient Measurements of High-Temperature Gas Flow
    with Thermocouple", Journal of Qinhua University, Vol. 1, 1978.

27. Ge Yongle, Diesel Engine Institute. Shanhai: "A Study of Temperature Field of Gas Turbine High-Temperature Parts", Third National Conference of Chinese Socity of Engineering Thermophysics, Paper 80-2002, 1980.

28. Wang Enhui, Wang Feng, Wang Zhu, Wei Fuqing, Beijing Aeronautics University, "Laser Holograph of Temperature Field in Boundary Layer", National Conference on Heat and Mass Transfer and Combustion, Paper 81-036, Huangshan, Anhui Province, Oct. 16-21, 1981.

29. F.K. Tsou, S.J. Chen, Shao-Yen Ko, "The Effect of Unsteady Main Flow on Film Cooling", to be presented at the JSME-ASME Thermal Engineering Joint Conference, Honolulu, Hawaii, U.S.A. March 12-16, 1983.

Heat Transfer to Turbine Blading

W. J. PRIDDY and F. J. BAYLEY
Thermofluid Mechanics Research Centre
School of Engineering and Applied Sciences
University of Sussex
Falmer, Brighton, Sussex BN1 9QT, England

ABSTRACT

This paper reports the most recent results of a continuing research programme studying the factors determining the distribution of heat transfer to
turbine blading. It is particularly concerned with the effects of mainstream
turbulence parameters, including intensity and frequency of disturbance. A
suggested correlation based on an earlier theoretical study (2) shows clearly
how different regions of the blade surface react to the perturbations in the
mainstream flow. It is clear that laminar-turbulent transition for the blade
boundary layer is crucial to this reaction and the large amount of heat transfer
data from the present programme is used to examine transition criteria in steady
flows.

NOMENCLATURE

c	Blade chord
C_p	Specific heat at constant pressure
D	Diameter of leading edge of circular cylinder
E	Eckert number, $U^2/2C_p(T_O - T_w)$
f	Frequency
F	Strouhal number, fL/U
G_θ	Goertler number $(Re_\theta \sqrt{(\theta/\nu)})$
h	Local heat transfer coefficient
I	Cumulative amplification factor, $\int \beta dx$
K	Acceleration parameter, $(\nu/u^2)(du/dx)$
L	Characteristic length
$\dot{m}$	Mass flow rate
M	Local Mach number
N	Rotational speed
Nu	Nusselt number
Nu_D	Nusselt number based on cylinder diameter
$\overline{Nu}$	Nusselt number in steady flow
Nu_x	Nusselt number based on surface distance
Pr	Prandtl number
r	Radius
Re_D	Reynolds number, with approach velocity and leading edge diameter
Re_x	Local Reynolds number
Re_2	Exit Reynolds number
Re_θ	Moment thickness Reynolds number
Tu	Turbulence intensity

$\overline{T_W}$	Surface mean temperature
T_O	Gas total temperature
U	Streamline velocity
$\overline{U}$	Surface mean velocity
u'	Streamwise velocity fluctuation
V_1, V_2	Cascade entry and exit velocities
X	Distance from turbulence generator
x	Surface distance from blade leading edge
α	Wave number (2π/wavelength)
β	Disturbance amplification factor
λ_x	Pohlhausen pressure gradient parameter $(\theta^2/\nu)(du/dx)$
θ	Boundary layer momentum thickness
ν	Kinematic viscosity
∞	Subscript for approach flow condition.

1.   INTRODUCTION

Many uncertainties still remain in attempting to predict precisely the distribution of heat transfer to gas turbine blading. These spring mainly from ignorance of the details of the flows which exist in the harsh engine environment and especially how these flows can be simulated in theoretical and experimental investigations to provide data for the designer. In particular, flows in engines are rarely steady and there is doubt about which parameters define an artificially turbulent flow and certainly which are the most critical in determining the true convective heat transfer rate to a blade surface. There is no doubt that amplitude or intensity of turbulence is highly significant and early work at Sussex showed also a clear and separable effect of frequency of oscillation, especially at very high intensities (1). Theoretical studies of the effects of flow oscillations on laminar boundary layers (2), which in a low turbulence mainstream exist over large parts of modern blade sections, had suggested that this might be the case. Recent experimental work at lower intensities and higher frequencies more appropriate to perceived engine conditions (3,4) has shown less clearly defined frequency effects. It has also revealed uncertainties about the nature of the flows over modern blade sections even when these are nominally steady, say with mainstream turbulence intensities of less than 0.5 per cent. In particular, it is crucial in determining the consequences of artificial mainstream turbulence to know whether the boundary layer in the corresponding steady flow was turbulent or laminar. It is quite clear that criteria for establishing the position and extent of laminar-turbulent transition are inadequate, and a substantial part of this paper uses the large amount of heat transfer data available from the present research programme to examine current transition criteria. This examination, together with the most recent unsteady flow data, shows also how important can be the development of the mainstream turbulence parameters through the blade passages in determining their consequences upon the distribution of heat transfer coefficient to their surfaces.

2.   EXPERIMENTAL MEASUREMENTS

2.1   Heat Transfer

Two separate procedures for measuring the distribution of convective heat transfer coefficient around a blade section have been developed and used at Sussex in the Thermofluid Mechanics Research Centre. The first method, and still necessary for testing metal blade sections at representative temperatures,

involves the measurement of blade surface temperature at a number of stations.
Numerical solution of the Laplacian conduction equations with internal boundary
conditions set by the cooling arrangements then yields the distribution of heat
transfer coefficients around the blade surface. This technique was developed by
Turner and is fully described in reference (5).

The second technique uses the blade section reproduced as a shell in a low
conductivity metal-plastic composite material, and measurement of the inner and
outer shell surface temperatures gives a direct measurement of the local rates
of heat transfer. This technique is much less demanding of computer time than
the first and is convenient for rapid testing of blade sections over a wide
range of varying conditions, as required in the present programme. Data
obtained by the two methods from a representative blade section are critically
compared, and the 'shell' technique described more fully in reference (6).

2.2   Turbulence Generation

As with the heat transfer measurements, two separate techniques have been
used in the Sussex research programme to generate mainsteam turbulence.
Throughout the programme use has been made of the conventional turbulence grids,
ranging from fairly fine gauzes to produce turbulence intensities around 5 per
cent up to a coarse mesh of orthogonal metal strips 5 mm wide and at 16 mm pitch
which yielded a nominal intensity of 17 per cent at the blade leading edges when
these were placed 90 mm downstream of the grid. These turbulence intensities
have for most of the programme been determined in the free jet downstream of the
generators using hot filament anemometry, but recently laser doppler anemometry
techniques have been employed with the blades in position.

The second procedure for generating mainstream turbulence was developed at
Sussex with the object of separately varying the characteristic frequencies and
amplitude of the turbulent perturbations. This procedure was thought more
closely to represent the conditions in a turbine in which these perturbations
are due, partly at least, to passing blade wakes and the work of Ishigaki (2),
previously referred to, had suggested that frequency of perturbation could be
significant in its effect on laminar boundary layers.

The device developed to achieve these objectives in essence comprises a
'squirrel cage' in which intensity is controlled by the diameter of the bars of
the cage, and its distance from the blade cascade, and the frequency, up to 10
kHz, by the rotational speed and the number of bars. A fuller description of
this device is given in references (1), (3), (4) and (6), together with its
operating characteristics. These showed that the rotating cage enabled the
separation of intensity and characteristic frequency, although the power density
spectra indicated a dominant frequency only at very high intensities, above 30
per cent r.m.s. Similar observations are reported by Wood (7) from experiments
in the flow downstream of large steam turbine blades. At lower amplitudes of
perturbation, although a frequency 'spike' at bar passing frequency could be
detected, its energy content as a fraction of the total was small, and quickly
dissipated in the flow at short distances downstream of the cage. These observations could be of considerable significance in representing the characteristics of the turbulent flows in a real turbine.

Both techniques of turbulence generation had effects upon the streamline
flow to the cascade of blades which should not be overlooked in interpreting the
consequent heat transfer data. The cages produced some distortion of the otherwise uniform velocity profile across the cascade entrance, which was affected by
the rotational speed. Generally, however, the flow over the mid span section of

the blades remained two-dimensional as shown by flow visualisation experiments referred to later in this paper. The static grids used as the alternative means of turbulence generation, on the other hand, had a tendency when used for intensities of the order of 15 per cent to induce from the wakes of the coarser elements a highly three-dimensional cellular vortex structure over the blade surfaces.

2.3   Results

Figure 1 shows some typical results obtained by the procedures described in this Section of the paper. Local heat transfer measurements from a 'shell' model of a rotor blade are given from tests with a cage driven at 10 000 and 20 000 rev/min, with corresponding characteristic frequencies in kiloherz of half these values since there were 30 bars in this cage. There is the small, but clear effect of cage rotational speed upon the heat transfer levels over the blade, an effect which is as evident over most of the blade profile as at the separately plotted leading edge.

Also shown are the observations from the same blade when the flow is perturbed by a static grid yielding the same nominal intensity of perturbation. The similarity of the heat transfer distributions is noteworthy.

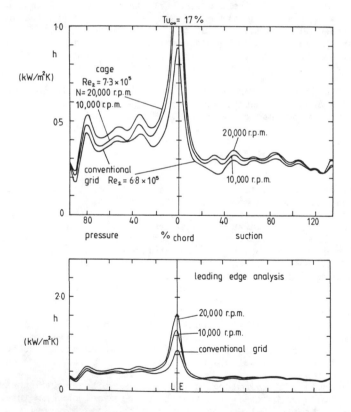

Fig. 1   Comparison of local heat transfer coefficients measured on the rotor blade downstream of the conventional grid and squirrel cage.

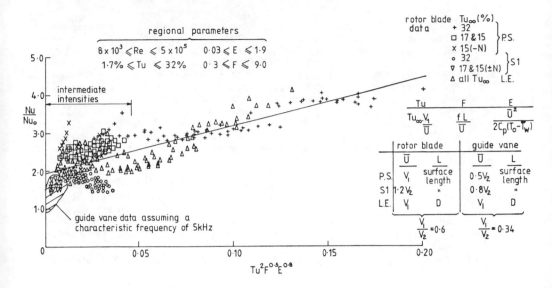

Fig. 2   Increase in surface mean heat transfer as a function
of turbulence: universal Ishigaki analysis (thin-shell data).

The general trend in all the perturbed mainstream programme is shown in Fig. 2. Here the Nusselt numbers in the ordinate are normalised by reference to the values measured under the corresponding conditions in the steady flow tests. To represent the heat transfer data the blades are divided into four separate regions, the leading edge radius (L.E.), the pressure surface (P.S.) and the upstream (S1) and downstream (S2) halves of the suction surface. Since laminar-turbulent transition always occurs before the start of S2, heat transfer rates over this region are little affected by the perturbations and the corresponding data are not plotted.

The abscissa of Fig. 2 is a parameter suggested by the theoretical work of Ishigaki (2), and which predicts the relatively small effect of frequency compared with amplitude or intensity as has been observed in most of our work. The resulting correlation is not discouraging although the scatter clearly conceals a number of factors which require further investigation. Certainly a deeper understanding is required than we have at present of the development of boundary layers on modern turbine blade sections, and especially how this development is modified by the presence of mainstream perturbations.

3.   BOUNDARY LAYER FLOW

Modern turbine blade sections with their high accelerations and sharp curvatures exhibit substantial regions of laminar flow in a steady mainstream. To determine the heat transfer rates over the whole surface, and especially its response to mainstream turbulence, requires reliable knowledge of the position and extent of the inevitable transition between the laminar and turbulent regimes. It is surprising therefore to find that there appears to be no universally applicable information in the huge amount of literature on the topic of transition to enable the blade designer reliably to predict its occurrence, or how it is affected by mainstream perturbations; nor, indeed, how to model these perturbations in the analysis of the complete boundary layer.

Daniels and Browne (8), for example, used five computer programs, each with a different turbulence model, to calculate heat transfer rates to gas turbine blades and compared the results with experimental work on cascades. They concluded that no advantages accrue from the use of the latest and most complicated turbulence models based on a full numerical solution of the governing differential equations. The major difficulties they reported were in the prediction of transition on the suction surface of their blade and the effect of free-stream turbulence on the laminar boundary layer. On the pressure surface, as well, agreement was poor; indeed, elsewhere in the literature there is evidence to suggest that the Taylor-Goertler instability system may well exist on the concave surfaces of blades but Daniels and Browne could not model this phenomenon.

The cascade work performed during the course of the present research programme has given the opportunity for the various theories and empirical correlations on boundary layer development to be tested against a wealth of blade heat transfer data. Some six different blade profiles have been tested at Sussex, five of them by the thin-shell technique. The heat transfer coefficients of all these profiles measured in the unperturbed stream have been used to analyse boundary layer behaviour and some specific examples to summarise the analysis follow.

As a beginning, the heat transfer coefficients have been compared with the simplest flat plate predictions for laminar and turbulent boundary layers and, at the geometric leading edge, with a well-established correlation for the forward stagnation point of a circular cylinder, as shown in Fig. 3. Such comparisons enable the nature of the boundary layer to be established.

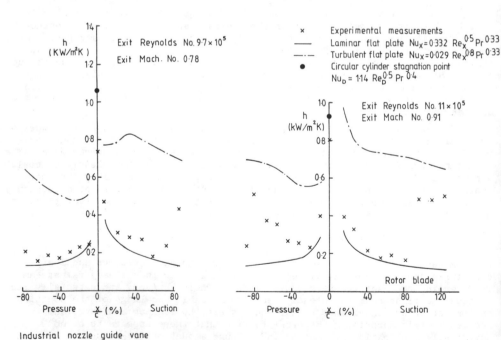

Fig. 3   Local heat transfer coefficients.

Agreement with the theory for a laminar boundary layer is not unreasonable over much of the two blades shown in Fig. 3 and a sharp transition on the suction surface of each is apparent. On the pressure side of the rotor blade the boundary layer would appear to have been in a transitional state over most of the surface. The necessary free-stream velocity profiles which enable the calculations to proceed have been derived from pressure-tapped aerofoils assuming one-dimensional isentropic flow.

With the computational techniques now available, it is expected that the heat transfer in a low turbulence stream (Tu < 1%) would be predicted with reasonable accuracy around the leading edge of a blade, and over those areas immediately downstream where the boundary layer is purely laminar. For the remainder of the blade the situation is far from satisfactory, even in this low turbulence flow. For this reason the analysis conducted at Sussex has primarily concentrated on the problem of boundary layer transition. As seen in Fig. 3, transition can occur on both sides of the aerofoil and influence significant areas of the heat transfer surface.

3.1   The Suction Surface

Considering first the suction surface, every single blade profile tested at Sussex has exhibited a sharp transition. In all cases, this sudden rise in the heat transfer curve occurs within the region of adverse pressure gradient and is preceded by a dip in the curve as may be seen, for example, in Fig. 3. Such a characteristic is indicative of boundary layer separation since by Reynolds' analogy the localised drop in heat transfer is accompanying the skin friction tending to zero. This effect was observed in the much earlier work of Walker and Markland (9). They varied the incidence of their blade and the position of the separation was found to be insensitive to changes from zero to progressively more negative angles. The Sussex blades have all been tested at zero incidence.

According to Dunham (10) an accepted criterion for the onset of laminar separation is $\lambda_x = -0.09$, where λ_x is the Pohlhausen Pressure Gradient Parameter. As can be seen in Fig. 4, the dip in heat transfer at separation does approximately coincide with the λ_x curve crossing the critical line and this has been observed for all blades tested. (The momentum thickness for this analysis was determined using the well known Thwaites equation.) At the highest Reynolds number in the cases shown, the transition has shifted rearward. Although the point where λ_x reaches the critical value is also further down the surface at higher flows the predicted movement is not in accord with observation and it appears that shock-boundary layer interaction may be determining the separation point.

Conclusive confirmation of the separation and almost immediate reattachment of the boundary layer, i.e. 'bubble' transition on the suction surface, was graphically provided by flow visualisation experiments using the Titanium Dioxide and oil technique of reference (11).

3.2   The Pressure Surface

On the pressure surfaces of turbine blades there are two opposing effects controlling the stability of the boundary layer. First, the concave streamlines tend to produce an inherently more unstable flow; second, the very strong favourable pressure gradient along the entire pressure surface of modern blades tends to stabilise the boundary layer. The former, destabilising influence can develop into a three-dimensional Goertler vortex system and available evidence suggests that the curvatures typical of turbine blades are sufficient to allow this; see, for example, the review of Winoto (12). In a theoretical study, Smith (13) imposed such a system of vortices on a laminar boundary layer and produced a solution to the equations of motion in the form of a stability chart, which has been applied to the Sussex data. The chart shows that Goertler vortices will develop when the Goertler number ($G_\theta = Re_\theta \sqrt{(\theta/r)}$) reaches the value 0.32, assuming a wavelength corresponding to the minimum of the neutral stability curve. Calculation of the Goertler number from the momentum thickness given by the Thwaites criterion, for all six Sussex profiles and the blade of Walker and Markland (9), has shown it to exceed the critical value at the very start of concavity downstream of the leading edge, in every case. This condition is realised, theoretically, for the full range of flows tested.

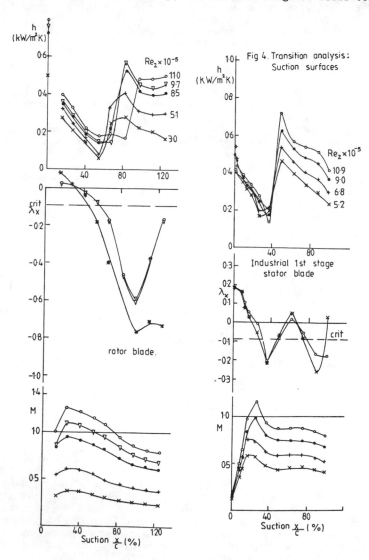

Fig. 4   Transition analysis: Suction surfaces

    If Goertler vortices do exist on the concave pressure surfaces of blades,
it then becomes necessary to know how they develop and eventually break down to
produce a turbulent boundary layer, under the restraining influence of the rapid
acceleration of the free-stream flow. The heat transfer coefficients measured
at Sussex have, on the whole, not exhibited transitional characteristics on the
pressure surface, although the rotor blade plot of Fig. 3 is one example where
they do. Smith in reference (13) suggests, on the basis of a very few test
cases, that an empirical criterion for the onset of turbulence is the attaining
of the value 9 by his 'cumulative amplification factor'. Fig. 5 summarises the
application of the procedures of reference (13) to the Sussex data, where it
will be seen that for the rotor blade Smith's correlation correctly predicts
transition at high Reynolds number. Further downchord, at low flow rate, the
predicted transition apparently did not occur, but in this case it will be
noticed in Fig. 4 that the free-stream acceleration, represented by the acceleration parameter K (= ν/u^2.du/dx), is somewhat more severe. In particular, the
value of K at low Reynolds number never reduces to below the value 2.5 x 10^{-6},
which is the figure normally associated with the relaminarisation of turbulent
boundary layers. Similarly, for the heat transfer data of Walker and Markland
at 20° incidence, the Smith transition prediction correctly shifts downstream
with reduction of Reynolds number, but at the lowest flow rate, K is substantially above the critical line and the expected rise in heat transfer is suppressed. Among the other blade profiles tested at Sussex, in many cases, the
acceleration parameter is above the relaminarisation value and transition has
not been observed. In some instances, K is below 2.5 x 10^{-6} but the Smith
criterion does not reach the critical value and, encouragingly, transition has
not been observed.

3.3   'Natural' Transition

    In addition to the Goertler vortex phenomenon and boundary layer separ-
ation, there is also the possibility of, so-called, 'natural' transition on the
pressure and suction surfaces of blades. Perhaps the best known empirical
correlation for the onset of this instability is that of Seyb (14), but the
Sussex data have been examined using a more recent equation due to Abu-Ghannam
and Shaw (15). Theirs is a modification of the correlation of Seyb, which
includes all available previous experimental results and numerous points from
their own very detailed hot-wire study of the boundary layer. Like its prede-
cessors the prediction for the start of transition is very sensitive to free-
stream turbulence up to Tu = 5%. For this reason it has not been possible to
apply it accurately to the Sussex blades because the local turbulence intensity
in the free stream has hitherto not been known. The analysis has shown, however, that on the suction surface, even for low level turbulence (as in the
approach flow where Tu = 0.5%) natural transition may well be expected near but
downstream of the point of separation (λ_x = -0.09). This explains why a short
bubble is formed rather than a lengthy circulation of detached flow, as seen in
the flow visualisation experiments.

    If the turbulence intensity in the approaching flow (Tu_∞ = 0.5%) is repre-
sentative of the local flow over the pressure surface, then the transitions
observed in 'steady flow' can only be attributed to the Goertler vortex phenomenon. Thus, for the pressure surfaces of all the blades the momentum thickness
Reynolds number (up to a maximum of 300 at the trailing edges as calculated by
the Thwaites procedure) did not approach the critical value for natural transition associated with mainstream turbulence intensity of one per cent. If
however there is amplification of the amplitude of perturbation as it is transported along the pressure surface to, say, three per cent, then the critical
Reynolds number will be reduced sufficiently for natural transition to occur.

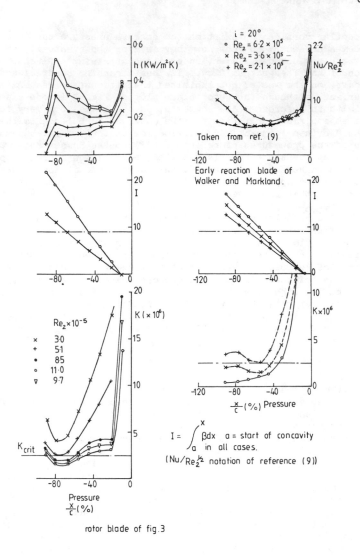

Fig. 5   Transition Analysis: Pressure surfaces

It is thus clearly important to establish how turbulence develops in strongly accelerating flows along a curved surface. For this reason the hot-wire anemometry system previously used in our work has been replaced by a forward scatter Laser-doppler optical anemometry system which as this paper is being written is just beginning to yield results. Highly significant, however, is an initial observation, as the laser beam is traversed towards the blade leading edges, which the hot wire cannot reach, of an amplification of the turbulence intensity from the approach mainstream value. This may be important for understanding of laminar-turbulent boundary layer transition, as discussed in this Section of the paper, as well as the varying consequences of the perturbations observed over the different regions of the blade as demonstrated in Section 2 and Fig. 2.

4.   CONCLUSIONS

The following principal conclusions may be drawn from the work reported in this paper:

(1)   A large amount of reliable and detailed heat transfer data has been gathered from the research programme, and is on computer file for continuing analysis. These data show a significant effect of free-stream turbulence on heat transfer, particularly to the pressure surface and leading edge of the blades. A nominal turbulence intensity of 17% trebled the heat transfer level over the pressure surface of a rotor blade compared to the unperturbed stream.

(2)   Under conditions of low turbulence (Tu = 0.5%) an analysis based on surface pressure measurements, as well as the heat transfer data, and backed by flow visualisation has shown that bubble transition occurs on the suction surfaces of seven blade profiles considered. The onset of separation complies with the criterion λ_x = -0.09, except when the flow becomes supersonic locally.

(3)   Consideration of the pressure surface for low, ambient turbulence level has revealed evidence of the existence of Goertler vortices from the very start of concavity on seven different blade profiles. A correlation which calculates the growth of the vortices along the pressure surface has successfully predicted the observed onset of transition where this has occured. Where the free-stream acceleration is, for the most part, above the value K = 2.5 x 10^{-6} transition on the pressure surface is suppressed.

(4)   Before the existence of Goertler vortices on blade pressure surfaces in cascades can be conclusively confirmed as the transition mechanism, the ambient turbulence level around the profile in the unperturbed stream will have to be measured. Should the fluctuations be found to amplify to levels above 3% intensity then correlations for natural transition may account for the observed local heat transfer rates. A laser doppler study of the development of turbulence through the blade channels is now under way to investigate this possibility.

ACKNOWLEDGEMENT

The authors express their grateful thanks for support in this work to Rolls Royce Ltd., GEC Gas Turbines Ltd., and the Science and Engineering Research Council, and to the Royal Commission for the Exhibition of 1851 for the award of a Fellowship to W.J. Priddy.

REFERENCES

(1)   Bayley, F.J. and   'The effect of free stream turbulence upon heat
       Milligan, R.W.   transfer to turbine blading', AGARD PEP CP229,
                              Paper 37, 1977.
(2)   Ishigaki, H.   'The effect of oscillation on flat plate heat
                              transfer', J.F.M., Vol. 47, 1971.

(3) Bayley F.J. and Priddy, W.J. — 'Effects of free stream turbulence intensity and frequency on heat transfer to turbine blading', ASME Jnl. of Eng. for Power, 80-GT-79, 1980.

(4) Priddy, W.J. — 'The effect of free-stream turbulence quantities upon heat transfer to turbine blading', U. of Sussex, D.Phil. Thesis, 1980.

(5) Turner, A.B. — 'Local heat transfer measurements on a gas turbine blade', I.Mech.E., Jnl. of Mech.Eng.Sci., Vol. 13, 1971.

(6) Bayley, F.J. and Priddy, W.J. — 'Studies of turbulence characteristics and their effects upon the distribution of heat transfer to turbine blading', AGARD PEP CP28, Paper 9, 1980.

(7) Wood, N.B. — 'Flow unsteadiness and turbulence measurements in the l.p. cylinder of a 500 MW steam turbine', I.Mech.E. Conf. on 'Heat and Fluid Flow in Steam and Gas Turbines', Paper No. 3, 1973.

(8) Daniels, L.C. and Browne, W.B. — 'Calculation of heat transfer rates to gas turbine blades', Int.Jnl. of Heat & Mass Trans. Vol. 24, 1981.

(9) Walker, L.C. and Markland, H. — 'Heat transfer to turbine blading in the presence of secondary flow', Int.Jnl. of Heat & Mass Trans., Vol. 8, 1965.

(10) Dunham, J. — 'Predictions of boundary layer transition on turbomachinery blades', AGARD AG164, No. 3, 1972.

(11) Nelson, W.C. (Ed.) — 'Flow visualisation in wind tunnels using indicators', AGARD AG70, 1962.

(12) Winoto, S.H. — 'Review of the literature on Goertler vortices', Imperial College London, Mech.Eng. Report, 1976.

(13) Smith, A.M.O. — 'On the growth of Taylor-Goertler vortices along highly concave walls', Quart.App.Maths., Vol. 8, No. 3, 1955.

(14) Seyb, N.J. — 'The role of boundary layers in axial flow machines and the predictions of their effects', AGARD AG164, No. 241, 1972.

(15) Abu-Ghannam, B.J. and Shaw, R. — 'Natural transition of boundary layers', I.Mech.E. Jnl. of Mech.Eng.Sci., Vol. 22, 1980.

Development of Computational Model for Full-Coverage Film Cooling

TAKESHI TAKAHASHI, NOBUHIDE KASAGI, and MASARU HIRATA
Department of Mechanical Engineering
University of Tokyo, Japan

MASAYA KUMADA
Department of Mechanical Engineering
University of Gifu, Japan

ITSUO OHNAKA
Department of Metallurgical Engineering
Osaka University, Japan

ABSTRACT

The target of the present study is to develop a simple model which can be used for the computational prediction of the temperature distribution within the full-coverage film-cooled (FCFC) wall. This paper summarizes the comparison and discussion of the experimental and numerical results recently obtained. Three computational models are proposed, while the numerical method is based on the Improved Inner Nodal Point Method. Boundary conditions are given of the heat transfer coefficients experimentally measured. The surface temperature distributions obtained by the combined heat transfer/heat conduction test are well reproduced by these computations and it is concluded that the heat transfer data of uniform wall-temperature condition can be extensively used for the general cases with heat conduction inside the FCFC wall. These models are also used to compute the transient response of the system under study and to clarify the effects of wall thickness and of wall thermal conductivity on the FCFC performance.

1. INTRODUCTION

Increasing attention has been directed towards advanced technology for high temperature gas turbines to achieve high thermodynamic efficiency for energy conservation. One of the most important problems involved must be development of an advanced cooling technique for the turbine components exposed to hot combustion gases. A feasible cooling scheme is the secondary coolant injection through a number of discrete holes provided on the surface to be protected. This has been termed as full-coverage film cooling (FCFC) and is currently under intensive studies [1-4].

Flow field and heat transfer in the near-hole region of FCFC are inevitably three-dimensional in contrast to most film cooling theories, which are well applicable to the region far downstream of injection. In addition, the heat transfer problem with FCFC becomes much complicated with the heat conduction effect inside the film-cooled wall. Therefore, further study should be needed to establish a reliable prediction method for FCFC.

A conventional procedure for treating film cooling heat transfer is based on the adiabatic wall temperature and the heat transfer coefficient without fluid injection [5]. The heat transfer coefficient in FCFC, however, might be considerably different from the value assumed for an ordinary turbulent boundary layer, because the boundary layer flow on the film-cooled wall is strongly disturbed by the coolant injection. Moreover the heat conduction inside the

439

wall would change the temperature field dependent upon the thermal properties of wall material.

An alternative procedure in the case of uniform wall-temperature has been proposed by Metzger [6], in which the heat transfer rate can be obtained for an arbitrary value of coolant/wall temperature ratio. Based on this concept, Stanford group [4] have reported the extensive data of Stanton number averaged over each row of injection holes under various experimental conditions. It is, however, uncertain that the heat transfer data of uniform wall-temperature condition could be applicable to the general cases with internal heat conduction.

In the present study, numerical computations of FCFC performance are performed by using three computational models, i.e., two kinds of two-dimensional models and a three-dimensional model. Boundary conditions are given of heat transfer coefficients obtained under the uniform wall-temperature condition [7], while the numerical method used is the improved Inner Nodal Point Method by Ohnaka [8,9]. The FCFC effectiveness reproduced by these computations are presented and compared with the heat transfer/heat conduction experiments by Kasagi et al. [10]. It is discussed whether the heat transfer data of uniform wall-temperature condition can be used for the problems of non-isothermal walls with internal heat conduction. The present models are also utilized to study the transient response of FCFC system as well as the effects of wall thickness and wall thermal properties on the FCFC performance.

2. NUMERICAL METHOD

A schematic of the FCFC plate with the Cartesian coordinates is shown in Fig. 1. The injection holes, which are slant by $\alpha°$, have been distributed in the staggered manner with prescribed hole pitches, p/d, s/d, in the streamwise and lateral directions, respectively. The secondary coolant at the initial temperature of T_2 impinges on the backside surface of the plate and flows through the discrete injection holes into the mainstream on the full-coverage film-cooled plate.

The numerical method used here is the improved Inner Nodal Point Method (improved INPM) developed by Ohnaka [8,9]. The principle of this method is shown in Fig. 2. From this figure and assumption of linear distribution of temperature between nodal points, the finite difference equation for conservation of energy of i-th element is derived as follows:

$$(\rho c_p V)_i \frac{\partial T_i}{\partial \tau} = \sum_k \frac{\lambda_k}{l_k} S_k (T_k - T_i) + \sum_k \frac{S_k}{1/h_s + l_g^i/\lambda_i + l_g^k/\lambda_k}(T_k - T_i) + \phi \qquad (1)$$

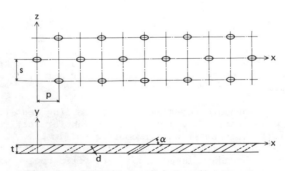

Fig. 1. FCFC Plate and Coordinates

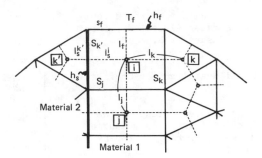

Fig. 2   Principle of the INP Method

The left-hand side of this equation is the net rate of accumulation of internal energy. The first term on the right-hand side is the net rate of heat in by conduction through side S_k according to Fourier's law. The second and third terms are the net rate of heat in through boundary sides and that generated by some heat sources, respectively.

The temperature of an element is evaluated at each nodal point. The nodal point is located at the circumcenter of the element or, if it does not exist, some approximate point is selected. The reason for this is to estimate accurately the local heat flux irrespective of real heat flow direction. Since the physical meaning of treatment is very clear without any mathematical difficulty, it is easy to apply this improved INPM to a variety of heat conduction problems with various kinds of boundary conditions [8,9]. The extension to three-dimensional problems is also very easy, although two-dimensional description is given here.

All the computations reported in this paper have been performed by HITAC-M200H at the University of Tokyo Computer Center. The computation time for unit time step is typically 0.1~0.2 second for the two-dimensional models (440 elements, implicit method), while 0.2~0.3 second for the three-dimensional model (2225 elements, explicit method).

3. SUMMARY OF EXPERIMENTAL WORKS

3.1 Heat Transfer/Heat Conduction Experiment

The combined heat transfer/heat conduction tests have been carried out by adopting brass and acrylic resin as a material for the full-coverage film-cooled wall. Since the thermal conductivities of these materials are 128 and 0.21 W/mK, the experimental results obtained must reflect the influence of thermal properties of wall material on the FCFC effectiveness. These tests have provided the basic data sets by which the evaluation of the computational models of FCFC should be made.

The 30°-slant injection holes are drilled on the plate of 25mm thickness in the staggered array with the two kinds of hole pitches, which are five and ten hole diameters in the streamwise and lateral directions. The freestream velocity has been changed to be 10 and 20m/s respectively in a blowdown wind tunnel, and the measurement of wall surface temperature has been made successfully with the aid of temperature-sensitive liquid crystal [11] illuminated by a monochromatic light of sodium lump. From these experiments,   quantitative data are obtained for the local and averaged cooling effectiveness for each material tested. For detailed description, see Kasagi et al. [10].

3.2 Measurement of Local Heat Transfer Coefficient

The local heat transfer coefficient on the full-coverage film-cooled wall has been measured under the uniform wall temperature condition by using the law of analogy to mass transfer, i.e., the technique of naphthalene sublimation. Both the geometrical shape of FCFC plate and the experimental condition are the same as those in the heat transfer experiment mentioned previously, while the naphthalene concentration in the secondary injection has been changed. Hence, the effects of the mass flux ratio, $M = \rho_2 u_2 / \rho_\infty u_\infty$, and the non-dimensional concentration of secondary injection, $\theta = C_2/C_w = (T_{20} - T_\infty)/(T_w - T_\infty)$, on the local mass transfer coefficient are quantitatively clarified. Fig. 3 shows typical examples of the contour diagrams of Stanton number which have been constructed from these precise measurements.

Since these data are equivalent to those under the condition of uniform wall-temperature, it is confirmed that the local Stanton number is a linear function of dimensionless secondary injection temperature as [4]:

$$St(\theta) = (1 + k\theta)St(0) \qquad (2)$$

Fig. 4 represents the experimental confirmation of the relationship of Eq. (2). This simple but usefull relationship is extensively utilized to evaluate the heat transfer rate in the present study as described later.

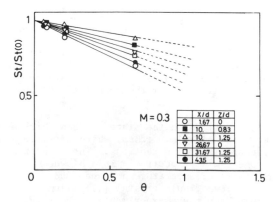

Fig. 4 Dependency of Local Stanton Number on θ

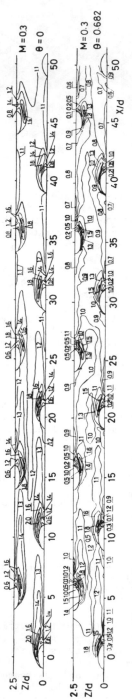

Fig. 3 Contour Diagrams of Constant Stanton Number on the FCFC Surface for $\theta=0$ and 0.682 (Numerical values in the diagram indicate $St \times 10^3$)

In order to enable the numerical computation of temperature distribution inside the FCFC plate, additional experiments have been performed of the measurements of the local heat transfer coefficients on the backside surface of FCFC wall and on the inner surface of an injection hole. For further information, see Kumada et al. [7].

4. COMPUTATIONAL RESULTS AND DISCUSSION

Based on the data of heat transfer coefficients given by the naphthalene sublimation technique, the temperature distributions within the FCFC wall of various thermal properties and thickness are reproduced numerically by using computational models presently proposed. These results are directly compared with those of the heat transfer/heat conduction tests. A separate report also includes additional results and discussion on this subject [12].

In all the present computations, a following assumption is made. When the local heat transfer coefficient on the FCFC wall is evaluated, the simple relationship of Eq. (2) is extensively used. At each time step of computation, the local values of heat transfer rate are calculated by using the ratio of local wall-surface temperature to that of right upstream injection with the assumption that Eq. (2) should hold even under the non-uniform wall-temperature condition. Then the temperature distribution at the next time step is calculated by the computational model. This sequence of procedure is repeated until the steady state is reached. This is based upon an intuitive speculation that in the case of FCFC the heat transfer rate should be predominantly affected by the secondary coolant injection at each row of injection holes in contrast to ordinary boundary layer flow, and that it should be practically insensitive to the upstream wall-temperature condition.

4.1 Two-Dimensional Slit Model

The two-dimensional slit model is schematically shown in Fig. 5, where the physical space is spanwise averaged. Temperature distributions are calculated in the case of $p/d=s/d=5$, $t/d=25/12$, $u_\infty=20$m/s and $M=0.3$. On each boundary of the model, $\overline{St}(0)$ and $\overline{k}$ in Eq. (2) are approximated by third order polynomials and by a linear distribution respectively, referring to the measurements [7]. The heat transfer rate inside the hole, St_H, is assumed to be constant and is the average on the whole inner surface of the injection hole, while that on the backside surface, $\overline{St}_B$, is given by the Aitken's interpolation method. The angle of injection hole realized in this model is 30.5°, while it is 30° in the heat transfer test [10].

Fig. 6 represents the isothermal diagrams of dimensionless temperature, $\eta=(T-T_\infty)/(T_2-T_\infty)$, for two typical values of $Bi=2hd/\lambda_B$, between the 5th and 6th rows of holes obtained by the slit model. In this figure, the distribution of local heat flux through each boundary is also shown. From these results, some noteworthy general features can be pointed out in spite of the simplicity of the present model. The calculated cooling effectiveness, $\eta_w=(T_w-T_\infty)/(T_2-T_\infty)$, shows a

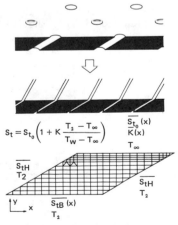

$$S_t = S_{t_0}\left(1 + K\,\frac{T_2 - T_\infty}{T_w - T_\infty}\right)$$

Fig. 5 2-D Slit Model

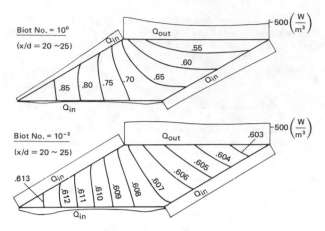

Fig. 6   Temperature Distribution Inside the FCFC Plate

considerable dependency on the thermal conductivity of the wall. The dimensionless temperature difference inside the plate of Bi=10^{-2} is about 0.01,
while that inside the plate of Bi=1 reaches more than 0.35. In addition, the
large temperature gradient normal to the upper surface can be observed in the
case of larger Biot number.

The cooling effectiveness calculated is in good agreement with the experiment for the brass plate, but not satisfactorily for the acrylic resin plate.
This is due to the fact that the geometrical configuration has been too much
simplified to reproduce the three-dimensional aspect of temperature field inside the wall of low thermal conductivity [10]. The cooling effectiveness calculated in the downstream region ($40 \leq x/d \leq 45$) is generally larger than that of
the experiment for both materials. The reason for this may be the heat conduction effect in the streamwise direction which cannot be taken into account by
the slit model.

4.2 Heat Source Element (HSE) Model

The HSE model shown in Fig. 7 is to consider the effect of longitudinal
heat conduction inside the FCFC plate, but avoid the detailed computation in

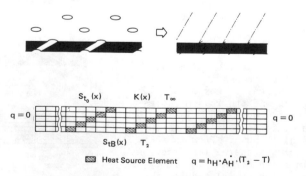

Fig. 7   2-D Heat Source Element Model

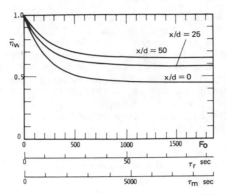

Fig. 8   Transient Response of Cooling Effectiveness

the region between injection holes. This model is again two-dimensional in the same manner as the slit model. Heat transfer on the inner surface of a hole is simulated by distributing an equivalent heat source in the corresponding element as follows:

$$q = h_H A_H^* (T_2 - T_c) \tag{3}$$

$$A_H^* = \pi dt / (ns\sin\alpha) \tag{4}$$

The boundary conditions are again given by the Stanton number data measured. Both $\overline{St}(0)$ and $\overline{k}$ are approximated as a linear function of x/d, and the distribution of $\overline{St}_H$ is represented by the linear connections of the data points, while $\overline{St}_B$ is assumed to be constant. The apparent injection angle in this model is 26.6°.

Fig. 8 shows the transient response of local cooling effectiveness of brass plate to a step change of secondary coolant temperature, T_2, at the three streamwise positions. The Biot number of brass plate is roughly the same as real gas turbine blades. Hence the abscissa is denoted by the three time scales, i.e., the dimensionless Fourier number, Fo=$\kappa\tau/d^2$, the real time in the experimental facility, τ_m, and the equivalent time in gas turbines, τ_r. The timespan predicted for the experimental setup to reach a steady state is quite in good agreement with the experimental time record of wall temperature.

The cooling effectiveness reproduced by this HSE model is shown in Fig. 9 with that of heat transfer/heat conduction experiment. In the case of brass plate the results with various backside heat transfer rates are also calcu-

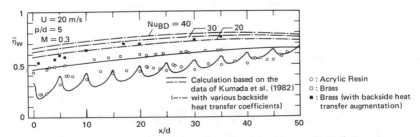

Fig. 9   Cooling Effectiveness by the 2-D HSE Model

lated, since in the experiment the data with the heat transfer augmentation on the backside surface of plate have been obtained. The cooling effectiveness of brass plate both with and without the backside heat transfer augmentation are reasonably well predicted, since the backside Nusselt number with augmentation must be slightly lower than 30 [12]. On the other hand, the pattern of cooling effectiveness distribution of acrylic resin does not seem to agree with that of experiment. This is again due to the fact that the geometrical boundary conditions have been too much deformed. However, as for the cooling effectiveness averaged over the length in between rows of holes, a fairly good agreement is obtained even for the acrylic resin.

The influences of thermal conductivity and thickness on cooling effectiveness are further investigated. The streamwise cooling effectiveness distribu-

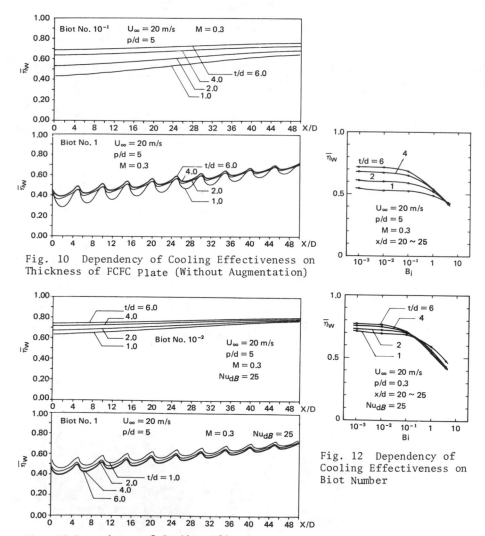

Fig. 10 Dependency of Cooling Effectiveness on Thickness of FCFC Plate (Without Augmentation)

Fig. 12 Dependency of Cooling Effectiveness on Biot Number

Fig. 11 Dependency of Cooling Effectiveness on Thickness of FCFC Plate (With Augmentation)

tions are represented for various plate thickness with and without the backside heat transfer augmentation as shown in Figs. 10 and 11, respectively. In the case of the Biot number of an order of 10^{-2}, the cooling effectiveness considerably increases with increasing the plate thickness at every streamwise position. This is because the hole-interior area, where the cooling performance is relatively high [7], is increased with the plate thickness. The effect of heat transfer augmentation also leads to a substantial improvement of effectiveness. In the case of Bi=1, the increase of plate thickness results in the increase of effectiveness in Fig. 10, but conversely the decrease in Fig. 11. This is a composite effect of the plate thickness and the heat transfer augmentation.

Fig. 12 summarizes the dependency of cooling effectiveness averaged over the streamwise distance of $20 \leq x/d \leq 25$ upon the Biot number. Without the heat transfer augmentation on the backside surface, the effectiveness is rapidly decreased with Bi beyond 10^{-1} when the plate thickness is relatively large. When the plate is relatively thin, the influence of Biot number is rather weak. With the backside augmentation, the effectiveness appears in a similar manner for various thickness plates. This implies the importance of internal cooling of turbine blade even in the case of FCFC.

4.3 Three-Dimensional Model

The two-dimensional models discussed in the previous subsections cannot simulate the three-dimensional aspects of FCFC. Hence the three-dimensional model as shown in Fig. 13 has been proposed. The FCFC plate is divided regularly into rectangular block elements for economical computation of three-dimensional temperature distribution inside the whole FCFC plate, avoiding heavy computation of detailed temperature field near the holes. On the side surface of the model except that corresponding to the injection hole, the heat flux normal to the wall must be zero according to the geometrical symmetry. Each spanwise intervals of the data points of surface heat transfer coefficient at every location of injection holes are enlarged uniformly so that the data points cover all the surface of the model. Inverse conversion has been made after the final computational result is obtained. In these computations, the values of Stanton numbers averaged over the surface of each element have been given as boundary conditions.

The cooling effectiveness diagrams for acrylic resin are compared between the computation and the experiment as shown in Fig. 14. The reproduced value and profile of cooling effectiveness agree reasonably well with the experimental results, but the computation gives the profile somewhat flatter than those of experiment. This is primarily due to the small number of elements in the spanwise direction (5 in the present computations). It is noted that the calculated cooling effectiveness of brass plate indicates a negligibly small change in the spanwise direction in good accordance with the experiment.

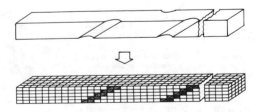

Fig. 13   Three-Dimensional Model

Fig. 15 shows a comparison of the spanwise averaged cooling effectiveness between the experiment and the computation by this three-dimensional model. The results both for the brass and acrylic resin plates are in excellent agreement in spite of the complex geometry of the present FCFC plate. In the computation by the 2-D HSE model shown in Fig. 9 there appear steep peak and valley at each injection hole when the Biot number is an order of 1 due to the overmuch simplification of the system under study. Therefore, it can be said that the three-dimensional model must be used for the prediction of FCFC performance with the wall of a relatively low thermal conductivity. The computation for the brass plate, however, can be reasonably well performed even by the 2-D HSE model because the remarkable heat conduction effect takes place. There is only a lttle difference between the result in Fig. 9 and that in Fig. 15.

5. CONCLUSIONS

Three computational models based on a simple principle have been presently proposed for the three-dimensional heat conduction problem associated with the prediction of FCFC performance. These models can also reproduce the time-dependent response of the system under study to an arbitrary change in temperatures of freestream and secondary coolant. It is concluded that the heat transfer data previously obtained under the isothermal wall condition can be extensively used as the boundary conditions for all of these models of non-isothermal condition. It is numerically demonstrated that the thermal conductivity and thickness of the FCFC plate have to be taken into account for a design of full-coverage film-cooled components in future high-temperature gas turbines.

With the Biot number less than an order of 0.01, the two-dimensional models can be used to predict the FCFC effectiveness in spite of the three-dimensionality of the system. With the Biot number of

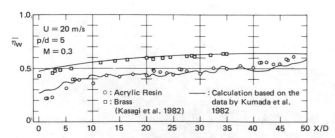

Fig. 15 Spanwise Averaged Cooling Effectiveness Reproduced by the 3-D Model

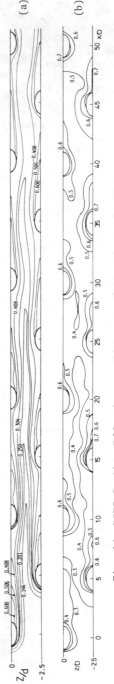

Fig. 14 FCFC Cooling Effectiveness Diagrams of Acrylic Resin Plate,
(a) Experiment, (b) 3-D Model ($p/d=8/d=5$, $t/d=25/12$, $u_\infty=20m/s$, $M=0.3$)

an order of 1, the whole distribution of cooling effectiveness along the FCFC plate may be reasonably well evaluated by the 2-D Heat Source Element model, although the 3-D model must be recommended for an accurate prediction. By the 2-D slit model, a qualitative evaluation can be made for the temperature distribution in the region between the rows of holes.

NOMENCLATURE

$A_H{}^*$; equivalent hole-interior area, Eqs. (3) and (4)
$Bi=2hd/\lambda_B$; Biot number
C ; concentration of naphthalene
c_p ; specific heat at constant pressure
d ; hole diameter
$Fo=\kappa\tau/d^2$; Fourier number
$k=(St(1)-St(0))/St(0)$; proportional constant
h ; heat transfer coefficient on the FCFC surface
h_B ; heat transfer coefficient on the backside surface of plate
h_f ; heat transfer coefficient, Fig. 2
h_H ; heat transfer coefficient on the hole-interiour surface
h_s ; heat transfer coefficient, Fig. 2
l ; distance between nodal points, Fig. 2
l_f l_s ; distance between a nodal point and a boundary, Fig. 2
$M=\rho_2 u_2/\rho_\infty u_\infty$; mass flux ratio
$Nu_B=h_Bd/\lambda$; Nusselt number
n ; number of elements in the y-direction, Eq. (4)
p ; streamwise hole pitch
q ; heat flux
$St=h/\rho_\infty u_\infty c_{p\infty}$; Stanton number
$St(\theta)$; Stanton number evaluated at the condition of $\theta=\theta$
$St_B=h_B/\rho_2 u_2 c_{p2}$; Stanton number
$St_H=h_H/\rho_2 u_2 c_{p2}$; Stanton number
$\overline{St_H}$; Stanton number averaged around the hole periphery
S,S_f ; surface area or side length of an element, Fig. 2
s ; lateral hole pitch
T ; temperature
T_f ; temperature of fluid, Fig. 2
t ; thickness of FCFC plate
u ; flow velocity
u_2 ; injection velocity evaluated far upstream of injection holes
V ; volume of an element, Fig. 2
x ; streamwise distance from the first row of injection holes
y ; distance normal to the wall
z ; lateral distance
α ; injection hole angle
$\eta=(T-T_\infty)/(T_2-T_\infty)$; local non-dimensional temperature
$\eta_w=(T_w-T_\infty)/(T_2-T_\infty)$; local cooling effectiveness
$\theta=(T_{20}-T_\infty)/(T_w-T_\infty)$; non-dimensional temperature at hole exit
κ ; thermal diffusivity
λ ; thermal conductivity of fluid
λ_B ; thermal conductivity of FCFC wall material
ρ ; density
τ ; time
ϕ ; heat generation per unit time, Eq. (1)

Subscripts

c ; evaluated on the element corresponding to the injection hole
i, j, k, k' ; nominal index of an element, Fig. 2

w ; evaluated on the FCFC wall surface
∞ ; evaluated at free stream condition
2 ; evaluated far upstream of injection hole
20 ; evaluated at hole exit

Superscript

$\overline{(\ )}$; spanwise averaged value except $\overline{St}_H$

REFERENCES

1. Metzger, D.E., Takeuchi, D.I. and Kuenstler, P.A., 1973, Effectiveness and Heat Transfer With Full-Coverage Film Cooling, ASME J. Engrg. Power, Vol. 95, pp. 180-184.

2. Mayle, R.E. and Camarata, F.J., 1975, Multihole Cooling Film Effectiveness and Heat Transfer, ASME J, Engrg. Power, Vol. 97, pp. 534-538.

3. Bergeles, G., Gosman, A.D. and Launder, B.E., 1981, The Prediction of Three-Dimensional Discrete-Hole Cooling Processes, Part 2: Turbulent Flow, ASME J. Heat Transfer, Vol. 103, pp. 140-145.

4. Crawford, M.E., Kays, W.M. and Moffat, R.J., 1975, Full-Coverage Film Cooling Heat Transfer Studies - A Summary of the Data for Normal-Hole Injection and 30° Slant-Hole Injection, Report HMT-19, Thermosciences Division, Mech. Engrg. Dept., Stanford University.

5. Goldstein, R.J., 1971, Film Cooling, in Advances in Heat Transfer, Vol. 7, pp. 321-379.

6. Metzger, D.E. and Fletcher, D.D., 1971, Evaluation of Heat Transfer for Film-Cooled Turbine Components, J. Aircraft, Vol. 8, pp. 33-38

7. Kumada, M., Hirata, M. and Kasagi, N., 1982, Studies of Full-Coverage Film Cooling (Part 2: Measurement of Local Heat Transfer Coeeficient), ASME Paper 80-GT-38, submitted to ASME J. Heat Transfer.

8. Ohnaka, I., 1980, Direct Finite Defference Method, in Innovative Numerical Analysis for the Engineering Sciences, Univ. Press of Virginia, pp. 555-565.

9. Ohnaka, I., 1979, Classification of Numerical Methods for Transient Heat Transfer Problem and Improved Inner Nodal Point Method (in Japanese), Iron and Steel, Vol. 65, pp. 77-86.

10. Kasagi, N., Hirata, M. and Kumada, M., 1982, Studies of Full-Coverage Film Cooling (Part 1: Cooling Effectiveness of Thermally Conductive Wall), ASME Paper 80-GT-37, submitted to ASME J. Heat Transfer.

11. Kasagi, N., 1980, Liquid Crystal Applications in Heat Transfer Experiments, Report IL-27, Thermosciences Division, Mech. Engrg. Dept., Stanford University.

12. Kasagi, N., Takahashi, T., Hirata, M. and Kumada, M., 1982, Studies of Full-Coverage Film Cooling (Part 3: Computational Models for the Prediction of FCFC Performance), submitted to ASME J. Heat Transfer.

Measurements of Film Cooling Effectiveness and Heat Transfer on a Flat Plate

H. KRUSE
Institut fur Antriebstechnik
Deutsche Forschungs- and Versuchsanstalt
fur Luft- und Raumfahrt
Cologne, FRG

ABSTRACT

Measurements on film cooling for a single row of holes as the basic blowing configuration are reported in this paper. The influence of the hole spacing to diameter ratio and the influence of the inclination of the holes to the surface is investigated on a flat plate at Reynoldsnumber Re $\approx$ 3x10^5 and mainstream temperature $T_o \approx$ 100 ^{o}C. Results are given in terms of film cooling effectiveness measured on an adiabatic model and in terms of heat transfer ratio for blowing and non blowing conditions on a heat transfer model at different positions and blowing rates.

1. INTRODUCTION

In recent years film cooling has become common in gas turbines. With film cooling compressor air is ducted to the turbine blades. After cooling the interior of the blades convectively the cooling air is injected through the blade surface. The cooling gas mixes with the hot mainstream and a film of cooler gas is formed which is effective over a certain region downstream from the point of blowing. Inclined or tangential slots with lips are most effective but high stresses within the blade and aerodynamic penalties have dictated rows of discrete holes. In addition the small coolant flows would require very small slots tending to deformation in a hot and corrosive environment; on the contrary there are no problems manufacturing discrete holes with larger diameter in equivalent rows. In the case of turbine blade cooling blowing out of a single row or numerous rows is of practical importance. The latter is commonly known as full-coverage film-cooling. Full-coverage film cooling offers the advantage of nearly two dimensional conditions within the mixing region and in this case successful calculation schemes have been developed. But downstream of single row injection the flow across the span of the film cooled wall is extremely nonuniform and because of the resulting threedimensional effects at least over a certain distance from the holes the numerical treatment of the flow is difficult. Therefore most of the studies of discrete hole cooling with blowing through a single row like the present study are experimentally based.

The heat transfer rate between the hot gas and a film cooled wall is dependent upon a large number of parameters such as the blowing rate, defined as the ratio of velocity and density of the coolant to mainstream velocity and density, the ratio of momentum, the boundary layer thickness relative to the hole diameter, the hole spacing to diameter ratio and the injection angle. In addition to these typical film cooling parameters there are general heat transfer parameters which

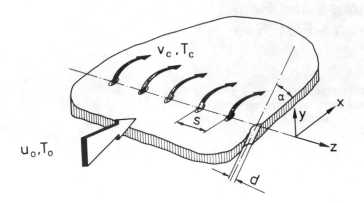

Fig. 1   Film cooling configuration and coordinate system

may change their influence due to film cooling like free stream turbulence,
pressure gradient and wall curvature.

The measurements reported in this paper were made on a simple flat plate
model for particular hole configurations with the hole spacing to diameter ratio
s/d and the streamwise inclination of the holes α (Fig. 1) in addition to the
blowing rate as the main parameters. The range of interest in practice for cooling turbine blades of those parameters is $1,5 \leq s/d \leq 5$ for the hole spacing
to diameter ratio, $10° \leq \alpha \leq 90°$ for the inclination and $0 \leq \varphi \leq 2$ for the
blowing rate. The objective of the studies is to provide some data for developing
empirical two dimensional computational methods which at least become valid
5 to 10 hole diameters downstream of the blowing position.

2.   HEAT TRANSFER WITH FILMCOOLING

The film cooling performance can be characterized in terms of a conventio-
nally used temperature difference between free stream-temperature T_0 and the
wall temperature T_w.

$$q_F = h \cdot A \cdot (T_0 - T_w)$$

but apart from other above mentioned parameters the heat transfer coefficient
then depends upon the dimensionless temperature ratio

$$\theta = \frac{T_0 - T_c}{T_0 - T_w}$$

T_c is the coolant temperature at the hole exit. Therefore for design purposes
the adiabatic wall temperature concept

$$q_F = h_F \cdot A \cdot (T_{wad} - T_w)$$

is in use where T_{wad} is the wall temperature downstream of the injection without any heatflux only with the effect of the coolant film. T_{wad} is defined in
terms of a film cooling effectiveness

$$\eta_F = \frac{T_o - T_{wad}}{T_o - T_c}$$

If this η_F is known for the film cooling configuration and blowing rate under consideration, in absence of additional information the heat transfer is often calculated assuming

$$\frac{h_F}{h_o} = 1$$

where h_o is the heat transfer coefficient for the unblown surface, but this approach is very inaccurate for short downstream distances from the holes particularly with high blowing rates.

In order to get information on this hypothetical heat transfer coefficient h_F it is necessary to measure under adiabatic and heat transfer conditions or at different temperature ratios , which allows to calculate either h_F or h_F and η_F. Fig. 2 shows the well known diagram which is the result of combining the above relations:

$$\frac{h}{h_F} = \frac{\theta^* - \theta}{\theta^*} \qquad \text{with } \theta^* = \frac{1}{\eta_F}.$$

3. EXPERIMENTAL TECHNIQUE

The measurements reported herein are taken from two models: first from an adiabatic flat plate giving the local film cooling effectiveness η_F and second from a heat transfer model with the same blowing configuration giving the heat transfer coefficient h (x) which leads to h_F (x) at $\theta = 0$, whereas h (z) is averaged by conduction. The adiabatic model was used for better distinction of local η_F -results in order to mark off those regions where two dimensional treatment is valid. The η_F -results presented are averaged in z-direction.

The adiabatic model is made from plastic material, which is interrupted by interstices in order to reduce heat conduction. The heat transfer model is made from plastic as the basic material too. It has ten copper blocks in streamwise direction individually cooled by internal air flows (Fig. 3). The film cooling measurements were carried out at 100 °C main gas temperature and about 30 °C to 50 °C coolant temperature and at about 60 m/s flow velocity. The total length of

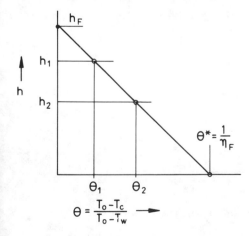

Fig. 2   Film cooling diagram

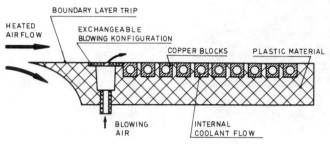

Fig. 3   Heat transfer model

the model is 200 mm. The boundary layer starts at a sharp edge and is then
tripped by a wire in order to get a defined turbulent boundary layer at the film
cooled section. For the tripped turbulent boundary layer the heat transfer without and with film cooling remains unchanged by free stream turbulence whereas
the processes in an untripped boundary layer starting at a profiled nose become
very complex as a consequence of turbulence influence in laminar boundary layers
and transition phenomena.

The blowing section is an exchangeable part of the surface. The hole diameter is $d = 2$ mm which leads to a realistic diameter to boundary layer thickness ratio. The representative boundary layer development along the plate without blowing is shown in Fig. 4.

The adiabatic wall temperature was measured by means of a very small thermo-couple probe looking like a hot wire probe. In a distance of about 0.1 mm
from the wall the temperature gradient at the adiabatic wall is still zero and
the measured sublayer temperature is equal to the local adiabatic wall temperature. The error is less than the sensitivity of the temperature measuring device.
The coolant temperature is measured by the same probe placed into the middle of
the jet in the plane of the surface. By means of the ten internal cooled copper
blocks ten heat fluxes at different x-positions averaged in z-direction can be
measured. The surface temperature is measured in the middle of the copper blocks
by means of an infra-red pyrometer calibrated by only one surface thermocouple. The mass flow and the inlet and outlet temperature of the individual
internal air flows measured by thermo-couples give the heat transfer rate. The
model is carefully isolated; remaining losses as well as the small heat fluxes

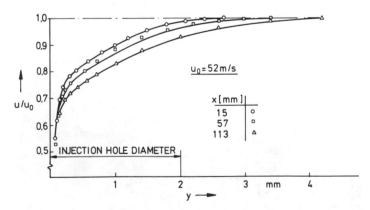

Fig. 4   Boundary layer development along the unblown wall

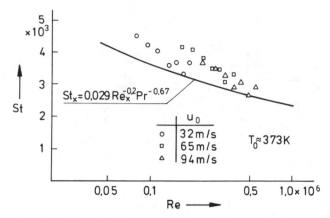

Fig. 5   Measured heat transfer without blowing

through the plastic material between the copper blocks are ignored.

   The accuracy of this simple stationary method for determining heat transfer
rates was checked for the wall without blowing with fully turbulent boundary
layer and compared with a standard turbulent flat-plate relationship. (Fig. 5)
The Reynolds number was varied by varying the free stream velocity and the position on the plate. The length x in Re_x was determined from a virtual origin
of the tripped boundary layer corresponding to the local boundary layer thickness
using emperical relations. This approach and perhaps other measuring errors lead
to some higher Stanton numbers than the standard relationship; but the objective
of this study was rather describing the relative changes caused by film cooling
than the determination of absolutely correct heat transfer rates.

4.   EXPERIMENTAL RESULTS

   In filmcooling applications there is a considerable temperature difference
and therefore density difference between the main gas and the injected coolant.
The measuring technique, the material of the model as well as the higher heat
losses with large temperature differences prevent temperature ratios of about
$T_0/T_c \geq 2$. The ratio of about $T_0/T_c \geq 1,2$ may be regarded as the main restriction
concerning the following results, but they may give some fundamental knowledge
of the mixing and heat transfer processes in the extremely nonuniform flow downstream of a row of injection holes. As one example the temperature distribution
in the mixing zone along the hole center line and at one cross section is shown
in Fig. 6 for the hole spacing s/d = 5 and the relatively high blowing rate

$$\varphi = \frac{\rho_c \cdot v_c}{\rho_0 \cdot u_0} = 2$$

   The pattern of constant temperature lines looks like that of single-hole
blowing. For computational treatment by means of two-dimensional methods which
may be sufficient for design purposes it is important to know under what conditions and at what distances the flow may be regarded as uniform. At the adiabatic wall a great deal of information on temperature distributions, depending on
hole spacing, inclination and blowing rate is available but presenting them
would be beyond the scope of this paper. The results given as follows are spanwise area averaged adiabatic wall temperatures in terms of film cooling effec-

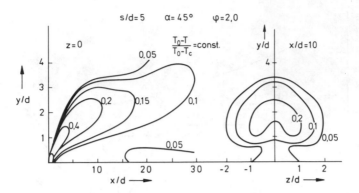

Fig. 6   Mixing of an injected coolant jet

tiveness. The averaged values are similar to those effective for the heat transfer model with the copper blocks and furthermore for realistic turbine blades
with much higher heat conductivity. In Fig. 7 the streamwise distribution of film
cooling effectiveness for a current single row blowing configuration with
$s/d = 2.5$ and $\alpha = 45°$ is shown. The curves below show the typical behaviour of
discrete hole injection: At low blowing rates ($\varphi = 0.5$) the jets are immediately
turned to the wall remaining more or less discrete jets attached to the wall over
a certain distance whereas with increasing blowing rate the jets tend to blow
off the wall where they interact before forming an equalized cooling film. The
upper curves show the potential of filmcooling using an inclined slot with the
width d. In Fig. 8 and 9 the film cooling effectiveness at two streamwise positions is plotted against the blowing rate. At a distance of 10 x d from the
blowing row (Fig. 8) the influence of the hole spacing to diameter ratio is small
with $s/d > 2,5$: the cooling film still consists of discrete jets blown off the
wall with spanwise secondary flow effects at higher blowing rates causing hot
sections immediately downstream of the holes. With $s/d = 1,5$ the jets nearly act
as a closed film without strong secondary flows. At greater distance ($x/d = 30$,

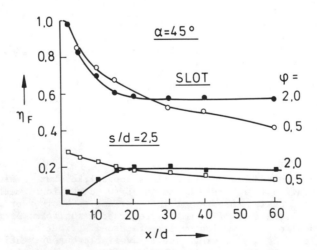

Fig. 7   Film cooling effectiveness downstream of a slot
          and a single row of holes ($s/d = 2,5$)

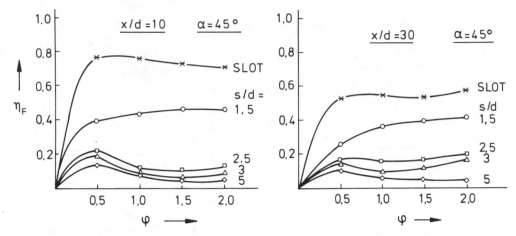

Fig. 8   Film cooling effectiveness for
        a slot and single row configu-
        ration with different spacing
        against the blowing rate
        10 x d downstream of injection

Fig. 9   Film cooling effectiveness
        with different spacing
        30 x d downstream of in-
        jection

Fig. 9) there is an almost continuous influence of the hole spacing because with
increasing distance x even jets with higher spacing to diameter ratios interact
forming an uniform film. With these considerations one had to keep in mind that
with increasing s/d the coolant mass flow per unit span decreases, but the
mixing and penetration phenomena can not be related to the absolute coolant
massflow without regarding the spacing and the injection angle. The complex
influence of downstream inclination on film cooling effectiveness is demonstrated in Fig. 10 and 11 for s/d = 2,5 at different distances: For shallow angle
injection (α = 10°) the coolant jets are attached to the wall and the flattened
jets cover extensive parts of the surface. For steeper injection with α = 45°
the film cooling effectiveness is less, the jets are less attached and inspite
of high effectiveness along the hole center line high cross flow gradients lead
to small averaged values. Normal injection results in penetrating a large region
of the main flow and as a consequence of crossflow the jets become unstable before or while bending into main stream direction. The spanwise temperature distribution is equalized by faster mixing. With 45°-injection the hole center line
values are higher but the averaged values are smaller than for normal injection.

    For thermal design of turbine blades both knowledge of film cooling effec-
tiveness and heat transfer coefficient is necessary. Relating the heat transfer
coefficient h_F with blowing (defined by T_{wad}) to the heat transfer coefficient
h_o at the same position without blowing one gets the ratio h_F/h_o, which tends
to unity for small blowing rates and large distances from the row. As Fig. 12
shows this is valid for commonly used spacing s/d = 2,5 and blowing rate
φ = 0.5. But increasing the blowing rate to φ = 2,0 results in considerable
deviations up to $h_F/h_o \approx 1.3$ at x/d = 25 for 45°-injection. Increasing film
cooling effectiveness by lowering the spacing leads to increasing heat transfer
coefficient at high blowing rates particularly at small distances in the region
between injection and the reattachment zone of the jets.

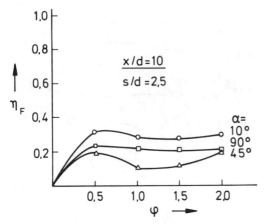

Fig. 10   Film cooling effectiveness
          with different downstream
          inclination at x/d = 10.

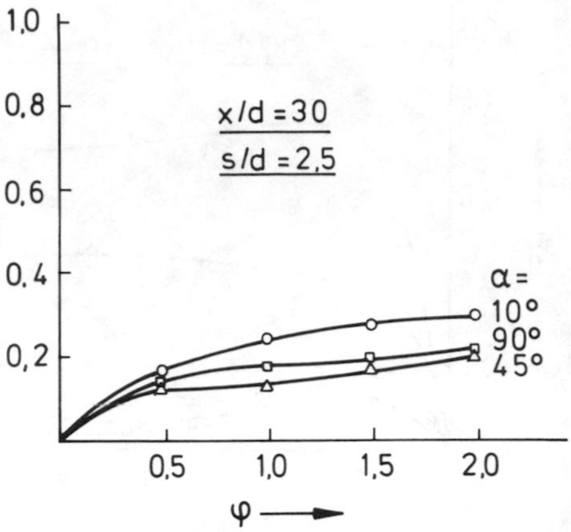

Fig. 11   Film cooling effective-
          ness with different down-
          stream inclination at
          x/d = 30.

    In Fig. 13 and 14 the cross-stream averaged heat transfer coefficient is
plotted against the blowing rate for different hole spacing to diameter ratios.
The injection angle is again α = 45°. With small hole spacing (s/d = 1,5) the
heat transfer coefficient increases rapidly as blowing rates are increased to
values greater than 0.5. With medium spacing (s/d = 2,5) small blowing rates
lower the heat transfer coefficient to values less than without blowing whereas
the ratio h_F/h_o exceeds unity at higher blowing rates. With large spacing
(s/d = 5) the crossstream averaged heat transfer coefficient just downstream
of 45°-injection is lower than that without blowing even at high blowing rates.
Fig. 15 and 16 demonstrate the influence of downstream inclination of the

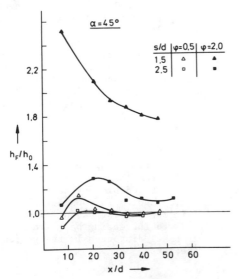

Fig. 12   Heat transfer ratio down-
          stream of a single row
          with different spacing
          and blowing rates

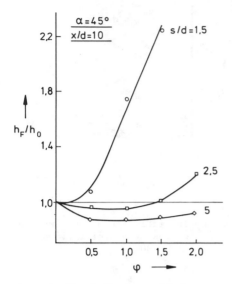

Fig. 13   The influence of hole
          spacing to diameter ratio
          on heat transfer ratio
          at x/d = 10

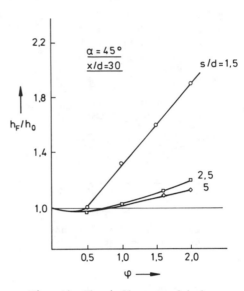

Fig. 14   The influence of hole
          spacing on heat transfer
          ratio at x/d = 30

holes: With $s/d = 2,5$ normal injection produces small heat transfer, to some
extent less than without blowing. Smaller injection angles produce higher heat
transfer coefficients: the more "tangential" the jets are the more the heat
transfer rate increases.

Most of these results are demonstrated without already having a complete
understanding. Many processes acting in the near wall flow disturbed by film
cooling injection may partially compensate each other and in some cases it remains yet unknown what influence prevails. The influence of some parameters is
obvious and confirmed under certain circumstances: Small hole spacing and/or
high blowing rates cause violent mixing and interaction of the jets with high
turbulence levels increasing the heat transfer rate. High heat transfer coefficients for shallow injection and low heat transfer rates for normal injection
are in agreement with the effects in thin respectively thick boundary layers.
The phenomenon of lowering the heat transfer by small blowing rates particularly just downstream of the injection may be due to smoothing the wall "hydrau-
lically". In this case the small normal velocity components of the almost laminar jets may lift off the sublayer whereas the effect of increasing the turbulence level by mixing and interaction is lower with small blowing rates.
Up to now these investigations were restricted to experimental work.

The use of existing calculation schemes and their further development may
allow parametric studies on film cooling which will be helpful to better understanding of the described phenomena.

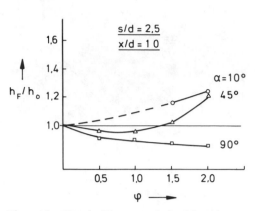

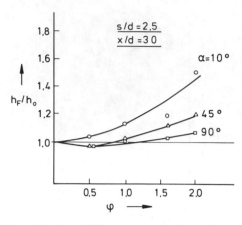

Fig. 15 The influence of inclination   Fig. 16 The influence of inclination
         on heat transfer ratio at   on heat transfer ratio at
         x/d = 10   x/d = 30

5. CONCLUSION

For different hole spacing to diameter ratio and different downstream inclination of a single row cross stream averaged film cooling effectiveness and
heat transfer ratios are presented. The results are plotted against distance
downstream of injection and blowing rate. The local effectiveness has been measured at an adiabatic flat plate model. By means of the known adiabatic wall
temperature and the heat transfer rate measured at a heat flux model with the
same aerodynamic conditions the heat transfer coefficient h_F has been determined. The film cooling effectiveness increases with decreasing hole spacing.
With regard to inclination there is a minimum between almost tangential (10°)
and normal blowing. The heat transfer coefficient too increases with decreasing
hole spacing. With small s/d there is a steep increase with blowing rates
greater than about 0.5 whereas at high spacing and small blowing rates the heat
transfer rate may be lower with film cooling than without blowing. Tangential
blowing causes high, normal blowing small heat transfer coefficients.

6. ACKNOWLEDGEMENT

The author wants to appreciate the assistance of Mr. Metzinger, who carried
out the tests.

7. NOMENCLATURE

A	Area of transfering surface	T_w	Wall temperature
d	Hole diameter	u_0	Free stream velocity
h	Heat transfer coefficient	v_c	Coolant injection velocity
h_0	Heat transfer coefficient without blowing	x	Streamwise coordinate
		y	Coordinate normal to the wall
h_F	Film cooling heat transfer coefficient	z	Cross stream coordinate
		α	Downstream injection angle
R_e	Reynoldsnumber	η_F	Film cooling effectiveness
s	Hole spacing	θ	Dimensionless temperature difference
St	Stanton number	ρ_c	Coolant density
T	Local gas temperature	ρ_0	Main gas density
T_0	Free stream temperature	φ	Blowing rate
T_{ad}	Adiabatic wall temperature		
T_c	Coolant temperature		

8. REFERENCES

Afejuku, W.O. Measured Coolant Distribution Downstream of Single and Double
Hay, N.   Rows of Film Cooling Holes,
Lampard, D.   ASME, 82-GT-144

Bergelen, G. Double-Row Discrete Hole Cooling: an Experimental and Numerical
Gosman, A.D. Study
Launder, B.E. Trans. ASME, Vol. 102, Apr. 1980

Brown, A.   Heat Transfer to Turbine Blades with Special References to the
Martin, B.W. Effects of Mainstream Turbulence
              ASME, 79-GT-26

Crawford, M.E. Full-Coverage Film Cooling
Kays, W.M.   Part I: ASME 80-GT-43
Moffat, R.J.   Part II: ASME 80-GT-44

Erikson, V.L.   Heat Transfer and Film Cooling Following Normal Injection
Goldstein, R.J. Through a Round Hole
                ASME, 74-GT-6

Kruse, H.   Investigation on Temperature Distribution Near Filmcooled
               Airfoils
               AGARD-CP 229-77

Kruse, H.   Fundamental Investigations on Effusion Cooling of Turbine Blades
               DLR-FB 77-39
               Translation: N78-24151
                            ESA-TT-469

Metzger, D.E. Evaluation on Film Cooling Performance on Gas Turbine Blades
Biddle, J.R.   AGARD-CP-73-71
Warren, J.M.

Ville, J.P.   Film Cooling and End Wall Heat Transfer in Small Turbine Blade
Godard, M.   Passages
Richards, B.E.   v. Karman Inst., Techn. Note 126, Febr. 1978
Sieverding, C.

Two-Phase Heat Transfer in Gas Turbine Bucket Cooling Passages: Part 1

J. C. DUDLEY, R. E. SUNDELL, W. W. GOODWIN,
and D. M. KERCHER
General Electric Company
Schenectady, N.Y., USA

ABSTRACT

Results are presented from an experimental program to study evaporative heat transfer to water in a rotating radial passage simulating a cooling passage in a gas turbine bucket. Part 1 of this paper presents results on the effects of coolant flow rate, steam flow, rotation number and passage geometry on the effective wetted area of a cooling passage. Part 2 evaluates the effects of flow rate, cooling passage length and tilt angle, centrifugal acceleration, pressure and wall wettability on the boiling transition power level which determines the thermal load limit for a passage.

NOMENCLATURE

a   -   passage radius

D   -   passage diameter

g   -   acceleration due to gravity

Q   -   power

R   -   radial distance from axis of rotation

T   -   temperature

U   -   velocity

W   -   volumetric water flow rate

X   -   exit steam quality

Z   -   passage length

$\overline{R}_e$ -   Reynolds number based on average film thickness

δ -   film thickness

ω -   angular velocity

ν -   kinematic viscosity

Γ -   volume flow rate per unit width

ρ   - density

Subscripts

w   - wall

sat - saturation

cu   - copper

sub - subcool

1.   INTRODUCTION

The combined cycle power generation system - with gas turbine generators
in the prime cycle and steam turbine generators in the bottoming cycle - has
long been identified as an excellent base load power generation system due to
its high efficiency (1,2). The trend towards coal utilization with the corresponding losses involved in converting coal to gases or liquids has introduced
the need for a new generation of highly efficient and reliable utility-size gas
turbines. Significant improvements in gas turbine performance can be attained
by increasing the firing temperature and pressure ratio (3). Also, coal derived
fuels may contain contaminants which can cause corrosion and deposition on
turbine components. The corrosive reactions are highly temperature dependent;
therefore, turbine surface temperatures must be kept low for corrosion protection.

The water cooling of gas turbine components has the potential for achieving both a high gas temperature and low metal temperature. The superior cooling
capacity of water maintains metal surfaces at approximately $1000^{o}F$ with firing
temperatures from 2600-$3000^{o}F$.

A particularly challenging technical problem in the design of a high-
temperature water cooled turbine is the cooling of the rotating turbine blades
or buckets. The water cooling of gas turbine buckets has been under consideration for well over 30 years (4). Many schemes have been experimentally tested,
but none applied commercially. Bayley and Martin (5) and Van Fossen and Stepka
(6) provide excellent reviews of these earlier efforts.

In the "open-circuit" cooling scheme considered here water travels through
near radial cooling passages beneath the bucket surface as shown in Figure 1.
The large centrifugal force (about 2×10^{4} times gravity for proposed turbines)
draws the water along the passage as a thin film, and the Coriolis force on the
water forces the film to one side of the passage. The heat sink is primarily
the water to steam phase change. The resultant mixture of water and steam
discharges freely at the tip of the blade.

The fluid dynamic and heat transfer phenomena in the cooling passages are
similar in some respects to those occurring in annular two-phase flow in
stationary systems. However, important differences exist due to the large
Coriolis and centrifugal forces in the rotating system. First, the liquid
film flow is the result of the centrifugal force rather than a pressure differential as in annular flow situations. Second, the Coriolis force on the
water film results in asymmetrical heat transfer around the passage similar
to two-phase flow in horizontal passages. Because of these effects, correlations commonly used for annular two phase flows are not applicable.

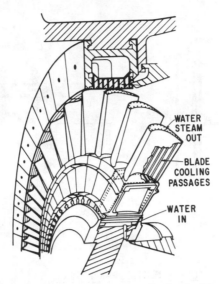

<u>Fig. 1</u> - Schematic View of the Blade Cooling System

The heat transfer performance of a turbine blade cooling passage is characterized by the film heat transfer coefficient around the passage perimeter and the boiling transition power level for the passage. The boiling transition power level, corresponding to a critical heat flux (CHF) condition, determines the thermal load limit for the cooling passage. If a nominal heat transfer coefficient is determined for the wetted fraction of the passage, the asymmetry in the heat transfer around the passage can also be represented as an effective fractional wetted area.

A previous paper has reported on experimental work to measure effective wetted areas and boiling transition power levels in rotating passages (7). This paper reports further experimental work with rotating passages simulating those in the proposed gas turbine. The effects of coolant flow rate, steam flow, rotation number and passage geometry on the effective wetted area are considered. A second paper (Part 2) discusses the effects of various parameters on the boiling transition power level for the passages.

2. <u>EXPERIMENTAL FACILITIES</u>

The turbine bucket cooling passages are simulated by rotating an electrically heated heat transfer specimen in two test facilities, shown in Figures 2 and 3. The range of test variables is summarized in Table 1. The test rig in Figure 2, called the turbine bucket simulator (TBS), consists of a two-armed rotor driven by an electric motor. The rotor spins within a steel enclosure vented to the atmosphere. High purity water is fed to the rotor through a feed system concentric with the rotation axis. The water enters a rotating annular trough which routes it to one of the two arms. The heat transfer specimen is located at the end of that arm. A trap located in the feed line upstream of the specimen prevents the backflow of steam up the arm. The water-steam mixture exits the specimen into the ambient air,

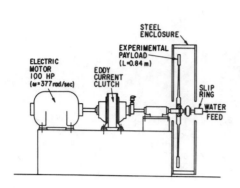

Fig. 2 - Turbine Bucket
Simulator Facility

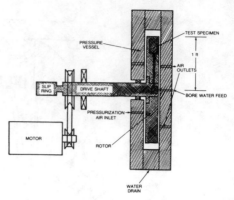

Fig. 3 - Pressurized Test
Facility

TABLE 1
EXPERIMENTAL PARAMETERS

Parameter	TBS[1]	PTF[2]	Visual Facility
R (inches)	33	12	8
D (inches)	0.063 - 0.116	0.063 - 0.116	0.375
Z (inches)	3.5 - 12.0	3.5 - 4.5	8
$\omega^2 R/g$	1×10^3 - 11.5×10^3	4.2×10^3	600
Flow (GPM)	0.01 - 0.04	0.01 - 0.04	0.20
Pressure (psia)	14.7	14.7 - 95	14.7
ΔT_{sub} (°F)	50 - 100	0 - 250	--
Flow Spreader Height (inches)	0.005 - 0.012	0.005 - 0.012	.037

(1) Turbine Bucket Simulator
(2) Pressurized Test Facility

Three pairs of arms permit testing, respectively, specimens with passage lengths up to 4.5 inches long, specimens with passages up to 12 inches long, and 4.5 inch specimens tilted in the plane of rotation at 0°, 10°, or 20° with respect to the radial direction. The various passage lengths and orientations bracket the bucket cooling passage geometries in the different stages of the proposed turbine.

In addition to this atmospheric rotating facility a pressurized test
facility (PTF) was used to investigate the effects of passage internal pressure
on heat transfer performance. The pressurized test facility, shown in Figure 3,
consists of an electrically-driven solid aluminum rotor which spins inside a
heavy-walled housing designed to operate at pressures up to 10 atmospheres.
The 4.5 inch long, electrically-heated heat transfer specimens are mounted in
a radial cavity in the rotor. The housing is pressurized with air; vents
and drains carry off air, steam, and water, and also control the vessel pressure.
A feed water system similar to that used in the Turbine Bucket Simulator facility
supplies the cooling water, except that the water must be pumped against vessel
pressure. An off-board preheater controls the inlet water temperature to the
facility.

An additional small bench-top facility was constructed to conduct visualization studies of the rotating flow phenomena. It consisted of a pair of
plastic or glass tubes rotating in the horizontal plane in an enclosure. Two
phase flow phenomena were simulated by using a cocurrent flow of water and air.
The passage was not heated in this facility.

The individual cooling passages in the buckets are modeled by heat transfer
specimens as shown in Figure 4. A cylindrical slug of copper (or aluminum in
the case of the 12-inch long specimens) is heated by 6 embedded resistance
heaters. The cooling passage runs along the axis of the slug. This cooling
passage is lined, in most of the specimens, with a 347 stainless steel or
IN-718 tube of 0.010 inch approximate thickness diffusion bonded to the copper.
Circumferential spreading promoters to counteract the Coriolis-induced asymmetry in the film flow were spaced periodically along some of the liners. The
interval between constrictions was 3/4 inch. Two geometries were tested. One
design consisted of square rings while the other geometry has a more contoured
restriction. A photograph of a sectioned channel liner with the contoured
restriction appears in Figure 5.

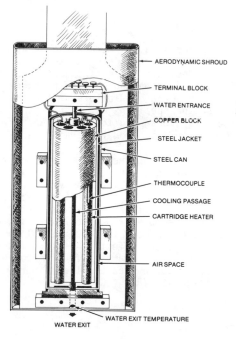

Fig. 4 - Heat Transfer
Specimen

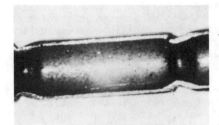

Fig. 5 - Interior of a Cooling Passage
with Contoured Flow Spreaders

Passages without liners were also tested and consisted simply of a hole drilled along the centerline of a copper or aluminum slug. These specimens permitted a measurement of the average wall superheat relative to the liquid temperature and formed the basis for estimates of the heat transfer coefficient.

The copper and aluminum slugs were encapsulated in thick-walled stainless jackets to limit creep deformation at the high g-levels. Air spaces between the jacket and a concentric can limited the heat losses to the environment.

Instrumentation on the specimens consisted of thermocouples embedded in 0.050 inch I.D. drilled holes at various positions in the copper or aluminum slug. Typically, thermocouples were located at two diametrically opposite mid-span positions outside the ring of heaters and also at three axial locations (0.75 and 1.0 inch from the root and tip end of the specimen, respectively, and at the mid-span location) within 0.040 inches of the cooling passage surface. The thermocouples were generally oriented such that they were in the plane of rotation and could be situated either at the leading edge or trailing edge by changing the direction of rotation. Water temperatures at the specimen inlet and exit were also measured.

3. WETTED AREA CALCULATION

The convection coefficient and wetted area parameters cannot both be determined explicitly from experimental data from a single specimen. This complexity arises because the flow does not wet the entire surface of the passage uniformly. It is believed that areas on the cooling passage are inter-mittently wetted by the deposition of steam-borne droplets or the changing paths of rivulets. Instrumentation on the specimens was insufficient to characterize this flow. Hence, the time averaged effective values of wetted area and convection coefficient were calculated from temperature data.

One procedure was to determine the wall superheat vs. power relation from measured temperatures in unlined specimens. Lined specimens with the same cooling passage geometry were assumed to have the same wall superheat-power relation, all other parameters being equal. For the lined specimens the low thermal conductivity of the liner caused significant temperature differences between the wall temperature (deduced from the power vs. wall superheat relation) and the measured specimen body temperature. These tem-perature differences were used to calculate the effective wetted area using a finite element thermal conduction computer program. Effective convection coefficients calculated from the deduced wetted area and the measured heat input were nominally 12,000 Btu/hr-ft^2-^{o}F at $\omega^2 R/g \sim 10^4$.

For specimens that did not have an unlined passage with duplicate geometry to permit use of the preceding method, the convection coefficient was
assumed to be 12,000 Btu/hr-ft^2-^{o}F and the difference between the measured
specimen temperatures and the saturation temperature of the water was used to
calculate the wetted area fraction. For each specimen a finite element thermal
conduction model was used to relate the measured temperatures to the effective
wetted area.

4. RESULTS

Results from a typical data scan (7) are shown in Fig. 6. The power to the
cooling channel is plotted as a function of the wall temperature on the Coriolis, or wetted, side of the passage for various flow rates. Rotation speed
and water flow are established and then the power is raised in increments
until boiling transition (or critical heat flux (CHF)) is reached. Boiling
transition is typified by a steadily rising copper temperature at constant
power. In the sensible heat region the data have slopes consistent with the
different flow rates. As the wall temperature increased above saturation the
slopes increase and for different flow rates the differences in wall temperature become small, indicating vaporization heat transfer. Finally, a CHF
condition is reached for each flow rate.

4.1 Wetted Area

Experimental information about the film heat transfer coefficient around
the passage and the effective wetted area was determined from thermocouple
data, visual tests and deposit patterns in the cooling passages. Effective
fractional wetted circumferences have been calculated from the heat transfer
data using the procedure described in the preceding section, generally at
power levels just prior to CHF. Because of variation of the wetted area
along the length of the passage, the above procedure gives a typical or
average value for the passage.

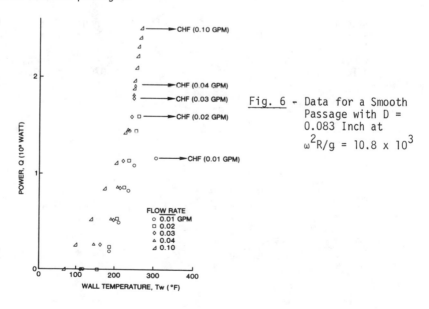

Fig. 6 - Data for a Smooth
Passage with D =
0.083 Inch at
$\omega^2 R/g = 10.8 \times 10^3$

Figure 7 shows the fractional coverage for smooth passages and passages with flow spreaders as a function of the dimensionless rotation number, $\omega\pi a^3/W$. The symbols for the data are described in Table 2. The rotation number correlation shows the influence of the channel diameter, the cooling flow rate and the rotational speed. The trend of decreasing coverage area with increasing rotation number is expected since the rotation number expresses the ratio of the Coriolis force on the coolant to the inertia force of the coolant.

Although substantial scatter exists in the data for passages with flow spreaders. it is apparent that the effective wetted area values for these passages are generally higher than those for smooth passages at any given rotation number. The periodically spaced spreaders interrupt the film and promote the redeposition of water on the otherwise dry portions of the channel wall.

A better indication of the asymmetry in the distribution of the water film around the passage is shown by wall temperature data taken earlier by Dakin (8). Figure 8 shows the results for a smooth passage specimen with a thermocouple welded on the back-side of a cooling passage liner providing a crude

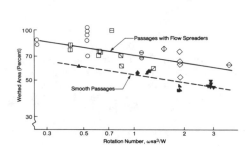

Fig. 7 - Effective Wetted Area for Passages With and Without Flow Spreaders. The Symbols are Explained in Table 2.

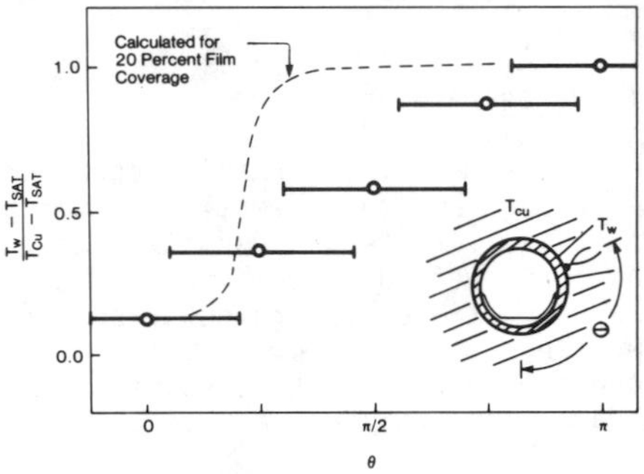

Fig. 8 - Circumferential Wall Temperatures from Dakin (8). $\omega^2 R/G = 11.5 \times 10^3$, D = 0.106 inch, Power = 1070 watts, 0.02 gpm

Symbol	D (Inches)	Flow Spreaders	W (GPM)	$\omega^2 R/G$	Facility
▨	0.073	Contoured	0.01	8.4×10^3	TBS
◩	0.073	Contoured	0.015	8.4×10^3	TBS
☐	0.073	Contoured	0.02	8.4×10^3	TBS
⊞	0.073	Contoured	0.03	8.4×10^3	TBS
⊟	0.073	Contoured	0.02	4.2×10^3	PTF
○	0.066	Contoured	0.02	10.8×10^3	TBS
⊕	0.066	Contoured	0.04	10.8×10^3	TBS
⊖	0.066	Contoured	0.01	10.8×10^3	TBS
◇	0.100	Rings	0.02	4.2×10^3	PTF
◈	0.100	Rings	0.025	4.2×10^3	PTF
⬦	0.100	Rings	0.01	4.2×10^3	PTF
◣	0.100	None	0.02	4.2×10^3	PTF
◢	0.083	None	0.02	4.2×10^3	PTF
▲	0.063	None	0.02	10.8×10^3	TBS
◆	0.083	None	0.02	10.8×10^3	TBS
▼	0.116	None	0.02	10.8×10^3	TBS

Table 2 - Symbols for Figure 7

heat flux gauge. The horizontal bars indicate the approximate width of the thermocouple cavity. On the wetted side of the passage ($\theta=0$) the thermocouple registers wall superheat, Tw-Tsat, of 22°F and on the dry side ($\theta=\pi$) temperatures close to the copper temperature (425°F). Calculations indicate that the heat flux falls off by 20 percent from $\theta=0$ to $\pi/4$ and by 50 percent from $\theta=0$ to $\pi/2$. Correspondingly, the temperatures indicate that the water film distribution and the local heat transfer coefficient decrease continuously from the Coriolis to the non-Coriolis side of the passage. This trend is similar to that observed in two-phase flow in horizontal passages (9). Hence, wetted area calculations using a nominal heat transfer coefficient give only nominal or effective wetted areas. For comparison, the dashed curve on Figure 8 shows the calculated wall temperature distribution if a continuous, uniform water film covered 20 percent of the passage wall with no heat transfer to the remaining 80 percent.

Visual tests (Figures 2 and 3 in Part 2 of this paper) also give evidence to the distribution of water around the passage perimeter. Observations from these tests show the film varying from a thicker narrow film on the Coriolis side to a distribution of droplets on the non-Coriolis side.

Finally, observations of deposit patterns inside cooling passages give evidence of the flow phenomena and water film distribution in the passage. In smooth passages a track left by the water film covered from 40 to 60 percent of the passage perimeter. These values are of the order of the effective wetted area values for smooth passages shown in Figure 7. Also, passages with deposits on the non-Coriolis side of the channel indicate a repetitive wetting and drying phenomena. This is evidence of the circumferential distribution of the water film measured above.

4.2 Steam Flow

High velocity cocurrent steam flow also has an effect on wetted area. Visual studies (Figures 2 and 3 in Part 2 of this paper) have demonstrated this phenomenon in passages with flow spreaders. Effective wetted areas have been calculated from temperature data at the root and tip locations of a 12 inch long aluminum specimen with a smooth passage. These values are shown in Figure 9. (The inlet water temperature for the 12-inch specimen was such that saturation conditions were reached in the first 1.5 inches of the passage at 2000 watts power.)

The wetted area is substantially larger at the tip where the steam velocity is larger even though less water is available there due to evaporation along the passage. Also, a trend of increasing wetted area with power and, therefore,

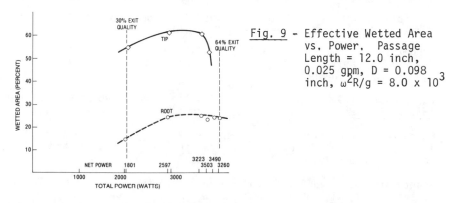

Fig. 9 - Effective Wetted Area vs. Power. Passage Length = 12.0 inch, 0.025 gpm, D = 0.098 inch, $\omega^2 R/g = 8.0 \times 10^3$

steam velocity is noted at the tip location prior to approaching a CHF condition. The trend is consistent with observations of stationary horizontal passages (10) which show an increase in wetted area with increasing steam flow. At CHF there is a substantial deterioration in the heat transfer characteristics at the passage tip.

5. CONCLUSIONS

   a. The effective wetted area ranges from 45% to 100% and correlates with rotation number.

   b. Turbulence promoters spaced periodically along the passage increase the effective wetted area and thereby the overall heat transfer performance.

   c. High velocity cocurrent steam flow increases effective wetted area and overall heat transfer performance.

ACKNOWLEDGEMENTS

These investigations were supported by the U. S. Department of Energy under the High Temperature Turbine Technology Phase II Program, Contract No. DE-AC01-76ET-10340 (formerly EX-76-C-01-1806) and Electric Power Research Institute under contract RP-234-3. The authors appreciate their support.

REFERENCES

1. Caruvana, A., Manning, G.B., Day, W.H., and Sheldon, R.C., (1978), "Evaluation of a Water-Cooled Gas Turbine Combined Cycle Plant", ASME Paper No. 78-GT-77.

2. Horner, M.W., Day, W.H., Smith, D.P., and Cohn, A., (1978), "Development of a Water-Cooled Gas Turbine", ASME Paper No. 78-GT-72.

3. Kydd, P.H., and Day, W.H., (1975), "An Ultra High Temperature Turbine for Maximum Performance and Fuels Flexibility", AMSE Paper No. 75-GT-81.

4. Schmidt, E., (1951), "Heat Transmission by Natural Convection at High Centrifugal Accelerations in Water-Cooled Gas Turbine Blades", Inst. of Mech. Engnrs., London, Section IV, pp. 361-363.

5. Bayley, F.J., and Martin, B.W., (1970-71), "A Review of Liquid Cooling of High Temperature Gas Turbine Rotor Blades", Inst. Mech. Engrs. Proc. 185, 18/71, pp. 219-227.

6. Van Fossen, G.J., and Stepka, F.S. (1978), "Liquid Cooling Technology for Gas Turbines: Review and Status", NASA TM78906.

7. Sundell, R.E., Goodwin, W.W., Dudley, J.C., Kercher, D.M., and Triandafyllis, J., (1980), "Boiling Heat Transfer in Turbine Bucket Cooling Passages", ASME Paper No. 80-HT-14.

8. Dakin, J.T., (1978), "Vaporization of Water Films in Rotating Radial Pipes", Int. J. Heat Mass Transfer, 21, pp. 1325-1332.

9. Butterworth, D. and Robertson, J.M., (1977), "Boiling and Flow in Horizontal Tubes", in Two-Phase Flow and Heat Transfer, D. Butterworth and G.F. Hewitt (eds.), Oxford University Press, Oxford, 1977.

10. Rounthwaite, C., (1968), "Two-Phase Heat Transfer in Horizontal Tubes", J. Inst. Fuel, 41, pp. 66-76.

Two-Phase Heat Transfer in Gas Turbine Bucket Cooling Passages: Part 2

**R. E. SUNDELL, J. C. DUDLEY, C. M. GRONDAHL,
and D. M. KERCHER**
General Electric Company
Schenectady, N.Y., USA

ABSTRACT

Results are presented from an experimental program to study evaporative heat transfer to water in a rotating radial passage simulating a cooling passage in a gas turbine bucket. This paper (Part 2) presents results on the effects of flow rate, cooling passage length and tilt angle, centrifugal acceleration, pressure and wall wettability on the boiling transition power level which determines the thermal load limit for a passage. Also, the two-phase flow phenomena in the passages are discussed. Part 1 of this paper evaluated the effects of coolant flow rate, steam flow, rotation number and passage geometry on the effective wetted area of a cooling passage.

1. INTRODUCTION

Water cooling of gas turbine components is an attractive means for designing for low metal temperature and very high gas temperatures. These two features will be important for a new generation of utility-size gas turbines burning coal derived fuels.

This paper presents further results from an experimental program to study heat transfer in the radial cooling passages of a water cooled gas turbine bucket. A boiling transition phenomenon (corresponding to a critical heat flux (CHF) condition) limits the cooling capacity of the radial passages. The effects of coolant flow rate, passage length and tilt angle, centrifugal force, pressure and wall wettability on this limit are considered. Part 1 of this paper examined the influence of flow rate, steam flow, rotation number and passage geometry on the effective wetted area for a cooling passage.

2. COOLING PASSAGE PHENOMENA

The two-phase flow phenomenon hypothesized for the cooling channel is shown schematically in Figure 1. This model is based on observations of co-current air/water flow in a bench-top flow visualization facility and on calculations and experimental heat transfer results to be described. The water travels along the wall as a thin film (of order 0.001 inch thick) at velocities up to 100 ft/sec and steam fills the passageway. The water film is drawn along the passage by the large centrifugal force (approximately 10^4 times gravity), and the Coriolis force (approximately 10^3 times gravity) on the water tends to force the film to the trailing surface of the passage, opposite the direction of rotation, resulting in asymmetrical heat transfer around the perimeter. Steam flows in the core of the channel at increasing

473

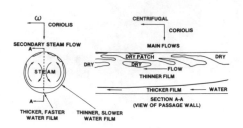

Fig. 1: Cooling passage phenomenology

Fig. 2: Flow of water exiting the
tip of a rotating glass
tube. A flow spreader is
located 7 diameters up-
stream of the tip. D=
0.375 inch, R=8 inch,
$\omega^2 R/g=600$, 0.2 gpm

velocities as it approaches the passage exit due to vaporization along the
passage. Steam velocities can approach sonic values at high exit steam
qualities and ambient pressures at the specimen tip, but are significantly
lower at the design operational pressure of 6.5 atmospheres. Also, the Coriolis force creates a secondary flow of steam as shown in Figure 1.

The film thickness and velocity were estimated from an expression taken
from the work of Dukler (1) who considered the analogous case of a vertical
falling film. Using the exponential distribution of eddy-viscosity proposed
by Deissler (2) he calculated the thickness of films falling under gravity.
His results give

$$\delta \left(\frac{g}{\nu^2}\right)^{1/3} = f(Re) \tag{1}$$

In applying this expression to the film in the rotating passage, the film
Reynolds number is defined in terms of the average film thickness

$$\overline{Re} = \frac{4\Gamma}{\nu} \tag{2}$$

and the gravitational acceleration is replaced by $\omega^2 R$.

Figure 1 shows the water film to consist of at least two zones. A relatively narrow and thicker high velocity film flows along the trailing Coriolis side of the passage and a thinner, slower, and possibly laminar film
exists on other wetted portions of the passage. The existence of this thinner film depends on the wettability of the passage wall, a parameter which
has a significant effect on heat transfer performance as will be shown later.
Dry patches may exist in the thin film region along with isolated wet
patches on other portions of the channel wall. Intermittently, rivulets of
liquid may form to return liquid to the Coriolis side.

The mechanism for wetting the anti-Coriolis side of the passage is still
a subject for further research as it is in two-phase flow in horizontal passages (3).   In the case of cooling passages with flow spreaders the liquid

film is interrupted at intervals along the passage and sprayed to the opposite side resulting in a more uniform distribution of water around the passage circumference. Also, droplets from the liquid spray can be entrained in the steam flow and redeposited on the anti-Coriolis surface.

These mechanisms are evident in photographs taken with the flow visualization facility operating with a rotating glass tube with a flow spreader located 7 diameters upstream of the tip. In Figure 2 the facility is operated without cocurrent airflow at the conditions indicated. A thicker, high velocity film exists on the Coriolis side of the passage and a spray of droplets emanates from the tip of the remaining perimeter of the passage as a result of the flow spreader which distributes the water around the passage circumference. In Figure 3 the facility is operated with cocurrent air/water flow at the conditions listed and, for this case, a fog of fine droplets entrained by the high velocity air is visible in the passage. Again, a thicker, high velocity film is evident on the Coriolis side of the passage and a uniform spray of droplets emanates from the tip of the remaining passage perimeter. Here, the deposition is enhanced by the entrained droplets.

As will be shown, heat transfer results in flow passages without flow spreaders also indicate wetting and vaporization heat transfer on a significant fraction of the passage perimeter. Where nucleation of steam bubbles is occurring in the film, eruption of the bubbles through the film provides a mechanism for transferring liquid droplets to the anti-Coriolis surface. Also, liquid droplets could be entrained from the liquid film by the high velocity steam flow and redeposited on the opposite surface. Air velocities of sufficient magnitude could not be attained in the visual facility to show this phenomenon in passages without flow spreaders.

Finally, the secondary vortices in the steam flow provide a mechanism for wetting the anti-Coriolis side of the passage by either drawing the liquid up the passage through viscous shear or redepositing entrained droplets on the passage wall. The redeposition mechanism is especially effective at flow spreader locations where a spray of droplets is formed. Also, the secondary flow is enhanced at the locations of the flow restrictors because of the substantial flow area reduction (up to 50 percent) and the corresponding stretching of the vortex lines.

Fig. 3:   Co-current flow of water and air exiting the tip of a rotating glass tube. A flow spreader is located 7 diameters upstream of the tip. D=0.375 inch, R=8 inch, $\omega^2 R/g$=600, 0.2 gpm, U_{air}=200 ft/sec.

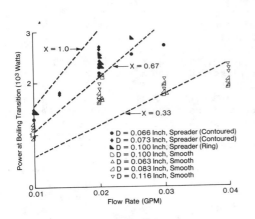

Fig. 4: Power at boiling transition for passages with
and without flow spread-
ers. $\omega^2 R/g = 4.2 \times 10^3$ to
10.8×10^3

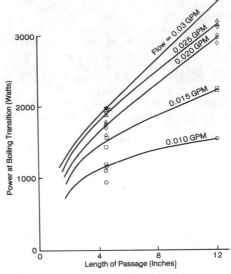

Fig. 5: Power at boiling transition
vs. passage length for
smooth passage. $\omega^2 R/g =$
10.8×10^3

Many of the above mechanisms have also been proposed to explain wetting
in two-phase flows in horizontal passages (3). Lacking a complete understanding of the precise mechanism for film spreading, it still appears that the
two-phase heat transfer phenomenon in the rotating passage can be represented
as one where a relatively narrow, thicker film on the Coriolis side of the
passage acts as a reservoir in supplying "make-up" water lost in the thin film
region either by evaporation or by Coriolis-driven flow returning liquid to
the trailing surface of the passage.

3. POWER AT BOILING TRANSITION

A boiling transition condition occurs in a cooling passage when an incre-
mental increase in power to the cooling channel results in a steadily rising
copper temperature. This is due to a deterioration (a substantial decrease)
of the nominal vaporization heat transfer coefficient in the passage. Test
results showing the effect of various parameters on the boiling transition
power level are now considered.

3.1 Boiling Transition Power Versus Flow Rate

Figure 4 shows the boiling transition power level for 4.15 to 4.5
inch long cooling passages with and without flow spreaders as a function of
flow rate. Passage diameters ranged from 0.063 inches to 0.116 inches, and
the flow spreader spacing along the passage was 3/4 inch. Both flow spreader
geometries are represented. Inlet water temperatures ranged from 0 to 150°F
below saturation conditions. Only a slight increase in boiling transition
power was noted as inlet water temperatures were reduced.

Cooling performance is a strong function of flow rate. However, there is a diminishing benefit above 0.02 gpm. Because of a limit on specimen heater power it was possible to obtain only one boiling transition data point for passages with flow spreaders at flows 0.03 gpm and above. Tests at flows above 0.03 gpm show that powers exceeding this data point can be reached without boiling transition. It is evident that passages with flow spreaders have superior cooling capacity relative to smooth passages.

A slight trend of increasing power with diameter is noted for the smooth passage at 0.04 gpm and for the flow spreader passages at flows 0.02 gpm and greater.

Lines of constant exit quality are shown in the figure to indicate the trend in the water evaporated with flow rate. The maximum possible quality attained in the passage exit prior to boiling transition decreases with increasing flow rate.

3.2 Boiling Transition Power Versus Passage Length

Boiling transition power levels were obtained for the 12-inch long (0.098 inch diameter) aluminum smooth passage specimen. The effect of cooling passage length on boiling transition is shown in Figure 5 by comparing these results with those from a smooth, 4.5 inch long (0.083 inch I.D.) specimen. Figure 4 indicates that the difference in passage diameter for the two specimens will have a negligible effect on the comparison. The observed increase in boiling transition power level is expected since for a given power and wetted perimeter fraction the heat flux would vary inversely with the passage length.

The effect of cooling passage length on boiling transition can also be shown by plotting the results for each specimen as shown in Figure 6. Lines of constant exit quality are included in the figure. Essentially all water is evaporated ($X = 1.0$) prior to reaching boiling transition for the 12-inch long passage at flows of 0.02 gpm and below. For this case boiling transition corresponds to a "dry-out" condition at the tip. These results can be compared with the results for the 4.5 inch long passage at 0.02 gpm where only 50% of the water is evaporated prior to reaching boiling transition.

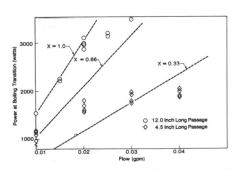

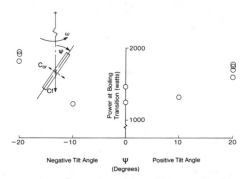

Fig. 6: Power at boiling transition for 4.5 and 12.0 inch long smooth passages. ω^2R/g = 10.8×10^3.

Fig. 7: Power at boiling transition vs. passage tilt angle. 0.01 gpm, $\omega^2R/g = 8.0 \times 10^3$, flow spreaders.

At flow rates higher than 0.02 gpm, boiling transition occurs prior to evaporating all the coolant in the 12-inch long specimen. The wetted area does not increase in proportion to the flow rate, and the larger flux necessary to evaporate the additional water is above the CHF limit for this flow configuration.

3.3 Boiling Transition Power Versus Tilt Angle and Centrifugal Force

Tests of the effect of passage tilt on boiling transition were performed in the turbine bucket simulator (TBS) with the arms which allow tilt angles of 0, +10, and +20 degrees in the plane of rotation. The 4.5 inch long test specimen had a $\overline{0}$.073 inch I.D. stainless steel liner with flow spreaders at 3/4 inch intervals.

A trend of increasing boiling transition power with tilt angle is noted in Figure 7 for $\omega^2 R/g = 0.8 \times 10^4$ and 0.01 gpm. Within the data scatter little change in boiling transition power occurs from 0 to +10 degrees tilt, but a substantial increase is noted from +10 to +20 degrees. In particular this trend is more pronounced for negative tilt angles. A sketch indicating tilt angle and the vector representation of the Coriolis and centrifugal accelerations is shown in the figure.

The most likely interpretation of this trend is that the onset of boiling transition is sensitive to the component of centrifugal force normal to the cooling passage wall. It is well known that in ordinary pool boiling, CHF increases with g-level (4,5). However, a more likely explanation associated with film flow is proposed in Section 4.

As was noted above, the trend of increasing boiling transition power with tilt angle is more pronounced for negative angles. This is consistent with the fact that, for negative tilt angles, the Coriolis and normal centrifugal

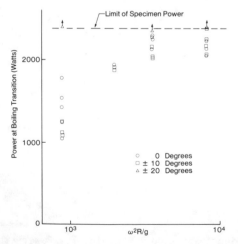

Fig. 8:   Power at boiling transition vs. centrifugal acceleration. 0.02 gpm, flow spreaders.

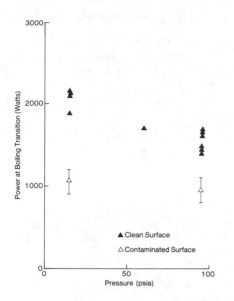

Fig. 9:   Power at boiling transition
          vs. vessel pressure for
          smooth passage. D=0.100 in.,
          0.02 gpm, $\omega^2 R/g = 4.2 \times 10^3$

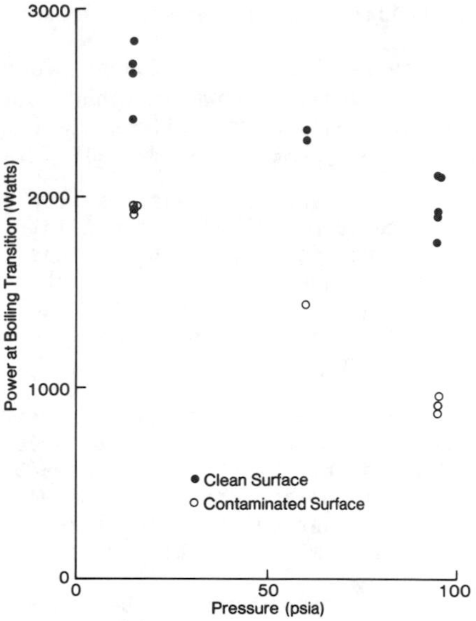

Fig. 10: Power at boiling transi-
          tion vs. vessel pressure
          for a passage with flow
          spreaders. D=0.073 in.,
          0.02 gpm, $\omega^2 R/g = 4.2 \times 10^3$

components are aligned while for positive angles they are opposed. Because of
the velocity dependence of the Coriolis force it should not be assumed that
Coriolis and centrifugal forces have quantitatively similar effects on boiling
transition levels. It is, however, reasonable to assume that their qualitative effect is the same.

     A trend of boiling transition power level with centrifugal acceleration,
$\omega^2 R/g$, for this specimen is shown in Figure 8. Data for all tilt angles are
shown for a flow rate of 0.02 gpm. It is noted that for tilt angles of 0 and
+10 degrees the boiling transition levels decrease abruptly as $\omega^2 R/g$ is reduced below 10^3. One possible interpretation of this observation is that below a particular rotational speed and film velocity flow spreaders are not as
effective in splashing the water film around the perimeter of the passage.
However, a similar trend in boiling transition has been noted with smooth passages. Also, data in Figure 7 in Part 1 of this paper would indicate increasing wetted area as $\omega^2 R/g$ is decreased. Another interpretation of the fall-off
in the boiling transition power is that the film is less stable and more likely
to break into rivulets at lower values of $\omega^2 R/g$. This phenomena is described
in References (6) and (7). Correspondingly, dry patches in the film caused
by bubble nucleation would remain and grow at lower accelerations. This
interpretation will be discussed further in Section 4.

     Figure 8 shows that the maximum specimen power was attained without
boiling transition for tilt angles +20 degrees for the range of $\omega^2 R/g$ values
tested. The superior heat transfer performance for these tilt angles is,
as above, associated with the large component of $\omega^2 R/g$ normal to the
cooling passage wall.

3.4   Boiling Transition Power Versus Pressure and Wall Wettability

The results for a 4.15-inch long specimen with a smooth IN718 cooling passage liner are shown in Figure 9. Also, Figure 10 shows the results for a 4.5-inch long specimen with a stainless steel liner with flow spreaders. Chamber pressures ranged from 15 psia to 95 psia.

Erratic and poor heat transfer performance was measured during early testing in the pressurized test facility. Borescopic inspection inside the cooling passages revealed that this was due to poor wetting characteristics on the cooling passage walls. The wettability was revealed by the shape of the meniscus associated with a water column in the cooling passage. It is suspected that silicone grease used on the front flange seal was the source of contamination.

A one-to-one correlation between heat transfer performance and passage wettability as revealed by the borescope was established for the smooth passage specimen. Results in Figures 9 and 10 show that surface contamination can affect boiling transition power levels by as much as a factor of two. With contamination the water likely travels along the wall in rivulets rather than as a film, drastically reducing the wetted heat transfer area.

A borescope could not be fitted in the smaller diameter passage with flow spreaders, but the high boiling transition power levels shown in Figure 10 were obtained after baking the specimen for one-half hour at 1000°F. This treatment would break down any silicone on the passage wall. Also, the cooling passage was purposely contaminated with silicone after a baking operation and the power levels returned to the low values indicated in the figure.

A significant pressure effect is noted in Figures 9 and 10 for both the contaminated and clean passage results. This is particularly true for the passage with flow spreaders. Steam velocities (and $\rho \bar{u}^2$) are reduced by a factor of 6 at the higher pressures and the effect of steam velocity on wetted area referred to earlier can, in part, explain this trend. Also, entrainment and redeposition at high steam velocities are much more prevelant in the passage with flow spreaders.

Figure 11 is a cross-plot between boiling transition power and the calculated effective wetted area. Wetted areas have been adjusted to represent the equivalent fractional wetted perimeter for a 0.100 inch I.D., 4.15-inch long channel. Passages with and without flow spreaders are represented, including both clean and contaminated surfaces.

The data show the expected trend of low boiling transition powers with low effective wetted area. The lowest values of power and wetted area are associated with contaminated passages. Also, the pressurized results generally lie below the results for 1 atmosphere and the bulk of the data corresponding to 95 psia indicate lower effective wetted areas. However, it should be noted that the wetted areas of a significant portion of the pressurized and 1 atmosphere data overlap, and yet higher boiling transition values are attained at 15 psia. This suggests another mechanism which reduces the boiling transition power at pressurized conditions. A likely cause is the higher wall temperature associated with the higher saturation temperature (325°F) condition at 95 psia. This could lead to an earlier Leidenfrost condition in the passage in which the water no longer wets the wall.

The curve through the data in Figure 11 indicates a power at boiling transition of 2900 watts for 100 per cent wetted area. This corresponds to a critical

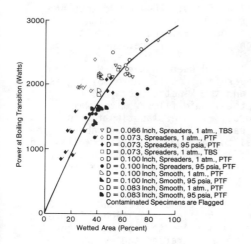

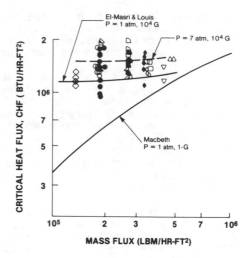

Fig. 11: Power at boiling transition vs. effective wetted area. 0.02 gpm, $\omega^2 R/g = 4.2 \times 10^3$ (PTF) and 10.8×10^3 (TBS)

Fig. 12: Critical heat flux. Data symbols and parameters are given in Fig. 11.

heat flux of 1.1×10^6 Btu/hr-ft^2. This value is approximately 75 percent larger than the corresponding CHF values measured in annular two-phase flows in stationary tubes at the same coolant mass flux and at atmospheric pressure (8).

With the exception of the contaminated specimens the boiling transition powers and effective wetted areas from Figure 11 are used to calculate critical heat fluxes which are plotted vs. mass flux coordinates in Figure 12. The mass flux is determined by using the total cross section of the cooling passage. The data is compared with the prediction of El-Masri and Louis (12) for the heat flux at CHF for thin films in wide rectangular channels and at high centrifugal accelerations. Values of the water flow at the exit of the passage per unit width of wetted perimeter are used to calculate CHF levels from the theory.

Because flow passage and film geometry assumed in the theory differs from that occurring in the experiment, good agreement of theory and experiment is not necessarily to be expected. Also, the significant effect of high velocity cocurrent steam flow on heat transfer performance noted in Part 1 of this paper is not accounted for by the theory. Nevertheless, it is interesting that the theoretical curves for 1 and 7 atmospheres fit nicely through the data. The data scatter is too large to detect any trend of CHF with pressure. A typical CHF correlation for stationary vertical tubes (8) at one atmosphere is shown for comparison.

Several mechanisms have been postulated for boiling transition in annular, two-phase flow (9). One of these is a mechanism observed by

Hewitt, et. al., (10,11) in studies of falling films. They noted that at powers just below CHF the film broke into rivulets forming dry patches resulting in local overheating. The dry patches developed, in some cases, when vapor bubbles formed and broke through the film surface.

Another mechanism, proposed by Tong (13) and used in the El-Masri and Louis model (12), is the boundary layer separation model. Here boiling is assumed to occur in the film prior to CHF and the kinetic energy of the vapor stream normal to the wall separates the liquid film from the wall. The boundary layer separation model also provides an explanation for the high levels of CHF attained in the tilted passages in Section 3.3 above. The large component of centrifugal force normal to the cooling passage wall would inhibit or delay separation of the liquid film from the wall; on the other hand, if the film were once separated the force would redeposit the droplets on the passage wall.

Both descriptions are consistent with copper temperatures and channel pressures measured in a specimen just prior to boiling transition. Figure 13 shows that as the power was increased to a level close to that at CHF the passage pressure increased with the added steam generation and eventually began to fluctuate, indicating either dry patch formation or the onset of film separation. At wall temperatures exceeding the Leidenfrost value heat transfer contact between the liquid and the surface is lost as indicated by the decreasing passage pressure and the rising temperature associated with a CHF condition.

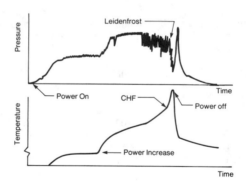

Fig. 13: Trace of specimen temperature
and passage pressure prior to
and at boiling transition

Using the information presented in the two parts of this paper water-cooled buckets have been designed for a 69 MW gas turbine firing at 2600°F (14). The design balances the need for low surface temperatures to prevent corrosion, the need to keep the heat load on each cooling passage below the power at which boiling transition occurs, the need to restrict strain levels, and the desire to obtain maximum efficiency from the gas turbine cycle.

5. CONCLUSIONS

1. Cooling capacity of rotating radial passages with partial channel water flow is limited by a CHF or boiling transition phenomenon.

2. The power at boiling transition increases with increasing coolant flow rate.

3. Cooling passages with flow spreaders have superior cooling capacity relative to smooth passages.

4. The exit quality and total power at boiling transition increases with cooling passage length.

5. Cooling passages tilted off-radial show increased power at boiling transition, especially when Coriolis and centrifugal force components are aligned.

6. Increased rotational speed (and centrifugal acceleration) increases heat transfer performance of radial cooling passages with flow spreaders.

7. Cooling passage pressure adversely effects heat transfer performance.

8. Contaminants which reduce cooling passage wettability reduce heat transfer performance by reducing wetted area.

ACKNOWLEDGEMENT

This investigation was supported by the U. S. Department of Energy under the High Temperature Turbine Technology Phase II Program, Contract No. DE-AC01-76ET10340 (formerly EX-76-C-01-1806) and Electric Power Research Institute under Contract No. RP 234-3. The authors are grateful for their support. F. W. Staub provided helpful consultation in identifying the influence of surface contaminants.

REFERENCES

1. Dukler, A. E. (1960). "Fluid Mechanics and Heat Transfer in Vertical Falling-Film Systems", Chem. Engr. Progr. Symp. Ser., 56, No. 30, pp. 1-10.

2. Deissler, R. G., (1951). "Analytical and Experimental Investigation of Adiabatic Turbulent Flow in Smooth Tubes", NACA TN-2138.

3. Butterworth, D., and Robertson, J. M., (1977). "Boiling and Flow in Horizontal Tubes", in Two-Phase Flow and Heat Transfer, D. Butterworth and G. F. Hewitt (eds.), Oxford University Press, Oxford.

4. Rohsenow, W. M., (1973). "Boiling", in Handbook of Heat Transfer, W. M. Rohsenow and J. P. Hartnett (eds.), McGraw Hill Book Co., New York.

5. Lienhard, J. H., (1981). A Heat Transfer Textbook, Prentice-Hall, Inc., Englewood Cliffs, p. 414.

6. Hartley, D. E., and Murgatroyd, W., (1964). "Criteria for the Breakup of Thin Liquid Layers Flowing Isothermally Over Solid Surfaces", Int. J. Heat Mass Transfer, 7, pp. 1003-1015.

7. Zuber, N., and Staub, F. W., (1966). "Stability of Dry Patches Forming in Liquid Films Flowing Over Heated Surfaces", Int. J. Heat Mass Transfer, 9, pp. 897-905.

8. Collier, J. G., (1972). Convective Boiling and Condensation, McGraw Hill.

9. Hewitt, G. F., (1977). "Mechanisms of Burnout, Pt. 2", in Two-Phase Flow and Heat Transfer, D. Butterworth and G. F. Hewitt (eds.), Oxford University Press, Oxford.

10. Hewitt, G. F., and Lacey, P. M. C., (1965). "The Breakdown of the Liquid Film in Annular Two Phase Flow", Int. J. Heat Mass Transfer, 8, pp. 781-791.

11. Hewitt, G. F., Kearsey, H. A., Lacey, P. M. C., and Pulling, D. J., (1965). "Burnout and Nucleation in Climbing Film Flow", Int. J. Heat Mass Transfer, 8, pp. 793-814.

12. El-Masri, M. A. and Louis, J. F., (1978). "On the Design of High-Temperature Gas Turbine Blade Water-Cooling Channels", ASME Paper No. 78-GT-29.

13. Tong, L. S., (1968). "Boundary-Layer Analysis for the Flow Boiling Crisis", Int. J. Heat Mass Transfer, 11, pp. 1208-1211.

14. Horner, M. W., (1982). "High Temperature Turbine Technology Program Phase II Final Report DOE/ET/10340-127, Contract No. DE-AC01-76ET10340", General Electric Gas Turbine Division, Schenectady, New York.

Critical Gas Turbine Blade Tip Clearance: Heat Transfer Analysis and Experiment

DAVID M. EVANS and BORIS GLEZER
Solar Turbines Inc.
Subsidiary of Caterpillar Tractor Co.
San Diego, Calif., USA

ABSTRACT

The gas turbine blade tip clearance control problem, as related to the relative transient thermal displacement of stator-rotor elements, is reviewed in detail. The methods of critical tip clearance calculation, based upon transient heat transfer between turbine disks, shafts, nozzle case and engine housing, are presented. Basic equations, assumptions and transient boundary condition selection, in support of the analyses, are discussed. Confirming and correlating engine generated test data are provided.

NOMENCLATURE

Symbols

A	area
C	specific heat
h	convection coefficient
K	thermal conductance
$K_{ia} =$	hA_i
$K_{ij} =$	kA_i/L_{ij}
$K_{iR} =$	$\varepsilon A_i \sigma \dfrac{(T_i^4 - T_a^4)}{(T_i - T_a)}$
k	thermal conductivity
L	characteristic length
t	time
T	temperature
V	node volume
α	thermal expansion coefficient
ε	emissivity
δ	tip clearance
$\Delta\delta$	tip clearance change
ρ	material density
σ	Stefan-Boltzmann constant

Subscripts and Superscripts

o	room temperature
2	compressor discharge
3	combustor outlet
5	power turbine inlet
a	atmosphere
B	bearing oil inlet
D	turbine disk
i, j	elemental nodes
I	instantaneous
max	maximum
min	minimum
NC	nozzle case
R	radiation
SH	turbine shaft
SS	steady state
TH	turbine housing
t	transient
th	thermal
r,q,s,z	number of elements in turbine component

1. INTRODUCTION

One of the more difficult problems confronting the gas turbine designer is that of maintaining control over turbine blade tip clearances, particularly with respect to unshrouded blades. Excessive tip clearance leads to excessive blade tip leakage and associated losses in engine performance.

Ideally, full load turbine blade tip clearance should be zero, from the viewpoint of performance, but transient rotor-stator displacement requires some additional clearance, in order to avoid tip rubs. No established analytical method is available to calculate critical blade tip transient clearance, because of the complex relationship between relative radial rotor-stator displacement. The problem becomes even more complicated for turbines with conical shaped stationary shrouds. In this case, axial rotor-stator displacement becomes a major factor in the net tip clearance change.

The changes that affect the engine operating clearance between the rotating turbine blade tip and the stationary shroud result from combinations of thermal and centrifugal displacements. Centrifugal displacement is a function of turbine rotor speed and can be calculated rather simply. Thermal displacement is a complex result of transient and steady-state thermal expansions and contractions. The critical tip clearance calculation, as a multicomponent function of transient turbine rotor-stator heat transfer and thermal displacement, presents a much more complicated task and is the main topic of this paper.

In addition to calculation methods required to establish critical turbine blade tip clearance, comprehensive experimental results have been obtained, using the Solar Centaur gas turbine engine, both in recuperated and simulated simple cycle configurations. The primary purpose of these engine tests was to obtain detailed experimental data required to correct boundary conditions and to calibrate the blade critical tip clearance thermal model. Some unusual experimental techniques were used in obtaining these test results.

2. CRITICAL OPERATING CONDITIONS

In order to establish the most critical period, in which turbine section blade tip rubs are most likely to occur, five operating conditions must be considered:

- Start transients, from a "cold" (effectively room temperature) condition
- Full load, steady-state operation, for which clearances are usually designed to a minimum
- Load transients, where sudden gas stream temperature changes result in different thermal responses of the component parts, due to differing thermal capacitances and heat transfer boundary conditions
- Shutdown transients, which include a rotor rundown period and a cooling period after the rotor stops
- "Hot" restart, after shutdown from a steady-state full load operating condition

Turbine tip clearance changes result from a combination of the following individual displacement components, which are highlighted in Fig. 1:

- Turbine blade radial thermal and centrifugal displacement
- Turbine nozzle case radial and axial thermal displacement
- Turbine tip shoe radial and axial displacement (usually represents very small values and can typically be neglected)
- Turbine disk radial thermal and centrifugal displacement (disk axial thermal displacement may be neglected, in regard to effects upon tip clearance)

- Turbine housing axial and radial displacement
- Turbine rotor assembly axial displacement

The need to consider axial displacements is particularly important for those turbine designs which feature a conical gas path shape, with flare at the blade tip. For this type of design, an axial relative displacement causes a tip clearance change which is proportional to the tangent of the turbine blade tip flare (half) angle.

Large circumferential temperature variations occur only in parts directly swept by the gas stream; e.g. turbine nozzle vanes and turbine tip shoes. These components, however, play a negligible role in determining blade tip clearance. The turbine nozzle case also exhibits a circumferential temperature variation, which is generally reduced to a negligible level, a result of a cooling design of even moderate quality. It may thus be assumed that the turbine tip clearance problem can typically be treated as two dimensional, which grossly simplifies the required thermal model.

The gas producer (compressor) turbine efficiency typically has a much more significant influence upon engine performance than the power turbine. Consequently, all subsequent material concentrates exclusively upon the gas producer turbine blade critical tip clearances, even though all basic principles apply equally well to the power turbine.

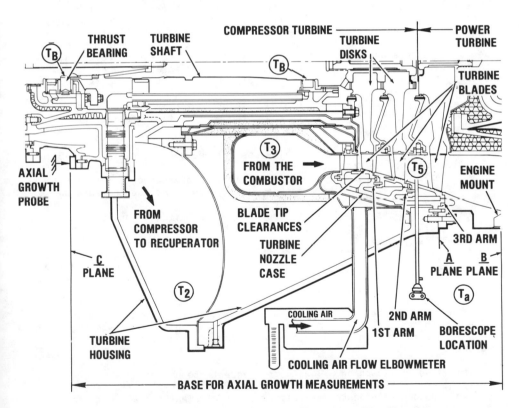

Fig. 1. Turbine Section Components Effecting Tip Clearance

Consider the tip clearance responses associated with the five previously listed engine operating conditions. During a fast engine start, radial growth of the turbine blades and disks, associated with centrifugal strains, occurs instantly upon rotor acceleration. Due to the fact that the turbine blade temperature time constant is extremely small, the blade radial thermal growth may also be considered to be essentially instantaneous. Turbine nozzle case heating and associated radial expansion provide a tip clearance increase. Axial nozzle forward thermal growth, from plane A (Fig. 1), also provides a tip clearance increase for engines with tip flared conical gas path shapes. Turbine disk radial thermal growth results in tip clearance reduction. Turbine housing axial growth, from mount plane B of Fig. 1, displaces the thrust bearing housing face (plane C of Fig. 1) forward, and thus pulls the turbine rotor forward. In the case of the turbine with a tip flared conical gas path shape, this axial displacement may produce a very significant reduction in tip clearance. The turbine nozzle case is restrained from radial expansion by the very rigid turbine housing (plane A of Fig. 1), for the typical design, and consequently this component has negligible influence upon tip clearance change. Turbine rotor axial expansion in the aft direction, from the thrust bearing (plane C Fig. 1), produces some tip clearance increase. A summation of the various displacement factors noted dictates the transient tip clearance value, during engine heating, after a start from a cold condition. The objective of each turbine design is to reach minimum (ideally zero) tip clearance during engine steady-state full load operation.

Sudden gas temperature and rotor speed drops, during engine shutdown, result in an instantaneous tip clearance increase from centrifugal force reduction and turbine blade cooling. At about the time that the rotor comes to a stop, the dominant factor in governing tip clearance is the fact that the nozzle case cools more rapidly than the turbine disks, due to the proximity of lower sink temperatures, resulting in continuing tip clearance reduction. At some point in time, after the rotor stops, the tip clearance reaches a minimum and then starts to increase, as a result of accelerated turbine disk cooling.

Considering all individual turbine displacement components, we can quite simply conclude that the "hot restart" is the most critical moment for potential blade rub, because at that moment:

- Turbine blades grow essentially instantaneously, resulting from both thermal and centrifugal strains
- Nozzle case, which has cooled partially after shutdown, responds to a gas stream temperature increase much slower than do the turbine blades
- Turbine disks grow instantaneously from centrifugal forces, resulting from engine acceleration, while maintaining about their maximum thermal growth for long periods after shutdown
- Turbine housing and rotor assembly axial thermal displacements, essentially compensate for each other during the cooling period after shutdown, but the turbine housing responds more rapidly during a start than does the rotor assembly

As a result of this combination of factors, a critical tip clearance is reached, with respect to a potential blade rub, during a "hot restart", if the turbine is designed with small (essentially zero) steady-state full load tip clearances.

This line of reasoning allows us to concentrate our efforts upon establishing a time to arrive at minimum (critical) tip clearance after shutdown, and determining the associated value of clearance at that time. Our problem has been grossly simplified, as it is necessary for us to consider only one of the five possible critical operating modes. Our problem has been further simplified as a result of the assumption that we can perform our thermal analyses in two rather than three dimensions, with acceptable accuracy.

3. TRANSIENT TEMPERATURE DISTRIBUTION MODEL

Our attention is now focused exclusively upon the period immediately following a shutdown from stabilized full load operation. The task we address ourselves to is determining the minimum turbine blade tip clearance, after shutdown, and further to predict the point in time at which this occurs. This was accomplished by constructing a computer model which describes the transient heat transfer events during this period, in the most accurate fashion that our data allow.

The first step in the preparation of the computer program, for the calculation of critical blade tip clearance, is to determine the transient temperature distribution within those turbine components which directly effect the result. Immediately after fuel cut-off, compressor speed and discharge pressure reduce very rapidly and, as a result, cooling airflow rates also decay very rapidly. As a consequence, it is reasonable to neglect the convective cooling of components, normally cooled under full load conditions. Relative small temperature differences exist between turbine disks and nozzle diaphragms, between the turbine rotor and bearing housings, and also between the nozzle case and surrounding parts, which allows us to neglect radiation heat transfer between these elements during engine cooling. The transient heat transfer between the turbine elements of concern, during the cooling period, is thus largely dictated by a combination of:

- Turbine disk conductance and natural convection
- Turbine shaft conductance and convection between cooling oil and the bearing surfaces and turbine shafts
- Nozzle case conductance and natural convection
- Turbine housing radiation, natural convection and conductance

Consider the transient heat transfer for an elemental volume of such a thermal system. The current time heat balance for such an elemental node i is determined by the equation:

$$K_{ij}(T_i-T_j) + K_{ia}(T_i-T_a) + K_{iR}(T_i-T_a) = 0 \tag{1}$$

The solution of the differential equation of heat transfer must satisfy the initial and also the boundary conditions, given by Equation 1, after each calculation interval (Δt). Using the method of finite differences, for the solution of the differential equation, we can write an expression for transient temperature distribution within our thermal model of n nodes:

$$T_i(t + \Delta t) = \frac{\sum\limits_{n} K_{ij}T_j(t)}{\sum\limits_{n} K_{ij}} \left[1 - e^{-\frac{\Delta t}{\rho_i C_i V_i} \sum\limits_{n} K_{ij}} \right] + T_i(t)\, e^{-\frac{\Delta t}{\rho_i C_i V_i} \sum\limits_{n} K_{ij}} \tag{2}$$

Solution of such a differential equation, arranged in terms of finite differences, is a relatively easy job for the computer.

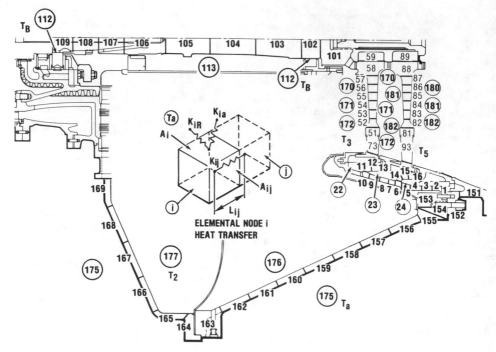

Fig. 2.   Recuperative Centaur Engine Turbine Section
Finite Element Thermal Model

Fig. 2 summarizes the simplified thermal model representing the turbine
section of the recuperative Centaur engine. Input data required by the model
include steady-state full load temperature distribution throughout the model,
time, and temperature dependent boundary conditions and alloy thermal and
transport properties. The time and temperature dependent boundary conditions,
which are easily obtained from engine measurements after shutdown, greatly
enhance the practicality of the results. The post shutdown measured boundary
conditions (Fig. 1) include the following:

- Compressor discharge air temperature (T_2)
- Bearing oil temperature (T_B)
- Gas producer turbine inlet (T_3) or power turbine inlet (T_5)
  temperature
- Engine surrounding ambient air temperature (T_a)

During the course of the computer solution of the system of equations
representing the thermal model, the temperature dependent heat transfer
coefficients were updated at each time interval.

4. THERMAL DISPLACEMENT MODEL

Consider now the preparation of the thermal displacement (expansion-
contraction) model, which utilized the output from the previously described
transient temperature distribution model. These two companion models form the

basis for the computer program whose purpose is to calculate the minimum turbine tip clearance which will occur after shutdown, following a sustained full load run, and the point in time when this event will occur. It is necessary to consider only those mechanical elements which directly effect tip clearance, a fact which helps ensure a model of reasonable size.

Expansion (contraction) of an elemental volume i, from room temperature to current temperature T_i, is proportional to the product of the thermal expansion coefficient and the difference between current and room temperatures. Let us say that the cold turbine tip clearance is δ_o and the near instantaneous tip clearance reduction during a cold start cycle, from blade and disk radial growth resulting from centrifugal forces, as well as blade heating, is $\Delta\delta_I$. The transient blade tip clearnace δ_t is then:

$$\delta_t = \delta_o \pm \Delta\delta_I + \Delta\delta_{th} \tag{3}$$

The minus sign before $\Delta\delta_I$ signifies that tip clearance is decreased during the engine start and the plus sign means that tip clearance is increased during shutdown. $\Delta\delta_{th}$ is the transient thermal tip clearance change which is calculated by the present program and is defined by:

$$\Delta\delta_{th} = \sum_r [\alpha_{NC} L_i^{NC} (T_i^{NC} - T_o)] + \sum_q [\alpha_D(-L_i)^D(T_i^D - T_o)] +$$

$$+ \sum_s [\alpha_{SH} L_i^{SH} (T_i^{SH} - T_o)] + \sum_z [\alpha_{TH}(-L_i)^{TH}(T_i^{TH} - T_o)] \tag{4}$$

The plus sign before the nozzle case and shaft node characteristic lengths (L_i^{NC} and L_i^{SH}), as well as the minus sign before the turbine disk and turbine housing node characteristic lengths (L_i^D and L_i^{TH}), indicate a positive or negative influence of node thermal displacement upon transient tip clearance. T_o is the datum for calculations performed with the subject program, and the thermal expansion coefficient (α) was treated as a variable.

The maximum calculated transient thermal tip clearance change ($\Delta\delta_{th}^{max}$), during engine cooling after shutdown, will determine the critical time for engine hot restart and consequently the permissible minimum cold tip clearance (δ_o^{min}) required to avoid blade tip rub:

$$\delta_o^{min} = \Delta\delta_I - \Delta\delta_{th}^{max} \tag{5}$$

Figure 3 clearly shows the physical significance of each of the variables in equation 5.

The same requirement determines the minimum tip clearance at steady-state full load conditions:

$$\delta_{SS}^{min} = \delta_o^{min} - \Delta\delta_I \tag{6}$$

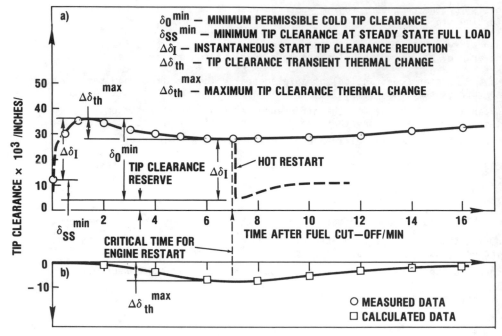

Fig. 3.   Comparison of Measured (a) and Calculated (b)
Tip Clearance Variations after Engine Shutdown

5. EXPERIMENTAL RESULTS

It has been established that the most probable point in time for a severe turbine blade rub will occur during a restart, subsequent to shutdown after a sustained full load run. Our plan encompassed the preparation of an analytical thermal model describing the events of interest, whose validity was ensured by calibration of the model with the following experimentally established boundary and initial conditions:

- Initial steady-state full load temperature distribution, within those turbine elements which have a major bearing upon tip clearance (obtained from previous tests)
- Local ambient temperatures, together with relevant gas and cooling airstream temperatures
- Turbine housing and turbine nozzle case temperature changes, during both rotor rundown and post shutdown cooling periods
- Turbine housing steady-state axial growth during start and shutdown transients
- Tip clearance changes, during engine cooling after shutdown, to determine the minimum tip clearance and the critical time interval to reach that minimum

The recuperated Centaur engine was equipped with detailed thermocouple instrumentation, in order to accomplish these objectives. An existing turbine nozzle case cooling air supply elbow was precalibrated and used for airflow measurments, in order to avoid pressure losses associated with standard flow measurment devices. A turbine housing axial growth measurement device, utilizing an inductance-type pickup probe, was mounted on the engine frame near the thrust bearing plane.

Tip clearance measurement was accomplished by using an air-cooled borescope (Fig. 4), in conjunction with an unusual photographic technique. The borescope was installed in the turbine section, through holes provided aft of the second turbine stage (Fig. 1), normally used for thermocouple probes. The 5 mm rigid focusing borescope, with a side view field of 20° and a fiberoptic light source, was furnished with a cooled outer shell. The cooled shell made it practical to use the borescope in environments to 900°F (482°C). This permitted observations immediately upon engine shutdown.

In order to get objective and precise tip clearance measurements, a rather simple and inexpensive method was used. Each of the photo frames was exposed twice, the first time while viewing ordinary graph paper and the second time with the field of view through the borescope. Total magnification of borescope and teleconverter was about 9. Fig. 5 presents a typical photograph taken through the cooled borescope. The upper part of the figure shows the portion of the blade and tip shoe seen through the borescope and the lower part is an actual picture taken, utilizing the teleconverter feature, upon the graph paper background.

The current tip clearance is the distance between the blade tip and the corresponding step on the tip shoe, which is immediately adjacent to the blade trailing edge. (Centaur tip shoes are stepped in the flow direction and thus have a saw-toothed appearance). To determine the minimum tip clearance value, during steady-state full load operation, some small cylindrical rub pins (0.125 in. or 3.175 mm in diameter) were installed in the tip shoes, in different circumferential positions (Fig. 5). Initial rub pin heights exceed the expected minimum steady-state full load tip clearance. Preliminary calibration of the reference dimensions of the shoe step height and the rub pin diameter provided a precise picture scale, sufficiently accurate to measure tip clearance to within ± 0.001 in. (± 0.025 mm).

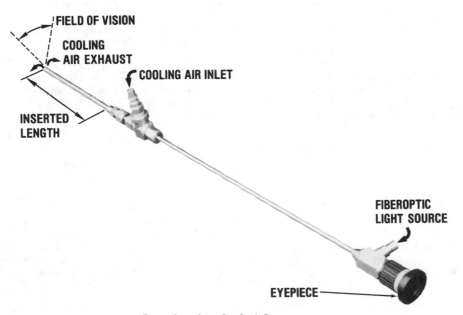

Fig. 4.   Air-Cooled Borescope

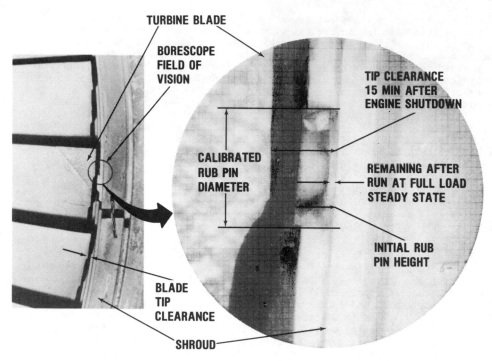

Fig. 5.   Typical Borescope Tip Clearance Measurement Data

The most important objective of the computer program was to predict
transient tip clearance during the cooling period, subsequent to engine
shutdown, at which time the critical (minimum) tip clearance occurs. Fig. 6
presents test data during the engine cooling period, 30 minutes after
shutdown. A sharp reduction in compressor discharge air temperature, as a
result of rotor run down, produces rapid nozzle case cooling (by 80°F or
45°C), soon after fuel cut-off. Tip clearance measurements during the cooling
period (Fig. 3) show a rather fast tip clearance increase, resulting from
sudden removal of the centrifugal load upon disk and blades during shutdown,
but also include some tip clearance reduction as a result of nozzle case
cooling. During the following hot restart, at the critical time (about six
minutes after shutdown), near instantaneous disk and blade centrifugal growth
will cause turbine blade rubbing, if the engine was designed for near zero tip
clearance at steady-state full load conditions. In our case, steady-state tip
clearance was 0.012 in. (0.3 mm) and critical tip clearance during hot restart
was about 0.005 in. (0.125 mm), according to rub pin height measurements.

Experimental data (Fig. 7) show turbine housing axial growth occurs very
rapidly during an engine start, responding primarily to rapid rises in
compressor discharge and turbine inlet temperatures. A significant turbine
housing axial growth produces an equal thrust bearing axial displacement,
yielding a corresponding tip clearance reduction for the turbine with a
conical gas path shape. A 0.120 in. (3 mm) recuperative Centaur thrust bearing
axial displacement measurement (Fig. 7) results in a tip clearance reduction
of 0.030 in. (0.75 mm), upon reaching steady-state full load. The necessity
for including axial turbine rotor displacements in a computer model, intended
for calculation of tip clearances in engines with conical flow paths, is thus
made manifestly clear.

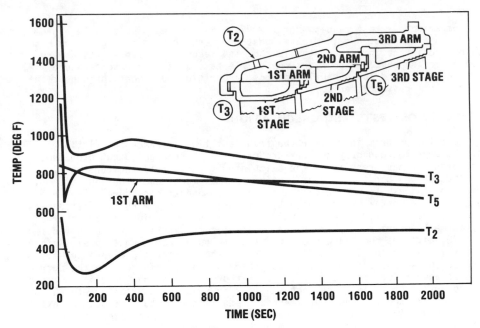

Fig. 6.   Recuperative Engine Cooling after Shutdown

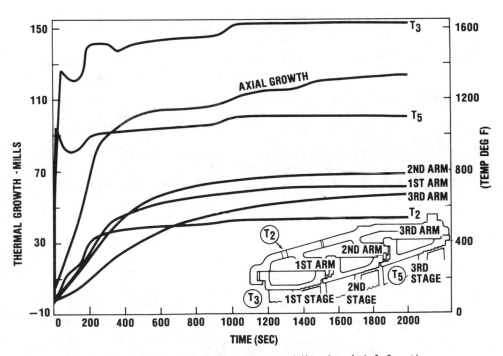

Fig. 7.   Transient Temperature and Housing Axial Growth
during Start from Cold Condition

Comparing our test results for the recuperative and simple cycle engines, during and after shutdown, indicated that the recuperative configuration has a distinct advantage in respect to tip rub. The recuperator represents a very large source of heat, which keeps the nozzle case at a higher temperature after shutdown, and consequently ensures larger tip clearances throughout the critical (hot restart) period.

6. RESULTS OF TESTS AND CALCULATIONS

Computer program input, for the transient tip clearance calculation, included measured boundary temperature conditions. Program output provided, among other things, results of the calculated transient temperature distribution through the turbine section thermal model and transient tip clearance changes, from each of the turbine components considered. Fig. 8 presents the computer program output, for an instant six minutes after engine shutdown, which includes both transient temperature distribution and thermal displacement information. The thermal element temperature nodes, numbered from 1 to 170, give the temperature distribution through the thermal model, which was specified in Fig. 2. Node 341 represents the current tip clearance increase (0.044 in.), from nozzle case thermal expansion; node 355 the current tip clearance reduction from turbine disk thermal expansion (-0.034 in.); node 365 the current tip clearance increase from turbine shaft axial growth (0.008 in.); node 375 the current tip clearance reduction from turbine housing axial growth (-0.026 in.); node 397 the current tip clearance reduction, from a combination of all of the factors above (-0.008 in.).

```
TIP CLEARANCE TRANSIENT ANALYSER      15:33:53 10-JUN-81  PAGE   18
           CRTB2TC: CENTAUR REC TURB 2 ST BLADE TIP CLEAR
           (COOLING OFF)   BG 5.81
   TIME HR        MINUTES      COUNT   PREV INC     NEXT INC     MIN RC
   9.9000E-02    5.9400E+00     187   4.2574E-04   5.2591E-04  2.1036E-03
                   CURRENT TEMP. AND THERMAL GROWTH FACTORS
     1            2            3            4            5
7.4944E+02   7.3808E+02   7.3525E+02   7.6381E+02   7.5974E+02

     6           14           15           16           51
7.5443E+02   7.3986E+02   7.1817E+02   6.7144E+02   8.8570E+02

    52           53           54           55           56
9.4162E+02   8.9911E+02   8.3273E+02   7.1508E+02   5.9137E+02

    57           58          101          102          103
4.9566E+02   4.5957E+02   3.6320E+02   2.9813E+02   2.5766E+02

   104          105          106          107          108
2.5085E+02   2.5011E+02   2.5070E+02   2.5154E+02   2.3995E+02

   153          154          155          156          157
6.4340E+02   6.3680E+02   6.4340E+02   5.2688E+02   5.1800E+02

   158          159          160          161          162
5.3380E+02   5.3469E+02   5.2976E+02   5.3632E+02   5.4390E+02

   163          164          165          166          167
5.6777E+02   5.7714E+02   5.8624E+02   5.8438E+02   5.9660E+02

   168          169          170          341          355
5.9407E+02   5.8953E+02   6.0000E+02   4.4377E-02  -3.4270E-02

   365          375          395          396          397
8.0060E-03  -2.6089E-02   1.0108E-02   1.8114E-02  -7.9752E-03
```

Fig. 8. Computer Output for Transient Tip Clearance Calculations

The results of the transient tip clearance calculations are presented in Fig. 3. A comparison of test data (Fig. 3) and the calculated results, shows very satisfactory agreement between 2 and 16 minutes after engine shutdown. Maximum tip clearance reduction of 0.008 in. (0.2 mm), for both analytical and experimental cases, was reached at approximately six minutes after shutdown. This result confirms the ability of the computer program outlined here, to calculate the transient tip clearance with reasonable accuracy, at that most critical time for probable turbine tip rubs.

7. CONCLUSIONS

From the results of our analytical and experimental studies, of the turbine blade tip clearance problem, the following conclusions can be drawn:

- For turbines with a flared blade tip, a combination of radial and axial thermal displacement must be considered.
- Engine restart, shortly after shutdown from steady-state full load, is defined as the most critical period for potential blade tip rubs.
- Some fairly unusual experimental methods, including a highly precise borescope tip clearance measurement technique, were essential for accurate post shutdown transient tip clearance measurements.
- The minimum permissible cold build tip clearance, needed to prevent rubbing, can be established by utilizing a transient tip clearance calculation computer program, based upon a two-dimensional turbine section thermal model. In order to provide an acceptable level of accuracy, the computer model for each turbine engine configuration must be calibrated with comprehensive test results, aimed primarily at establishing actual temperature boundary conditions.
- Comparison of the recuperative and simple cycle engines shows an advantage of the recuperative configuration with respect to a potential tip clearance rub problem; i.e., the heat exchanger thermal mass ensures larger tip clearances throughout the critical engine rundown and hot restart periods.

Effects of Initial Crossflow Temperature on Turbine Cooling with Jet Arrays

L. W. FLORSCHUETZ and D. E. METZGER
Mechanical and Aerospace Engineering
Arizona State University
Tempe, Ariz., USA

ABSTRACT

Two dimensional arrays of circular air jets impinging on a surface parallel to the jet orifice plate are considered, with the objective of modeling
the impingement cooled mid-chord region of gas turbine airfoils. Heat transfer and flow distribution results for the case without an initial crossflow
approaching the array, previously obtained under a NASA sponsored study, are
first reviewed. As an extension of this work, the effect of an initial crossflow approaching the array is being investigated. The effect of initial
crossflow on the jet array flow distribution is discussed, followed by a presentation of early results for the effects of both the flow rate and the
temperature of the initial crossflow on the jet array heat transfer characteristics. The three-temperature nature of the problem is emphasized and the
significance of the effects on cooling system design is discussed.

1. INTRODUCTION

The cooling of gas turbine engine components has become established as an
important aspect of turbine engine design. The most common cooling schemes
involve the use of air cooling, where a portion of the compressor discharge is
diverted around the combustor to be used directly as a heat exchange medium.
In this way the temperatures and temperature gradients of components exposed
to the hot gas stream can be reduced, thereby extending service life at a
given performance level.

Modern high performance engines use 20 percent or more of the compressor
discharge flow for cooling purposes. The design of such engines requires
great care so that the performance improvement to be derived from operating at
higher temperatures is not more than offset by the cycle and aerodynamic penalties associated with compressing and using the cooling air. In order to do
rational and confident design, the designer must have access to detailed accurate information on the flow and heat transfer characteristics of cooling
schemes in use or under consideration.

The most critical areas in the engine from the viewpoint of thermal
exposure are the first-stage airfoils, both stator vanes and turbine blades.
The stationary first stage vanes, situated immediately downstream of the burner experience the highest gas temperatures, including 'hot streaks' of several
hundred degrees above the mean temperature associated with combustor pattern
non-uniformities. The first stage blades, although experiencing lower

relative velocities and a rotational averaging of the combustor pattern, are subject to the additional complications and stresses of rotation.

For both these airfoil sets, the external heat load around the airfoil surface is very non-uniform. The situation depicted in Figure 1 is typical. The leading edge region experiences very high external heat transfer coefficients. These decrease quickly but usually grow again in the midchord region, particularly on the suction side of the airfoil.

The large external heat loads require an internal cooling scheme with high heat transfer coefficients between the cooling air and inner surface of the airfoil. An impingement cooled arrangement is often the choice because of the high heat transfer coefficients possible and the capability of placing jets in patterns dictated by the external thermal loading. This flexibility in jet placement can be advantageous not only in the chordwise direction, but also in the spanwise direction to reflect, for example, the burner pattern in the radial direction. Figure 2 shows a typical midspan arrangement of jets. Note that the jets are constrained to exit in the chordwise direction; so the accumulated jet flow from upstream rows acts as a crossflow to downstream jet rows in the array. The drop-off in external load behind the leading edge eliminates the need for new cooling jets in this region and the leading edge coolant flows around to to become a separate, or initial, crossflow to the midchord jet array.

Despite the complications involved in fabricating airfoils with inserts to provide a jet plenum, the jet array remains an attractive cooling scheme for the midchord region for the reasons stated above. It is potentially a much better match to the spanwise and chordwise distribution of external heat loads than a multipass spanwise cooling flow arrangement. These multipass designs are often subject to either overcooling or undercooling problems in the turn regions at the airfoil root and tip.

Over the past several years, Arizona State University (ASU) has engaged in an extensive NASA sponsored study of the flow and heat transfer characteristics of multiple jet arrays of the type depicted in the midchord region of Figure 2. The early work in this study was directed at cases where an initial crossflow is not present. It should be recognized, however, that crossflow is always present downstream of the first row, whether or not a separate initial crossflow is imposed. The geometry of the airfoil application dictates that all of the jet flow will exit in the chordwise direction toward the trailing edge. This fact has stimulated much of the prior work on the effects of crossflow on confined jets, as typified by References [1-8].

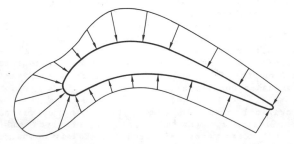

Fig. 1 Example of airfoil external heat
        load distribution

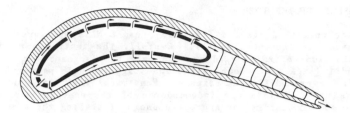

Fig. 2 Example of impingement cooled airfoil

More recently, the ASU study was expanded to consider the effects of initial crossflow, first with equal crossflow and jet temperature, and now with crossflow temperature elevated above the jet temperature. The latter case is of considerable importance. The designer is often faced with an initial crossflow temperature which is substantially above the jet flow because of heat pickup in the leading edge region. Confident design can be achieved only if the designer knows the proper effective coolant temperatures and heat transfer coefficients to use in the region where the initial cross-flow penetrates into the jet array. To date, there is virtually no information in the literature to help the designer answer these questions. The purpose of the present paper is to present the first results of a study on the effect of initial crossflow temperature on jet array performance. These first results have been obtained for a single, but typical, jet array arrangement over a range of cross-to-jet flow rate ratios important in practice. The arrangement chosen has a uniformly spaced inline array of ten spanwise jet rows in the chordwise direction. The chordwise spacing between jet rows (x_n/d) is 5; the spanwise spacing between adjacent holes in a given row (y_n/d) is 4. The spacing between the jet plate and impingement surface (z/d) is 2. Figure 3 depicts the arrangement and nomenclature.

Before proceeding, it is appropriate to review some of the previous results of the study. Examples of these results are presented in the next section mainly for the $x_n/d=5$, $y_n/d=4$, $z/d=2$ (5,4,2) geometry.

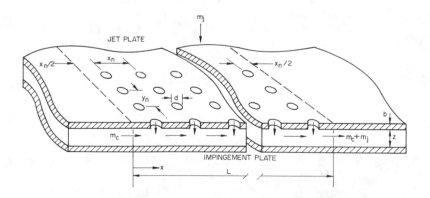

Fig. 3 Test geometry and nomenclature

2. PRIOR RESULTS OF THE STUDY

The prior published results from the present NASA sponsored study have been documented in References [6-11]. Overall, the testing has encompassed various combinations of x_n/d, y_n/d and z/d over the ranges $5 \leq x_n/d \leq 15$, $4 \leq y_n/d \leq 8$, and $1 \leq z/d \leq 6$. Both inline and staggered hole patterns have been investigated. Several different sizes of geometrically similar configurations have been used to ascertain the independence of the non-dimensional results from jet size and to confirm the appropriateness of scaling these results down to the prototype size common in turbine engine applications.

Figure 4 shows some typical chordwise Nusselt number profiles, acquired with a jet array sized to yield a chordwise resolution of spanwise-averaged Nusselt number equal to one chordwise jet hole spacing. These results are all shown for the same value of mean jet Reynolds number, $\overline{Re}_j = 10^4$. Tests were typically conducted over the range $5 \times 10^3 \leq \overline{Re}_j \leq 2 \times 10^4$, but the relative behavior of different array geometries was found to be essentially independent of the $\overline{Re}_j$. Figure 4 shows that the chordwise profiles of heat transfer coefficients can vary significantly with z/d, especially with close spanwise spacings. Nevertheless, the mean values remain relatively insensitive to z/d over the range covered.

Figure 5 shows more detailed results acquired with the $x_n/d=5$, $y_n/d=4$ array arrangement of primary interest in the present work. Here, tests were conducted with a geometrically larger array where resolution of Nusselt number profiles to one-third the chordwise hole spacing was possible. For convenience, these results are normalized with the corresponding array mean values which are shown on the figure for $\overline{Re}_j = 1.5 \times 10^4$.

The periodic variation of heat transfer coefficient is clearly evident in these higher resolution results. It should be noted that much larger amplitudes of the periodic variation have been observed and presented [7,9]. In general, the tendency is for the amplitudes to increase as x_n/d is increased. The general trend for all array arrangements is for the amplitude of the periodic variations in Nu to be reduced downstream in the array by the

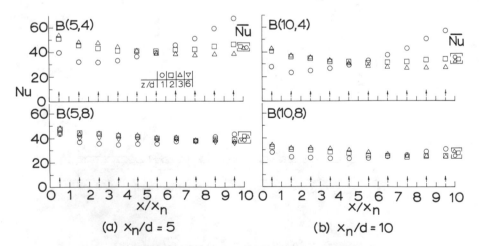

Fig. 4 Typical chordwise Nusselt number profiles
without initial crossflow

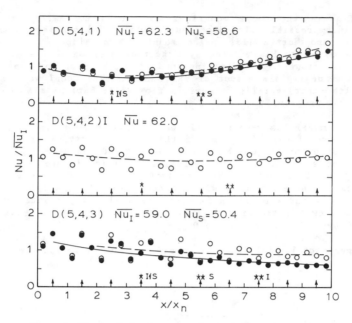

Fig. 5   Typical higher resolution Nusselt number
profiles without initial crossflow

influence of crossflowing air from the upstream jets.

The $x_n/d=5$, $y_n/d=4$ geometry is typical in that the general chordwise trend of Nu can be either increasing, decreasing, or constant, depending on the value of z/d. This is true for either inline or staggered arrangements of jet holes, and both are shown in Figure 5. Inline arrangements have been found to always yield Nu values that are equal to or greater than the corresponding values with staggered arrangements.

The measured chordwise Nusselt number profiles (smoothed over the periodic variations) have been correlated with individual spanwise row jet Reynolds numbers and crossflow-to-jet velocity ratios [8]. The correlation is in terms of individual row Nusselt numbers relative to the Nusselt number of the first row, Nu_1.

$$Nu/Nu_1 = 1 - C(x_n/d)^{n_x}(y_n/d)^{n_y}(z/d)^{n_z}(G_c/G_j)^n \qquad (1)$$

where

$$Nu_1 = 0.363(x_n/d)^{-0.554}(y_n/d)^{-0.422}(z/d)^{0.068}Re_j^{0.727}Pr^{1/3} \qquad (2)$$

and the constants in (1) take the following values:

	C	n_x	n_y	n_z	n
Inline	0.596	−0.103	−0.380	0.803	0.561
Staggered	1.07	−0.198	−0.406	0.788	0.660

G_c/G_j and Re_j in the above equations have been evaluated from a one-dimensional incompressible flow distribution model [8] similar to those used by Dyban [12] and Martin [13]. The model was developed by assuming the discrete hole array to be replaced by a surface over which the injection is continuously distributed. It also included assumptions of constant discharge coefficient and negligible effect of wall shear. The model was verified by comparison with experimentally determined flow distributions also obtained as part of the test program.

This basic model has recently [11] been extended to include the presence of an initial crossflow. For some of the array geometries of interest the effect of crossflow on the discharge coefficient and the effect of wall shear were found to be significant when an initial crossflow is present. Figure 6* shows these flow distribution characteristics for the (5,4,2) geometry for which the heat transfer results with initial crossflow have been obtained. This figure shows the comparison between measured and predicted values for both the jet flow distribution ($G_j/\bar{G}_j$) and the cross-to-jet mass velocity ratio (G_c/G_j) over the nominal range $0 \le m_c/m_j \le 1.0$.

The experimental distributions were obtained from chordwise channel

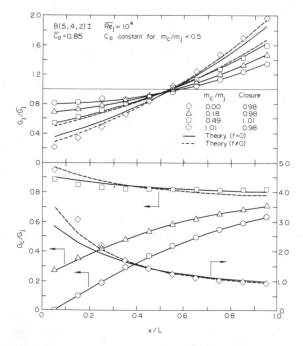

Fig. 6 Typical jet array flow distribution
with initial crossflow

*In Figs. 4 and 5, the chordwise displacement (x) is normalized as x/x_n. In Fig. 6 and subsequent figures it is normalized as x/L. Since these geometries have ten spanwise rows of holes, the two scales differ precisely by a factor of ten.

pressure traverses accomplished with static pressure probes inserted from the open downstream end of the flow channel [11]. The closure obtained for the experimental mass balance in each test is listed on Figure 6. This closure value is the ratio of the sum of the individual spanwise row jet flow rates to the total jet flow rate measured by a standard orifice.

The solid curves in Figure 6 are based on the theoretical model with f=0 (i.e., neglecting wall shear effects). For m_c/m_j < 0.5, the local crossflows are not high enough to affect the discharge coefficients and the solid curves for these cases are from a closed form solution. For m_c/m_j > 0.5, local crossflows are high enough to decrease discharge coefficients in some rows, and the solid curves represent numerical solutions. Numerical solutions including the effect of wall shear (f≠0) are shown by dashed curves for those cases where the effect is noticeable relative to the corresponding f=0 case. The ability of the analysis to predict the measured flow distributions is quite good. This is the case not only for the (5,4,2) geometry discussed here, but also for the other tested geometric arrangements as well over the range $0 \leq m_c/m_j \leq 1.0$. This validated flow model provides the foundation necessary to explore the chordwise heat transfer behavior when m_c/m_j > 0.

3. FORMULATION AS A THREE-TEMPERATURE PROBLEM

The situation of jet array cooling with an initial crossflow temperature differing from that of the jets can be viewed as a three-temperature problem. This is a convection heat transfer situation where the surface heat transfer is to a fluid that is in the process of mixing from two different sources at two different temperatures. The best known example of a three-temperature situation is film cooling. In film cooling it is well known that the interaction of a secondary fluid stream with a primary stream affects not only the heat transfer coefficient, but also the value of the reference fluid temperature which drives the heat flux. In the most simple terms (Figure 7):

$$q = h \ (t_{aw} - t_o) \tag{3}$$

where t_{aw} is the adiabatic wall temperature and is embodied in a non-dimensional effectiveness:

$$\eta = (t_m - t_{aw})/(t_m - t_f) \tag{4}$$

The heat fluxes for jet array impingement with an initial crossflow can also be written as in Eq. (3), but t_{aw} is now expressed as the non-dimensional adiabatic wall temperature (effectiveness) in terms of t_j and t_c (Figure 8):

$$\eta = (t_j - t_{aw})/(t_j - t_c) \tag{5}$$

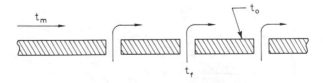

Fig. 7   Film cooling as a three-
temperature problem

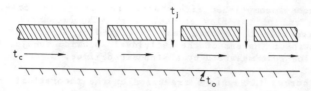

Fig. 8   Jet array impingement cooling with initial
crossflow as a three-temperature problem

In cases of practical interest in turbine cooling, the temperature differences
between t_c, t_j and t_o will be large; and the values of t_c and t_j can be easily
established. In the present experiments the temperature differences are small
(10-30K). This situation requires care in establishment of the appropriate
values of the cross and jet flow temperatures. In the following section the
experimental apparatus and test procedures will be briefly described.

4.   EXPERIMENTAL APPARATUS AND PROCEDURES

A cross-sectional view of the testing arrangement is shown in Figure 9.
Details of the basic facility have been described previously [6,9]; therefore
only a brief description will be given here. For the testing with initial
crossflow there are two plenum chambers. Each chamber is supplied with laboratory compressed air individually metered with standard orifices. The inlet
line to the initial crossflow plenum is equipped with an electric resistance
heater which allows the initial crossflow temperature to be variably elevated
above the jet temperature. The jet plates, each with ten spanwise rows of
holes, are interchangeable as are the spacers which fix the channel height
(z).

The jet hole diameter used in the initial crossflow tests reported here
was 0.254 cm. The channel width (span) is 18.3 cm. The initial crossflow
development length upstream of the array is 24.1 cm. The channel length

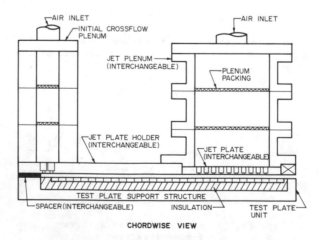

Fig. 9   Initial crossflow test unit assembly

downstream of the last spanwise jet row is 4.5 cm.

The heat transfer test plate unit consists of a segmented copper test plate with individual segment heaters, necessary thermal insulation, and a support structure. The segmented design provides for control of the chordwise thermal boundary condition at the heat transfer surface, as well as for determination of heat transfer coefficient and effective driving potential profiles in the chordwise direction. There are 31 segments: 19 upstream of the first jet row, one under each of the 10 jet rows, and two downstream of the last jet row.

Tests were conducted by first establishing the total jet flow rate so that $Re_j = 10^4$. Second, the desired relative amount of initial crossflow was established with the line heater operating so that the temperature in the initial crossflow plenum was above the temperature in the jet plenum. Segment heater power inputs, segment temperatures, and flow conditions are recorded after steady-state conditions are reached with all individual segment heaters adjusted to achieve an isothermal surface boundary condition.

Two separate sets of tests, each with three different heater power levels, are required to determine the chordwise profiles of η and h for a given flow ratio m_c/m_j. First, with the initial crossflow geometry, but with $m_c=0$, tests are conducted to determine t_j, the characteristic temperature for the jet flow alone, for each segment. These tests are conducted with three different power levels and a linear least squares fit to the three sets (q, t_o) for each of the segments under the jets is used to determine the appropriate t_j for each segment from $q=h(t_o-t_j)$.

Second, a similar set of three different power level tests are conducted with the heated initial crossflow present. A linear least squares fit is then used to determine both h and t_{aw} for the segments under the jets from $q=h(t_o-t_{aw})$ and the three sets (q, t_o). Note that only two sets (q, t_o) are necessary to determine the two unknowns h and t_{aw}; the third set provides additional confidence to the fit.

The initial crossflow temperature, t_c, was characterized as the adiabatic wall temperature of the initial crossflow at the entrance to the array (one-half a streamwise hole spacing upstream of the first spanwise row, see Fig. 3). For the conditions of these tests, this adiabatic wall temperature was essentially identical to the mixed-mean stagnation temperature of the initial crossflow, which was determined from the measured initial crossflow plenum temperature combined with an energy balance over the initial crossflow channel. The test procedure is described in more detail in [14].

5. RESULTS AND DISCUSSION

Figure 10 shows the chordwise profiles of η for the (5,4,2) inline array geometry for three values of m_c/m_j. The η values all span almost the entire range from $\eta=1$ (initial crossflow dominates) to $\eta=0$ (jet dominates). However, the penetration of the initial crossflow temperature influence into the array is a strong function of the flowrate ratio. Midway through the array, η varies from about 0.2 to 0.8, depending on m_c/m_j.

Figure 11 shows the corresponding heat transfer coefficients (in terms of Nusselt numbers). These results show a similar strong influence of m_c/m_j. High values of m_c/m_j suppress the high jet heat transfer coefficients over a large fraction of the array.

The practical implication of these results is of considerable importance. For example, in a highly cooled first stage vane like that shown in Figure 2, t_c is often several hundred degrees above t_j. Typical values are $t_o = 1800F$ (1256K), $t_j = 900F$ (755K) and $t_c = 1100F$ (866K). The results of Figures 10 and 11, if converted into heat fluxes, imply that local cooling rate predictions within the array can easily be in error by 100% or more, depending on

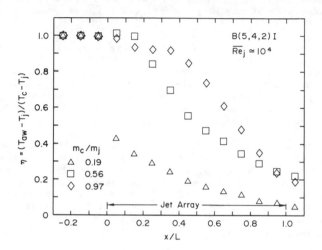

Fig. 10   Chordwise distributions of the normalized
          adiabatic wall temperature (η) for jet
          array impingement with initial crossflow

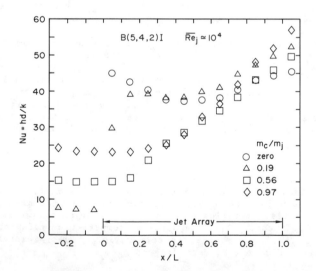

Fig. 11   Chordwise distributions of Nusselt number
          for jet array impingement with initial
          crossflow

the designer's guess. There is evidence that this level of design uncertainty exists in practice, and that premature failures of impingement cooled airfoils have been the result. With better information available on the effects of initial crossflow, it should be possible to make significant improvements in design and to further develop the full potential of impingement cooled gas turbine vanes and blades.

NOMENCLATURE

C_D	discharge coefficient
d	jet hole diameter
f	wall friction coefficient
G_c	crossflow mass velocity based on channel cross-sectional area
G_j	jet mass velocity based on jet hole area
h	convective heat transfer coefficient averaged across span
k	fluid thermal conductivity
L	streamwise length of jet array
m_c	initial crossflow mass flow rate
m_j	total jet array flow rate
Nu	Nusselt number averaged errors span, hd/k
Nu_1	value of Nu at first upstream spanwise jet row when Mc=0
Pr	Prandtl number
q	surface heat flux
Re_j	jet Reynolds number, $G_j d/\mu$
t_{aw}	adiabatic wall temperature
t_c	characteristic temperature of initial crossflow air
t_f	injection temperature in film cooling situations
t_j	characteristic temperature of jet air
t_m	mainstream temperature in film cooling situations
t_o	surface temperature
x_n	chordwise jet hole spacing
y_n	spanwise jet hole spacing
z	channel height (jet plate-to-impingement surface spacing)
η	effectiveness, Eqs. (4) and (5)
μ	dynamic viscosity
ρ	fluid density

REFERENCES

1. Kercher, D.M., and Tabakoff, W., ''Heat Transfer by a Square Array of Round Air Jets Impinging Perpendicular to a Flat Surface Including the Effect of Spent Air,'' ASME Journal of Engineering for Power, Vol. 92, No. 1, Jan. 1970, pp. 73-82.

2. Metzger, D.E. and Korstad, R.J., ''Effects of Cross Flow in Impingement Heat Transfer,'' ASME Journal of Engineering for Power, Vol. 94, 1972, pp. 35-41.

3. Gauntner, J.W., Gladden, H.J., Gauntner, D.J., and Yeh, F.C., ''Crossflow Effects on Impingement Cooling of a Turbine Vane,'' NASA TM X-3029, March 1974.

4. Bouchez, J.P. and Goldstein, R.J., ''Impingement cooling From a Circular Jet in a Crossflow,'' International Journal of Heat and Mass Transfer, Vol. 18, 1975, pp. 719-730.

5. Saad, N.R., Mujumdar, A.S., Abdel Messeh, W., and Douglas, W.J.M., ''Local Heat Transfer Characteristics for Staggered Arrays of Circular Impinging Jets with Crossflow of Spent Air,'' ASME Paper 80-HT-23, 1980.

6. Metzger, D.E., Florschuetz, L.W., Takeuchi, D.I., Behee, R.D., and Berry, R.A.,''Heat Transfer Characteristics for Inline and Staggered Arrays of Circular Jets with Crossflow of Spent Air,'' <u>ASME Journal of Heat Transfer</u>, Vol. 101, 1979, pp.526-531.

7. Florschuetz, L.W., Berry R.A., and Metzger, D.E., ''Periodic Streamwise Variations of Heat Transfer Coefficients for Inline and Staggered Arrays of Circular Jets with Crossflow of Spent Air,'' <u>ASME Journal of Heat Transfer</u>, Vol. 102, 1980, pp. 132-137.

8. Florschuetz, L.W., Truman C.R., and Metzger, D.E., ''Streamwise Flow and Heat Transfer Distributions for Jet Array Impingement with Crossflow ,'' <u>ASME Journal of Heat Transfer</u>, Vol. 103, 1981, pp. 337-342.

9. Florschuetz, L.W., Metzger, D.E., Takeuchi, D.I., and Berry, R. A., <u>Multiple Jet Impingement Heat Transfer Characteristic</u> - <u>Experimental Investigation of Inline and Staggered Arrays with Cross-flow</u>, NASA Contractor Report 3217, Department of Mechanical Engineering, Arizona State University, Tempe, January 1980.

10. Florschuetz, L.W., Metzger, D.E., and Truman, C.R., <u>Jet Array Impingement with Crossflow--Correlation of Streamwise Resolved Flow and Heat Transfer Distributions</u>, NASA Contractor Report 3373, Department of Mechanical Engineering, Arizona State University, Tempe, January 1981.

11. Florschuetz, L.W., and Isoda, Y., ''Flow Distributions and Discharge Coefficient Effects for Jet Array Impingement with Initial Crossflow,'' ASME Paper 82-GT-156, April, 1982.

12. Dyban, E.P., Mazur, A.I., and Golovanov, V.P., ''Heat Transfer and Hydrodynamics of an Array of Round Impinging Jets with One-Sided Exhaust of the Spent Air,'' <u>International Journal of Heat and Mass Transfer</u>, Vol. 23, 1980, pp. 667-676.

13. Martin, H., ''Heat and Mass Transfer Between Impinging Gas Jets and Solid Surfaces,'' <u>Advances in Heat Transfer</u>, Vol. 13, Academic Press, New York, 1977, pp. 1-60.

14. NASA Contractor Report, Department of Mechanical Engineering, Arizona State University, Tempe (in press).

STEAM TURBINES

Some Heat and Mass Transfer Problems in Steam Turbines

MARKO MAJCEN
Faculty of Mechanical Engineering
and Naval Architecture
University of Zagreb
Zagreb, Yugoslavia

ABSTRACT

Problems connected with heat and mass transfer influencing steam turbines design, efficiency and proper operation are discussed. Phenomena in the wet steam region with theoretical explanations, experimental methods and installations for defining of liquid phase flow in low pressure steam turbine parts are described.

1. INTRODUCTION

Although the basic principle of gas and steam turbines does not differ essentially, detailed knowledge of heat and mass transfer problems and their solutions is more important for the successful design of gas turbines.

This can be clearly seen from the number and proportion of papers dealing with heat and mass transfer in rotating machinery such as gas or steam turbines.

The previous statement may also be confirmed by the fact that a usable steam turbine dates back to 1883, while the first gas turbine appeared approximately 55 years later.

The reason for this delay can also be explained by the fact that the gas turbine efficiency is very much dependent on the properly selected aerodynamic characteristics in the flow apparatus of the compressor and turbine. Considerably higher temperature levels are required at the inlet of gas turbine for obtaining acceptable efficiency. On the other hand any incorrect assumption on the value of thermal stresses may lead to thermal distortion. Sudden change of load levels and, especially, fast starts of gas turbines increases even more the fact under discussion.

Although the lowest maximum temperature limit at gas turbine inlet used today is even markedly higher than the upper temperature limit of a steam turbine cycle, further increase of inlet temperatures is still expected in gas turbine field. In steam turbine field significant improvements due to the increase of temperature at turbine inlet are not expected.

At the same time some improvements resulting from reduced harmful influence of moisture content in low pressure turbine stages could be expected.

The aim of this paper is to summarize the facts which point out this stand-

513

point, and at the same time raise an interest for theoretical and experimental work in the field of two-phase steam flow in steam turbines.

2. STEAM CONDITIONS AT INLET AND OUTLET OF STEAM TURBINES

Main factors of increased steam turbine plant efficiency are steam parameters the temperature T_o and pressure p_o at the inlet, and pressure p_c at the outlet. By analyzing the changes in the applied scheme of Rankin cycle regenerative feed water heating and reheating as well as the aerodynamic influences of steam flow apparatus are not taken into account.

In case of the pressure in condenser, for instance $p_c = 4$ kPa is considered to be constant, the information about isentropic heat drop H_o and thermal efficiency η_{Ho} of the Rankin process, depending on the temperature T_o and pressure p_o of steam at the turbine inlet, can be obtained from on Fig. 1.

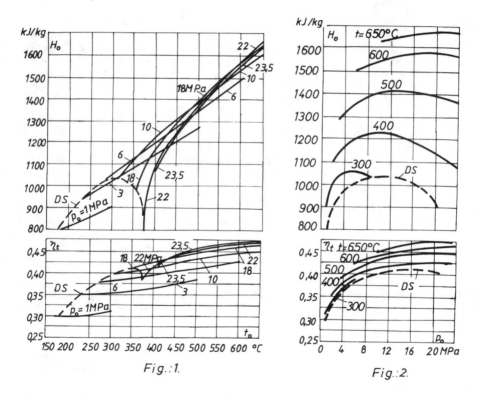

Fig.:1.   Fig.:2.

The increase of temperature T_o exerts positive influence on the heat drop H_o and η_o rise and they are the most important magnitudes for energy conversion.

Influence of the pressure increase p_o for the same process is presented in Fig. 2. However, the existence of the optimum pressure for each temperature level and the fact that to a higher temperature level higher optimum pressure belongs is obvious.

From Fig. 3., which represents a part of h-s diagram, we can notice that
at selected constant pressure p_c at the turbine outlet, the pressure increase
of p_o leads to the increase of heat drop H_o, until the point where tangent on
the temperature line becomes parallel with isobar p_c =const. represented in the
saturated region by a straight line. So, at least at first approximation, the once
chosen temperature T_o anticipates pressure p_o selection at turbine inlet.
Selection of temperature T_o depends on technological development, price,
availability and reliability of the parts of steam generator, turbine parts,
piping system and other material in use.

Maximum temperatures and belonging pressures were the limiting factors
during certain periods, which resulted in the following levels of the steam
turbine building development.

Until 1925/26 parameters at the inlet were about p_o =2,5 MPa, and 325-375°C.
Then front-connected turbines appeared with p_o =5,0 MPa and T_o =450°C, followed
by experimental plants with 9,0 MPa/411°C and 18,1 MPa/420°C. After 1930 the
plants with 10-11 MPa and 480-500°C were applied.

Comparing these realized parameters on the turbine inlets of some existing
plants, with optimal values on Fig. 2. it can be noticed that for the chosen
temperatures pressures in use are always lower than the theoretical optimal ones.
In other words, the designers chose that pressure which was assosiated with
lower heat drop, than maximal available. The reason for this lies in the fact
that the expansion in the turbine will lead to the region where steam is too wet
in the low pressure stages of the steam turbine, as it known from experience,
in case when expansion starts from the theoretically optimal parameters. This
takes place even under real conditions taking into acoount entropy increase
resulting from the imperfection of working medium and friction in the flow apparatus. The plants realised with inlet pressures chosen by experience gave,
with lower heat drop and its associated lower wetness on the turbine outlet,
better actual efficiency of the steam turbine plant. Negative influence of increased wetness will result not only in lowering the turbine efficiency, but
also in damaging the low pressure turbine blades and other parts of low pressure turbine because of errosion effect of dynamic forces, caused mostly by
steam droplet impacts. This effect appears due to steam droplet impacts and associated dynamic forces.

2.1 Reheated Steam

The application of the Rankin process with numerous bleeding for feedwater
heating and with reheating leads to the increase of turbine plant efficiency.
Reheated steam, with correctly chosen pressure and temperature has a positive
effect by enabling higher mean temperature at which heat is supplied. The conditions at turbine inlet can be moved to the left in h-s diagram, in order to
be closer to the optimal starting point. Inspite of this the outlet conditions
can be found even in the region of lower moisture content than if the reheating
was not applied, see Fig. 4.

The first plant with reheated steam appeared at the end of the 1930's
with the power of 50 MW and with 11 MPa/475°C. After 1945, due to the development of austenitic steel, the building of some plants that operated with temperatures at turbine inlet of 600°C or even 650°C was possible. Inlet pressures
varied even more, so that from 25 high pressure turbines built till 1960 some
were even built for overcritical pressures up to 30.0 MPa at rated, and up to
33.0 MPa at maximum loading. The majority of these turbines operated with pressures from 15.0 MPa to 20.0 MPa at the turbine inlet, developing power from 3 MW

for back pressure turbines to 275 MW for condensing turbines /1/. In this period and, the previous one important research had been done in the field of heat transfer, metallurgical tests and analyses of combined thermal and mechanical stresses in the hot high pressure turbine parts.

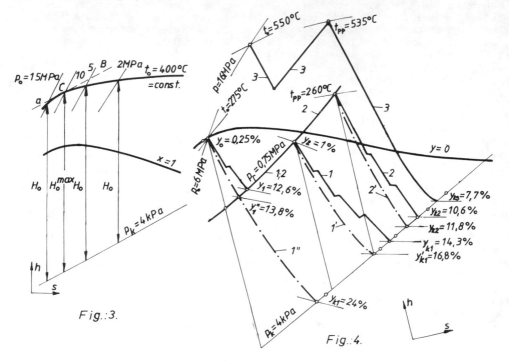

Fig.:3.

Fig.:4.

Usually, resuperheating appears widely in heat schemes, a large plants making possible the increase of η_{th} even as far as to 4%. Plant efficiency can be further improved by introducing double resuperheating which enables further increase of η_{th} for only 1-1.5%. The introduction of double reheating leads to a greater complexity of the plant. According to this the prices of turbine and steam generator parts increase and the reliability and availability decrease, so that economical effect can be anuled. Therefore this solution was applied in a limited number of units in USA and Europe. However, double resuperheating at this moment does not promise any remarkable progress in the actual exploitation of steam turbine plant.

2.2 The Influence of Other Factors

After 1960 the return to lower temperatures can be noticed. Good reason for it could be found in the costs increase of the materials which are required in connection with increase of inlet temperatures. On these high temperature levels the yield point of available materials is rapidly lowered, and at the same time the resulting creeping becomes evident. The temperatures above 535°C ask for the use of higher priced austenitic steel, alloy consisting of strategical materials. The poor heat conduction of this steel causes in this case also a price increase because of complicated welding procedures. Above all cracks occur frequently because of temperature asymmetry during sudden load changes. It is considered

/2/, /3/ that the price of the condensing steam turbine is increased for more
then 10% if, instead 500°C temperature of 535°C, and for further 7-9% if temperature of 565°C applied.

Especially for large power units, losses that occur because of unpredicted
shutdown soon take away collected savings, therefore after 1960 designers turned
to design turbines in many cases with the inlet temperature of 535°C and rarely
with 565°C. In the first case ferritic material and in the second case austenitic
materials have been used.

2.3 The Increase of Power Unit Size

The increase of power unit size, reduced number of low-pressure outlets,
number of casings and shafts i.e. saving of material and machining costs are the
main goals in turbine design. Turbines with unit-size from 60 to 100 MW were built
in Europe at the begining of 50's. There were the initial designs for further
development. In USA in that period 200 MW sets were built. In the second half of
the 50's in USA the unit-size of about 420 MW was attained while in Europe with
USSR included, the power unit-size was between 200 and 300 MW. The problems of
electric generator cooling led to the design of twin-shaft turbine of cross-com-
pound type. At the beginning of 60's these kinds of turbine were produced rated
from 550 to 600 MW, and at the beginning of 70's turbines from 800 to 900 MW were
introduced. Tandem compounded turbines grew with the lower rate, so that 500 MW
unit were found in that period. Today after 1100 MW type, twin-shaft turbines
reached the 1300 MW, and after the 800 MW type single-shaft turbine of 1200 MW
power unit-size is already in operation.

The nuclear power plants appeared in
1953 and the installed turbines in these
early plants were of small power unit-size.
It was soon found out that for the construction of a nuclear power plant with unit-size
under 600 MW, economic justification could
be hardly found. Nowadays nuclear power
plants have a turbine power unit-size from
900 to 1350 MW.

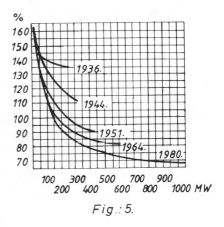

Fig.: 5.

Reduction of the specific price per
installed kWh depending on the enlargement
of power unit-size for oil fired classical
power plant is shown in Fig. 5. At the same
time by using twice as large unit, crew ex-
penses are cut down by half.

The tendency of building steam turbine
plants of as large unit-size as possible
seems to be justified, and is actually widely found.

2.4 Differences in Design Approach

Identical power unit-size of certain turbines does not include the same design and dimensional concept for each unit. Producers in USA, Japan and Korea,
because of the characteristics of their electric network system, are mostly
building turbines with the rotational velocity of $60 \, s^{-1}$. In their turbines identical mass of rotor blade will involve 1.44 time higher centrifugal forces in

comparision with forces found in turbines of $50\,s^{-1}$ rotational velocity. It can be expected that with the same materials and required knowledge,for the 50 Hz systems units with higher power unit size may be constructed. This is the case with gas turbine design and construction. With steam turbines for the $60\,s^{-1}$ electric network system this inconvenience is in a great number of cases partially compesated for by higher inlet pressures. The increase of steam velocity through the last blade row, has the greatest influence on the steam flow quantity, and thus on the unit-size. Another way for increasing the steam flow capacity is possible with the increase of pressure p_c in the condenser. Both actions have negative influence on the turbine efficiency. The possible increase of low-pressure turbine outlet numbers, brings about the opposite effect, because they increase power unit-size end efficiency with the same blade length. This is widely used by manufacturers of both 50 and 60 Hz types, but it also leads to the remarcable rise of the unit price.

Nowadays there exists an outstanding tendency to reduce the number of outlets as much as possible, although from the rise of fuel price the opposite tendency could be expected. The already mentioned increase of pressure p_c is frequently related to the limited cooling possibility due to cooling water shortage, especially in the industrially developed regions. Therefore the specific volume of steam is reduced. The quantity of steam flow, when using the same blade length, i.e. through the same cross-section of the last wheel outlet, rises, and the developed turbine power rises proportionally inspite of the reduction of the total heat drop.

There is an other way, which is quite efficient but less economical for achieving unit-size of a very large power, apply of the half-speed design, used especially for nuclear power plant steam turbines in 60 Hz network system.

2.5 The Influence of Moisture Content

The expansion lines for different types of turbines are represented in Fig.4. The first initial state is related to the nuclear power plant. Here the expansion must be interrupted at some mean pressure in point 1, and an outside moisture separation up to the point 2 or superheating up to point 3 is to be carried out, in order not to exceed the normally permitted limit of the moisture content. This is moisture content of 14% in extreme, and 10% in acceptable case. Considering steam turbines on fossil fuel and resuperheated steam, outlet steam moisture content of 7% is considered to be usual. Although it is not drawn in Fig. 4., the moisture content at outlets of turbines without the reheating (fossil fuel) could be expected in the 10 to 12% range.

It is obvious from expansion line for the nuclear turbine that internal separation in some stages takes place, so that the stepwise h-s diagram readings refer to the reduced flow quantity, and no more to one kilogram of steam carried to the inlet.

For some determined turbine parts during different operating conditions, essential deviations from thoese obtained by calculations moisture content for given rated power can be expected. This appears for example by lowering the pressure p_c in a condenser at varying climatic conditions, where a drive during the winter period even more aggravates the effects due to greater wetness of the steam on the outlet.

What is the amount of theoretically possible improvement of efficiency, if all negative influences of wetness could be totaly excluded? At an existing 900 MW fossil fueled turbine, arbitrarily the wetness losses appear in the last and the fore last stage. If the losses due to wetness could be entirely eliminated, internal efficiecy of this turbine could be improved for about 1%. Equivalent losses are far more important for turbines in nuclear power plants, altough that here the peripheral velocity of some wheels are considerably lower. Here they are present mostly in all, and not only in the last two stages.

An additional consequence of loss reduction in the wet steam flow will be the reduction of damages caused by erosion on a flow apparatus and on other turbine parts. The reduction of the harmful influence of wet steam flow, will lead to direct improvement of internal turbine efficiency, additional lower distortion, especially on rotor blades and slower growth of losses, caused by erosion effect in time. Advanced erosion causes damage on aerodynamical profile of rotor blade, resulting in the growth of aerodynamical losses. All these losses cannot be avoided, but as far as the turbines of large unit-size are concerned, even a small proportional savings will represent a significant absolute amount. The decrease of losses caused by wetness, could be attained only when the laws and rules valid for wet steam flow were better known, and when counter-measures for eliminating harmful influences of wet steam on the energy conversion in steam flow apparatus were discovered.

Nowadays the steam turbine designer in his effort to improve successfulness of his constructions expect significant help from the development and research departements and scientific institutions, in the field concerned with wet steam presence and its understanding, especially in the design of the last stage turbine vanes and blades.

3. RESEARCH IN THE WET STEAM REGION

Research in the wet steam region can be divided into two main fields. The first deals with research in order to define the rules valid for wet steam behavior, conditions for its occurence, states and distinctive forms where the presence of liquid phase is manifested. The second research field is related to the problems of erosion and corrosion caused by wet steam flow in steam turbines.

3.1. Formation of Wet Steam and Its Behavior

The necessity of studying phenomena arising in wet steam flow has appeared since the very first beginning of steam turbine design. The great number of turbines built in this pioneer period operated with inlet steam conditions located in the vicinity of saturation line. Thermodynamic calculations were carried out under the hypothesis of steam expansion under equilibrium conditions. Observing these turbines during operation, differences between results received by estimation and experimental measurements were indicated. Already Stodola /4/, has stated that the events connected with the realised wet steam expansion, differ from those expected at expansion at equilibrium conditions.

The classical thermodynamics approach is based on fundamental hypothesis that the conditions of the wet steam flow are changing quasistatically and that a liquid phase is present in a fine dispersed condition, equally distributed in amount and direction.

Later development in steam turbine design is based on a more accurate knowledge about events connected with the wet steam expansion. This phenomenological approach to the thermodynamic events, is based on the first two thermodynamic laws as well as on Kelvin's theorem on temperature scale. Such an approach does not require either examination of molecular medium structure or

defining the influences of interior mechanisms of examined phenomena. Macroscopic
conditions are discussed, thermodynamic analyses are limited to the examination
either of equilibrium systems and processes, or to processes developed in conditions insignificantly different from instantaneous equilibrium ones /5/, /6/.

During expansion from superheated to saturated region, subcooling and supersaturation of steam occur. Depending on expansion rate and heat drop at the first
moment during the passage through saturated region steam condensation does not
appear. Instead of condensation the subcooling of steam is realized. The steam
further behaves as a superheated steam, reaching the lower temperature from the
corresponding temperature of saturated steam. Stodola /4/ reports about experiments where the temperature of subcooled steam was for ΔT=15 to 25 K lower than
that of saturated steam at existing pressure. Recent results report even of ΔT
from 30 to 45 K. /5/.

Reaching the Wilson's point, the steam starts to condensate. Geometrical
places of Wilson's points obtained under the same conditions form Wilson's line.
With different expansion rates, different Wilson's lines are obtained and all are
placed under the saturation line in Wilson's area.

Wilson's lines for defined expansion rate $\dot{p}$ is approximately parallel
with saturation line, but with the increase of $\dot{p}$ this line is more and more
distant from the saturation line.

$$\dot{p} = -\frac{1}{p}\frac{pdp}{dt} = \frac{\varkappa}{\varkappa-1}j \, . \quad (1) \quad \text{whence} \quad j = \frac{1}{h}\frac{hdh}{dt} \, . \quad (2)$$

h = enthalpy, p = pressure, t = time and $\varkappa$ = exponent of isentropic expansion.

Although the subcooled steam properties can be approximately determined by
extrapolation of superheated steam properties, the subcooled steam structure is
different from that of the superheated steam. During steam subcooling coagulation constantly occurs. Under certain conditions coagulation could pass to a
condensing nucleus and condensation process will start. The enthalpy of subcooled steam differs from that of the superheated steam enthalpy. Specific volume
of generated wet steam is greater than specific volume of corresponding subcooled
steam because of developed liquid phase. At the same time the temperature of the
equilibrium process is higher because the admission of latent heat due to condensation. Available work done during expansion of subcooled steam is lower if
compared with work done from expansion at equlibrium.

The above mentioned facts are valid only if the starting point of steam
expansion is in the superheated region. In a number of cases examined expansion
in steam turbines starts in the saturated region, with two-phase medium, consisting of gaseous and liquid phase.

The liquid phase is composed of water droplets, which move with the velocity
c' that differs from gaseous phase velocity c''. The velocity difference influences the wet steam conditions from the beginning of expansion. But even if
this starting velocity difference is neglected, the gaseous and liquid phase expansion, between the same initial pressure p_1 and final pressure p_c, will result with considerably higher acceleration of gaseous phase, comparing with the
acceleration of wet steam created during the liquid phase expansion. This is
valid especially if steam deceleration and droplets acceleration caused by mutual
friction are neglected. Acceleration of droplets created in the gaseous phase
expansion is taken into account, because they arise with condensation of already accelerated subcooled steam.

With described expansion of wet steam, subcooled steam with arising liquid phase in clusters, mist, fog or droplets, was not obtained. Instad of this, here is a mixture of subcooled steam in which liquid phase arises later, moving with high velocity c'', with a mixture of droplets and evaporating steam moving with much lower velocity c', obtained at the nozzle outlet.

If the velocity difference at nozzle inlet is taken into account, which could be important, exact definition will be difficult to obtain. However, by obtaining experimentally necessary relations for calculations of occurences in turbine flow apparatus, it can be seen that velocity c' in a broad range is approximately 0.1 to 0.25 of the velocity c''. For example for a pressure drop from 0.2 MPa to 0.1 MPa velocity c'=0.15 c''.

Diagrams and tables of thermodynamic properties are made for equilibrium state and conditions of two-phase medium, and they assume equal gaseous and liquid phase velocity. They do not take into consideration the energy which is consumed during liquid phase acceleration. In reality this part of energy consumed due to friction increases the entropy of the system. When in the saturated region diagrams and tables are used this energy consumption should be taken into account in relation on the liquid phase quantity on the nozzle inlet.

If dry saturated steam at inlet condition expands as subcooled steam, losses connected with droplets acceleration do not exist. In this case higher steam velocity would be expected. But the velocity is lowered because of subcooling. Considering these both phenomena, the real gaseous phase velocity does not reach the one, which would be realized by expansion in equilibrium conditions. The phenomena already mentioned, even treated in a very simple way point out to the complexity of analysis and evaluation of these occurences, even if they appear in rectilinear nozzle at steam exspansion in wet steam region.

The complexity of these occurences is even greater due to the very short time period of only a few mili seconds in which they appear. Tiny particles flow in a quantity of millions, they change the form and exchange the heat in these short periods of time. /7/.

In spite of these facts, these phenomena have been widely described in literature and examined experimentally.

3.2 Kinetic of the Condensation Process

Interdependency among radii ξ of droplets in thermodynamic equilibrium with surrounding gaseous phase and partial steam pressure is expressed by the Kelwin's equation /5/:

$$\ln \frac{p''}{p_s} = \frac{2 G}{\xi \varrho' RT}$$
(3)

where is p''-gaseous phase pressure, p_s-saturation pressure over flat dividing surface between phases at given temperature T, T-temperature required for thermodynamic balance between the phases, different from the temperature at which the heat transfer occurs. R-gas constant for steam, ϱ' liquid phase density, and G surface tension.

Thermodynamic theory of phase changes deals with the condensation process only in the conditions of balance between the original and the forming phase. The time sequence of the condensation process is not included in this analysis.

Similar dependability between temperature and droplet radii ξ is given by equation:

$$\ln \frac{T_s}{T''} = \frac{2\sigma}{r \xi \rho'} \tag{4}$$

where is T_s-saturation temperature over flat surface dividing the phases at given pressure, T''-gaseous phase temperature, r-heat transfered between phases.

The Kelvin equation is based on the supposition that the thermodynamic phase potentials are in balance with the influence of capilarity forces at given constant temperature and constant pressure.

Taking into account the supposition that surface dividing phases is flat, the Clausiuss - Clapeiront equation, representing conditionally the situation of droplet with infinite radius, can be applied:

$$r = T(v'' - v') \frac{dp}{dT} \tag{5}$$

where is r heat transferred between phases, T-temperature, v'' - gaseous phase volume, v' - liquid phase volume, p-pressure.

Using equation (3) for droplet radii ξ =1μ and equation (5) for the calculation of the phase balance curves for steam will give approximately the same result. This is in favor of the supposition that droplets with radii $\xi > 1$ can be treated as liquid with flat surface with steam above.

For given conditions of state it is possible to calculate the critical droplet radii ξ_{cr} by means of the Kelvin equation.

$$\xi_{cr} = \frac{2T\sigma}{r \rho' \Delta T} \tag{6}$$

where is $T = T_s - T''$ - subcooling rate, $T = T' = T''$ wet steam temperature.

In case when the droplet radii are smaller than critical, liquid phase thermodynamic potential grows at given conditions. In time this metastable state is changed into the stable state because of inevitable equalization of thermodynamic potentials of phases which causes the droplets to evaporate.

In case when droplet radii are larger than critical, the liquid phase termodynamic potential diminishes and thus due to the equalizing process, the droplets grow with further condensation.

In order to obtain information about the condensation process, it is necessary to know the deviations from the stable state, at which the liquid phase occurence and growth are possible. For this purpose the thermodynamic methods must be extended by molecular - kinetic theory of phase occurence.

If the gaseous phase molecular free path length is proportional to the droplet dimensions of the liquid phase contained in wet steam, molecular free path mean length and radii of the present droplets with their dimensions, have essential influence on the condensation process.

The mean length of molecular free path $\bar{\lambda}$ is given by:

$$\bar{\lambda} = \frac{\mu''}{p} \sqrt{\frac{9\pi RT}{8}} \tag{7}$$

The rate between the mean length and droplet diameter is known as Knudsen's number

$$Kn = \frac{\bar{\lambda}}{2\xi} \tag{8}$$

On the basis of this Knudsen's number, it is possible to define the approach to the condensation process as macroscopic, in cases in which $Kn < 0.01$. In this case the gaseous phase behaves in relation to the droplets as a continuum. For $Kn > 4.5$ the gaseous phase behaves to the droplets as gas with free molecules and this should be considered as a microscopic state. Unfortunately there is a wide range between $Kn = 0.01$ and $Kn = 4.5$, which represents the so called transition state. This range is widely present at low pressures in the last stages of steam turbine. Here the temperature field is rapidly changing, the heat transferred between phases should be extracted, thus droplet temperature in the condensation process should be higher than the gaseous phase temperature. The various temperatures should be defined: T'-droplet temperature, T''-gaseous phase temperature, T_s - saturation temperature of droplets of certain radius, T_s-saturation temperature for liquid phase in condition of a flat liquid surface.

It is important to mention at this point, that in case of very small droplets, the temperatures T'' and T_s could differ to a great extent.

With sufficient accuracy the supposition can be established /5/, /10/, that the droplet temperature T' and saturation temperature T_s calculated from equation (3) will not differ for the given rate p_s'/p_s.

The droplet growth velocity depends on the temperature difference $T' - T''$ and Knudsen number splits the velocity range into two parts. When $Kn > 1/2$ the growth velocity could be expresed by:

$$\frac{d\xi}{dt} = \frac{c_p'' \, \rho''}{r \, \rho'} \sqrt{\frac{R T}{2}} \quad \frac{T' - T''}{1 - \frac{2}{r\rho'\xi}} \quad \frac{1}{\frac{c'' - T''}{r(1 - \frac{2\sigma}{r\rho'\xi})}} \tag{9}$$

where Cp is specific heat of gaseous phase at isobaric conditions.

If $Kn < 1/2$ then:

$$\frac{d\xi}{dt} = \frac{\lambda''(T' - T'')}{r \, \rho' \, \xi} \tag{10}$$

where λ is heat conductivity coefficient of gaseous phase.

In the case when $Kn < 1/2$, the droplet grouth is proportional to the time t at the begining of the condensation. When $Kn > 1/2$ the growth is proportional to time square root $\sqrt{t}$.

At the beginning of the condensation process the condensation nucleii are molecular clusters, small droplets or impurities existing in gaseous phase /7/. The velocity of the condensation nucleii formation can be expressed as the sum of nucleii created in one cubic meter of the flowing medium per second. This can be calculated from Frankel's equation /11/:

$$I = \delta_I \, e^{-\frac{4\pi\sigma}{3kT''}\xi_{cr}^2} \, \frac{1}{\rho'} \, \left(\frac{p''}{kT''}\right)^2 \, \frac{2\mu\sigma}{\pi N_A} \tag{11}$$

where is k - Boltzman's constant, N_A - Avogardo's number, μ nuclear mass and δ_I correction factor.

The velocity defined in such a way depends on steam pressure and subcooling degree ΔT, which also influences ξ_{cr}. In the range of smaller subcooling the amount of created condensation nucleii is negligibly small, thus there is nearly no condensation. The number of droplets rapidly grows in the region of high subcoolings. In Fig. 6., this dependency is shown by the line 1 for pressure 10 kPa and for the pressure of 1 MPa with the line number 2. The broken line shows the influence of changing of surface tension which changes with droplet radii growth. This causes intensification of the condensation process even at small subcooling.

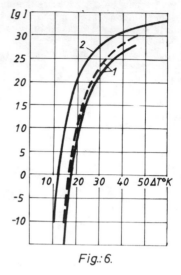

Fig.: 6.

In real conditions, at the begining of expansion, the steam can contain a certain amount of moisture .

During expansion in vanes of various shapes, the velocities of different values and the various grades of dispersion of the liquid phase in different directions with regard to dry steam velocity can be achieved. The droplets can lose stability and desintegrate.

Weber's number defines the stability characteristic,

$$W_e = \frac{d(c'' - c')\, \varrho''}{\sigma} \qquad\qquad (12)$$

where d-droplet diameter, $c'' - c'$ - velocity difference between dry steam and liquid phase.

Depending on the flow conditions the critical value of Weber's number is in the range of 9 to 18 (12). If $We < We_{cr}$, the droplets are stable, under the condition, that the difference between droplet and dry steam velocity, causing the droplets desintegration, is established in a time sequence, which is comparable with a sequence of the droplet free standing vibration frequency. If, on the contrary, the actuating time of dry steam phase on the droplet is very short, it is not obligatory that the droplet disintegration will occur even when $W_e > W_{cr}$.

In conditions of supersonic and transonic flow the non-stationary shocks are present. The periodical non stationary characteristic of this process dictates the pulsation of flow conditions with the frequency of 500 to 2000 Hz. This leads to very intesive heat transfer and creates condensation jump. Thus, the condansation jumps theory does not explain the physical nature of the process /5/, the condensation does not often occur suddenly, so that in this case subcoling is also partially present.

4. WET STEAM FLOW IN STEAM TURBINE FLOW APPARATUS

The exemplified approach to the phenomena of wet steam flow includes a number of simplifications. A more detailed description, connected with the flow in turbine cascades, requires the application of two phase fluid flow dynamics, i.e. the continuity equation, the conservation of impulse and energy equation, and the equation of gaseous phase state, in addition to formerly quoted equations

for the velocity of condensation nucleii (11) and the velocity of droplets
growth.

This interlinked system of equations can be solved by means of a computer
for a given turbine cascade geometry /13/.

However, the question still remains, how much the obtained results differ
from the real situations that are realized during the real two-phase fluid flow.

In each of the turbine cascades operating under the saturation line, the
liquid phase is present either in the shape of very fine dispersed droplets,
i.e. mist, fog or droplets or even in a thin stream of liquid which flows on
the cascade profile surface or free in intermediate spaces.

The dispersity of the liquid phase is characterized by the size of droplets,
and the mean diameter of present droplets is taken as the measure of dispersity.

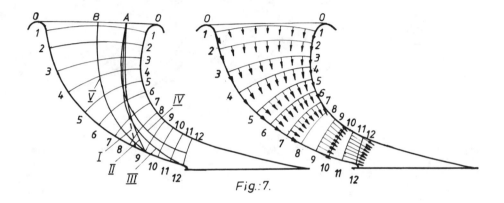

Fig.:7.

The equation valid for one dimensional flow cannot be here succesfully applied. The part of the liquid phase content, which is in the shape of relatively
large droplets, has the trajectories essentially different from the steam phase
stream lines, Fig.7. All this can be particularly seen in the relative flow at
the rotor cascade inlet and at the folloving stator cascade inlet respectively
in Fig.8. Fig.7. shows a) the steam velocity field and b) the liquid phase velocity field and trajectories. Line I shows the path of droplets of 100 μ radii,
line II droplets of 10 μ radii, line III that of 1 μ radii and IV and V a path
of droplets with 0.5 μ radii.

Supposing that at stator cascade inlet, the velocities of both phases are
equal (c'=c") and with ideal direction of inlet angle α_0, the inertia forces
will bring the droplets across the steam phase stream lines in contact with
concave surfaces of profiles. The droplets will gather on these surfaces. The
steam phase dynamic forces will continuously, while other arriving droplets will
discontinnously separate them from these surfaces and bring them in the stream
of the steam phase. The secondary flow will stimulate the gathering of the
liquid phase in the corners of cascade channels. The major part of the liquid
phase will thus be found in the vortexes of the profile leaving and trailing
edge wakes. Fig.9. shows the flow of droplets: 1-in wakes, 2-and 3-decollated
from profile surface and 4 reflected from water layer on profile surface. The
largest droplets are contained in stream 1 and 2 and they have the smallest

velocities. The distribution of these droplets over the vertical cascade cross-
-section depends on cascade geometry and secondary flow conditions. (14).

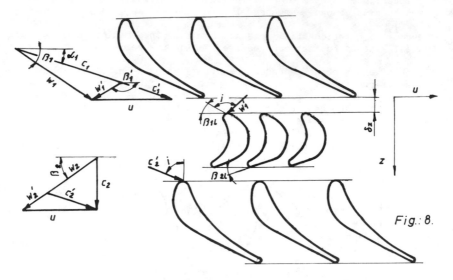

Fig.: 8.

Part of the practical problems dealing with the droplets flow can be
solved, under the limitation on action of aerodynamical forces only, and on supposed spherical shape of droplets. For different ranges of the Reynold's number
it is possible to establish for certain practical cases, quite simple equations
describing the droplet flow in the steam turbine flow apparatus spaces of interest /5/, /15/.

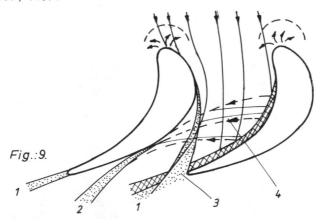

Fig.: 9.

The particular characteristics of two-phase flow through the vanes can be
divided in the folowing way:

a) Wet steam expansion with subcooling changeable not only in axial direction, but also perpendicular to the axis and along the height of the channel.
b) Previously discribed phenomena of droplets growth or evaporation, while the trajectories are different from steam phase stream lines.
c) The formation of the liquid layer of different shapes and thickness on profile surfaces and channel thorus surfaces.

In addition, friction and heat transfer as well as mass transfer between the phases, are also present in the flow through the cascade channels.

This results in constant change of local and integrated conditions on the cascade inlet and outlet thus changing the flow coefficient and losses which considerably differ from those expected when superheated or saturated steam flows through the same cascade.

Behind the stator cascade, under the influence of size of axial clearance, the effect of relative radial elevation, especially at small outlet angls α_1 is present. The local moisture content is higher towards the external circumference. Correct knowledge of this phenomenon is neccesary for the future design of efficient internal moisture separation devices.

It is important to know here the sliding coefficient, i.e. the earlier mentioned difference in phase velocities, represented as the ratio between the liquid phase velocity c', and steam phase velocity c". This coefficient can be obtained for mean values of c' by calculation.

The evaluation of experimental results by means of losses analysis shows that this coefficient is up to 50% smaller than if obtained by calculation for mean values of c'. This is explained by the fact that the droplets in the profile wake have velocities much smaller than the mean velocity c' and that approximate estimated droplet dimensions has substantial effects, neglected in theoretical calculation.

The droplets of different diameters d are unequally distributed along cascade pitch t, and the distribution is different for different pitches. In Fig.10, along the pitch on one side the large droplets of 125 - 175 μ can be observed. They originate from the profil trailing edge. Further on, the smaller droplets of 60-80 μ are present. The size of the droplets grows again to the previously mentioned dimensions in the vicinity of one third of the pitch length, due to droplets leaving the back side of the profile. Further on, once again, the smaller droplets are present, and at the end of the channel pitch again the larger droplets, coming from the concáve profile surface, can be recognised.

The cascade with a small pitch will show a smaller difference in droplet size, while the cascade of greater pitch, characterised with light-through passages, will show one more jump in droplet size, due to passage of droplet directly from the profile back. In the case of small pitch, this jump does not exist because these droplets will be stopped and absorbed in the liquid layer existing on the concave surface of the next profile.

Fig.10 (2) shows one of the experimental analysis on the cascade of profiles, tested at moisture content y=0.06 and with pressure ratio ε =0.75.

At the rotor cascade inlet the negative effect of wet steam flow in steam turbines is noticable. As it can be seen from Fig.10, due to the essential difference between velocities c' and c", it is imposible to obtain favorable steam phase flow without collision and at the same time avoid heavy collision from liquid phase particles with rotor blades. Large droplets coming from profile leaving edges collide at the wide angle with rotor blades, due to smaller velocity. ·

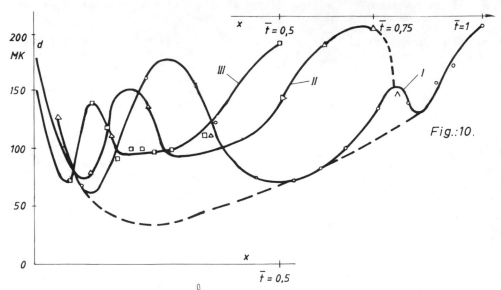

Fig.:10.

The sliding coefficient ν and cascade reaction degree ϱ_T are of greater influence to theoretical maximal collision velocity component $w_{sh\,max}$ for roughly dispersed droplets. This velocity will occur at stage reaction ratio $\varrho_T = 0.5$ and angle $\alpha_1 = 90°$. The value of this collision component will be:

$$w_{sh\,max} = (1 - \nu)\,u \tag{13}$$

where u is peripheral blade velocity.

The finely dispersed droplets have the velocity vector which is approximately the same in value and direction with the steam phase velocity vector.

The droplets flow in the rotor cascade is under the influence of inertia, friction, aerodynamic and gravity forces. For roughly dispersed droplets which came into contact with rotor blade surfaces, the forces of inertia due to high circumferential velocities are many times greater than gravity and even aerodynamic forces. Thus, the latter forces can be neglected in the analysis of droplets trajectories in rotor cascade.

The rotor blade surface can be substituted by narrow flat stripes, tangential to the profile, which form different angles β to the circumferential velocity plane, to enable the analysis of droplets trajectories under the influence of inertia forces with approximate friction influence /5/. Friction forces are taken as proportional to droplet velocity, and inertia forces include centripetal and Coriolis acceleration.

At the moment of impact between droplets and the rotor blade surface, very complex local phenomena take place. The angle β of the supossed stripes relative to the plane of circumferential velocity u, is of great influence on the character of droplet trajectories. That is the reason why this method can be used only conditionally for determining the initial sliding component $\overline{w}_{xo}$ of the relative droplets velocity w'_1.

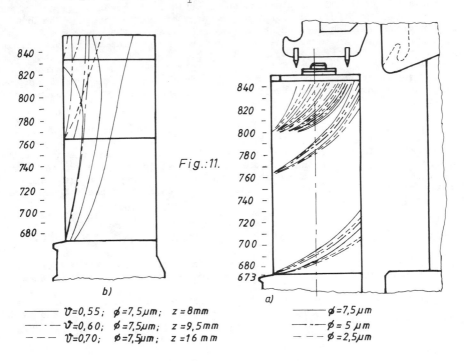

Fig.:11.

b)

———— $\mathcal{V}=0{,}55$; $\phi=7{,}5\mu m$; $z=8mm$
—·—— $\mathcal{V}=0{,}60$; $\phi=7{,}5\mu m$; $z=9{,}5mm$
———— $\mathcal{V}=0{,}70$; $\phi=7{,}5\mu m$; $z=16\ mm$

a)

———— $\phi=7{,}5\mu m$
—·—— $\phi=5\ \mu m$
———— $\phi=2{,}5\mu m$

At the rotor cascade inlet, where the angle $\beta < 90°$ some droplets may be rejected in the direction opposite to steam flow, towards the former stator cascade. Sometimes this can lead to the erosion of stator blades. Repetition of such collisions from rotor to stator blades and vice versa increases the amount of mechanical losses. If $\beta \approx 90°$ the Coriolis forces are perpendicular to the rotor blade surfaces, thus the droplets trajectories are dependent on the velocity component w_{xo} magnitude i.e. from initial conditions. At the outlet of the rotor blade profile $\beta < 90°$, the droplets are directed towards the trailing edge due to the Coriolis forces influence. This effect increasees in the vicinity of the trailing edge. This phenomenon is greatly emphasized in case of twisted blades. Due to the requests of the three dimensional flow the angle β incrcases from the rotor blade root towards the tip. Fig. 11. /16/ shows experimental results of tracing the traectories on a) impulse stage, with changeable sliding coefficient $\mathcal{V}$ and changeable axial clearance z and b) reaction stage with change in droplet diameter.

The velocity growth rate for $\Delta w'_x$ (the axial sliding component on the previously supposed stripe which is placed under angle β on the rotor plane) and for $\Delta w'_r$ (the radial component on the same plane) are of the same sign for $\beta > 90°$ and of opposite sign for $\beta < 90°$. During the translation in radial direction the radial velocity component w'_r increases, thus increasing the effect of the Coriolis forces. Here the rate between $\Delta w'_r$ and circumferential velocity u is of great importance.

The Coriolis forces tend to separate the droplets from the blade surface at the concave side of the profile.

In the space behind the rotor cascade there is the mixture of droplets of various sizes, velocities and direction of motion. The greatest amount of moisture drops off from trailing edges in the shape of large droplets. The droplets come from the trailing edge under the angle similar to the cascade angle β_2, and certain angle γ to the cascade diameter. The rest of the droplets which were carried by the steam flow and did not even come into contact with the cascade surface, or were rejected from it, are also present behind the rotor cascade. The aerodynamic forces tend to force them in the direction of the steam flow, but the angles of their velocities could be quite different from cascade outlet angle β_2.

It is very difficult to establish the laws of droplets behaviour in this zone, although it is of great interest to attain this knowledge because of possible design solutions for moisture separation. Draining away the separated moisture can diminish the harmful influence of liquid phase on the next turbine stage. The efficiency of separation is dependable on droplet radial elevation and the ratio of this elevation to the traveled path of droplets behind the rotor blades cascade.

5. MOISTURE LOSSES

Moisture losses consist of thermodynamic losses and brakeage effect losses. These losses are additional to the losses in the simple single phase flow.

The real wet steam expansion process differs from the equlibrium expansion process. Different heat drops, h_o in the equilibrium and h_T in the expansion process with subcooling define the magnitude of thermodynamic losses Δh_{sc}.

$$\Delta h_{sc} = h_o - h_T \tag{14}$$

From here the thermodynamic loss coefficient is obtained.

$$\xi_{sc} = \frac{\Delta h_{sc}}{h_o} = 1 - \frac{h_T}{h_o} \tag{15}$$

Depending on the expansion ratio $\varepsilon = p_1/p_2$ in the range of 1 to 5, this coefficient has values from 0 to 0.08.

Different dynamic losses appear due to mutual action between the steam phase and the droplets. The largest part of these losses is spent on roughly dispersed droplets, especially in the space between cascades, while acceleration losses of the fine dispersed droplets can be included in the velocity coefficient f. The coefficient of dynamical losses is defined as:

$$\zeta_d = \frac{\Delta h_d}{h_o} \tag{16}$$

where Δh_d is energy expended on droplet acceleration. This can be expressed with simplification as:

$$\Delta h_d = \vartheta y_R c_1^2 \tag{17}$$

by means of sliding coefficient, moisture content in the shape of roughly dispersed droplets y_R, and flow velocity c_1.

The brakeage effect losses appear as the result of impact between droplets and the rotor cascade. The magnitude and the sign of the brakeage force N_b can be defined by equation:

$$N_b = \int_{r_a}^{r_b} (u_1 c'_{1u} dM_{B_1} - u_2 c'_{2u} dM_{B_2}) \qquad (18)$$

where M_{B_1} and M_{B_2} is that part of the liquid phase mass flow which consists of roughly dispersed droplets, at cascade inlet end outlet. This part cause the brakeage effect. Values of internal and external radii of cascade bordering cylindrical surfaces are represented by r_a and r_b. With a lot of simplifications as $M = M_{B_1} = M_{B_2}$, and the supposition that the droplet velocity on the outlet c'_{2u} equals the peripheral circumferential velocity u_p, further that $u_1 c'_{1u} = \text{const}$ (equal circulation stage) and with as previous $c'^p_1 = c_1$, the equation (18) after integration changes to:

$$N_B = M_B (\sqrt{} u\, c_{1u} - u_p^2) \qquad (19)$$

The brakeage loss coeficient is:

$$\zeta_B = \frac{N_B}{M_B h_o} = 2 y_B \left(\frac{u_p}{c_o}\right)^2 - \frac{\sqrt{} u c_{1u}}{c_o} \qquad (20)$$

It is obvious that increase in sliding coefficient Ψ will diminish the brakeage loss, but at the same time, the dynamic loss spent on acceleration of droplets will be greater. For this reason, that two losses should be analysed together.

The efficiency coefficient of the turbine stages which operate in the wet steam region can be expressed as:

$$\eta_{wet} = (1 - y_B) \eta_A - (1 - y_B) \zeta_{sc} + (1 - y_B) \zeta_d + \zeta_B \qquad (21)$$

where the η_A is the efficiency coefficient of the steam phase flow with fine dispersed droplets.

Dynamic and brakeage losses are mechanical losses. They are under the strong influence of turbine stage axial clearances and the cascade reaction ratio ζ_T. They are increasing with increased ratio of u/c_o.

For practical purposes the coefficients a_1 and a_2, widely used, can be introduced as:

$$a_1 = \frac{\Delta \eta_{wet}}{y} \qquad \text{and} \qquad a_2 = \frac{\Delta \eta_{wet}}{y\, \eta_A} \qquad (22) \text{ and } (23)$$

In this case the wet steam efficiency coefficient is:

$$\eta_{wet} = \eta_A - a_1 y = \eta_A (1 - a_2 y) \qquad (24)$$

This evaluation of losses due to the wet steam flow is also given under simplification needed. The impact of droplets causing brakeage effect was treated as nonelastic one, which is true only to a limited extent. The elastic impact, taking place in reality, causes up to two times greater brakeage, depending on the distance of the impact from the blade tip. Theoretical results differ from the experimental as well as from the real ones. The values of a_1 and a_2 are strongly influenced by turbine stage design characteristics, flow conditions and operating regimes. More detailed determination of these coefficients, which are of great importance for turbine designers, requests large, complicated, and expensive experimental research.

6. RESEARCH METHODS

Experimental research methods dealing with wet steam flow can be divided into direct methods, with measuring in the wet steam flow and indirect methods, with measuring on the sample taken from the wet steam flow.

The methods used for determination of droplet sizes and distribution are based on optical, electrical or inertial properties of the wet steam flow.

Light scattering or laser beam application enables the analyses of droplet sizes, amount and distribution. A wide application is expected from holografic method.

The print method is based on inertia principle. A small plate covered with soft material layer is exposed to the impact of droplets in the wet steam flow, thus the prints are registered at the places of impact. The plate is exposed in the wet steam flow by means of a special probe which enables the change in exposition time, from one thousandth from a second up to a few seconds. For the proper probe orientation the auxiliary probe for the droplets flow distribution along the blades is used. The analysis of prints gives the droplet number per unit area, droplet sizes and distribution.

The next point of great importance is the determination of the wetness or the moisture coefficient y. The range of interest is between y=0 and 0.20 with a wide variation of pressures, temperatures and velocities of wet steam flow. The applicable methods are based on the throttling effect, electrical calorimetrics, measuring of electrical properties, radioactive tracers and radioactive beams. The methods, based on measuring of electrical properties are suitable for the determination of the droplet flow, local wetness coefficient, average wetness coefficient and local liquid layer thickness.

All mentioned methods have certain imperfections and limited field of application. To take samples where necessary is particularly complicated. Anyhow the constant improvement and application of more sophisticated equipment gives hope that the laws of wet steam flow in steam turbine stages will be better known, which will undoubtely lead to improvements in design and longer life of steam turbines.

7. EXPERIMENTAL RESEARCH INSTALLATIONS

The problems connected with wet steam flow in steam turbines are so complex that they must be investigated on experimental installations in conditions close to reality. The purpose of such research is:
   a) to determine the basic characteristics of cascades and turbine stages in different flow conditions,
   b) the improvement of turbine flow apparatus, and
   c) the improvement of internal separation arrangements and other protection

methods against erosion of blades, vanes and other steam turbine parts.

The profile cascades are investigated on stationary models, where modifications are easily made. There the details of flow patterns, droplet formations, liquid layer formation and two-phase flow are investigated. There is as another possibility, research based on the similarity criteria of phase transfer.

Experimental turbines enable the research of characteristics of one or more interconnected stages. Conditions created in these turbines are more or less similar to conditions in acctual turbines in operation. They are the main source of information of stator vanes influence on development, dispersity and motion of liquid phase particles during rotor blades rotation, especially with variable circumferential velocity.

During model testing care should be taken on similarity conditions: the coefficient of compressibility $\mathcal{H}$, Prandtl's, Reynold's and Mach number Pr, Re and Ma, expansion ratio ε , velocity ratio u/c_o, fan ratio, turbulence grade and in nonstationary conditions from Strouhal number Sh too.

It is evident that fulfilment of all these conditions at once is impossible. On an already constructed turbine variations in rotational velocity, due to variation of u/c_o ratio, variation of inlet conditions, due to realisation of different moisture coefficient y, and different Reynold's numbers due to outlet pressure variation is possible.

Because of economical reasons application of reduced scale models is preferable. Other reasons for such solutions are:
a) dimensions, velocity of rotation and steam conditions of tested turbine stages,
b) existing capacities of steam and cooling water sources,
c) dimension of area on which the experimental plant is to be located, and
d) quality level of measuring technique.

Testing the models connected with design of large power unit size $50\,s^{-1}$ range turbines, experimental turbines are running with outlet pressures p_c from 0.004-0.008 MPa, with coefficient of moisture content y of 0.11-0.14, rotor blade lenghts from L_R from 800 to 1400 mm with fan factor λ_R from 0.33 to 0.42 /17/.

Experimental turbines can be constructed with equal blade lenghts, blade roots on constant wheel diameter or with constant peripheral blade diameter.

The use of steam as a working medium in experimental turbines has many advantages enabling a wide range of variations of Reynold's and Mach number, while at the same time tests can be performed nearest to real conditions, following the tendency to identical values of coefficient of compressibility and Prandtl's number Pr.

On the other hand, choice of steam as a working medium introduces more complicated experimental installation, more complicated measuring instruments as well as the use of more sophisticated measuring methods. For these reasons combined usage of air and water is preferable. In this case problems connected with realisation of dispersion of water droplets in air as working medium arise.

Experimental turbines can be built with one or more turbine stages, with one or two shafts. The advantage of single-stage turbines is their simplicity, but turbines with more stages can realise conditions that are more similar to those in constructed multi stage steam turbines while in opperation. With the

	D	M1	M2	M3	M4	M5
$\dfrac{d_{Mi}}{d_D}$	1	1	$\dfrac{1}{\sqrt{m}}$	$\dfrac{1}{\sqrt{m}}$	*	**
$\dfrac{y_{Mi}}{y_D}$	1	1	1	1	1	**
$\dfrac{Re_{Mi}}{Re_D}$	1	1	$\dfrac{1}{m}\dfrac{p_{M2}}{p_D}$	$\dfrac{1}{m}\dfrac{p_{M3}}{p_D}$	$\dfrac{1}{m}\dfrac{p_{Mt}}{p_D}$	$\dfrac{1}{m}\dfrac{p}{p}\cdot(1-y_D)$
$\dfrac{M_{Mi}}{M_D}$	1	1	1	1	1	1
$\dfrac{1}{m_n}=\dfrac{n_{Mi}}{n_D}$	1	1	m	m	m	$\dfrac{a_{M5}}{a_D}\,m$
$\dfrac{1}{m_Q}=\dfrac{Q_{Mi}}{Q_D}$	1	$0{,}14 \div 0{,}18$	$\dfrac{0{,}14\div0{,}18}{m}\dfrac{p_{M2}}{p_D}$	$\dfrac{0{,}14\div0{,}18}{m^2}\dfrac{p_{M3}}{p_D}$	$\dfrac{0{,}12\div0{,}16}{m^2}\dfrac{p_{M4}}{p_D}$	$\dfrac{0{,}12-0{,}16}{m^2}\dfrac{p_{M5}}{p_D}\cdot(1-y_D)$
$\dfrac{1}{m_N}=\dfrac{N_{Mi}}{N_D}$	1	$0{,}1 \div 0{,}15$	$\dfrac{0{,}1\div0{,}15}{m^2}\dfrac{p_{M2}}{p_D}$	$\dfrac{0{,}08\div0{,}12}{m^2}\dfrac{p_{M3}}{p_D}$; $\dfrac{0{,}02\div0{,}04}{m^2}\dfrac{p_{M3}}{p_D}$	$\dfrac{0{,}02\div0{,}04}{m^2}\dfrac{p_{M4}}{p_D}$	$\dfrac{0{,}02-0{,}04}{m^2}(1-y_b)\cdot\dfrac{p_{M5}}{p_D}\left(\dfrac{a_{M5}}{a_D}\right)^2$
Scheme of installation		N_{M1} ; p_D ; Q_{M1}	N_{M2} ; p_{M2} ; Q_{M2}	$N_{M3,1}$; $N_{M3,2}$; p_{M3} ; Q_{M3} ; N_{M31}	N_{M4} ; p_{M4} ; Q_{M4}	N_{M5} ; p_{M5} ; Q_{M5} ; $a_{M5} > a_D$
Geometric scale	1:1	1:1	1:m	1:m	1:m	1:m
Turbine type	D	M1	M2	M3	M4	M5

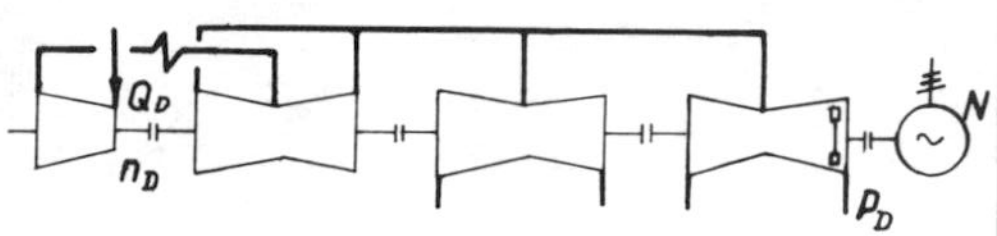

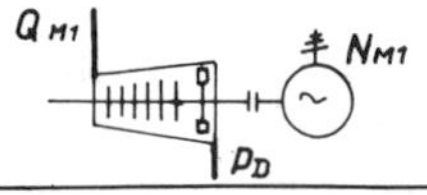

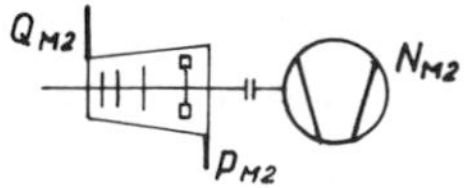

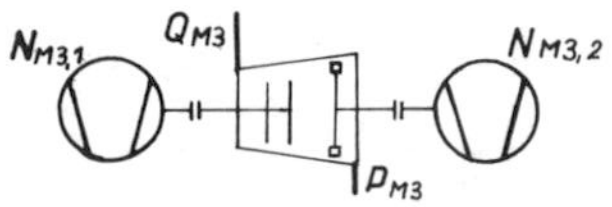

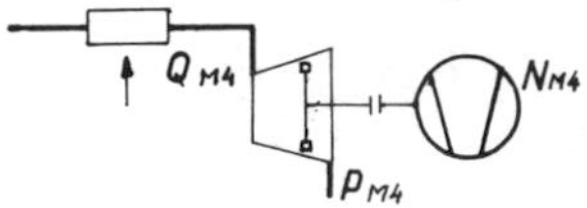

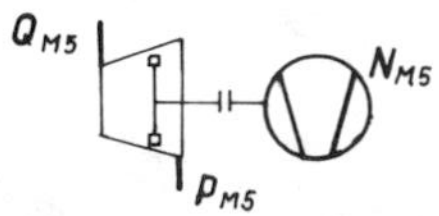

* Essentially different dimensions and of dropplets dispergivity
** Superheated steam

Fig.:12.

use of the two-shaft turbines it is possible to realise the desired inlet condition at examined stage inlet, at different rotational velocities. Inconvinient
levels of critical number of revolutions connected with overhang of some turbine
wheels, diminish sometimes their preference. Fig. 12.shows schematic presentation
of typical layouts of five experimental turbines, with presentation of influence
of geometrical scale m coefficient of pressure scale $m_p = P_D/P_M$, and relevant
ratios for similarity criteria $(Ma_D/Ma_{M_i}, Re_D/Re_{M_i}, y_D/y_{M_i})$.

Turbine M_1 is in fact a low pressure part of a real steam turbine in
m=1:1 scale. Due to large power unit size this unit delivers   large quantity
of energy during testing and must be connected to an electric generator for
economical reasons. The large steam generator, large condenser and the large
turbine itself, are the reason while high costs are connected with the use of
such an instalation. A more convenient turbine is M_2, where m>1, connected
with hydraulic brake. The two shaft turbine M_3 which is more expensive, includes earlier mentioned advantages and disadvantages. Turbines M_4 and M_5 are
the simplest single stage units. In the simple turbine M_5 experiments with wet
steam cannot be realized. But, however, if an additional outside wet chamber is
connected the transformation of this turbine in M_4 type is possible. In this
way the wet steam flow, similar to this in real turbines, can only approximately
be realized.

Fig. 12. represents simultaneously the limiting values of coefficients for
power scale $m_N = N_D/N_{M_i}$, flow scale $m_Q = Q_D/Q_{M_i}$, and rotational velocity scale
$m_\omega = \omega_D/\omega_{M_i}$.

Unfortunately the dispersity scale $m_d = d_D/d_M$ will always be different from
$m_d = 1$, immediately when the rotational velocity scale is different from one, m≠1.

After extensive analyses in the Jugoslav steam turbine factory "JUGOTURBI-
NA" in Karlovac, one three stage, one shaft experimental steam turbine, shown
in Fig. 13. and 14. was built and installed. Until present only experiments with
superheated steam have been performed. Experiment with wet steam are in preparation. /18/.

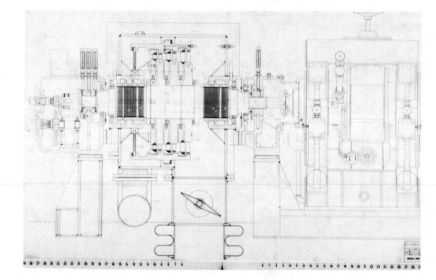

Fig.13.

Since complete similarity of experiments during tests on these turbines cannot be performed, /19/, it is generally necessary to make numerous tests and measurements, which will differ from real conditions, with insufficient precision of measurement, which will lead to inadequate interpretation of achieved results.

For this reason designers of turbines built for exploitation, include intuitive solutions into their design thus paying their major attention to moisture separation /19/, either by annular channels placed in front or behind the stators and rotors, or by suction of moisture through openings on the leading edge, concave or convex side and even on the stator vanes leaving edges. Simultanoeusly, the designers stimulate extraction of roughly dispersed liquid phase parts, due to centrifugal force, through the radial grooves on rotor blade surfaces. At the same time these grooves protect the blade surface from disastrous erosion effects. /20/.

Fig. 14.

At variable load conditions the mentioned openings, even carefully designed, can cause undesirable effects.

8. CONCLUSION

In the first part of this paper it was shown that the momentary development of steam turbines depends to a lesser extent on the research achievements in heat transfer region at high temperatures on the turbine inlet.

On the other hand, much is expected from the research whose aim is a better understanding of the principles in the field of heat and mass transfer in low pressure turbine parts. This is not only in aim of improving steam turbine efficiency, but also because of avoiding or diminishing harmful, of the here not dealt with, erosion and corrosion consequences.

Besides, the paper does not either deal with one of the primary field of interest of today's dynamic control study, i.e. the discovery of interrelationship of temperature changes between the rotational parts under control and the temperature changes on stationary parts accessible to measurements.

This problem from the field of heat transfer and heat conduction should be treated separately.

9. REFERENCES

/1/   Scheffezyk, H. Dampfturbinen, KWV - Fach aufsätze '76 - KWV - Mülheim - Ruhr.

/2/   Ščegljaev, A.V. 1976. Parovie turbini, izdanie pjatoe, dopolnenoe i podgo-
        tovlenoe k pečati prof. B.M. Trojanovskim, Energija, Moskva.

/3/   ... 1978. Design principles of very large steam turbines, BBC Brown, Boveri
        and Company, Baden, Switzerland.

/4/   Stodola, A. 1910. Die dampfturbinen, Verlag von Julius Springer, Berlin.

/5/   Kirilov, I.I. and Jablonik, R.M. 1968. Osnovi teorii vlažnoparovih turbin,
        Mašinostroenie, Lenjingrad.

/6/   Benediktov, V.D. 1969. Turbini i reaktivnie sopla na dvuhfaznih potokah,
        izdateljstvo Mašinostroenie, Moskva.

/7/   Gyarmathy, G. 1962. Grundlagen einer Theorie der Nassdampfturbine, Disser-
        tation ETH Juris-Verlag, Zürich.

/8/   Moore, M.J. and Sieverding, C.H. 1976. Two phase steam flow in turbines
        and separators, Hemisphere - McGraw - Hill Book Company, London.

/9/   Durst, D., Tsiklauri, G.W., Afgan, N.H. 1979. Two-phase momentum heat and
        mass transfer, Hemisphere, Washington.

/10/   Jablonik, R.M., Markovič, E.E. and Aljtušev, L.E. 1965. Dviženie kapelj v
        mežvencovih zazorah parovih turbin, Energetika, No. 10.

/11/   Selznev, L.J. 1975. Teoretičeskie i eksperimentaljnie isledovanja obrazova-
        nja kondenzirovanoj fazi v protočnih častjah turbin, IV Konference: Parny
        turbiny velkeho vykonu, Plzen.

/12/   Saltanov, G.A. 1972. Sverhzvukovie dvuhfaznie tečenija, izdateljstvo Višej-
        šaja škola, Minsk.

/13/   Kosjak, J.F. 1978. Paroturbinie ustanovki atomnih elektrostancij, Energija,
        Moskva.

/14/   Dejč, M.E. and Filipov, G.A. 1968. Gazodinamika dvuhfaznih sred, Energija,
        Moskva.

/15/   Kirilov, I.I., Ivanov, V.A. and Kirilov, A.I. 1978. Parovie turbini i pa-
        roturbinie ustanovki, Mašinostroenie, Lenjingrad.

/16/   Kirilov, I.I. and Nosovickij, A.I. 1975. Procesi kondenzacii i separacii
        vlagi v turbinih stupnjah, IV Konference: Parny turbiny velkeho vykonu, Plzen.

/17/   Kačer, V. 1975. Conception of research turbine design, IV Konference: Par-
        ny turbiny velkeho vykonu, Plzen.

/18/   Majcen, A. 1980. Neke metode ispitivanja utjecaja vlažne pare u niskotlačnim
        dijelovima turbina - Master thesis - FSB - University of Zagreb.

/19/   Trojanovski, B.M. 1979. O vliami vlažnosti na ekonomičnost paroviih turbin
        - Teploenergetika, Moskva.

Wet Steam in Turbines

G. DIBELIUS, and K. MERTENS
Institute for Steam and Gas Turbines
of the Technical University Aachen
Aachen, FRG

ABSTRACT

The presence of droplets in the flow through low pressure parts of conventional steam turbines or throughout nuclear steam turbines causes additional losses and erosion damage. They depend very much on the droplet concentration and their size distribution. Both quantities can be influenced by the
circumstances at the very onset of condensation. For the study of these phenomena a wet steam experimental turbine has been set up. The condensation process can be followed by means of a light scattering probe detecting number,
size and velocity of the droplets in the wet steam flow. The experimental
results are compared with the generally accepted theory that is based on
basically one dimensional steam nozzle experiments. The comparison indicates
clearly that more three dimensional flow effects have to be embodied in the
theory to match experimental evidence.

1. INTRODUCTION

Expansion into the wet steam region is a thermodynamic necessity for
ordinary steam cycles and constitutes the only possibility for light water
nuclear power plants as long as superheating cannot be accomplished. In both
cases the presence of droplets in the steam flow causes additional losses and
erosion damage, if from the original mist flow with droplet sizes of tenths of
micrometers by various mechanisms the droplet size has been increased by two
orders of magnitude. This is the reason, why for some time much attention has
been paid to the formation of mist and to the transformation mechanisms to
greater droplets in steam turbines.

2. CLASSICAL THEORY

The theory for the condensation in steam Turbines |1, 2| was developed
on the basis of calculations and experiments for the one dimensional flow of
steam expanding in nozzles into the wet steam region. In this case the flow
is basically one dimensional and ends up with supersonic velocities. The
process is described referring to Fig. 1: When the saturation line is crossed
during a rapid expansion no condensation occurs immediately as it does in
equilibrium; the steam continues to behave like superheated steam in that so
called metastable region, where it is subcooled or supersaturated.

With increasing subcooling there is a tendency that random accumulations
of molecules, so called nuclei, become stable, since the so called critical

539

radius of nuclei decreases with increasing subcooling. At that point a sudden, spontaneous formation of many stable nuclei occurs succeeded by a rather rapid growth to mist droplets |Fig. 2|. The heat rejected by this condensation process diminishes the subcooling to a small residue accompanied by an increase in entropy, temperature and pressure and a decrease in velocity of the remaining vapor. In a steam nozzle this takes place in the supersonic part of it. Only in this case the enthalpy change causing the expansion is big enough for creating condensation, when the steam is slightly superheated at the nozzle entrance as indicated by the expansion lines A - E in Fig. 1.

Since the nucleation rate depends very much upon the degree of subcooling, within this theory it is very important how fast the expansion takes place: With rapid expansions subcooling still increases during the formation of mist droplets until it is reduced by the condensation process (Fig. 2); therefore a very large number of nuclei is generated that can grow only to a limited size of mist droplets making up for the deficit of condensate. With slow expansions the heat rejection outweighs any further subcooling resulting in a smaller number of larger mist droplets.

Up to now the condensation in turbines has been treated much the same way applying the before mentioned principles to the flow as described by the properties averaged across the blade channel (Fig. 3). Of course, there are differences due to the pressure drop changing between high values within the blade rows and very small values in the gap in between. Therefore, all quantities relevant for condensation exhibit corresponding changes and do not change smoothly as for the nozzle flow (Fig. 4).

There is still another difference in comparison to the nozzle flow: In the nozzle growth of droplet size is only possible by further condensation according to the degree of subcooling and maybe due to some agglomeration. On the contrary, in a turbine part of the mist droplets are deposited on the blade and wall surfaces, forming rivulets along the surface nourishing water accumulations near the trailing edge of the blades. When they are big enough they will be entrained by the surrounding flow and torn apart to smaller droplets by shear forces until they become stable under the prevailing flow conditions. These mechanisms have to be taken into account since they are of utmost importance to erosion and to loss phenomena caused by the presence of these droplets.

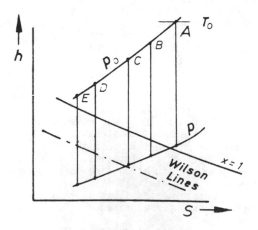

Fig. 1. Expansion in a steam nozzle from various inlet temperature

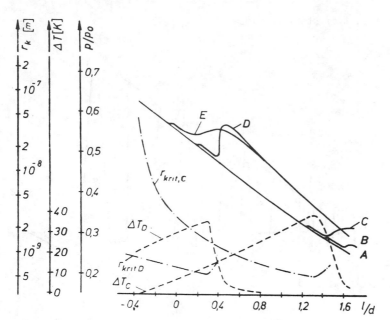

Fig. 2.   Formation of droplets in a steam nozzle

r_{krit} critical radius of stable nuclei,   ΔT subcooling,
p/p_0 pressure ratio

It has to be pointed out, that for the transformation processes of droplet size just described the original droplet size distribution of the mist
droplets is very important: The smaller the droplets are, the more easily
they follow the steam path, the less rivulets are formed along the walls,
which will be entrained as secondary droplets in the wet steam flow. Hence,
expansion rate in the nucleation zone appears according to this theory a
decisive quantity for all that follows from there onward.

3.   WET STEAM EXPERIMENTAL TURBINE

On the basis of the theoretical model a wet steam experimental turbine
(Fig. 5) has been installed at the Institute for Steam- and Gas Turbines of
the Technical University Aachen: In a first axial turbine stage the steam
is subcooled from a slightly superheated state, yet without any considerable
condensation. The stage is followed by an unbladed circular channel, the
contour of which can be modified by plastic inlays for different expansion
rates that are kept constant in each case within that channel. In this way,
it was thought to influence the spontaneous condensation occuring in the
channel. Finaly, the fate of the droplets within a second turbine stage
operated independently of the first stage can be observed. It is obvious that
the phenomena occuring in this type of an experiment are much closer to what
is happening in an actual turbine than to the flow through a nozzle.

Steam is supplied by the Universitys power and heating plant either from
the live steam side via a pressure and temperature reducing installation

Fig. 3. Low pressure part of steam turbine

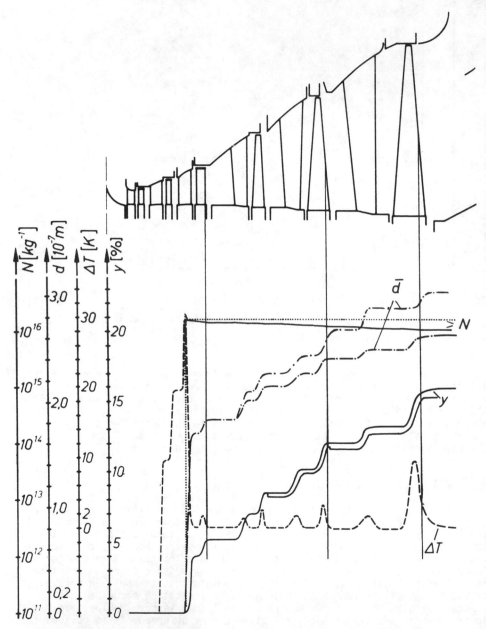

Fig. 4. Condensation in low pressure part of steam turbine. N droplet concentration, $\bar{d}$ mass averaged diameter, y wetness fraction, ΔT subcooling

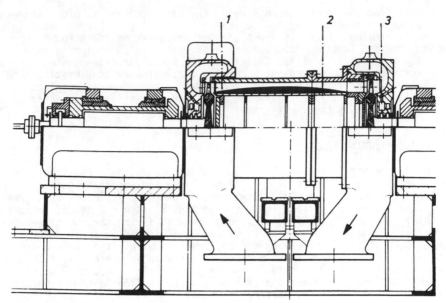

Fig. 5. Wet steam experimental set up

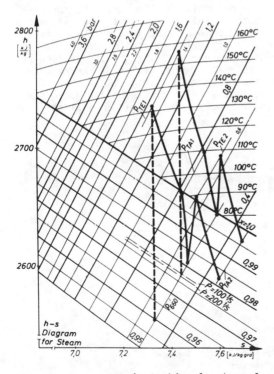

Fig. 6. Expansion in turbine 1, channel and turbine 2

or directly from the low pressure side of the back pressure turbine. The experiment then may be conducted in the following way (Fig. 6): For a chosen inlet pressure (variable between 0.5 and 4 bar) of the entering steam the temperature will be decreased in steps from values in the superheated region to values just above the saturation line. Therefore, the expansion extends step by step deeper into the two phase region. According to the theory condensation should start in the second stage, would occur at lower steam temperatures in the condensation channel moving toward the front end and is finaly expected to take place already in the first stage.

4. LIGHT SCATTERING PROBE

For determining where and how condensation occurs the light scattering probe developed by the Institute |3, 4| has been used detecting number, size and velocity of the droplet in the two phase flow (Fig. 7). Radial traverses can be taken at the 3 stations indicated on Fig. 5. It works on the following principle: The scattered light of single droplets flying through a very small observation volume V is measured. It is optically formed by the intersection of the images of the apertures B_1 and B_2. Droplets within this volume are illuminated by the primary light beam of an Argon laser via a glassfibre, the aperture B_1 and the objective O_1. According to their size they scatter light into different directions as a result of refraction, diffraction and reflection. Receiving this scattered light within a fixed range of scattering angles (in this case $90^0 \pm 20^0$) a unique relation between scattered light intensity and droplet size is given under certain boundary conditions. The scattered light is transferred via the objective O_2, the prism U, the aperture B_2 and a flexible glassfibre to a photomultiplier. Here the light signals are converted into voltage signals; they can be evaluated in a multi-channel-analyzer in terms of number of droplets in different classes of droplet size. For that purpose an extensive computer program applying the calibration relationship and correcting for some systematic errors is used. The results are immediately plotted in form of a histogram indicating the number of droplets per unit of massflow in the previously defined classes for various droplet sizes (f.e. Fig. 9). In addition the pulse width gives an indication of the droplet velocity.

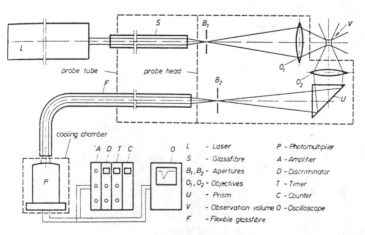

Fig. 7. Principle of a light scattering probe

Summing up the masses of all droplets liquid mass flow density can be computed. Relating it to the total steam flow density steam wetness fraction results.

5. RESULTS

5.1 Onset of condensation

Fig. 8 shows a typical test result where the total droplet concentration at the entrance and the exit of the condensation channel is plotted against the isentropic enthalpy drop between the saturation pressure and the pressure at the channel entrance as a measure of how far the expansion extends into the equilibrium wet steam region, for short "expansion depth". Lowering the turbine inlet temperature droplets are first detected at the channel exit with an increasing concentration. For higher expansion depths droplets with increasing concentration can be observed at the channel entrance. Qualitatively this corresponds to the theoretical understanding. However, if compared with calculations condensation should only start at much greater expansion depths and should lead to smaller numbers of droplets. Only with very deep expansion into the wet steam region condensation should start within the first stage at much higher expansion rates leading to a much higher droplet concentration.

Fig. 9 indicates the droplet size distributions in form of histograms for the two measuring conditions indicated in Fig. 8. They are much wider than expected from theory. In fact, with the light scattering probe as it has been applied so far the detection of very small droplets is limited to about 0.1 μm. Therefore, up to now it is not known how much further the droplet size distribution extends into the region of very small droplets.

5.2 Influence of the expansion rate

Fig. 10 and 11 plotted for the same variables as the previous ones exhibit measurements at the channel exit for different expansion rates of $\dot{P} = 60$ s^{-1} and 200 s^{-1}. There are only very small differences noticable for these two expansions rates. According to the theory, as outlined before there should be large differences in droplet concentration as well as in droplet size.

The discrepancy between the test results and the so far generally accepted theory can only be resolved if some processes are embodied in the theory that are typical for the three dimensional and generally unsteady flow in turbines in contrast to the more or less one dimensional flow in a nozzle.

5.3 Influence of pressure and temperature differences within the blade channel

In this connection it is obvious that the pressure and temperature is unevenly distributed in the flow field around the blades: For example the subcooling on the suction side of the blades will be higher than on the pressure side (Fig. 12). Therefore, on the suction side local condensation has to be expected earlier than calculations based on average temperatures indicate. Those droplets will spread out and become nuclei for condensation in the entire flow field.

Inclusion of this influence into the theory brings it closer to the experimental results, but does not account for the entire difference.

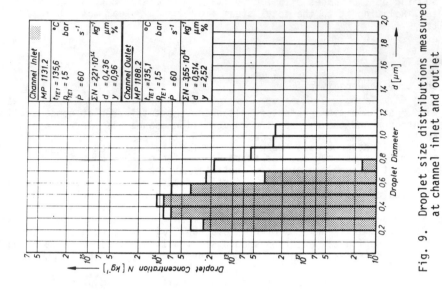

Fig. 9. Droplet size distributions measured at channel inlet and outlet

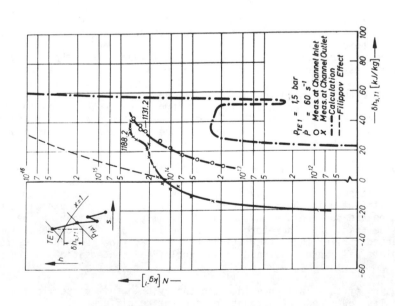

Fig. 8. Droplet concentration measured at channel inlet and outlet

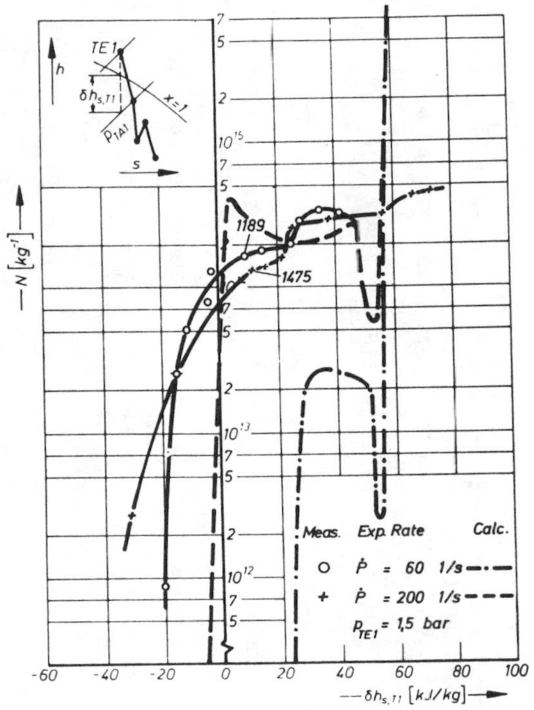

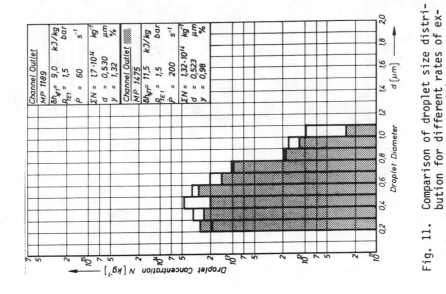

Fig. 11. Comparison of droplet size distribution for different rates of expansion (see Fig. 10).

Fig. 10. Droplet concentration measured at
channel outlet for constant pressure
and different rates of expansion

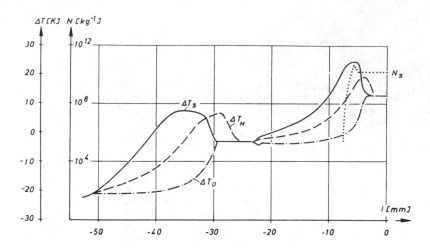

Fig. 12.   Subcooling ΔT and droplet concentration N at sucdion side, midle
           stream line and pressure side of first turbine stage

5.4 Filippov effect

There are certainly more local phenomena accounting for dropping temperatures in the flow field below the average. For instance the wake behind the trailing edge of the blades consists of a vortex line each one rotating in the opposite direction to the neighboring vortex |5|. In the center of each vortex a lower temperature has to be expected than in the surrounding flow. According to Filippov |6| this has the effect of much earlier condensation leading to a higher droplet concentration as indicated in Fig. 8.

An experimental check of this effect is possible by triggering the light scattering probe by a signal synchronous to the rotating wheel and activating it only for a short duration. Therefore, droplets can be observed in a small section of the circumference. When this "time window" is shifted the droplet concentration can be observed along the circumference and in particular in the neighborhood of a wake. These experiments are under way.

5.5 Influence of steam impurities

The results change when the experimental setup is supplied either by live steam that has been converted to lower temperatures and pressures or by exhaust steam from the back pressure turbine (Fig. 13, 14). There seems to be some preseeding with condensation nuclei in the case of the converted steam inspite of the fact, that the steam supply line after the reducing station to the turbine is approx. 30 m long. Hence with a degree of superheating of at least 10 K all spray droplets should have been evaporated. However, if impurities are present, these might act as condensation nuclei.

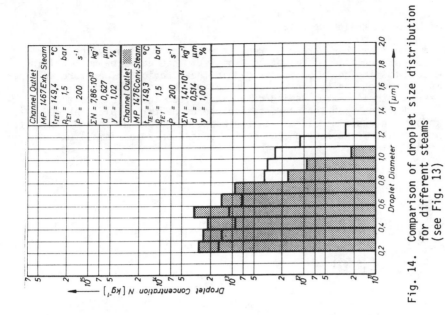

Fig. 14. Comparison of droplet size distribution for different steams (see Fig. 13)

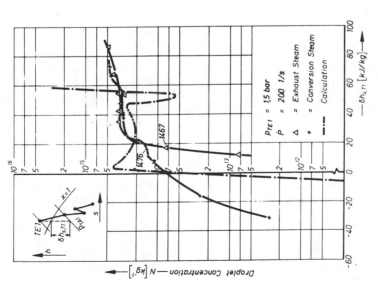

Fig. 13. Droplet concentration measured in exhaust and conversion steam for constant pressure and expansion rates

5.6   Influence of residence time

In a real flow there are many regions with reduced flow velocity or prolonged residence time. It might be expected, that equilibrium conditions for the condensation could be approached in these regions. Therefore an experiment was run without inlays in the condensation channel resulting into an almost zero expansion rate. As indicated in Fig. 15 there is almost no difference in comparison with the condensation phenomena taking place at 200 s^{-1}. This can only be explained if there is preseeding before or in the first stage and hence, further condensation can be influenced only to a small extent by the rate of expansion.

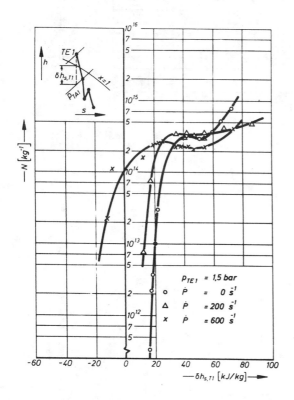

Fig. 15.   Droplet concentration measured in exhaust steam at the channel outlet for constant pressures different rates of expansion

REFERENCES

|1| Gyarmathy, G. 1962. Grundlagen einer Theorie der Naßdampfturbine. Zürich: Juris-Verlag.

|2| Kotake, S. and Glass, I.I. 1978. Survey of Flows with Nucleations UTIAS Review, No 42.

|3| Marx, P.P. 1970. Streulichtmeßsonde zur Tropfengrößenspektrometrie insbesondere in Naßdampfturbinen. Dissertation RWTH Aachen.

|4| Ederhof, A. 1977. Bestimmung der Flüssigkeitsbeladung mit Tropfen von Zweiphasenein- bzw. -mehrkomponentenströmungen mittels Streulichtmessungen. Dissertation RWTH Aachen.

|5| Lawaczeck, O. and Heinemann, H.-J 1975. Von Karman Vortex Streets in the Wakes of Subsonic and Transsonic Cascades. AGARD-CP-177, 28-1-28-13 Unsteady Phenomena in Turbomachinery.

|6| Filippov, G.A., Nekker, R. and Seleznev, L.I. 1977. Calculation of the Formation of Moisture in Turbine Flow Sections. Teploenergetika (Thermal Engineering) 24, 7, 7-14.

Steam Condensate Droplet Evolution: Experimental Technique

ROBERT A. KANTOLA
Turbine Business Group
General Electric Company
Lynn, Mass., USA

ABSTRACT

Condensation of water vapor from moist air or steam has been an active research area for over eighty years. A large portion of this research has been
concentrated on the condensation of steam in turbines, where the moisture induced efficiency losses and possibilities of erosion damage impose a lower
limit on the expansion of the steam so that no more than a 10 to 12% moisture
content is reached. In order to accurately assess and reduce the mechanical
and thermodynamic losses (which can be as high as 4-8 percentage points of efficiency), the thermodynamic conditions at nucleation onset and the rate of
the subsequent droplet growth must be accurately known. The information available prior to this study has, for the most part, resulted from measurements in
rather short, convergent-divergent nozzles resulting in excessively high expansion rates and interactions between the supersonic flow field and the condensation front. To overcome these objections to this prior "state-of-the-art"
data, a subsonic technique for investigating steam condensation has been developed. This technique uses a non-steady expansion wave caused by the blowdown of a steam filled tube to provide a controlled, but rapid, expansion of
slightly superheated steam.

To sense the onset of nucleation and the evolution of the condensate cloud
characteristics, droplet size and number density, three different optical techniques were developed. For the very small droplets (0.1 to 0.2μm) a side
scattering technique is used. For the larger droplets (0.2 to 1.5μm), forward scattering and attenuation techniques were used.

The most dependable system is the attenuation method, which uses laser
light at two wavelengths to determine the droplet size and number density time
histories. In conjunction with the optical measurements, fast response pressure transducers are used to determine the steam pressure during this transient
test. Through the use of high speed digital recorders (with sampling intervals as low as 0.1 microseconds) to capture the data and common time bases an
accurate determination of the thermodynamic conditions at onset as well as
detailed description of the droplet growth rates has been obtained.

With this technique, droplet diameters less than 0.2 micron have been
measured. The quality of the data is very high with the technique sensitive
enough to be able to sense the difference between using triple-distilled feedwater and demineralized triple-distilled feedwater. The demineralized

---

Robert A. Kantola was formerly with General Electric Corporate Research
and Development.

feedwater produced much more orderly droplet growth histories. These tests represent the first evidence of the effects of dissolved contaminants on the steam condensation process.

1. INTRODUCTION

Condensation occurs in the low pressure stages of fossil-fired steam turbines and throughout nuclear-powered steam turbines. The moisture losses and erosion damage resulting from the condensed water droplets can be reduced, however, once the mechanics of spontaneous nucleation of the supersaturated steam in the turbine environment are quantified. The appearance of moisture in a steam turbine occurs at subcooled steam conditions near what is commonly called the Wilson line. In order to accurately assess and reduce mechanical and thermodynamic losses, the thermodynamic conditions at nucleation onset and the rate of the subsequent droplet growth must be accurately known.

Prior data available to tackle this problem (References 1 through 6) have been largely inadequate. Past onset data have been taken for the most part, in rather short, convergent-divergent nozzles, resulting in excessively high expansion rates and with condensation occurring usually in supersonic flow, where interactions between the condensation front and the shock patterns contaminated the results. In a turbine the flow is subsonic, except for the last stages, thus limiting the usefulness of the data. With the apparatus used in this program, the primary objections raised about the currently available data have been overcome. The expansion rate can be changed independently of starting pressure, and all flow conditions are subsonic.

Relevant available data on condensate droplet size are also very sparse. The majority of the available data are concentrated at pressures of only a few atmospheres. Most of this data has also taken in convergent-divergent nozzles, and therefore suffers from the same limitations as the onset data.

In all of the prior studies, scant, if any, attention has been paid to the effects of steam purity. Although minute levels of impurities (10 ppb) can produce 10^{12} molecules of contaminant per cm^3, none of the references listed in the Bibliography have reported any measurements of the steam purity used in their tests.

It is the goal of the work reported herein to provide an experimental facility capable of producing data on nonequilibrium steam condensation in its simplest form that is without the effects of turbulence, sheared flows, temperature gradients and the other effects previously mentioned. The goal is to measure the size growth of the droplets and the droplet number density from onset to the later stages of condensation.

2. EXPERIMENTAL

2.1 Steam Condensation Facility

The basis for the work reported herein is established by the need to locate the Wilson point at thermodynamic conditions appropriate for steam turbines, and to obtain data on droplet size and number density. The concept developed in these studies for producing the desired wide ranges in the steam pressure, temperature, and expansion rate is based upon the phenomenon of an expansion wave traveling through slightly superheated steam. The basic gas dynamic process corresponds to the driver section of a shock tube or shock tunnel (7). In order to relate this concept to these studies, it has been designated as the nucleation tube.

Conceptually, the nucleation tube (Figure 1) consists of an externally heated tube which is permanently closed at one end and sealed with a rupture diaphragm at the other. Immediately ahead of the diaphragm is a smaller dia- meter orifice plate which can be used to reduce the cross sectional area of the tube. A closed dump tank (connected to the downstream side of the diaph- ragm) is maintained at a low enough pressure to cause choking at the orifice when the diaphragm is ruptured. The nucleation tube is initially filled with slightly superheated steam upstream from the rupture diaphragm.

When the diaphragm is ruptured, a nonsteady expansion wave centered at the orifice/diaphragm station propagates through the steam. As illustrated in Figure 1 the expansion wave reflects from the closed end of the tube and pro- cesses the steam a second time as it travels toward the orifice. When the reflected wave returns to the orifice, further reflection and transmission wave processes occur. Duplication of the controlling parameters corresponding to a particular wet stage operation is accomplished by independently varying the following:

o   Location of the sidewall observation point relative to the closed end
    of the tube
o   Length of the tube
o   Ratio of the tube and orifice cross sectional areas
o   Initial conditions (temperature and pressure) of the steam in the tube

The axis of the viewing ports and the centerline of the pressure-transducer port, all lie in the same plane. This plane is normal to the tube centerline. The windows are flush with the inside walls of the test section, which is necessary in a transient test to eliminate expansion and compression waves that would result with either a protrusion or a recess.

The nucleation tube is surrounded by an insulated furnace consisting of multizoned main tube heaters, guard heaters, and control equipment. Viewing ports in the furnace walls are aligned with the test section windows. Lenses of the optical instrumentation system form the "windows" of the furnace wall ports.

The steam generator is a vertical electric boiler that can provide steam up to 1000 psia and 560°F. To remove any impurities, the boiler feedwater (triple distilled water) is circulated through a demineralizer (Sybron/ Barnstead Nanopure). To monitor water purity an in-line sodium monitor (Orion Research Model No. 151102) and a conductivity meter (Sybron Barnstead Model No. D2770) are used.

2.2 Optical Systems

To measure the properties of the evolving droplet cloud, a multiplicity of measurements was provided. This was felt to be necessary as the wide varia- tion in test conditions (pressure and expansion rate) and the general lack of knowledge of the droplet sizes and droplet number densities did not allow a focus on a given set of experimental parameters.

To first place this task in perspective, a brief description of the drop- let cloud evolution will be given. According to nucleation theory, the nucleation onset is defined as the point at which the subcooling of the steam allows stable clusters of water molecules to form and grow. During the initial phases of nucleation, the droplet sizes are so extremely small that very little condensate is generated and the sub-cooling continues as the expansion pro- ceeds. Since the rate of droplet generation and growth rate of existing

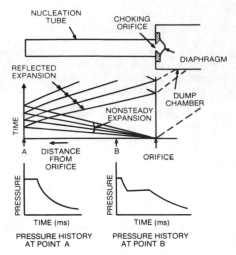

**FIGURE 1. NUCLEATION TUBE CONCEPT FOR
STEAM CONDENSATION STUDIES**

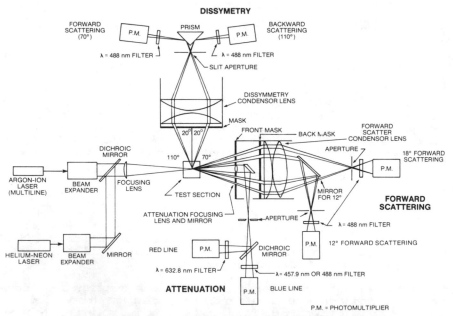

FIGURE 2. OPTICAL ARRANGEMENT—STEAM CONDENSATION

droplets increases rapidly with subcooling, a polydisperse droplet cloud is
generated. Once the amount of condensed droplets reach a level that the transfer of the latent heat of condensation from the droplets warms the surrounding
steam, the subcooling will peak and then decrease. This peak in the subcooling
defines the end of the nucleation period and while the steam temperature approaches the droplet temperature (equilibrates), condensation on the existing
droplets continues. This description is at best a relatively simplified picture of the actual events but it does give some flavor to the task at hand.

Three separate optical techniques are used to define the properties of
the condensate droplet cloud. It should be noted here that all three techniques are based on analytical predictions of the light scattering from either
a single droplet or from a mono-disperse droplet cloud. These techniques include attenuation, forward scattering, and side scattering, as shown in
Figure 2. In order to provide the necessary spread of wavelengths in the illumination beam, two separate lasers are used: an argon ion laser is used to
emit a blue line at a wavelength of either 457.9 or 488 nm, and a helium-neon
laser emits a red line at a wavelength of 632.8 nm. After passing through
beam expanders, the two beams are mixed with a dichroic mirror. This mirror
reflects the red line and transmits the blue line. The combined beam is then
focused by a diffraction-limited lens with a focal length of 80 mm. This convergent beam then passes through a 1/2 inch thick synthetic sapphire window,
with the waist of the beam located at the nucleation tube centerline. The
highly illuminated waist is used as the scattering volume for the forward and
side scattering techniques. There are two exit windows: one coaxial with
incident beam and one 90 degrees (vertical) to the beam. All the windows are
anti-reflective coated, thereby reducing the light scattered from the window
surfaces. Scattered light would raise the background level and reduce the
measurement scheme's sensitivity.

An optical table placed underneath the nucleation tube provides a mounting
surface for all the optical components. A channel beam supports the nucleation tube independently of the optical table, as shown in Figure 3. Motion of
the tube then doesn't effect the alignment of the optical systems and cause
spurious signals. The following paragraphs describe and discuss each measurement method used.

Attenuation - A single attenuation measurement determines a product of
three quantities: the scattering efficieny (K), the droplet number density (n),
and the droplet diameter squared (d^2). To separate the droplet diameter (d)
from the number of droplets per unit volume (n) requires at least two different
wavelengths. These wavelengths should have a reasonable separation. The wavelength separation requirement is the reason for using two lasers, as it is
difficult to achieve such a large separation with a single laser.

The attenuation technique differs from scattering techniques in that it
measures light removed by scattering or absorption from the incident beam that
traverses the test section. It integrates the events that occur along the
beam. The diverging beam existing from the test section is incident of a
small achromatic lens. This lens refocuses the beam. Directly behind the
attenuation focusing lens, and before the focal point, is a small, flat elliptical mirror. This mirror reflects the beam at a right angle, through a tube.
The beam's focal point occurs outside the main housing that contains both the
attenuation and forward scattering collection lenses. A 1.3 mm circular
aperture is located at the focal point of the focused beam and serves to
eliminate scattered light from entering the system. A dichroic mirror (located past the aperture) separates the beam into the two parts. Light with
wavelengths greater than 500 nm is reflected, that with less than 500 nm is

FIGURE 3. TWO LASER-OPTICAL ARRANGEMENT

transmitted. The particular wavelengths used to determine droplet size and number density are selected by placing interference filters immediately in front of the two photomultipliers. Each of the photomultipliers then responds to the attenuation of a different wavelength. The magnitude of the ratio between the two measured values of scattering efficiency then becomes a function of the droplet size, as shown in Figure 4 for several different wavelength combinations. Pendorf's (8) tabulated values of scattering efficiency were used in these predictions.

One of the main advantages of using the attenuation technique in a transient test is that it is self-calibrating. Before the diaphragm is broken, the illumination is a maximum and the attenuation due to the onset of droplets is measured from the maximum. This eliminates the need for a reference beam.

Forward Scattering - The forward technique is a single-wavelength two-collection angle method. By collecting light at two angles close to this forward direction, changes in the lobe patterns of the scattered light can be directly related to the droplet diameter. This technique is described in Reference (9). The collection angles are defined by two masks in front of the condenser lens. A spherically corrected anti-reflection coated lens is used to refocus the droplet scattered light from the two (one degree wide) light cones with 12 degree and 18 degree half angles. A flat elliptical mirror is used to reflect the inner core of light, through 90 degrees, to a defining aperture.

After passing the aperture, the cone of light is incident on an interference filter-photomultiplier combination. This allows the outer cone to remain coaxial with the incident beam and be directed to a different aperture-filter-photomultiplier combination. The lowest possible level of background light is attained by separating the cones of light and incorporating independent adjustments of the spatial filtering.

The Mie scattering theoretical predictions (10) were used to determine the expected optical output from the configuration. The prediction is based on the scattering of plane-polarized light from a single spherical water droplet located in an infinitely wide uniform illumination field. The incident scattered light is integrated over the mask openings, and yields a ratio of captured light flux (R_d) to the droplet diameter, as is given in Figure 5.

In contrast to the attenuation method, the forward scattering technique requires calibration of optical paths in the individual channels by using a known scatter. The calibration defines the effects of light loss and photomultiplier sensitivty.

Side Scattering - In an attempt to view single droplets, a side scattering method was implemented. Rather than use a single measurement at 90 degrees to the incident beam, a two angle system is used. The addition of the second channel allows the ratio of outputs to be compared and therefore eliminating the problem of locating the droplet in the non-uniform illumination field of the scattering volume.

A brief description of this technique, called dissymmetry, is given in Reference (11). This method gathers light along two rays placed symmetrically fore and aft of the incident laser beam's perpendicular plane, and on the same plane that contains the beam's axis and its polarization vector. To ensure that both rays emanate from the same spot, a single set of condenser lenses are used to collect and refocus the droplet scattered light. A viewing angle 20 degrees from the vertical was selected as a compromise angle. This angle is a compromise between droplet sensitivity and the amount of the scattering

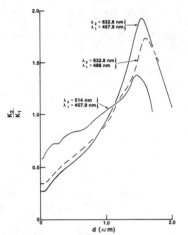

FIGURE 4.   ATTENUATION RATIO VS DROPLET DIAMETER

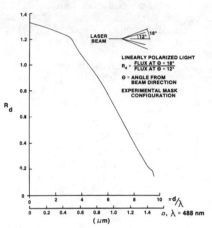

FIGURE 5.   RATIO OF LIGHT FLUX AT FORWARD
SCATTERING ANGLES VS DROPLET
DIAMETERS

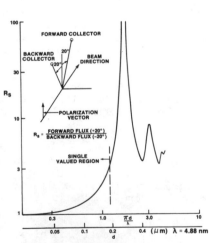

FIGURE 6.   DISSYMMETRY RATIO (+ 20°/-20°)
VS DROPLET DIAMETER

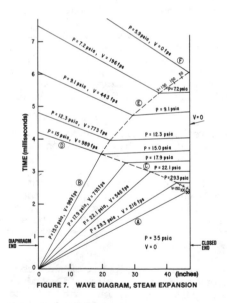

FIGURE 7.   WAVE DIAGRAM, STEAM EXPANSION

volume ("slice") that must be viewed. An adjustable split aperture serves to define the width of the "slice" and after passing through the aperture, an inverted prism deflects the fore and aft scattered rays into their respective photomultipliers. Interference filters are used to eliminate any other light sources. Again, the Mie (10) scattering theory is used to predict the relation between the ratio of forward to backward scattered light and the droplet diameter, as shown in Figure 6. The detection range is limited to less than $0.2\,\mu$m due to the multi-values of the function at greater sizes.

2.3 Pressure and Expansion Rate Measurements

Piezoelectric pressure transducers, similar to those used in shock tunnel research for many years, meet the fast-response requirements demanded of the pressure instrumentation for these tests. The high temperatures encountered in the furnace surrounding the nucleation tube dictated the use of pressure transducers with electrical connections and lead wires that could survive approximately 290°C. High-temperature transducers, manufactured by Kristal Instrumentation Corporation (Model Z4946A), were used in these studies.

2.4 Temperature Measurement

Determination of the thermodynamic state in the test section can be divided into two parts, before condensation onset and after. Before condensation, the steam is in a supersaturated state and, because it is a single phase system, knowing the pressure and temperature is sufficient. However, once condensation commences, the resulting two phase mixture requires that the moisture content must be known as well as either the pressure or the temperature. This is because the constant pressure lines in a Mollier chart, in the wet region, are also lines of constant temperature.

Determination of the steam temperature is very difficult. The measurement must be made at the wall to avoid triggering the steam nucleation and therefore, it is necessary to infer the temperature in the center of the test section. For the portion of the decay leading up to onset, it is customary to assume an isentropic process. Then only the measurement of pressure is required to determine the thermodynamic state. When the condensation process reaches equilibrium again, only the pressure will be required provided the moisture is known. Therefore, the starting thermodynamic point and the ending point can be determined with reasonable certainty by measuring the pressure and moisture content. For these reasons, no direct or indirect measure of the steam temperature is being attempted, except for the initial values.

3. EXPERIMENTS

3.1 Basic Approach

A short discussion on the basic aspects of the experiment will be given before discussing experimental techniques. The experimental apparatus consists of a long segmented tube with a 1 inch square bore. A test section, with provisions for four windows, can be positioned at several different locations along the tube due to the modular construction of the entire apparatus. High purity saturated steam is admitted into the nucleation tube which is maintained at a temperature corresponding to 20°F of superheat. Before heating up the tube, the optical systems are aligned and calibrated with neutral density filters and a known scatterer. After heating the tube, alignment is rechecked due to the close proximity of the optics to the tube and its heaters. The test is initiated by bursting the diaphragms, which causes an expansion wave to run toward the test site.

To provide a greater understanding of this wave expansion phenomenon, a closer look at the nature of the expansion process that occurs in the nucleation tube will be taken here. Details about the pressure and velocity changes expected in the test are obtained by the use of an analytical prediction of the transient. The steam will be approximated by a perfect gas and it will be assumed to be undergoing an isentropic expansion. An analytical technique called the "method of characteristics" can be used to predict this one-dimensional unsteady expansion process. This technique is described in standard texts such as Reference (12), and will not be discussed further here. To illustrate some of the properties of the expansion, the typical results of the calculation procedure are shown in Figure 7. The full lines shown on Figure 7 are lines of constant pressure. While the process is continuous, the method of characteristics uses a piecewise approximation to represent this. During the initial portion of the expansion fan, between lines A, B and C , waves of only one family exist and the lines of constant pressure are also lines of constant velocity. These lines are inclined with respect to the abscissa, and as such, neighboring points in the tubes undergo different pressure histories. It is quite the different case, in the region between lines C, E and the closed end of the tube. Here the interaction between the initial waves and waves reflected from the closed tube end cause the isobars to be lines of nearly constant time. Now neighboring points in the tube are subject to nearly equal pressure histories, and subsequently should have very similar condensation droplet growth. The test site in the work reported herein is located 1.9 inches from the closed end. Nucleation occurs nearly midway, time-wise, between line C and line E. This is important as the test site is fixed in space and due to the particle velocity (about 50 fps for conditions of Figure 7), droplets drift into and out of the test site. Therefore, the droplets seen at the start of nucleation are not those observed (some 1/2 milli-second later) near the end of the observed interval of droplet growth. However, since the pressure history is very nearly equal for all neighboring points, one must conclude that the droplet growth rates are also nearly equal and this convection effect is quite small. Also equally important are the absence of any steam density or drop density gradients which would tend to bend the laser beam causing it to miss the aperture and therefore yielding an erroneous attenuation signal.

The pressure decay in this wave is very rapid and after arrival of the wave nucleation onset usually occurs within a 2 ms time interval. After onset the droplet growth transient lasts up to 1.5 ms. To capture this information, three transient recorders are used. Delays are set in such a manner that the data sampling occurs only in the time frame of interest. This aspect of catching the event "on the fly" requires experience and careful selection of delay times and pressure trigger levels. By recording two optical channels on the same transient recorder as the pressure signals, a very accurate measurement of the onset pressure can be achieved by observing when the optical signals start to change. By assuming an isentropic expansion up to the point of onset, the thermodynamic conditions at onset can be determined. A typical x-y plot of these traces is seen in Figure 8. The values used in the calculations of onset conditions are taken from the stored digitized signals residing in the transient recorder's memory. As can be seen in Figure 8, the sudden change in the attenuation signals are a very accurate indication of onset. The other optical outputs also give a good indicator of onset, as can be seen from Figure 9.

3.2   Experimental Results - Nucleation Onset

With the pressure known, a Mollier chart can be used to locate the equilibrium wetness at nucleation onset (Wilson line), as shown in Figure 10. The

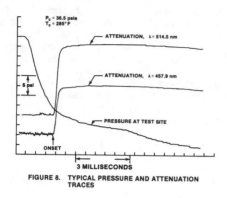

FIGURE 8. TYPICAL PRESSURE AND ATTENUATION TRACES

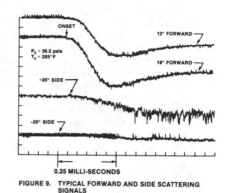

FIGURE 9. TYPICAL FORWARD AND SIDE SCATTERING SIGNALS

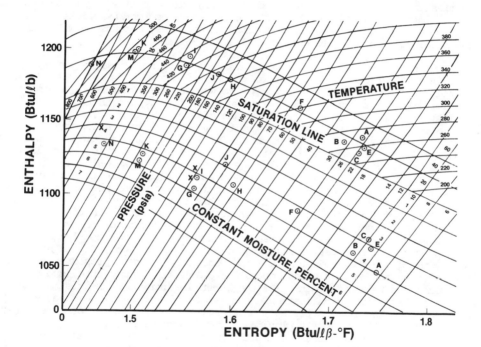

NOTES: DIMERALIZED WATER, X · INDICATES
EARLY ONSET, EXPANSION RATES (s⁻¹)
ARE AS FOLLOWS: A(470), B(375), C(409),
(410), F(385), G(562), H(430), I(868), J(553),
K(633), M(572), N(632). 42″ NUCLEATION
TUBE LENGTH

FIGURE 10. EQUILIBRIUM WETNESS AT NUCLEATION ONSET (WILSON LINE)

initial starting point is also located and the expansion rate at onset is given.
To locate the onset point the supersaturated vapor is assumed to expand as a
perfect gas with the properties of the initially superheated steam. By calculating the isentropic enthalpy drop the enthalpy at onset can be used along
with the measured pressure to establish the onset location on the Mollier diagram. This procedure was used previously by Yellot, et al (13). In some cases
an "early" onset is noted, that is the optical signals start changing at a
very slow rate before the sharp change at normal onset. The start of this slow
change is designated as the "early" onset. The general trend of the data indicates that supersaturation increases with the pressure at onset.

3.3   Experimental Results - Droplet Growth

Attenuation - In the course of testing it became clear that the attenuation method was the most reliable technique and would be the main source of
information. The discussion of the problems with the two scattering systems
will be taken up later and we will continue with the discussion of the attenuation method. For the attenuation method, only the voltage at the no light
transmission condition is necessary to determine droplet size. This data is
recorded immediately after each test with the same recorder by blocking the
light input. The quantity of data sampled in each test depends on which transient recorder is used, with either 1024 and 4096 samples.

With such large amounts of data, averaging over many samples is employed
to eliminate the effects of random background noise and to "smooth" the data.
Initially, it was hoped to use two separate wavelengths produced by a single
argon-ion laser to avoid the added complexity of integrating another laser into
the tightly packed optical set up. The use of a single laser system was aban-
doned when it was realized that, unless the wavelength separation was increased,
the ability to reliably sense droplets of less than 0.5 μm in diameter would be
in doubt. After changing to the two laser operation, as seen in Figure 11, an
increased sensitivity of the method was evident. Also shown on Figure 11 are
the steam pressure and the equilibrium wetness based on an isentropic expansion.
With this technique, droplets less than 0.2 micron have been measured, as seen
in Figure 11. One particular advantage of this technique over the convergent-
divergent nozzle studies is that the expansion process occurs in time rather
than space. By high speed sampling, a detailed description of the droplet
growth history can be obtained instead of a somewhat sketchy description that
can be provided by the spatial sampling of the nozzle method. In the nozzle
studies, to change the expansion rate independently of onset pressure requires
different nozzles for each expansion rate. In the nucleation tube, one need
only change the relative position of the burst diaphragm and the test site.
Likewise the location of nucleation onset is identified as the point at which
the attenuation and the scattering signals move away from their baselines. The
sudden change in these signals provides a very accurate location of the point
of onset. This is to be contrasted with the prior method of locating the
pressure change due to release of latent heat. In fact, as both signals are
available in the test data, it can be seen that this pressure rise can be very
subtle and usually only noticeable well after nucleation has started. Due
to the low signal-to-noise ratio in this very early part of the transient,
however, it is impossible to determine reliable droplet sizes until somewhat
later in the transient.

Effect of Contaminants - A heuristic insight into the effect of contaminants can be gained by comparing two series of tests: one with triple-distilled
feedwater, and a second that used triple-distilled water that was additionally
circulated through the Barnstead Nano-Pure demineralizer. Figure 12 displays
the typical droplet diameter growth curves of the triple-distilled water test.

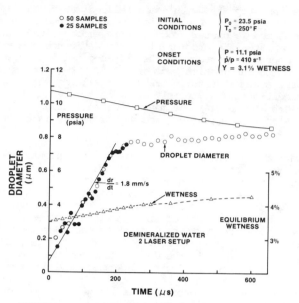

FIGURE 11. DROPLET GROWTH, PRESSURE AND
EQUILIBRIUM WETNESS VS TIME

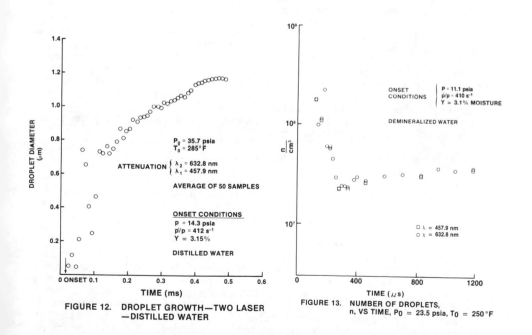

FIGURE 12. DROPLET GROWTH—TWO LASER
—DISTILLED WATER

FIGURE 13. NUMBER OF DROPLETS,
n, VS TIME, $P_0 = 23.5$ psia, $T_0 = 250\,°F$

These tests were then repeated with demineralized water and Figure 11 shows typical results. Compared to the earlier results, the droplet diameter changes with demineralized feedwater produced were smoother and exhibited, in most cases, a definite plateau in droplet size. The evidence presented here suggests that the impurities present in the triple-distilled water tests had a definite effect on the nucleation process. A 10 ppb concentration of NaCl in 100 psia saturated steam results in 1.2×10^{12} salt molecules per cubic centimeter. This concentration is probably equal to or greater than the number density of droplets at the start of measurable nucleation. The effect of dissolved impurities on the nucleation process is clearly in need of further investigation.

Forward Scattering - The forward scattering techniques has not produced results as promising as attenuation. Forward scattering consistently yields clear signals with repeatable trends. However, when the ratio of the two signals is taken and all the calibration factors are applied, the ratio is not always in the proper range. To resolve this, one can resort to scaling the data from some base, such as the peaks of the individual traces which occur when the lobes of the scattering patterns cross the mask openings. These positions can be identified with a certain monodisperse droplet size and then used to readjust the ratio, R_d, to scale the data. At times this worked and at other times it would not. The scaling factor, likewise, did not remain invariant.

On occasion the forward scattering results were in good agreement with the attenuation results at small droplet sizes and predicted somewhat smaller droplet sizes than the attenuation signal above $1 \mu m$. The most probable cause of the divergence of the two results, after $1 \mu m$, is multiple scattering as the droplet cloud becomes more opaque. The forward scattering method is sensitive to the lobe structure of the scattered light, while the attenuation method is sensitive only to the amount of light scattered.

Side Scattering - Likewise, the side scattering or dissymmetry technique has had very limited success. The technique suffers with the scaling problem encountered in the forward ratio method. Also, it is very difficult to keep the scattered light beam's focal points centered in the slit aperture. To maintain centering, a much wider aperture than anticipated was used. The combination of a wide aperture and a high droplet number density, results in an inability to characterize single droplets. The principal function of this technique is to help locate the onset of nucleation by feeding into the same recorder as used for the pressure data.

3.4 Experimental Results - Droplet Number Density

The number of water droplets per unit volume can be directly calculated from either of the two individual attenuation measurements, once the drop diameter is known. Implicit in this is the assumption that the droplets are all of the same size, i.e., monodisperse. Using the blue laser line ($\lambda = 457.9$ nm) and the droplet diameters as shown in Figure 11 (for demineralized water), a droplet number density transient has been calculated and is displayed in Figure 13. The general form of the result is a rapid decrease in number density immediately after nucleation onset. If the droplet cloud is polydisperse, then some alteration in these trends could be expected. To gauge the polydispersity of the droplet fog, the attenuation levels from both wavelengths can be used. From the work of Dobbins and Jizmagian (14), it is known that when the scattering coefficient is integrated over a polydisperse droplet size distribution the "effective" coefficient will be changed. Since the original monodisperse droplet scattering coefficient was used to determine the functional relation between droplet size and the ratio of the attenuation levels, then use of both wavelengths will predict the same droplet number density if the droplet fog is truly monodisperse. If the droplet fog is polydisperse, the two calculations of the number density will be different. In Figure 13 the results of such a calculation can be seen. The red line (λ = 632.8 nm) predicts a slightly lower droplet number density. This slightly lower droplet number (as predicted by the red line absorption) is consistent with a droplet size distribution that is biased to the smaller sizes. The difference here is quite small - about 4% initially to 2% later in transient. The sensitvity of the attenuation method to variations in the width of the polydisperse distribution function is needed to quantify these results, however.

4.   CONCLUSIONS

A technique for measuring non-equilibrium steam condensation using a nonsteady expansion wave to provide a controlled, but rapid, expansion of slightly superheated steam has been developed. The onset of droplet formation and its subsequent growth are determined by measuring the attenuation and scattering of laser light. With this technique, droplets less than 0.2 micron have been measured. The location of nucleation onset is identified as the point at which the attenuation and the scattering signals move away from their baselines. The sudden change in these signals provide a very accurate location of the point of onset. The quality of the data taken on this project is very high and the technique is sensitive enough to be able to sense the difference between using triple-distilled feedwater and demineralized triple-distilled feedwater. These tests represent the first evidence of the effects of dissolved contaminants on the steam condensation processes. Since a two-color attenuation technique is used, the droplet diameter and the droplet number density history have both been determined.

ACKNOWLEDGEMENTS

The research contained in this paper was supported by the Electric Power Research Institute under Contract RP735-1. The author would like to point out that the experimental results stated in this paper are in large part due to the talent and diligent effort of Frank Bowden. The author would also like to thank Dr. Murray Penney for his very necessary consulting on the optical design of the droplet sizing systems. Lastly, but not leastly, the author is grateful for the fine work done by Roger Johnson on the design of the nucleation tube test section and associated hardware.

REFERENCES

1.   Gyarmathy, G. and Lesch, F., "Fog Droplet Observations in Laval Nozzles and in an Experimental Turbine," Proc. Inst. Mech. Eng. (London), Vol. 184, pt. 3G(III), pp. 29-36, 1969-1970.

2.   Petr, V., "Measurement of an Average Size and Number of Droplets During Spontaneous Condensation of Supersaturated Steam," Proc. Inst. Mech. Eng. (London), Vol. 184, pt. 3G(III), pp. 22-28, 1969-1970.

3.   Krol, T., "Results of Optical Measurements of Diameters of Drops Formed Due to Condensation of Steam in a Laval Nozzle," (in Polish), Trans. Inst. Fluid Flow Mech. (Poland), No. 57, pp. 19-30, 1971.

4.   Saltanov, G.A., Seleznev, L.I., and Tsiklauri, G.V., "Generation and Growth of Condensed Steam at High Pressure: First Experimental Results," Proc. of the Institution of Mechanical Engineers, Vol. 187, 1973, p.192.

5. Gyarmathy, G., Burkhard, H.P., Lesch, F., and Siegenthaler, A., "Spontaneous Condensation of Steam at High Pressure: First Experimental Results," Proceedings of the Institution of Mechanical Engineers, Vol. 187, 1973, p. 192.

6. Smith, A., "Wilson Point Experiments in Steam Turbines," Third Conference on Steam Turbines of Great Output, September 24-27, 1974, at Gdansk, Poland.

7. Glass, I.I. and Patterson, G.N., "A Theoretical and Experimental Study of Shock Tube Flows," Journal of the Aeronautical Sciences, Vol. 22, No. 2, February 1955, pp. 73-100.

8. Penndorf, R. (1965), "New Tables of Mie Scattering Functions for Spherical Particles," Geophysical Research Paper No. 45, Part 6: Air Force Cambridge Research Lab, Bedford, Massachusetts.

9. Gravatt, C.C., Jr., "Real Time Measurement of the Size Distribution of Particulate Matter by a Light Scattering Method," Journal of the Air Pollution Control Association, Vol. 23, No. 12, 1973, p. 1035.

10. Mie. G. (1900), Ann. Physik, 25, 377.

11. Winkler, E.M., "Section G - Condensation Study by Absorption or Scattering of Light," Physical Measurements in Gas Dynamics and Combustion, Vol. IX, Princeton University Press, Princeton, New Jersey, 1954, pp. 289-306.

12. Shapiro, A.H., "The Dynamics and Thermodynamics of Compressible Fluid Flow," Vol. II, Ronald Press, 1954.

13. Yellot, J.I. and Holland, C.K., "The Condensation of Flowing Steam, Part I: Condensation in Divergent Nozzles," Transactions of the American Society of Mechanical Engineers, Vol. 59, 1937, p. 171.

14. Dobbins, R.A. and Jizmagian, G.S., J.Opt. Soc. Amer., Vol. 56, p. 1345, 1966.

Prediction of Blade Erosion in Wet Steam Turbines

GIANNI BENVENUTO and MICHELE TROILO
Istituto di Macchine
Facoltà di Ingegneria
Università di Genova, Italy

ABSTRACT

The paper considers the problem of the liquid phase motion in wet steam turbines and presents a simplified mathematical model that, starting from the drop trajectory evaluation in the region between fixed and moving blades, is applied to calculate the impact points, where the drops hit the blade, and the component of the impact velocity normal to the blade surface.

On the basis of such results, it is possible to identify the blade regions where the erosion is likely to take place. The erosion threat can be then quantitatively predicted by means of correlations which take into account the effect of the normal impact velocity, of the impinging droplet size and of the erosion resistance of the material.

An extensive calculation of the specific erosion rate has been carried out for a typical wet steam turbine rotor, by varying the drop initial conditions and the stage axial clearance, with the aim to detect the relevant parameters and to give criteria to reduce the erosion threat since the early stage design.

1. INTRODUCTION

The problem of blade erosion in wet steam turbines is regarded as a limiting factor for further possible developments in large output turbine design. This explains why studies on the subject, both theoretical and experimental, are still developed in many research centers in the world. Although the main events leading to erosion in steam turbines are known, a better knowledge of the phenomena involved and of the relative importance of the different parameters, both geometrical and fluid-dynamic, that influence the steam turbine during its operation, can be useful not only to give a more complete understanding of the problem but also to conveniently suggest the design solutions that could be adopted in order to reduce the erosion threat.

The purpose of the present paper is to give a contribution to the solution of such problems, by means of a theoretical analysis of the liquid phase motion in a low pressure stage, assuming a simplified mathematical model, that nevertheless takes into account the different behaviour of the liquid and gas phases and the effect of the characteristic three dimensional flow inside the turbine. The model is applied to the quantitative prediction of the erosion threat for a

569

typical blade of a low pressure turbine rotor, by means of a procedure which requires the computation of the impact points, where the water drops hit the blade,
and of the normal component of the impact velocity.

2. CAUSES OF THE EROSION

It is known ([1]) that the blade erosion in turbine stages operating with
steam in the high moisture content region, is caused by the high impact velocity
of comparatively large drops on the moving blade surface. The creation of micro-
subsurface cracks by each impact leads after a time to the formation of a mosaic
that ultimately allows pieces of the blade material to fall out. The sequence of
events leading to this end can be briefly summarized as follows. A small fraction
of the liquid phase present in the steam flow in form of fine water drops (having
a size of less than about 1 μm) is deposited on blading and other surfaces
through impact and diffusional processes. The water collected by stationary blades flows over them towards the trailing edge as liquid film or rivulets under
the influence of the steam drag. The water rupture at the trailing edge is generally believed to occur in two distinct phases: the first one in which the water
detachement gives rise to the formation of large drops (primary atomization) and
a second one in which such drops, at a certain distance from the trailing edge,
are further on disrupted by the steam flow into smaller fragments (secondary atomization). It may be observed that large drops (up to 1500 μm diameter) formed
by primary atomization, can be stable only in the low velocity zone of the nozzle
wake. The secondary atomization otherwise can take place in flow regions near the
wake, where the steam velocity, relative to the drops, is large enough to produce
their disruption.
    The water droplets are then accelerated by the gas phase in the anulus be-
tween stator and rotor in such a way that the smaller is their size, the greater
will be their acceleration and viceversa. The collision velocity and hence the
erosive action will be higher for the drops of greater size because of their inability to follow the gas phase in the rotor channels.
    It seems confirmed ([2])by experiments and investigations of several authors
that the dimensions of water drops dangerous as regards the erosion are limited
in a relatively narrow range. Drops less than 20-40 μm diameter should not be
dangerous and, on the other hand, drops grater than 300-400 μm diameter could
not be stable in the stream and should be fragmented before reaching the rotor.
Besides droplet dimensions, other factors influencing the extent and depth of the
blade erosion are:
- the peripheral velocity of the rotor blades, that substantially affects the
drop impact velocity;
- the steam density, whose influence on the Weber number and Reynolds number is
determinant for the development of drop atomization and acceleration processes;
- geometrical characteristics of the stage design, such as the axial gap between
the vane and the following blade, or the particular conformation of the stator
blade trailing edge, that can affect the life history of the water drops.

3. TWO-PHASE FLOW SIMPLIFIED MODEL

The basic assumption under which the two-phase flow is described in this paper, is that the presence of droplets does not affect the gas flow field. As shown in [8] , the treatment of the gas flow field without interaction terms due to the particles, is correct for small values of the particle volume fraction. For the case considered here, this parameter is less than 10^{-4} (see [4]), allowing the "single particle" calculation scheme to be adopted.

Following this approach, the problem of the motion of the suspended phase in wet steam turbines is reduced to the solution of the drop motion equation, written in vectorial form:

$$m \frac{d\bar{c}}{dt} = \bar{F}_D + \bar{F}_g + \bar{F}_m \qquad (1)$$

where: m droplet mass, $\bar{c}$ absolute velocity, t time, $\bar{F}_D$, $\bar{F}_g$, $\bar{F}_m$ external forces due respectively to aerodynamic drag, to gravity and to droplet mass variation. Expressing the mass of the droplet, supposed of spherical shape, as:

$$m = \frac{\pi D^3}{6} \rho_1 \qquad (2)$$

and the right hand side forces as:

$$\bar{F}_D = - \frac{1}{2} \rho_g \, c_D \, \frac{\pi D^2}{4} \, w \, \bar{w} \qquad (3)$$

$$\bar{F}_g = \bar{g} \, (\rho_1 - \rho_g) \, \frac{\pi D^3}{6} \qquad (4)$$

$$\bar{F}_m = - \frac{dm}{dt} \, \bar{w} \qquad (5)$$

where c_D is the sphere drag coefficient and $\bar{w}$ the relative velocity between the water drop and the gas phase:

$$\bar{w} = \bar{c} - \bar{c}_g \qquad (6)$$

the equation (1), divided by the mass of the drop, takes the form:

$$\frac{d\bar{c}}{dt} = - \frac{3}{4} \frac{\rho_g}{\rho_1} \frac{c_D}{D} w \, \bar{w} + \bar{g} \, (1 - \frac{\rho_g}{\rho_1}) - \frac{3}{D} \frac{dD}{dt} \, \bar{w} \qquad (7)$$

As shown in previous works ([3,4,5]), the gravity force is small as compared to the drag force and usually can be ignored.

For checking the importance of the reaction force due to the drop mass variation, expressed by the third right hand side term of the above equation, a simple calculation has been made, based on the Gyarmathy ([9]) droplet growth equation:

$$\frac{dD}{dt} = \frac{4 \lambda}{\rho_1 L} \frac{\Delta T \, (1 - D_{crit}/D)}{D \, (1 + 3.18 \, Kn)} \qquad (8)$$

in which the variation of the drop diameter is written in function of the steam undercooling ΔT and of the ratio between the critical and the actual drop diameter. Results obtained by solving the system of the two equations (7) and (8), taking into account the dependence of the critical diameter on the steam thermodynamic variables according to the Kelvin–Helmholtz formula:

$$D_{crit} = 4\sigma/(\rho_1 \, R \, T_g \, \ln(p/p_S)) \qquad (9)$$

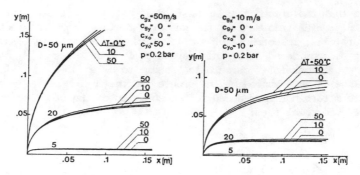

Fig. 1 - Trajectories of drop subjected to mass variation

are presented in fig. 1. Here water drops of different initial diameters are supposed to be injected at right angle in a uniform stream of steam. The trajectory
variations due to different degrees of steam undercooling are evidentiated. It
can be seen that, even with small values of velocity and high degrees of undercooling, the trajectory variations, for distances corresponding to the path traveled in a turbine stage, are small, so that the influence of the drop mass variation can in general be disregarded.

Then equation (7), retaining as relevant only the aerodynamic force, reduces
to the form:

$$\frac{d\overline{c}}{dt} = - Q\, c_D\, w\, \overline{w} \tag{10}$$

being

$$Q = \frac{3}{4} \frac{\rho_g}{\rho_1} \frac{1}{D} \tag{11}$$

For a three-dimensional study of the liquid phase motion in turbine it is
convenient to write eq.(10) with respect to a fixed cylindrical frame, in which
r, z, ϑ are the radial, axial and tangential coordinates respectively (see fig.2):

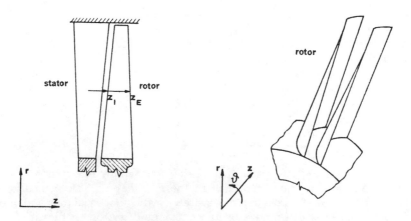

Fig. 2 - Sketch of a turbine stage in the meridional plane and three-dimensional view of two adjacent rotor blades with the proper cylindrical reference frame.

$$\frac{dc_r}{dt} = - Q\, c_D\, w\, w_r + \frac{c_\vartheta^2}{r} \qquad\qquad \frac{dr}{dt} = c_r$$

$$\frac{dc_z}{dt} = - Q\, c_D\, w\, w_z \cdot \qquad\qquad\qquad \frac{d\vartheta}{dt} = \frac{c_\vartheta}{r} \qquad (12)$$

$$\frac{dc_\vartheta}{dt} = - Q\, c_D\, w\, w_\vartheta - \frac{c_\vartheta\, c_r}{r} \qquad\qquad \frac{dz}{dt} = c_z$$

The gas velocity field in the turbine, that must be known in the whole region of interest in order to integrate the system (12), has been considered till now a simple free-vortex flow in which the circulation is destroyed inside the rotor at a fixed rate, reaching zero at the outlet, while tha axial velocity remains constant: the small deviation between the assumed flow and the real flow encountered in turbine -apart from two-dimensional near the blade leading edges, and non stationary or secondary effects, that anyhow are ignored also in through flow calculations- is widely compensated by its very easy analytical description:

$$c_{gr} = 0 \qquad\qquad c_{gz} = \text{const.} \qquad\qquad c_{g} = \frac{\Lambda(z)}{r} \qquad (13)$$

being $\Lambda(z)$ the circulation defined by:

$$\Lambda = c_{g\vartheta H} \cdot r_H \qquad\qquad (z < z_I)$$

$$\Lambda = c_{g\vartheta H} \cdot r_H \cdot \frac{z - z_E}{z_I - z_E} \qquad (z_I < z < z_E) \qquad (14)$$

$$\Lambda = 0 \qquad\qquad\qquad (z < z_E)$$

The system (12) can be solved by means of a Runge-Kutta numerical procedure for a set of initial conditions, taking into account the proper dependence of the drag coefficient from Reynolds and Knudsen numbers (see e.g. [4]) and calculating the gas flow velocities along the integration path with the simplified formulation (13).

In previous works ([3,4,5]), in order to get acquaintance with the basic behaviour of the drop trajectories in turbine, an infinitely thin blade has been considered, whose shape, in the relative reference frame, is given by the equation:

$$\vartheta_R = \vartheta_{RI} - \frac{c_{g\vartheta H} \cdot r_H\, (z_E - z_I)}{2\, r^2\, c_{gz}} \left[\left(\frac{z_E - z}{z_E - z_I}\right)^2 - 1 \right] + \frac{\omega}{c_{g\vartheta}} (z_I - z) \qquad (15)$$

obtained by integration of the tangency condition between relative gas velocity and blade profile:

$$\frac{dz}{r\, d\vartheta_R} = \frac{c_{gz}}{c_{g\vartheta} - \omega r} \qquad (16)$$

As an example of these previous calculations, fig. 3 shows the trajectories and the points of impact of 100 μm water drops, supposed to leave the stator at different blade heights and angular positions. The two cases presented show the influence of the initial drop velocity on the trajectory shape and on the impact point locations.

In the present work we have extended the calculations to a real rotor blade,

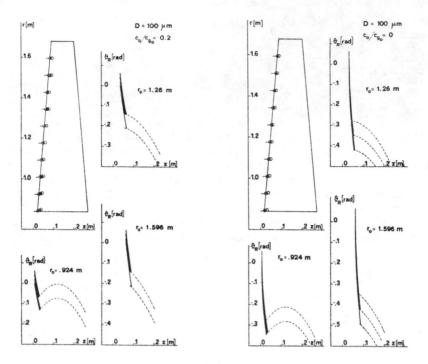

Fig. 3 - Drop trajectories and points of impact in the (z,r) and (ϑ_R,z) planes; dotted lines are traces of blade sections at the radius of impact

with the aim to improve the adequacy of the theoretical model and to verify the possibility of a quantitative evaluation of the blade erosion.

4. EVALUATION OF THE EROSION THREAT

A quantitative evaluation of the erosion threat requires, for given drop diameters and initial conditions, the calculation of the impact points where the drops hit the blade and of the normal component of the impact velocity. The total volume of water impinging on the unit blade surface element per unit time should also be known.

4.1. Impact Point Prediction

The point of impact of water drop impinging on a rotor blade can be considered as the intersection of the drop trajectory, calculated in the relative reference frame, moving with the rotor, and the blade surface. The problem can be treated writing the drop trajectory in parametric form (with the time as parameter):

$$r = r(t) \qquad\qquad z = z(t) \qquad\qquad \vartheta_R = \vartheta_R(t) \qquad\qquad (17)$$

together with the equation of the blade surface:

$$\vartheta_S = \vartheta_S(r,z) \tag{18}$$

and searching for that point of the trajectory for which:

$$\vartheta_R(t) = \vartheta_S(r(t),z(t)) \tag{19}$$

For the numerical solution, an approximated procedure has been adopted till now. Since the trajectories, determined by integration, are known as a series of points, as well as the blade surface, at a given radius $r(t)$, linear interpolation has been made between two adjacent points, both of the trajectory and of the blade surface. The approximation thus introduced is shown in fig. 4.

4.2. Calculation of the Normal Component of the Impact Velocity

From a mathematical point of view, the normal component of the impact velocity w_i is given by:

$$w_{in} = \bar{w}_i \cdot \bar{n}_S \tag{20}$$

being $\bar{n}_S$ the unit vector perpendicular to the blade surface in the point of impact. In the relative frame, if the blade surface is given in the form:

$$f_S(r,z,\vartheta_R) = 0 \tag{21}$$

the vectors $\bar{w}_i$ and $\bar{n}_S$ in eq. (20) can be expressed as:

$$\bar{w}_i = w_{ir} \bar{i}_r + w_{iz} \bar{i}_z + w_{i\vartheta} \bar{i}_\vartheta \tag{22}$$

$$\bar{n}_S = \frac{1}{\Delta} \left(\frac{\partial f_S}{\partial r} \bar{i}_r + \frac{\partial f_S}{\partial z} \bar{i}_z + \frac{1}{r} \frac{\partial f_S}{\partial \vartheta} \bar{i}_\vartheta \right) \tag{23}$$

where:

$$\Delta = \sqrt{\left(\frac{\partial f_S}{\partial r}\right)^2 + \left(\frac{\partial f_S}{\partial z}\right)^2 + \left(\frac{1}{r} \frac{\partial f_S}{\partial \vartheta}\right)^2} \tag{24}$$

For computational purposes, some approximations can be introduced in the calculation of w_{in}:
- the radial component w_{ir} in eq. (22) can be neglected, being small in comparison with w_{iz} and w_i: previous calculations ([4]) have shown in effect that w_{ir} becomes important only near the blade hub, where erosion is harmless;
- in eq. (23) the term $\partial f_S/\partial r$ can be neglected with respect to the others,

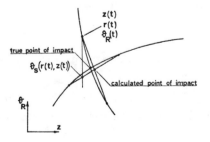

Fig. 4 - Impact point calculation.

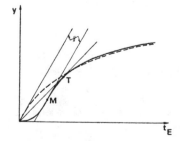

Fig. 5 - Erosion versus time curve.

owing to the characteristic shape of the blades in the axial turbomachinery: the consequent overestimation of w_{in} compensates the errors due to the first approximation.

4.3. Calculation of the Erosion Rate

Following the classical approach reported in [6],[7], an empirical correlation is defined:

$$t_E = 4.89 \cdot 10^{11} \; y \; \frac{k}{q_B} \; w_{in}^{-4.92} \; D^{-1.69} \; \exp(0.25y/y_T) \qquad (25)$$

where:

t_E time for erosion depth y, (s)

y erosion depth (m)

y_T erosion depth for maximum mean erosion rate (see fig. 5)

k numerical factor, depending on the material

q_B water volume impinging the unit blade surface in the unit time

Referring to fig. 5, the slope of the curve (25), dotted line, in the point $t_E = 0$, is equal to the maximum slope of the true erosion curve (solid line), in the point M. So the maximum erosion rate is given by:

$$u_{E_{max}} = (\frac{dy}{dt_E})_{t_E=0} = 0.205 \cdot 10^{-11} \; \frac{q_B}{k} \; w_{in}^{4.92} \; D^{1.69} \qquad (26)$$

In the present work, whose aim is the study of the influence, on the erosion threat, of design and operation parameters, the following non-dimensional quantity will be considered:

$$\frac{u_{E_{max}}}{q_B} = 0.205 \cdot 10^{-11} \; \frac{w_{in}^{4.92} \; D^{1.69}}{k} \qquad (27)$$

Indeed, the non linearity of the erosion correlations prevents from superimposing the effects of drops of different diameters and velocities. Conversely, the maximum erosion rates for different drop diameters, give directly useful informations on the measures to be adopted to reduce the erosion threat. Anyway, the calculation of the absolute erosion depth, by means of (25), requires the knowledge of the amount of impinging water, which is till now hard to estimate. Nevertheless, if the water leaving the stator blades as droplets is supposed -as reasonably- to be seen from the rotor as uniformly distributed along the pitch, at least the water distribution along the blade surface can be evaluated. Indeed the water volume impinging on the unit blade surface in the unit time may be written:

$$q_B = w_{in} \qquad (28)$$

The same water amount leaving the stator may be written, per unit cross section, at the stator trailing edge, and per unit time, as:

$$q_S = w_z \qquad (29)$$

being w_z the axial component of the drop velocity in the section considered. Then one has:

$$q_B = q_S \; \frac{w_{in}}{w_z} \qquad (30)$$

and (27) may be referred to a water volume constant over the pitch, introducing the quantity:

$$E = \frac{u_{E_{max}}}{q_S} = 0.205 \cdot 10^{-11} \frac{w_{in}^{4.92} D^{1.69}}{k} \frac{w_{in}}{w_z} \quad (31)$$

This way the relative variation of the maximum erosion rate along the blade surface can be evaluated, so evidentiating the blade regions subjected to higher erosion threat.

5. DETAILED ANALYSIS OF A ROTOR BLADE

The calculation procedure already outlined has been applied to the erosion prediction of a typical turbine rotor operating in wet steam. In fig. 6, a three dimensional sketch of the blade is reported, where the five sections considered for the erosion calculation are also indicated. The drop diameters have been chosen, according also to the Weber criterion, as $D = 50, 100, 200$ μm; the influence of the axial gap between stator and rotor, and of the slip between drops and gas phase at the leaving has been studied.

Results here presented are relative to cases with $c_o/c_{go} = .1, .5$ and axial gap = 1, 2, 3 cm. In fig. 7, sample trajectories are plotted, for the considered blade sections, to show differences mainly due to the blade velocity increase towards the tip. In fig. 8, the influence of the slip and axial gap at a selected radius is shown; while the axial gap seems to have a poor influence on the trajectory, the slip plays an important role, in that a remarkable reduction of the normal component of the impact velocity and a spreading of the impact points in a wider region are observed, when the initial drop velocity is closer to the gas velocity.

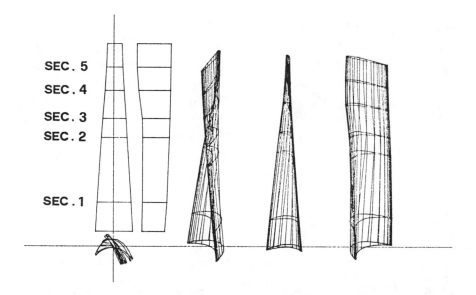

SEC. 5
SEC. 4
SEC. 3
SEC. 2

SEC. 1

Fig. 6 - Three-dimensional sketch of the analyzed blade.

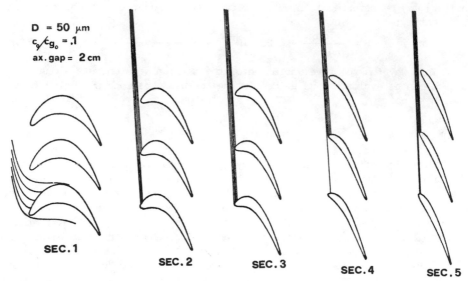

Fig. 7 - Drop trajectories at different blade heights.

In fig. 9 and 10 the function E defined in (31) is plotted in polar-logaritmic coordinates around the profile sections considered. The following comments describe the situation, in agreement with the basic erosion phenomena knowledge:
- the erosion rate increases towards the blade tip;
- the erosion rate increases for large drop diameters;
- a higher initial drop velocity reduces the erosion rate, that anyhow affects a wider blade surface;
- the axial gap, in the range considered, does not seem to influence appreciably

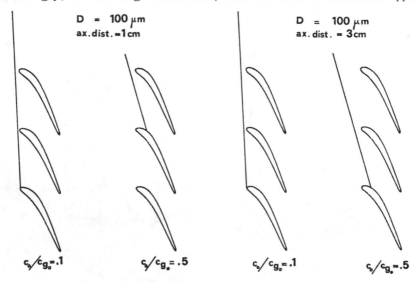

Fig. 8 - The influence of the initial velocity on the drop trajectory.

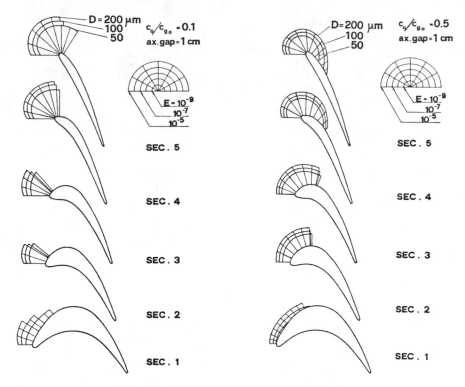

Fig. 9 - Calculated erosion rates.

the erosion rate, except for hub sections, because the drop relaxation time is too high to admit major changes in the trajectory within the concerned space.

Remembering that the absolute erosion depth in function of the time can be calculated from the E function -plotted in figs. 9,10- once q_s and k are known, the results obtained may be directly used as a relative erosion threat criterion, for they allow a quantitative comparison of the erosion threat for different operating or design conditions.

6. CONCLUSIONS

By means of a calculation procedure, based on some simplifying assumptions, the erosion rates on a typical turbine blade, operating in wet steam, have been obtained, in function of drop size, initial slip and axial gap.

A relative erosion threat criterion has been established, which seems to be well useful, as long as data on water content in actual situations, and material erosion resistance indexes are not yet available.

The analysis carried out till now indicates a low initial slip as an important means to keep the erosion rate at comparatively low levels, so encouraging those efforts directed towards the erosion threat reduction by means of wake dynamization techniques.

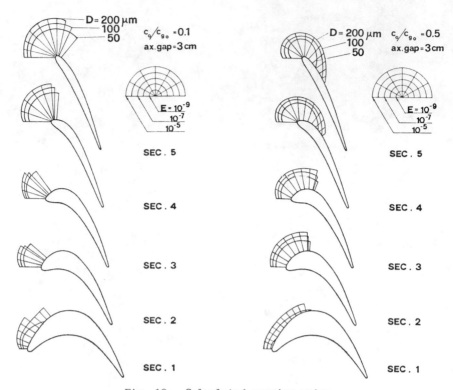

Fig. 10 - Calculated erosion rates.

SYMBOLS

		Subscripts	
c	absolute velocity, coefficient	B	blade
D	drop diameter	D	drag
E	specific erosion rate	E	erosion, exit
F	force	g	gas phase
k	material resistance index	i	impact
i	unit vector	I	inlet
L	latent heat of vaporization	l	liquid
m	mass	n	normal
n	normal unit vector	r	radial component
p	pressure	R	relative
q	specific volume flow rate	S	surface, saturation
r	radial coordinate	z	axial component
R	gas constant for the steam	ϑ	tangential component
t	time	o	initial
T	temperature		
u	erosion rate		
w	relative velocity		
y	erosion depth		
z	axial component		

ϑ tangential coordinate
λ steam thermal conductivity
$\wedge$ circulation
ρ density
σ interfacial surface tension
ω angular velocity

REFERENCES

1. G.C.Gardner: "Events leading to erosion in the steam turbine" - Proceedings of the I.M.E. 1963-64, vol. 178, p. 153.

2. M.J.Moore, K.H.Sieverding: "Two-phase steam flow in turbines and separators"- Hemisphere Publishing Corporation, 1976.

3. O.Acton, G.Benvenuto, M.Troilo: "On the droplet motion in two-phase flow a- round single and cascaded airfoils" - VI Conference on Fluid Machinery, Budapest, 1979.

4. G.Benvenuto, M.Troilo: "Prediction of drop trajectory and points of impact in wet steam turbines" - Conference on Gas Borne Particles, I.M.E., Oxford, 1981.

5. G.Benvenuto, M.Troilo: "Previsione dell' erosione di rotori di grandi turbi- ne a vapore" - XXXVI Congresso Nazionale A.T.I., Viareggio, ottobre 1981.

6. F.J.Heimann: "Toward quantitative prediction of liquid impact erosion" - ASTM Special Technical Pubbl. 474, Philadelphia, 1969.

7. J.Krzyzanowski, Z.Szprengiel: "The influence of droplet size on the turbine blading erosion hazard" - ASME Journ. of Eng. Pow., vol. 100, p. 561, 1978.

8. G.Rudinger: "Fundamentals and applications of gas particle flow" - Agardo- graph N. 222, 1976.

9. G.Gyarmathy: "Condensation in flowing steam" in "Two-phase steam flow in turbines and separators" - Hemisphere Publishing Corporation, 1976.

Comments on the Influence of Trace Elements on Nucleation Rates of Steam

WOJCIECH STUDZINSKI, RICHARD A. ZAHORANSKY,
SIGMAR L. K. WITTIG, and GÜNTER H. SPIEGEL
Institut fur Thermische Strömungsmaschinen
Universität Karlsruhe [T.H.]
7500 Karlsruhe, FRG

ABSTRACT

Homogeneous binary nucleation theory is applied in analyzing the influence of impurities on the nucleation in low pressure end steam turbine stages. In modeling turbine flow, experiments were performed in an expansion tube. Water samples of different quality from operating steam power plants were used in comparing the experimental and theoretical results. It is shown that observed phase transitions at relatively low supersaturation - frequently cited as responsible for corrosion fatigue - are not induced by water impurities such as silicic acid or metal components of typical concentrations. In addition, fluid mechanical effects are to be considered in explaining the condensation phenomena in steam turbines during normal operation. Calculations show, however, that organic substances such as acetic acid can cause a shift of the nucleation zone.

1. INTRODUCTION

The relatively high supersaturation of pure steam, which is predicted by homogeneous unary nucleation theory and realized under ideal conditions in laboratory experiments such as in nozzle expansion flow or in expansion fans is not found in real turbine flow. Observed phase transitions in low pressure end stages of condensation turbines at very low supersaturation are generally explained by the influence of impurities in the steam due to water conditioning, metal components and other trace elements [5,8]. As a result corrosion fatigue of blades and bucket corrosion have been explained by binary enrichment in the nucleation zone [1,2,18].

The purpose of this paper is a critical analysis of the influence of these impurities on the nucleation rate by applying homogeneous binary nucleation theory based on the classical theory. The main emphasis is directed towards characteristic systems typical for turbine flow. Generally, silicic acid is a major impurity during normal operation of the power plant whereas organic acids are found in the feed water after the occurance of leakages - in the condensor for example [1,9,18]. In addition to the classical steady state nucleation theory the relaxation time of the binary nucleation process is to be considered in de-

583

scribing the turbine flow with characteristically high cooling rates.

For an analysis, clarification and verification of the theoretical models, experiments were performed using water samples of drastically differing quality expanded in the high pressure section of a shock tube. A major advantage of this technique is that the composition of the samples is well defined and that a nearly isentropic expansion is obtained up to the onset of nucleation, thus resembling approximately ideal turbine flow.

2. THEORETICAL MODEL

In analogy to unary condensation, three major regimes are found during homogeneous binary phase change:

- nuclei are formed after passing saturation up to the collapse of the subcooled state within the so-called Wilson regime

- growth and change of composition of the stable nuclei

- equilibrium condensation of the vapor on the surface of the droplets.

Of major importance in the present context is the first regime which determines particle concentration and size distribution. Growth of the binary droplets as a function of the expansion rate is observed in the second regime as described earlier by us [21]. Here the droplet composition will soon approach its equilibrium state. It should be noted, therefore, that the enrichment at the onset of nucleation is not only determined by the difference in the fugacities of the relative components but primarely by the surface tension. In estimating the enrichment of impurities within the droplet, knowledge of the composition of the nucleus on one hand and the phase equilibrium on the other are sufficient as limiting values. In principle, the time dependent droplet composition can be obtained by a binary droplet growth model which has been discussed by us in another context [21].

During fast expansion, the equilibrium composition, however, is quickly obtained. In typical Laval nozzles, for example, the characteristic path length of a nucleus in reaching equilibrium is of the order of a few millimeters.

Size and composition of the critical nucleus are determined by the conditions of metastable equilibrium. It can easily be shown that the saddle point on the energy surface built by the free enthalpy difference $\Delta G(n_1, n_2)$ of the nucleus with n_1 and n_2 molecules of the binary fluid from the vapor phase is determined as follows:

$$-\frac{\partial \Delta G}{\partial n_1}\bigg|_{n_2} = 0 \qquad (1a) \qquad \frac{\partial \Delta G}{\partial n_2}\bigg|_{n_1} = 0 \qquad (1b)$$

$$\left(\frac{\partial^2 \Delta G}{\partial n_1 \partial n_2}\right)^2 - \frac{\partial^2 \Delta G}{\partial n_1^2} \cdot \frac{\partial^2 \Delta G}{\partial n_2^2} > 0 \qquad (2)$$

The free enthalpy difference for isothermal nucleus formation from the vapor phase ΔG is

$$\Delta G = n_1 \Delta \mu_1 + n_2 \Delta \mu_2 + 4\pi r^2 \sigma + (p - p_s) \cdot (n_1 + n_2)v \quad (3)$$

Here $\Delta \mu_i$ is the difference of the molecular chemical potentials between gas and liquid bulk phase for component i, r is the droplet radius, $\sigma(x_i, T)$ the surface tension, p the total pressure of the binary mixture, $p_s(x_i, T)$ the saturation pressure, $v(x_i, T)$ the specific molecular volume of the mixture, T the temperatur, and x_i the mole fraction.

Terms containing the chemical potential describe the change of free enthalpy during condensation of n_1, n_2 molecules. The third term in Eq.(3) represents the energy necessary to form the surface whereas the last term is of minor importance and can be neglected.

Various techniques can be applied in solving Eqs. (2a) and (2b). In the following it is reduced to a nonlinear equation for the composition of the nucleus and an explicit expression of the droplet radius:

$$\frac{n_2}{n_1} = \frac{b_2}{b_1}\zeta\left[1+\frac{n_2}{n_1}\right]^{1-\zeta} \qquad (4)$$

$$r = 2\sigma(x_2,T)v(x_2,T)/(kT\ln S) \qquad (5)$$

where $\qquad b_i = a_i/\gamma_i = p_i/p_{si}^{\ominus}\gamma_i \qquad (6)$

$$\zeta = \frac{v_2 + 1.5x_1 v \partial \ln\sigma/\partial x_2}{v_1 + 1.5x_2 v \partial \ln\sigma/\partial x_1} \qquad (7)$$

$$S = S_1^{x_1} S_2^{x_2} \qquad (8)$$

$$S_i = p_i/(p_{si}^{\ominus}\gamma_i x_i) \qquad (9)$$

$$x_i = \frac{n_1}{n_1 + n_2} \qquad (10)$$

Here p_i is the partial pressure of component i, $p_{si}^{\ominus}$ the saturation pressure of the pure component i, γ_i the activity coefficient, v_i the partial molecular volume and S_i the supersaturation.

For small values of the mole fraction x_i the partial vapor pressure in equilibrium p_{si} for one component can also be described by Raoult's law whereas the other partial vapor pressure saturation state is obtained from Henry's law:

$$p_{s1} = p_{s1}^{\ominus} x_1 \tag{11}$$

$$p_{s2} = K_H x_2 \tag{12}$$

K_H is Henry's constant.

For steady state conditions, the binary nucleation rate can be calculated from kinetic theory [6,17]:

$$I = C \exp[-\Delta G^*/(kT)] \tag{13}$$

The star denotes the value of the saddle point.

The frequency factor C can be derived following Reiss [17]:

$$C = \frac{\beta_1 \beta_2}{\beta_1 \sin^2\phi + \beta_2 \cos^2\phi} (N_1 + N_2) Z 4\pi r^{*2} \tag{14}$$

$$\text{with } \beta_i = \frac{p_i}{(2\pi m_i kT)^{1/2}} \quad \text{the collision rate,} \tag{15}$$

m_i the molecular weight and N_i the molecular concentration in the vapor phase.

$$Z = \sqrt{-P/Q} \quad \text{is the Zeldovich factor} \tag{16}$$

$$P = \frac{\partial^2 \Delta G}{\partial n_1^2} \cos^2\phi + 2\frac{\partial^2 \Delta G}{\partial n_1 \partial n_2} \cos\phi \sin\phi + \frac{\partial^2 \Delta G}{\partial n_2^2} \sin^2\phi \tag{17}$$

$$Q = \frac{\partial^2 \Delta G}{\partial n_1^2} \sin^2\phi - 2\frac{\partial^2 \Delta G}{\partial n_1 \partial n_2} \cos\phi \sin\phi + \frac{\partial^2 \Delta G}{\partial n_2^2} \cos^2\phi \tag{18}$$

ϕ is the angle between the n_1-ordinate and the projected direction of the main nucleation path across the saddle point. It can be determined according to

$$tg\phi = (d_A - Rd_B)/2 + [(d_A - Rd_B)^2/4 + R]^{1/2} \tag{19}$$

where R is the correction factor proposed by Stauffer (e.g.[15]):

$$R = \beta_1/\beta_2 \qquad (20)$$

$$d_A = - \frac{\partial^2 \Delta G}{\partial n_1^2} \Big/ \frac{\partial^2 \Delta G}{\partial n_1 \partial n_2} \quad , \quad d_B = - \frac{\partial^2 \Delta G}{\partial n_2^2} \Big/ \frac{\partial^2 \Delta G}{\partial n_1 \partial n_2} \qquad (21a) \; (21b)$$

An important consideration which has been neglected so far in the nucleation theory presented is hydratation which is to be expected in mixtures with acids. In analyzing this effect it can be shown that the nucleation is kinetically accelerated and simultaneously the partial pressure of the free acid molecule is reduced leading to a decrease of the nucleation rate. The latter thermodynamic effect of hydratation dominates [19]. Application of our model not considering hydratation, therefore, will lead to a somewhat higher nucleation rate.

In deriving Eq. [13] for the nucleation rate infinitely slow expansion was assumed. The relaxation time required to attain the steady state value for I can be estimated by applying a formally simple expression [20]:

$$\tau = 3(n_1^* + n_2^*) \; \frac{L}{D^* \ln S^*} \qquad (22)$$

with $D^*(x_2) = 4\pi r^{*2}\beta_1\beta_2[x_2^2\beta_1 + x_1^2\beta_2]^{-1}$ $\qquad (23)$

Function L is a multiple integral term, which fortunately is of minor dependance on parameters of the nucleus. For supersaturation applicable in technical systems, L will vary between one and two. It, therefore, is possible to calculate the time dependent nucleation rate I_i knowing τ:

$$I_i = I[1 - \exp(-t/\tau)] \qquad (24)$$

3. APPLICATION TO PRACTICAL SYSTEMS

As shown by various recent statistics, the most important problem with respect to turbine reliability seems to be corrosion fatigue in the transitional stages from superheated to wet steam flow [11]. Acetic acid, sulfuric acid and chlorides have been mentioned as potential sources in inducing this phenomenon [1,2, 11,18]. Furthermore it has been argued that these impurities are responsible for a relatively early onset of nucleation at low supersaturation [5,8,14]. This is in contrast to available experimental data obtained under laboratory conditions.

Although multicomponent systems are found in real turbine flow the theoretical model presented in chapter 2 is only capable of describing binary mixtures. The present paper, therefore, is directed to describe the isolated influence of selected mixture components on the homogeneous condensation process.

	Rheinhafen Power Plant Karlsruhe	Power Plant EC2 Gdansk Poland	Ion Exchanger	Public Water Supply	German Standard(VGB)
Extraction Place	Condensate Pump	Condensate Pump			
Conductivity μS/cm	0.11	0.60	15	600	< 0.2
SiO_2 mg/kg	0.010	0.026	36.52	12.72	< 0.02
$Fe+Fe_2$ mg/kg	0.010	0.018	0.029	0.042	< 0.02

Tab. 1   Composition of water samples studied

In general two groups of impurities can be found in the feed water and condensate of conventional power plants. Primarely silica and metal components as well as water conditioners are found during normal operation. Table 1 shows the composition of feed water samples from two power plants in comparison with water from a laboratory ion exchanger and from public water supply. In addition, the German standards for power plant operation are found in column 5.

The second group consists of components which are due to component failure such as condensor leakage. Typically, organic components and their derivatives such as acetic acids are found.

Of predominant interest is silica an additive to feed water under normal operational conditions. In general this is the primary impurity. The solubility of amorphous SiO_2 and quartz in water vapor is a function of temperature and pressure. In contrast, for the liquid phase, it is a function of the temperature and allotropic structure. In liquids, solution of SiO_2 leads to a chemical reaction or hydratation as follows [12]:

$$(SiO_2)_x + 2H_2O \rightleftharpoons (SiO_2)_{x-1} + Si(OH)_4 \qquad (25)$$

The formed orthosilicic acid $Si(OH)_4$ transforms by hydratation and polymerisation by surpassing the solubility. The resulting sol and gel will lead to coatings on turbine blade surfaces, for example (see [9]).

Detailed information on SiO_2 solubility especially at relatively low pressures and temperatures which is of interest for condensation steam turbines can be found in the literature [9,12]. In knowing the solubility limits it is possible to calculate the partial vapor pressure and the composition in the liquid phase. Therefore, Henry's constant is determined. Fig. 1 illustrates the solubility of silicic acid in water.

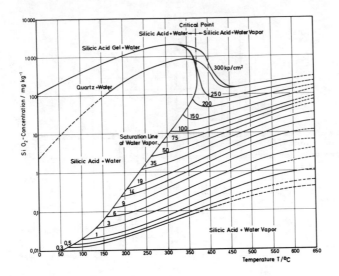

Fig. 1 Solubility for water/silica mixtures

The density of water/Si(OH)$_4$ solutions is derived from measurements [7]. Major difficulties, however, are experienced in determing the surface tension. Values for pure water are assumed.

Considering the second group, acetic acid aqueous solutions have been studied. Typical concentrations of 0.03 mg/kg steam and 0.29 mg/kg condensate have been observed [1]. The phase equilibrium for water/acetic acid mixtures as well as the saturation pressure of the pure acid, the density of the solution and the surface tension can be found in literature [4,10,22].

4. EXPERIMENTAL SETUP

The five water samples, shown in Table 1, with silica concentration from relatively high to extremely low levels have been investigated using conventional shock tube techniques. The experimental setup is shown in Fig. 2.

The vapor is expanded from the superheated state beyond the dew point into the two phase region within the expansion fan of the high pressure section. The driver section is preheated to ensure that the states for onset of nucleation are comparable with those of condensation turbine.

The tube is equipped with several measurement planes along its axis. Thus, by altering the distance of the observation station, the expansion rate can be varied. Quartz pressure transducers were used to record the fast expansion. For known initial conditions, the temperature can be calculated as the expansion is nearly isentropic up to the phase transition. The onset of nucleation is detected by photomultipliers monitoring the extinction of laser beams passing through the tube perpendicular to its axis.

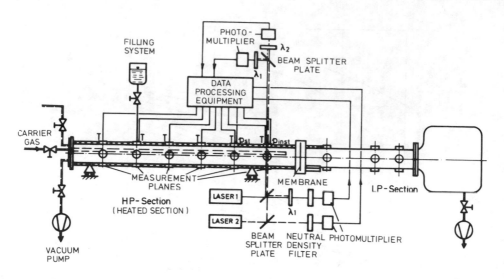

Fig. 2   Schematic of experimental setup

Calibration of the experimental technique and the feasibility of
the binary nucleation theory have been demonstrated by studying
mixtures such as n-propanol/ethanol and water/ethanol [23].

The test fluid is evaporated and superheated within the tube
itself. To ensure a nearly isothermal nucleation process - one
major assumption of the nucleation theory - it is necessary to
dilute the vapor by a carrier gas, nitrogen for example. A concentric perforated inner tube of small diameter provides intensive
and homogeneous mixing.

5. RESULTS AND DISCUSSION

Starting from initial temperatures in the range of approximately 40°C and partial steam pressure of 5750 Pa to 7070 Pa,
actual nucleation temperatures were close to room temperatures
(274°C to 277°C). This is shown in Fig. 3 which illustrates the
- somewhat surprising - experimental results:it can be seen that
even in varying the SiO_2 concentration two to three orders of magnitudes, it is 0.01 to 36.5 mg SiO_2/kg water, no remarkable effects
on the state of nucleation is found. It should be noted, however,
that the solubility of silicic acid in the vapor phase is limited
to 0.01 mg/kg steam apparently at a temperature of 40°C. As can
be seen, all 26 runs which were performed during the present test
series compare well with previous measurements in pure water vapor.
Deviations due to SiO_2 addition could not be isolated. Also other
additives such as Fe, Fe_2, calcium and others (see Table 1) are of
minor influence. The small scattering band of 26 experimental data
shown in Fig. 3 derives from differing expansion rates which have
their origin in the variation of the initial total pressure from
2.2 bar to 5.5 bar.

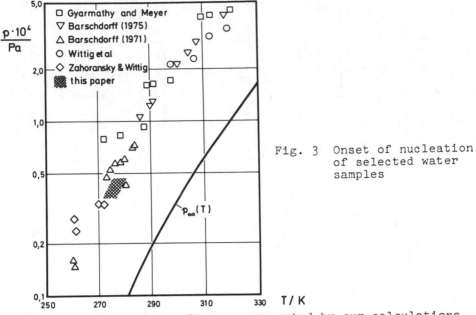

Fig. 3 Onset of nucleation of selected water samples

The experimental results are supported by our calculations which also demonstrate that the "unary" phase change is not affected by impurities of concentrations typical for feed water and even for public water supply. In determining the condensation parameters, following the binary nucleation theory, leads to difficulties for mixtures where one component is extremely diluted. For example, the Zeldovich factor as shown in Eq.(16) will be zero in approaching the unary limit. From a physical point of view this must be incorrect and results from the continuous treatment of the binary nucleation process. A plausible explanation, however, is that low concentrations will have a minor influence on the kinetics of the nucleation reactions. In a first approximation the unary Zeldovich factor can be applied [15]. In contrast, the thermo-dynamic consequences on the free enthalpy difference of binary nucleus formation ΔG is to be considered as few molecules in the nucleus can lead to a pronounced reduction.

The results of the calculations are summarized in Table 2 for water/orthosilicic acid and Table 3 for water/acetic acid solutions. For the $H_2O/Si(OH)_4$ system, saturation concentrations in the vapor phase of orthosilicic acid is used. As a typical concentration for the water/acetic acid mixture 30 ppb is used as known from power plant operation [1].

In Tables 2 and 3 y is the mole fraction of the relevant impurity in the vapor phase and x the mole fraction in the nucleus. In addition the binary nucleation rates are compared with those for unary water nucleation. Also, the critical radii and the relaxation time is shown.

T K	y 10^{-11}	x 10^{-8}	S_1 1	S_2 10^4	p_1 10^4 Pa	p_2 10^{-7} Pa	r 10^{-9} m	I $m^{-3}s^{-1}$	I_{H2O} $m^{-3}s^{-1}$	τ 10^{-7} s
275	6.7	0.11	6.0	3.8	0.42	2.7	0.66	$6.0\ 10^{16}$	$5.8\ 10^{15}$	1.3
275	7.5	0.3	5.0	1.3	0.35	2.7	0.74	$6.5\ 10^{11}$	$6.3\ 10^{11}$	2.1
275	9.2	1.1	4.0	0.35	0.28	2.7	0.86	$5.9\ 10^{4}$	$5.6\ 10^{4}$	3.8
300	1.4	0.12	6.0	3.8	2.1	3.0	0.58	$2.4\ 10^{22}$	$2.3\ 10^{22}$	0.2
300	1.7	0.35	5.0	1.3	1.8	3.0	0.65	$4.8\ 10^{19}$	$4.6\ 10^{19}$	0.3
300	2.1	1.3	4.0	0.35	1.4	3.0	0.75	$8.1\ 10^{14}$	$7.9\ 10^{14}$	0.6

Tab. 2 Calculated cluster parameters for water/silicic acid mixtures

T K	y 10^{-10}	x 10^{-8}	S_1 1	S_2 10^{-2}	p_1 10^4 Pa	p_2 10^{-6} Pa	r 10^{-9} m	I $m^{-3}s^{-1}$	I_{H2O} $m^{-3}s^{-1}$	τ 10^{-7} s
275	3.2	3.4	5.0	6.5	0.35	1.1	0.58	$7.3\ 10^{21}$	$6.3\ 10^{11}$	1.3
275	3.9	2.3	4.0	9.5	0.28	1.1	0.67	$2.1\ 10^{19}$	$5.6\ 10^{4}$	2.5
300	0.65	0.7	5.0	6.5	1.8	1.2	0.53	$2.1\ 10^{25}$	$4.6\ 10^{19}$	0.2
300	0.84	0.5	4.0	9.5	1.4	1.2	0.62	$3.4\ 10^{22}$	$7.9\ 10^{14}$	0.4

Tab. 3 Calculated cluster parameters for water/acetic acid mixtures

Enrichement effects are obvious in comparing the mole fractions in the vapor phase y and in the nucleus x. For water/ orthosilicic acid the distribution coefficient x/y for the nuclei varies approximately between 20 and 620 and for the system water/ acetic acid from 60 to 110. The absolute contribution of the diluted component in the cluster, however, is still extremely low at these small partial pressures p_2. It seems to be impossible for that reason that this small concentration will lead to a corrosive fluid. In addition, the droplet composition will reach its equilibrium value extremely fast. Further enrichment with the diluted constituent appears to be impossible for growing droplets, as in thermodynamic equilibrium the distribution coefficient x/y is lower than in the metastable equilibrium (Tables 2 and 3).

Furthermore it should be pointed out that the cluster concentrations calculated with a purely formal binary procedure can lead to unrealistic consequences with one cluster containing only fractions of one molecule of the diluted component. If the influence of the surface tension is not considered, the nucleation rate, therefore, will not deviate drastically from the unary value as shown for the system H_2O/SiO_2.

In contrast, the nucleation rate of the system water/acetic acid differs appreciably from the unary rate. This effect contributes primarely to the strong decrease of the surface tension even by addition of only small amounts of acid in the solution (compare our results on alcohol/water solutions [23]). The calculations indicate an accelaration of the nucleation process for this solution and thus a reduction of the supersaturation. From a physical point of view, however, these mathematical results are somewhat doubtful:the surface tension for small clusters will be higher as the used bulk value as discussed earlier [16]. Consequently, the nucleation rate will be reduced. Furthermore, the calculated number of acid molecules is smaller than one. To confirm, therefore, the results with a higher nucleation rate, the models applied have to be refined.

The time lag τ for the systems studied are nearly similar to those of pure water nucleation which is in the range of a microsecond (Tables 1 and 2). This behaviour is a consequence of the small mole fraction x of the impurities in the clusters as can be seen in Eq. (23).

As the main result of the present study it has to emphasized, that impurities alone at small concentrations should not be of major influence in altering the condensation characteristics. For steam turbine flow, other sources must be held responsible for "early" condensation, i.e. low supersaturation. Threedimensional effects, for example, with excessive flow speeds and other phenomena in cascade and bucket flow can lead to high local supersaturations and isolated high cluster formation with subsequent heterogeneous vapor condensation. These questions, however, at present are awaiting additional detailed analysis.

ACKNOWLEDGEMENTS

Thanks are due to Dipl.-Ing. Bundschuh, Director, and Dipl.-Ing. Zickwolf from Rheinhafen Power Plant (RDK), Badenwerk AG and Dipl.-Ing. S. Marcinkowski from the Institute of Fluid Flow Machinery, Gdańsk, for the analysis of the water samples.

The work was supported by a grant from the A.v.Humboldt Foundation to Dr inż. W. Studziński from the Institute of Fluid Flow Machinery Polish Academy of Sciences in Gdańsk.

REFERENCES

1. Bodmer, M. 1977.
    Brown, Boveri Mitt. 6, 343

2. Bohnsack, G. 1978.
    VGB Kraftwerkstechnik 58, 373

3. Brunner, R. 1976.
    VGB Kraftwerkstechnik 56, 9o

4. D'Ans Lax 1967.
    Taschenbuch für Chemiker und Physiker, VI, 823
    Springer Verl.

5. Dibelius, G., Mertens, K. Sept. 17/18, 1981
    in:Tagungsbericht Koll. "Prozesswärme aus HTR", RWTH Aachen.

6. Flood,H. 1934.
    Z. Phys. Chem. A 170, 286.

7. Gmelins Handbuch der Anorganischen Chemie, 1959.
    Silicium, Teil B, 448, Verlag Chemie GMBH.

8. Gyarmathy, G. 1976.
    in:Two-Phase Steam Flow in Turbines and Separators, 1.
    Eds. M.J. Moore and C.H. Sieverding, Hemisph. Publ. Corp.

9. Heitmann, G. 1964.
    Mitt. der VGB 90.

10. Hirata, M.;Ohe, S., Nagahama, K. 1975.
    Computer Aided Data Book of Vapor-Liquid Equilibria
    Kodansha Ltd. Elsevier Sci. Publ. Comp.

11. Höxtermann, E. 1979.
    VGB Kraftwerkstechnik 59, 952.

12. Iler, R.K. 1979.
    The Chemistry of Silica, J. Wiley & Sons.

13. Kreitmeier, F., Schlachter, W. and Smutny,J.1980.
    VDI Ber. 361, 201.

14. Marcinkowski, S., 1979.
    Institute of Fluid Flow Machines, Gdańsk, Private Communication.

15. Mirabel, P. and Clavelin, J.L. 1978.
    J. Aerosol Sci. 9, 219.

16. Mirabel, P. and Katz J.L. 1977.
    J. Chem. Phys. 67, 1697.

17. Reiss, H. 1950.
    J. Chem. Phys. 18, 840.

18. Schieferstein, U. and Schmitz, F. 1978.
    VGB Kraftwerkstechnik 58, 193.

19. Shugard, W.J., Heist, R.H. and Reis, H. 1974.
    J. Chem. Phys. 61, 5298.

20. Wilemski, G. 1975.
    J. Chem. Phys. 62, 3772.

21. Wittig, S., Zahoransky, R. and Studziński, W. 1981.
    Eds. G. Adomeit and H.-J. Frieske, VDI-Verlag.

22. Wright, E.H.M., Akhtar, B.A.A. 1970.
    J. Chem. Soc. (B), 151.

23. Zahoransky, R.A. and Wittig, S.L.K. 1982.
    in:Shock Tubes and Waves, 682
    Eds. C.E. Treanor and J.G. Hall, State Univ. of New York
    Press, Albany.

Latter Stage Condensing Turbine Rotor Blade Trailing Edge Erosion

BRANKO STANIŠA and MILENKO DIČKO
SOUR Jugoturbina
47000 Karlovac, Yugoslavia

ABSTRACT

This paper indicates the latter stage rotor blade trailing edge erosion damage investigation results of many greater power turbines in use. The basic characteristics and conditions of the wet steam flow, in the trailing edge part of a low pressure greater power turbines, due to which latter stage blade trailing edge erosion occurs, are listed in this paper. Subsequently, the actions for reducing and protection of latter stage condensing turbine rotor blade trailing edges were introduced for the purpose of erosion protection.

1. INTRODUCTION

In the condensing turbines usually occurs a more significant damage of latter stage rotor blade leading edges working in the wet steam area. Rotor blade leading edge erosion damage causes as well as most frequently used erosion protection methods in the greater power condensing turbine use are indicated in the earlier works [1,2,3]. However, besides the leading edge erosion in the greater power condensing turbine use, a more significant latter stage rotor blade trailing edge erosion damage could occur. In some cases a need for replacement of those blades appears due to the latter stage rotor blade trailing edge erosion damage, and in some cases the damage caused blade fracture and a significant turbine defect [4,5].

Currently, a greater attention is given to this problem and many theoretical and experimental investigations are carried out [6,7,8]. However, up to now, the rotor blade trailing edge erosion is still not well enough understood and investigated. In order to solve the greater power condensing turbine rotor blade trailing edge erosion problem, a detailed theoretical and experimental investigation of the wet steam flow in the low pressure trailing part, by a smaller work load, erosion damage experimental investigations as well as the observation and analysis of the erosion process in turbines in use are needed.

Since, a greater number of greater power turbines is daily built into the Yugoslav electro-power industry, a growing interest

595

for the erosion problem and its effect on the life and work reliability of latter turbine stages is evident. During the turbine
repair, besides the leading edge erosion, a significant latter
stage rotor blade trailing edge erosion is evident on some greater
power turbines. This appearance caused a need for a more detailed
analysis and theoretical investigations for erosion cause explanation, its effect on the turbine work relaibility and actions
to be taken for reduction and protection of latter stage rotor
blade trailing edge from erosion damage. The results of such
investigations are listed in this paper.

2.   THE INVESTIGATION RESULTS OF THE ROTOR BLADE TRAILING EDGE
      EROSION ON TURBINES IN USE

      In order to obtain an insight of erosion and possibility to
observe further erosion process for the purpose of relaibility
increase in turbine work, the condition of erosion damage is taken
and the anlysis of latter stage rotor blade trailing edge erosion
damage on many greater power turbines installed in Yugoslav electro-power industry is carried out.

      In the Thermo-electro power plant "Nikola Tesla" Obrenovac
a check and analysis were carried out in 1981 on the condensing
turbine erosion, power 308,5 MW, block 6 Alsthom Atlantique to
BBC technology, by the turbine erosion after 10080 working hours.
A significant latter stage rotor blade trailing edge erosion was
noted during the check. The example of the trailing edge erosion
damage is shown on Fig. 1. Erosion damage on the shown blade
trailing edge convexed side is 600 mm long from the root to the
top, wide up to 10 mm and deep up to 1 mm. Erosion damage, as
shown on the figure, has a form of the horizontal grooves and
notches regarding the blade lenght. The latter stage rotor blade
length is 990 mm. All the remaining latter stage rotor blade
trailing edge erosion damages were 550 to 600 mm long from the
root to the top, wide 6 to 10 mm and deep 0,5 to 1 mm. A more
significant erosion damage was noted on the blades whose trailing
edges due to the bad manufacturing or installment outstand up to
1,5 mm from the stepped grate usual order.

      A significant latter stage rotor blade trailing edge erosion
was noted on the turbines of the same type with power of 308,5 MW,
block 4, during their repair in 1981 by erosion check after 23004
working hours when they were put to work from the cold condition
15 times. The trailing edge erosion damaage was long up to 620 mm
from the root to the top of the blade, wide up to 12 mm and deep
up to 1,3 mm. A more significant blade erosion damage was noted
on this turbine in which blade trailing edge outstands from the
stepped grate usual order. Blades located directly behind those
which outstand had a much lower trailing edge erosion rate in
regard to the other.

      On the basis of rotor blade trailing edge analysis and given
data of these two turbines of the same type, with the power of
308,5 MW and approximately same working conditions, it was noted
that the blade trailing edge erosion damage is higher on a turbine
which was longer in use. We could conclude from this that the
erosion on these turbines progresses during the use. In order to

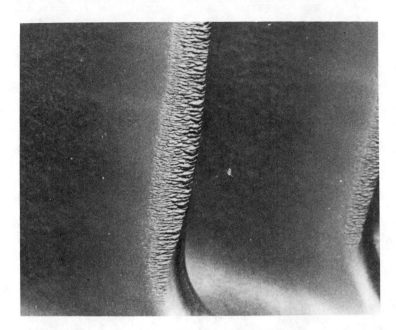

Fig. 1. Latter stage rotor blade trailing edge erosion on a
         turbine, power 308,5 MW, after 10080 working hours

follow erosion progress process, the highest latter stage rotor
blade trailing edge erosion damages were measured. On the basis
of such data, during the upcoming checks of erosion damage, it
will be possible to determine the erosion speed. Knowing the speed
it will be possible to determine whether or not it is necessary
to carry out the additional safety precautions from the additional
erosion damages. It is recommended that the further erosion process speed is determined as well as the damage effect on the blade
strength, since they do not have the vibration damping wire, so
they have a small damping rate. On the same type turbine, with
the power of 305 MW, block 3 a latter stage rotor blade fracture
occured in 1980 after 17790 working hours; this blade had a significant trailing edge erosion damage.

       Trailing edge erosion damage check and analysis results of
the latter stage rotor blades which are part of condensing turbines of IMZ production with the power of 210 MW, and which are
installed in Yugoslav electro-power industry, showed that the
highest erosion rate is present on the Thermo-electro power
plant "Tuzla", block 4 turbine. The latter stage rotor blade
trailing edge erosion damage example of this turbine after 55000
working hours is shown on Fig. 2. It could be seen with the help
of a ruler that the trailing edge convexed side erosion damage
is wide from 10 to 12 mm and deep up to 1 mm. The erosion damage
is long 300 mm from the root to the top of the blade. Erosion
damage has a form of the horizontal grooves and notches in regard
to the blade length. The remaining latter stage rotor blade
trailing edge erosion damage was long from 200 to 300 mm from

Fig. 2.   Latter stage rotor blade trailing edge erosion on a
          turbine, power 210 MW, after 55000 working hours

root to the top, wide 6 to 12 mm and deep 0,5 to 1 mm.   The length
of the blade is 765 mm.   The erosion damage was also noted on the
latter stage rotor blade rivet head part on a disk in a rotating
direction and on balancing weights.   The rivets are damaged in
a form of horizontal grooves, 4 mm long and deep to the disk surface.   Balancing weights had a needle shape damage 3 mm deep.
Basic geometric and thermodynamical characteristics of latter stage rotor blade trailing edge erosion in turbines, power 210 MW,
are given in the previous work [9].

        A less significant latter stage rotor blade trailing edge
erosion was noted on the same type of turbines, power 210 MW,
installed in other thermo-electrical power plants.   For example,
in the Thermo-electro power plant "Nikola Tesla" Obrenovac the
latter stage rotor blade trailing edge erosion damage of turbines,
power 210 MW, block 1, after 53800 working hours, was long up to
450 mm from the root to the top, wide up to 5 mm, deep up to 0,4
mm.   All blades were damaged from erosion to an approximately same
rate.   However, due to a very significant leading edge damage [1],
all latter stage rotor blades were replaced during the regular
repair in 1978.

        The check and erosion analysis results of condensing turbines,
power 125 MW, Jugoturbina & Zameck's production, as for example
in Thermo-electro power plant "Trbovlje", showed that the latter
stage rotor blade trailing edge erosion has almost stopped.   Since

the erosion check by the turbine repair in 1978, until the check
in 1981, an unsignificant increase in latter stage rotor blade
trailing edge erosion was noted. The turbine was put to work in
1969. Measured erosion damage was long 300 mm, from the blade
root to the top, wide up to 3 mm and deep 0,2 to 0,4 mm. The
blade length is 535 mm. On the basis of a very slow latter stage
rotor blade trailing edge erosion further progress, by the almost
same working conditions as before the check, it could be assumed
that here, as on the leading edges, the erosion process is in the
third slowed down stage [10].   It was noted during the check that
the erosion present on the latter stage rotor blade rivet head in
a form of 3 mm long, 0,5 mm deep grooves and balancing weights on
the disc in a form of 2 mm deep needle.

     Check results and erosion analysis on many greater power tur-
bines in use also showed that the latter stage rotor blade trailing
edge erosion does not occur on all turbines, regardless of a longer
work under reduced load. For example, a turbine, power 100 MW,
IMZ production in Thermo-electro power plant "Tuzla" worked for
a longer period of time under reduced load, however the latter
stage rotor blade trailing edge erosion was not evident. Latter
stage rotor blade leading edge erosion was very high, so that the
stellit plates were replaced on leading edges and a latter stage
reconstruction was carried out in order to reduce erosion [3].

     The latter stage rotor blade trailing edge erosion was not
present on a turbine, power 320 MW, Ansaldo production in Thermo-
electro power plant "Rijeka" after 149 testing hours in 1978,
17 times were put to work from cold and 2 times from warm conti-
tion and a larger portion of work taken when the turbine is not
loaded. However, due to such running, a more significant latter
stage rotor blade trailing edge erosion occured in regard to a
small number of working hours. All leading edges were damaged.
on the end parts of the blades due to the poor erosion of perforated and dotted 0,3 to 1 mm deep form. On the blade No. 15, a
leading edge damage was noted in a form of 1,5 mm deep notch.
By the erosion check during the turbine repair in 1980, the
trailing edge erosion nor the further latter stage rotor blade
leading edge erosion progress were not present.

3.   THE ROTOR BLADE TRAILING EDGE EROSION CAUSES AND ITS EFFECTS
     ON TURBINE WORK RELIABILITY

     Latter stage rotor blade trailing edge erosion checks and
analysis on the above mentioned turbines as well as on many greater
power condensing turbines in use showed that the erosion appears
and develops on turbines which have been for some time on a reduced
load and often let to work. By comparison of the rotor trailing
edge to leading edge erosion it was concluded that they have same
nature and cause. The investigation showed that the only possible
cause of the latter stage rotor blade trailing edge erosion is the
streaming on the steam flow and a large water drop striking against
their surface by backward and whirling wet steam flow, usually occuring during the reduced turbine load and work with no load. The
source of larger water drops could be moisture in wet steam flow,
moisture shown on the leading housing surfaces, steam from different watter supplies, to the condensor leading part, moisture

from sealing steam and moisture from first row cooling moisture
pipes. Besides that, additional moisture source on many turbines
could be trailing housing cooling fixture, included in the reduced
load regime and no load rebime. Additional moisture source could
be steam cooling fixture launched in the condensor leading part.

The rotor blade trailing edge erosion area mainly depends
upon the turbine trailing housing construction and latter stage
rotor disc, as well as length of work in the regime with wet steam
backward flow. Investigations of turbines in use showed that
this area usually is 30 to 65% of the rotor blade length from
root to the top and 10 to 15 mm wide. Erosion damage depth mainly
depends on length of turbine work under reduced load and on launch-
ings to work.

In order to obtain an idea about the reduced load effect on
steam flow in latter stages and trailing housings, many experimental testings have been recently carried out on models in the
experimental turbines and actual turbines in use [11,12] . The
results of such investigations showed that by a turbine reduced
load regime, especially when turbine is not loaded, flow segregation occurs, usually accompanied by the whirling and backward
steam flow in the trailing housing with a high rate of negative
reaction on the blade root. Fig. 3 shows the investigation
results of the area where the flow segregation appears lenghwise
of the blades $\bar{L}_{seg} = l_{seg}/l_2$ and vibration amplitude increase
$\bar{A}_{seg} = A_{seg}/A$ in dependance upon the volume steam flow by the
alternated and nominal work rebime $\bar{G}v_2 = Gv_2/(Gv_2)_n$ for the
greater power turbines [6,13] .

It could be seen from the diagram that the flow segregation
mainly depends on the steam volume flow, i.e. turbine work regime.
Ba load reducing the steam seg egation is extended from the root
lenghtwise of the blade and extends itself in the backward flow
and whirling flow. For example, for the value of $\bar{G}v_2 = 0,12+0,15$
which suits turbine no load regime or very small load, by vacuum
in condensor 88+96 %, the flow segregation area amounts $\bar{L}_{seg}=0,7+$
0,75. Here, a very intensive backward and whirling steam flow
occurs, latter stage working as ventilator. Backward flow speed
from the turbine trailing housing in the blade proximity could
amount to 100 to 120 m/s [6] . It has been determined by measuring that the diameter of the largest water drops which strike
against rotor blade trailing edges usually amounts to 40 to 50 μm.
The investigations showed that the backward flow starts by load
reducing to approximately 35% of the nominal turbine power by the
nominal pressure of by the nominsl power if the pressure in the
condensor is increased 3 times in regard to the nominal value[14].
Pressure increase in condensor could cause, for example that the
great amount of steam is allowed to come out of the roundybout
pipe to its leading part through the cooling fixture. Here, a
pressure increase could occur behind the latter stage and backward
steam flow in the first not-regulated extraction.

It could be seen from the diagram, also, that the reducing
of the volume steam flow $\bar{G}v_2 = 0,6$ causes a significant latter
stage rotor blade amplitude vibration increase to $\bar{A}_{seg} < 2,5+2,8$.
Blade vibration increase causes adequate increase of their dynamical stress [15,16] . The increased dynamical stresses in the

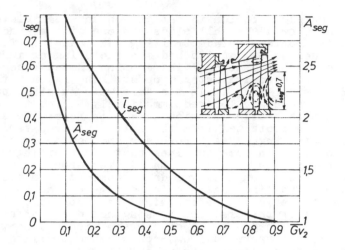

Fig. 3. Flow segregation area dependance $\bar{I}_{seg}$ and
vibration increase $\bar{A}_{seg}$ from the volume stream
flow $\overline{Gv}_2$ in the greater power turbines

no load regime or small load regime and erosion damages of the
latter stage rotor blade trailing edges significantly reduce turbine work reliability. The trailing edge erosion damages in a
form of grooves and notches could be the places where the fracture
starts, which by material fatigue at the increased dynamical
stresses could cause blade fracture and turbine defect. Those
problems of backward steam flow, erosion, dynamical stresses and
material fatigue have not been yet well enough studied and understood. For the purpose of work reliability increase detailied
intergal wet steam flow studies, vibration increase and latter
stage turbine blade trailing edge erosion under reduced load are
needed. On the basis of these intergal studies it is necessary
to determine the allowed reduced load work regime for the turbines
sensitive on erosion, in regard to the maximum allowed erosion
damage of the trailing edges and allowed dynamical stresses.

On the basis of flow analysis in many greater power turbines
and trailing edge erosion damages a scheme of the backward steam
flow in the condensing turbine, power 210 MW, trailing edge is made,
at the no load regime or at the small load regime, Fig. 4.a. It
could be seen from the flow scheme that the moisture with backward
flow streams on latter stage rotor blade trailing edges from the
right hand side of the trailing housing. Backward flow speed
depends upon the negative reaction size on the blade root, and in
a limited quantity upon the amount of moisture on the surfaces of
the trailing housing. Due to this backward flow and larger water
drop striking against the latter stage rotor blade trailing edge
back part surface, their erosion occurs, shown in the scheme with
the help of lines. When the turbine is put to work, the sealing
steam temperature could be low and such it could be the cause for
the additional moisture and rotor blade trailing edge erosion.

Trailing edge erosion investigation and analysis on many turbines in use showed that all blades in one disc are not equally damaged from erosion. Blade trailing edge outstand from the stepped grate usual order, due to bad manufacturing or installment could cause greater erosion. The erosion increase could be explained by more intensive water drop striking against the trailing edge surface similar as it is the case with outstanding leading edges [2]. The blade located immediately behind the outstanding ones have a much smaller erosion damage in regard to others, since they have to be protected from the intensive drop striking on the outstanding trailing edges. However, independently of this, it has been observed on some turbines that the trailing edges of certain blade groups are less damaged from erosion, and some are more damaged even though they didn t outstand from stepped grate usual order. Trailing edge erosion damage on the rotor blade alternating groups could be explained by drops which fall from the upper part of the trailing housing and don t affect all the blades equally. Sheded drops on certain disc and part places affected by steam backward flow, could cause trailing edge erosion damage on the rotor blade alternating groups.

The carried out moisture movement analysis in the trailing housing of many condensing turbines enabled determination of moisture drop source, which together with backward flow is under reduced load striking against rotor blade trailing edges. On the basis of such analysis a scheme of moisture flow in the condensing turbine trailing housing upper part, power 210 MW, Fig. 4. b. The scheme also indicated the movement paths of the drops sheded from the latter stage disc, under the turbine work on the reduced load regime, as well as the trailing edge erosion damage, rotor blade rivet head and balancing weights. It could be seen from the scheme that the larger moisture drop source, which due to the backward flow strike against the trailing edges and cause their erosion could be drops sheded from the trailing housing walls, stiffening ribs, aerodynamical inserts, constructional forming, then the moisture coming down the trailing housing walls as well as the moisture from the sealing steam. The moisture drops coming down the trailing housing walls, moisture drops from the sealing steam and moisture drops from the steam condensing in clearance between the disc and trailing housing, due to the suction activity are sheded from the disc and part and strike along the rotor blade trailing edge. The above mentioned moisture, due to the suction activity of the disc and backward flow, causes rotor blade rivet head erosion as well as the erosion on the balancing weights on the disc.

Checkings and analysis of the greater power turbine in use under the no load and reduced load rebime showed that the latter stage rotor blade trailing edge erosion does not appear in all turbines. The analysis showed that the drop inlet angle in regard to the trailing edge convexed side [17] had a great influence on the erosion appearance. It has been showed that under smaller inlet angles the erosion did not appear. It could be explained by the angles that the vertical component of the water drop striking speed against the blade trailing edge surface is smaller than the speed under which erosion starts.

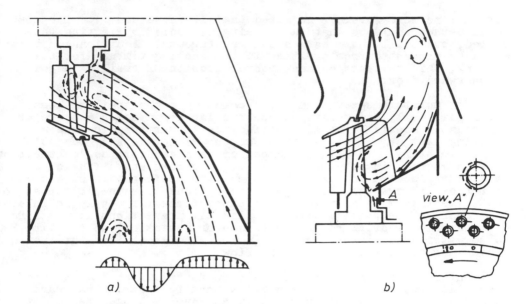

Fig. 4. Wet steam flow scheme in the trailing housing of the
turbine with the power of 210 MW, under small load

a) housing lower part, b) housing upper part

4.   ACTIONS FOR REDUCTION AND PROTECTION OF ROTOR BLADE TRAILING EDGES FROM EROSION

The action for reduction and protection of latter stage rotor
blade trailing edge protection on the greater power turbines from
erosion could be divided in two groups: active and passive. Active protections consists of projecting accounting and use actions
for reduction or prevention of the backward steam flow and increase
of its moisture in the trailing housing of a turbine. The passive
protection consists of actions for reduction of erosion activity
under the backward damp steam flow.

Active protection actions. It could be suggested that one
of the actions for the protection purposes could be that under
projecting and calculation of greater power turbines a greater
stage of reaction is chosen on the root and a greater temperature
drop at the latter stage. The reduced work regime area will be
extended in this way without segregating the flow at the blade
root and backward flow. The latter stage turbines need to be aero-
dynamically projected in such a way that a wide area of work regime
without flow segregation could be enabled. Improve cooling fixture
construction of the trailing housing under reduced load, chose the
most suitable position, optimum number of places and water injection
angle, the fixtures should be projected with nozzles in such a way
that the water disperses in the smallest possible drops, which are
not very dangerous for the erosion damages. The optimal quantity
of the injected water must be determined as well in regard to the
maximum allowed temperature of the trailing housing, in such a

way that no more water is injected that it is necessary. The most
ideal method would be that the housing is cooled by letting the
cooled, but still a bit warm steam run through. Additional moisture
would be prevented from creating in the cooling fixture for the
housing, this moisture causes greater erosion of trailing edges
by backward flow.

For moisture seperation which comes down the trailing housing
upper part walls and moisture from the labyrinyh seals a segregating chamber should be located behind the latter stage on the
trailing part of the seals, this chamber will take away the collected moisture. In [17] a suggestion is given for the construction
of such segregating chamber. Trailing housings and diffusion
implements located in the chambers must be projected in such a way
that they improve the steam going conditions from the latter stage
and prevent flow whirling. On all pipes for water supply in inlet-
ing condensor part it is needed to install preventions in the form
of water flow rectifiers in the condensor direction, so that it
makes water drop lifting in trailing housing and latter turbine
stage impossible for the backward flow.

On the basis of explotation action for trailing edge erosion
prevention it is not to allow turbine work regime with backward
flow or not to allow such work to extend over a longer period of
time. Frequent stops should be avoided as well as putting the
turbine to work. Putting the turbine to work and turbine load to
approximately half the nominal power should be carried out without
extensive hold back. The turbine stopping shall be carried out
without hold back as well. Power reduction during the night hours
due to the need reduction for the electrical power, shall be carried
out only to the backward flow limitation. In the case when the
pressure in the condensor is larger than what the nominal is, for
example, in the summers due to the cooling water tremperature increase or due to the transaction to work with cooling towers, reduced quantity of cooling water, cooling pipe clogging and other
causes, the allowed area of power reduction is smaller. In the
greater power extraction turbines used for heating purposes, the
reduction shall be carried out in such a way that enough steam is
going to the condensor, which will ensure the work without the
backward and whirling flow in the turbine trailing housing.

Passive protection actions. A one of basic protection actions
against erosion is suggested as material selection for latter stage
rotor blade manufacture. This material must be very resistant to
the erosion damage. For the purpose of reducing outstanding
trailing edge greater erosion from the stepped grate usual order,
due to bad manufacturing or installment it is suggested that the
rotor blade trailing edge outstanding is not higher than 0,3 mm.
One of the erosion reducing actions is suggested as hardening, i.e.
resistance increase of the latter stage rotor blade trailing edge
to the erosion, reheating or covering the trailing edge convexed
side of the blade with the protection layer of material which is
high resisting to erosion. For the purpose of the trailing edge
erosion protection it could be suggested that the angle selection
of the latter stage rotor blade trailing edge installment is
carried out. This angle should be such that the erosion will
not appear. The formula for determining this angle is given in [18]

CONCLUSION

The investigations showed that on some greater power turbines in use a more significant latter stage rotor blade trailing edge erosion appeared due to the backward flow of the wet steam under the reduced load. Due to the amplitude increase in vibration and dynamical stresses, under backward and whirling steam flow, rotor blade trailing edge erosion damage could significantly put reliable turbine work to danger. Trailing edge erosion damages in a form of grooves and notches under the increased dynamical stresses by the material fatigue could cause blade fracture and great turbine significant defect.

It is suggested that for each turbine type endangered by increased erosion of the rotor blade trailing edges, a work regime with reduced load under which the backward steam flow starts is determined. For the purpose of erosion reduction and reliability increase a longer turbine work is not allowed in regime with backward steam flow. For the turbines where a longer work is impossible to avoid on the reduced load it is suggested that on the basis of the intergal erosion investigation processes and dynamical stress increase allowed work regime is determined in regard to the maximum allowed trailing edge erosion damage and acceptable blade life.

It is suggested that the turbines on which an increased rotor blade trailing edge erosion is noted are not stopped and put to work very often. At the same time, the putting to work and stopping shall be carried out without unnecessary holdings on the reduced loads. For the purpose of trailing edge erosion reduction, it is needed that the cooling fixture checks of the trailing housing under reduced load are made as well as pipes for water and steam letting to the leading condensor part checks. It is concluded that they are the causes of the additioanl steam moisture which under the back flow causes erosion, it is necessary to carry out their rectification and possible reconstruction.

REFERENCES

1. Staniša, B. 1979. Utjecaj strujanja vlažne pare na eroziju turbinskih lopatica. Strojarstvo, Vol. 21, (3/4) pp. 161-169.

2. Staniša, B. and Dičko, M. 1981. Ispitivanje raznih površinskih zaštita turbinskih lopatica od erozije. Strojarstvo, Vol. 23, (3) pp. 141-148.

3. Staniša, B. 1981. Utjecaj odvođenja vlage ispred rotorske rešetke zadnjeg stupnja turbine 100 MW na smanjenje erozije. VI jugoslavenski simpozij termičara, Bled.

4. Kuličihin, V.V. and Tažiev, E.I. 1981. O povyšenii nadežnosti raboty lopatok poslednih stupenej parovyh turbin, Električeskie stancii. (7) pp. 24-27.

5. Hauke, W. 1967. Schaufelbrüche durch Erosion an den Laufschaufelaustrittskanten von Dampfturbinendstufen, Maschinenbautechnik, Vol. 16, (1) pp. 25-29.

6. Lagun, V.P. and Simoj, L.L. and Nahman, J.V. and Semenov, J.E.
    and Naftulin, A.B. and Kolokolcev, V.M. 1977. Erozija vy-
    hodnyh kromok raboč̌ih lopatok poslednih stupenej parovyh tur-
    bin. Teploenergetika (10) pp. 12-17.

7. Kuličihin, V.V. and Tažiev, E.I. and Soloveva, O.V. and Tru-
    bulov, M.A. 1978. O nekotoryh pričinah erozii vyhodnyh
    kromok rabočih lopatok poslednih stupenej parovyh turbin.
    Teploenergetika (5) pp. 16-19.

8. Javelskij, M.B. and Šilin, J.P. 1981. Erozija vyhodnyh
    kromok rabočih lopatok poslednih stupenej parovyh turbin
    i meroprijatija po ee ustraneniju. Energomašinostroenie
    (10) pp.11-15.

9. Staniša, B. 1981. Erozija rotorskih lopatica turbina snage
    210 MW instaliranih u našoj zemlji. Šesto savetovanje o ter-
    moelektranama Jugoslavije, Sarajevo, knjiga3, pp. 1005-1026.

10. Staniša, B. 1981. Ispitivanje utjecaja struje vlažne pare
    na brzinu erozije glatke i erodirane površine turbinskih
    lopatica. Strojarstvo, Vol. 23, (5) pp. 245-252.

11. Štastny, M. 1976. Some Problems of a Flow Through the Low-
    Pressure Section of a Large-output Steam Turbine. IMP PAN
    (70-72) pp. 637-654.

12. Šnee, J.I. and Ponomarev, V.N. and L.N. Bystrickij. 1977.
    Eksperimentalnoe issledovanie častičnyh režimov raboty tur-
    binnyh stupanej. Energomašinostroenie (11) pp. 10.14.

13. Kuličihin, V.V. and Tažiev, E.I. and Ljudomirskij, B.N. and
    Antonov, E.I. and Pervušin, S.M. 1976. Vlijanie ekspluata-
    cionnyh režimov and kolebanie rabočih lopatok poslednej stupe-
    ni teplofikaionnoj turbiny, Električeskie stancii (7) pp.
    17-18.

14. Štasny, M. 1980. Betriefverhältnisse einer Dampfturbine
    mit Endstufenventilation. Energietechnik, Vol. 30, (5)
    pp. 176-180.

15. Klebanov, M.D. and Jurkov, E.V. and Šapiro, G.A. 1978. Vlija-
    nie rashoda para i davlenija v kondensatore na vibraciju ro-
    bočih lopatok poslednej stupeni teplofikacionnoj turbiny,
    Električeskie stancii (6) pp. 22-24.

16. Klebanov, M.D. and Jurkov, E.V. 1979. Vlijanie režima raboty
    na dinamičeskie naprjaženija v rabočih lopatkah poslednej stu-
    peni teplofikacionnoj turbiny. Električeskie stancii (10)
    pp. 30-33.

17. Faddeev, I.P. 1974. Erozija vlažnoparovyh turbin. Mašino-
    stroenie, Leningrad.

18. Smetanin, A.G. and Semenov, J.E. 1979. Nekotorye harakteri-
    stiki erozionnoj nadežnosti vyhodnyh kromok rabočih lopatok
    poslednih vlažnoparovyh stupenej turbin. Tr. Vses. teplotehn.
    NII (21) pp. 78-81.

ROTATING HEAT PIPES AND THERMOSYPHONS

Rotating Heat Pipes

PAUL J. MARTO
Department of Mechanical Engineering
Naval Postgraduate School
Monterey, Calif., USA

ABSTRACT

The basic concept of the rotating heat pipe is presented along with its
operating principles and thermal limitations. Various applications are described. The existing heat transfer literature pertinent to this device is
reviewed, and appropriate analyses and/or correlations are provided.

1. INTRODUCTION

The rotating heat pipe is a closed device to transport thermal energy either radially or axially in rotating machinery components. It relies upon the
evaporation and condensation of a small amount of working fluid, and thus may
be appropriately referred to as a rotating two-phase thermosyphon. Unlike ordinary heat pipes, which rely upon a capillary wicking structure to return the
condensate to the evaporator (or an ordinary thermosyphon which relies upon
gravity), the rotating heat pipe contains no wick and relies, instead, upon
centrifugal forces to return the condensate. This general concept and ter-
minology was first formulated by Gray [1] in 1969 in the United States. Quite
remarkably, earlier that same year, a specific application of the device was
independently proposed in a patent application by Hoffmann and Fries [2] in
Germany.

1.1 Principles of Operation

Figure 1(a) shows the simplest form of a rotating heat pipe designed to
transfer thermal energy axially. It consists of an evacuated hollow shaft rotating about its axis in which a small amount of liquid has been placed. At
sufficiently high rotational speeds, the liquid will cover the inside walls
of the rotating shaft as a thin annulus. Heat added to one end of the shaft
will evaporate some of the liquid (causing the film to get thinner), and the
vapor generated will flow to the colder end. Heat removed from this end of
the shaft will condense the vapor (causing the film to thicken). The condensate will flow back to the evaporator end due to the differential change in
film thickness and the resulting hydrostatic pressure change in the film.
This design, although noted for its simplicity and low fabrication costs,
leads to relatively poor thermal performance due to the resultant thick films
in the condenser end. A more effective means of returning the condensate to
the evaporator is therefore shown in Figure 1(b) where the shaft is shown to
have a slight internal taper. A slight taper of only 1 or 2 degrees allows
centrifugal acceleration to pump the condensate back to the evaporator. This

609

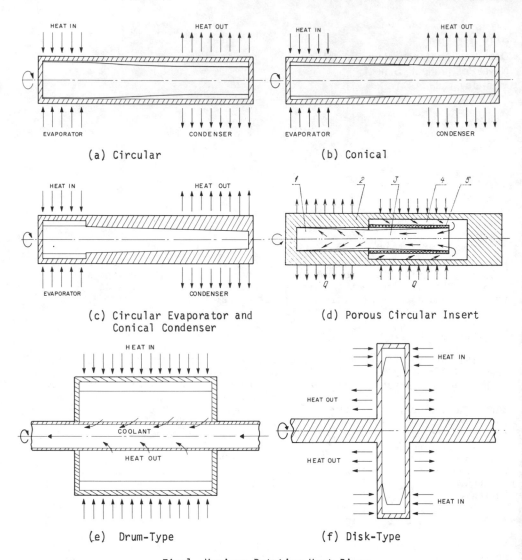

Fig.1. Various Rotating Heat Pipes

results in much thinner films and more effective condensation heat transfer. Another improvement is shown in Figure 1(c) where the evaporator end has an abrupt change in diameter in order to allow the liquid to collect as a uniform annulus. This additional modification makes the rotating heat pipe less sensitive to the amount of working fluid used, and permits uniform heat addition in the evaporator with nucleate boiling occurring within the annulus. Vasiliev [3] has proposed the use of a porous cylindrical insert in the evaporator as shown in Figure 1(d). In this case, centrifugal acceleration forces the condensate through the pores within the wall of this insert and the condensate sprays onto the evaporator surface in numerous fine droplets, causing high evaporator heat transfer coefficients. Axial heat transport may also be established by mounting heat pipes eccentric to the axis of rotation [4, 5].

In addition to the above-described heat pipes, two proposed schemes for radial transport of energy are shown in Figures 1(e) and 1(f). The device shown in Figure 1(e) permits heat to flow from a large cylindrical drum-type surface whereas Figure 1(f) shows a device configured as a rotating disk-type surface.

1.2 Operating Limits

The thermal performance of a rotating heat pipe (or two-phase thermosyphon [6]) is very effective due to the resulting high fluxes that are possible with evaporation and condensation heat transfer, and the nearly isothermal transport of thermal energy as latent heat in the vapor space. As with conventional heat pipes, however, there are various limitations to its thermal performance. These limits are imposed by the critical nucleate boiling heat flux in the evaporator, the entrainment of condensate by the counter-current flow of vapor, sonic vapor flow conditions or the ability to condense vapor in the condenser. These limitations are drawn schematically in Figure 2. It should be realized that the exact location and shape of these limit curves will depend upon the heat pipe geometry, the type of working fluid used, the presence of noncondensable gases in the vapor space, and heat pipe rotational speed. It is important to realize that as these conditions change, the limit curves will shift in relation to one another, providing different limitations on thermal performance. In general, however, for ordinary geometries and working fluids, it appears that the major limitation to rotating heat pipe performance is caused by the ability to remove heat in the condenser end [7].

2. APPLICATIONS

Numerous applications for the rotating heat pipe concept exist wherever there is a rotating part and the need to transfer heat. Gray [1, 8] described various applications including a scheme to cool high speed drills as shown in Figure 3. In this scheme, the working fluid transfers heat axially from the hot tip of the drill bit to a region along the shaft where it is removed by a coolant flowing through an annular gap. Vasiliev [9] described various energy recovery applications which occur in gas turbine engines, rotating drum-type dryers, and centrifugal systems used for heating granular material or pasteurizing milk and other food products. Figure 4 shows a cross sectional representation of a heat pipe heat exchanger proposed for use in a radial turbocompressor [9]. The main part of the system is a circular centrifugal heat pipe to which are attached two stages of turbine and compressor blades.

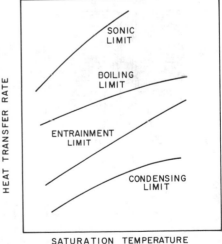

Fig.2. Operating Limits of Rotating Heat Pipes

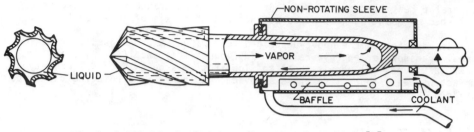

Fig.3. Drill Bit Cooled by a Rotating Heat Pipe [8].

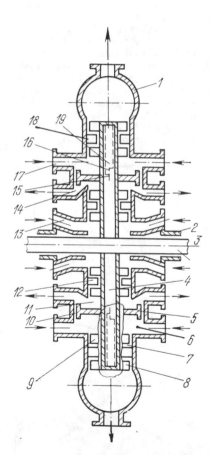

Fig.4. Turbocompressor with a
Centrifugal Heat Pipe
[9].

Hot gases enter at the first line of turbine
blades (9) and expand radially outward, giving
up part of their thermal energy to the heat
pipe working fluid. Some of this fluid eva-
porates and flows radially inward to where it
condenses on the interior walls of the compressor section thereby preheating the incoming air.
The possibility of using liquid cooling (i.e.,
two-phase closed thermosyphons) to reduce turbine blade tip temperatures has been under study
for many years [10] and current research findings in this important area have been reported
elsewhere at this Symposium.

Perhaps the most promising application for
rotating heat pipes is in cooling electric
motors and generators. It is widely recognized
that modern electric machines must be effectively cooled to permit large loadings and
compact size [11]. Two-phase heat transfer
can therefore be utilized to flatten temperature profiles within the rotor and stator. In
1970, Fries [12] described his use of a hollow-
shaft, rotating heat pipe to cool electric motors. Since then, he and co-workers have received several patents on this concept in the
Federal Republic of Germany [13, 14]. Figure
5 shows his concept schematically. The shaft
is hollow with a uniform inside diameter.
Cooling occurs at both ends of the motor casing using fan blades (19) and/or fins (16). In
1976, Oslejsek and Polasek [15] reported some
experimental results that were obtained with an
electric motor containing a rotating heat pipe.
Their heat pipe had a conically-shaped condenser section and a stepped evaporator as
shown in Figure 6. The heat pipe was finned
on the condenser end to lower the thermal resistance to the cooling air. With this configuration, they found a significant reduction
in rotor temperature and a flatter temperature
profile when the heat pipe was installed. Several manufacturers produce heat pipe cooled DC
motors for commercial use [16, 17].

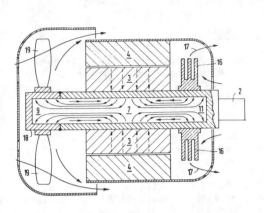

Fig.5. Electric Motor Cooled
      by a Circular Rotating
      Heat Pipe [13].

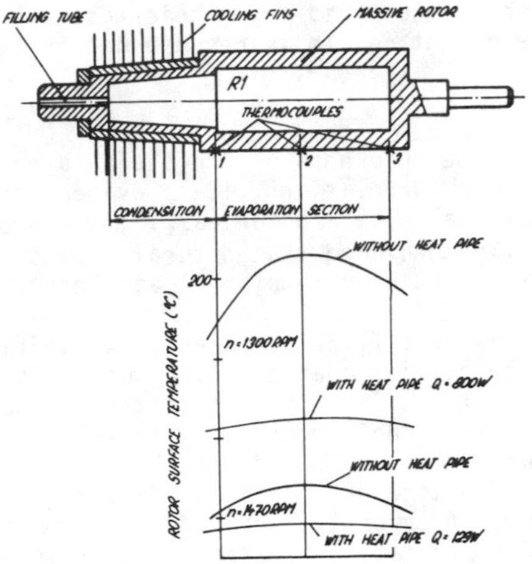

Fig.6. Induction Motor with Heat Pipe
       R1 and Temperature Profile on
       Its Surface [15].

Rotating heat pipes have also been proposed to cool automotive brakes [18]
and to provide a rotating isothermal oven for spectroscopic analysis of various vapors [19].

3. EVAPORATOR HEAT TRANSFER

Heat transfer in the evaporator section of a rotating heat pipe may occur
either with nucleate boiling in thick liquid layers, or with evaporation from
thin liquid films.

3.1 Nucleate Boiling

The effect of acceleration on nucleate pool boiling heat transfer has been
studied extensively by various investigators such as Merte and Clark [20], Costello and Tuthill [21], Adelberg and Schwartz [22], Gray, Marto and Joslyn [23],
Marto and Gray [24], and Körner [25]. As a result, it is now well known that
in the low heat flux region, high acceleration diminishes the number of active
nucleation sites such that little or no bubble nucleation occurs. Instead,
very strong natural convection currents are induced with evaporation occurring
primarily from the free surface. The natural convection data of Marto and Gray
[24] in a rotating cylindrical boiler (102 mm diameter by 51 mm high) were correlated by the equation of Fishenden and Saunders [26]:

$$Nu = 0.14 \ Ra^{1/3} \tag{1}$$

where the characteristic length in the Nusselt and Rayleigh numbers was the
boiler diameter. Körner [25] obtained data using a similar rotating boiler
(145 mm diameter by 52 mm high) and recommended the correlation:

$$Nu = 0.133 \ Ra^{0.375} \tag{2}$$

in which the fluid layer thickness is used as the characteristic length. A more recent result is recommended by Vasiliev and Khrolenok [27]:

$$Nu = 0.75 \ Ra^{1/4} \tag{3}$$

where the thickness of the fluid layer is the characteristic length. In each of these correlations, it is evident that the Nusselt number is proportional to rotational acceleration $(a/g)^S$, where $0.25 < s < 0.375$. As a result, it may be readily concluded that natural convection heat transfer can be substantially enhanced at high rotational acceleration.

As heat flux increases and significant bubble nucleation occurs, the nucleate boiling heat transfer coefficients appear to be practically independent of acceleration. This is shown in Figure 7 which is reproduced from Marto and Gray [24]. Notice that at low fluxes, the data agrees reasonably well with the natural convection correlation of eq'n (3). At high fluxes, however, the data converge together and it is difficult to ascertain the exact effect of acceleration. Apparently, the effect of increased buoyancy of the bubbles caused by high acceleration is offset by the decreased number of active nucleation sites. The data of Merte and Clark [20] and Costello and Tuthill [21] even show a crossover trend in that at high heat fluxes, an increase in acceleration decreases the heat transfer coefficient. This effect, however, if it does exist, is only minor.

The high flux behavior, as described above, continues up until the critical heat flux is reached. Costello and Adams [28] have shown that for water the critical heat flux is proportional to $(a/g)^{1/4}$. This result is in agreement with the Zuber-Kutateladze prediction [29]:

$$q''_{max} = 0.13 \ \rho_v^{1/2} h_{fg} [g(\rho_f - \rho_v)\sigma]^{1/4} \tag{4}$$

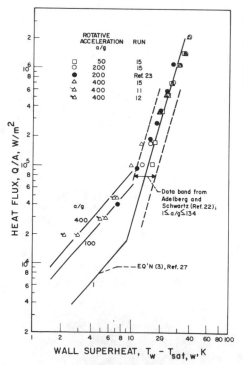

Fig.7. Heat Transfer Data in a Rotating Boiler at High Accelerations [24].

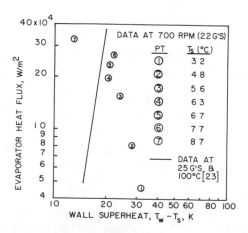

Fig.8. Evaporator Heat Transfer in a Rotating Heat Pipe [32].

and with correlations for other fluids [30, 31].

Figure 8 shows some evaporator data for water taken by Marto [32] in a rotating heat pipe whose evaporator was 79 mm in diameter and 102 mm long. The rotational speed was 700 RPM (corresponding to a/g = 22 at the test surface). The data were taken in the nucleate boiling regime, as observed through a pyrex glass end-window, with a liquid annulus approximately 5-6 mm thick. The data of reference [23] for a rotating boiler at 100°C and a/g = 25 is shown for comparison. During the run, seven data points were obtained for increasing heat flux (and therefore, increasing operating pressure and saturation temperature within the vapor space). As shown in Figure 8, as the saturation temperature increases, the wall superheat decreases. Evaporator heat transfer coefficients therefore increase with applied heat load and operating pressure. This result is in agreement with the well-known pressure effect upon saturated nucleate pool boiling [33].

3.2 Thin Film Evaporation

Evaporation in the presence of very thin liquid films where no bubble nucleation takes place may be treated in a similar way to laminar film condensation. Roetzel [34] described laminar film evaporation in a fast rotating drum having a curved, conical inside surface (Figure 9), and presented a theoretical solution for the case of uniform heat flux. He calculated a wall profile such that the sum of the film resistance (which increases in the flow direction) and the wall resistance (which decreases in the flow direction) remained constant. He included a numerical example which calculated the curvature of the evaporator wall in a 0.5 m diameter by 1.0 m long drum rotating at an angular velocity of 30 radians per second. Assuming a film thickness of 0.1 mm at the beginning of the flow path, he calculated an overall evaporation heat transfer coefficient (including conduction across a copper wall) of 3101 W/m^{2}K. Roetzel [35] has also examined film evaporation from slowly rotating disks which dip into a liquid bath and drag a thin liquid film around. He used this result to predict very high heat transfer rates in a rotary plate evaporator. Dakin [36] studied vaporization of thin water films in rotating radial pipes similar to those proposed for use in gas turbines. He obtained evaporation heat transfer coefficients greater than those predicted by the model of Dukler [37]:

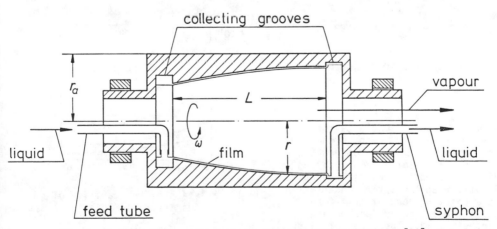

Fig.9. Cooling Drum with Curved, Conical Inside Surface [34].

$$h = k \left(\frac{\omega^2 R}{\nu^2}\right)^{1/3} \left[\left(\frac{1}{3Re}\right)^{1/3} + 0.032\ Re^{0.23}\right] \tag{5}$$

where the average film Reynolds number Re is approximated by:

$$Re \simeq Q_{in}/\ 2\pi r\ F\nu \tag{6}$$

Equation (5) predicts heat transfer coefficients as large as 150,000 W/m²K for $\omega^2 R/g$ equal to 11,500. The solution is dependent on F, the wetted fraction of the passage circumference, which may be very difficult to predict.

A problem associated with thin film evaporation is the ability to keep the evaporator wall wetted at all times. This condition depends on the amount of liquid used within the heat pipe, and is discussed by Daniels and Al-Jumaily [38]. They concluded that for the optimum operation of a rotating heat pipe, there is a unique fluid charge for a particular heat flux and rotational speed. Khrolenok [39] analyzed the evaporation performance in a rotating heat pipe with a porous insert within the evaporator zone to generate a fine mist of droplets impinging upon the heated surface (See Figure 1(d).). He studied the movement of the condensate along a conically shaped porous wall, and arrived at an expression for the film thickness $\delta(x)$ along the insert as a function of distance from the end of the adiabatic section of the pipe $(L_c + L_t < x < L_c + L_t + L_b)$:

$$\delta(x) = \left(\frac{6K\ \cos\phi}{7H\ \sin^2\phi}\right)^{\frac{1}{2}} \frac{\left[(R_0 + (L_c + L_t + L_b)\sin\phi)^{7/3} - (R_0 + x\ \sin\phi)^{7/3}\right]^{\frac{1}{2}}}{(R_0 + x\ \sin\phi)^{2/3}} \tag{7}$$

where K = permeability of porous insert,
          H = thickness of porous insert, and
          ϕ = half of the internal cone angle.

From this equation, he determined the permeability K of the insert which would guarantee that the heated surface remains wetted for a given heat flux. Figure 10 shows the dependence of the permeability on the heat flux and the rotational speed.

4. CONDENSER HEAT TRANSFER

4.1 Rotating Condensation Models

The literature contains numerous studies of film condensation in rotating systems, and much of this information is applicable to rotating heat pipes. Sparrow and Gregg [40] analyzed laminar film condensation on a cooled, isothermal, rotating disk situated in a quiescent vapor. A similar analysis for condensation on a rotating cone of half-cone angle ϕ was made by Sparrow and Hartnett [41]. These results

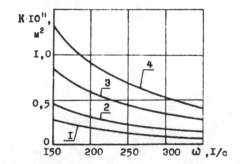

Fig.10. Relation of Permeability of Porous Insert to Rotational Speed and Heat Load (1- 1 KW, 2- 2KW, 3- 5KW, 4- 10KW) [39].

showed that the heat transfer coefficient is independent of position along the condenser wall. For high Prandtl number fluids, Sparrow and Hartnett arrived at an approximate expression for the heat transfer coefficient:

$$Nu = \frac{h}{k}\left(\frac{\nu}{\omega \sin \phi}\right)^{\frac{1}{2}} = 0.904 \left(\frac{Pr}{C_p(T_{sat}-T_w)/h_{fg}}\right)^{\frac{1}{4}} \qquad . \qquad (8)$$

Nandapurkar and Beatty [42] measured condensation heat transfer coefficients on a rotating disk. Their results for methanol, ethanol and Freon-113 were about 25 percent below the theoretical prediction of eq'n (8) (with ϕ = 90 degrees), and they attributed this discrepancy to the omission of vapor drag in the Sparrow and Gregg theoretical model. In a later study therefore, Sparrow and Gregg [43] treated the effect of vapor drag and found that the effect was quite small for ordinary fluids, leaving the discrepancy between theory and experiment attributable to other causes. Dhir and Lienhard [44] studied laminar film condensation on isothermal, axisymmetric bodies in a non-uniform gravity field and included numerous examples of condenser geometries. Their solution, when applied to a rotating disk, agrees exactly with eq'n (8) above.

In 1969, Ballback [7] performed a Nusselt-type analysis for film condensation on the inside of an isothermal, rotating, truncated cone, of half angle ϕ. He found that the heat transfer coefficient varies with position along the condenser. His results for the average heat transfer coefficient may be put in the form:

$$Nu = \frac{h_m L}{k} = 0.904 \left(\frac{\omega^2 L^2 R_o^2}{\nu^2} \cdot \frac{Pr}{C_p(T_{sat}-T_w)/h_{fg}}\right)^{\frac{1}{4}} G(\beta) \qquad (9)$$

where $G(\beta) = \left|\frac{(1+\beta)^{8/3}-1}{\sqrt{\beta}\ (2+\beta)}\right|^{3/4}$, and $\beta = \frac{L \sin \phi}{R_o}$.

Equation (9) was later confirmed to be a special case of the work of Dhir and Lienhard [44]. In order to solve for the local condensate thickness, and hence local heat transfer coefficient, Ballback assumed that the slope of the condensate film $d\delta/dx$ is much less than tan ϕ. This last assumption is not valid for very small half-cone angles, and limits this analysis to truncated cones which are not too slender. In fact, eq'n (9) predicts zero heat transfer when ϕ = 0 (i.e., condensation on the inside of a rotating cylinder). Nimmo and Leppert [45, 46] however, in studying laminar film condensation on finite horizontal plates, have shown that even when the body force is normal to the condensing surface, a finite amount of heat transfer can occur if condensate drainage is permitted over the edges of the condensing surface. Their results can be applied to condensation on the inside of a rotating cylinder (ϕ = 0) and are approximated by:

$$Nu = \frac{h_m L}{k} = 0.82 \left(\frac{\omega^2 L^2 R_o^2}{\nu^2} \cdot \frac{Pr}{C_p(T_{sat}-T_w)/h_{fg}}\right)^{1/5} \qquad . \qquad (10)$$

Marto [47] analyzed laminar film condensation on the inside of a slender, rotating, truncated cone whose geometry is shown schematically in Figure 11(a). He allowed the slope of the condensate film $d\delta/dx$ to be of the same order of magnitude as tan ϕ, and numerically integrated the resulting energy equation to

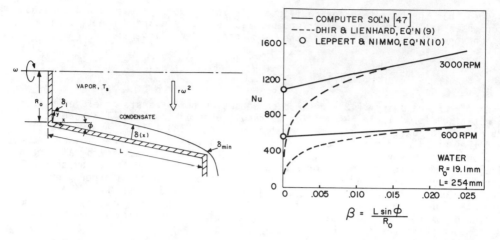

Fig.11(a). Model of Laminar Film Condensation on a Rotating,
Truncated Cone [47].

Fig.11(b). Comparison of Film Condensation Models [47].

find the heat transfer rate and average Nusselt number. Figure 11(b) compares
his results for a specific geometry to those of Dhir and Lienhard (eq'n (9))
and Nimmo and Leppert (eq'n (10)). Notice that the computer solution agrees
with the Nimmo and Leppert solution for the rotating cylinder when $\beta = 0$,
whereas eq'n (9) predicts an average Nusselt number of zero. It is perhaps
surprising, however, that eq'n (9) accurately predicts the average Nusselt
number even down to very small half-cone angles ($\beta \approx 0.02$). The prediction is
even better at high rotational speeds.

   In studying fast, rotating, paper-drying drums, Roetzel [48, 49] investigated laminar film condensation on a curved conical surface while maintaining
a uniform heat flux. He showed that a significant increase in heat transfer
rate can result in going from a cylindrical, rotating drum to one with curved
conical walls, and derived a useful result for the average Nusselt number which
is applicable to rotating heat pipes:

$$Nu = \frac{h_m L}{k} = \left[\left(\frac{4-\alpha}{3} \right)^{31/7} \cdot (G \cdot A)^{31/21} + (0.86A)^{31/28} \right]^{7/31} \tag{11}$$

In the above expression,

$$G = \left(\frac{k_w}{k_f} \right)^{3/4} \cdot \left(\frac{1}{A} \right)^{1/4} \quad , \qquad A = \frac{\omega^2 \bar{R} \rho h_{fg} L^2}{6 \nu q''} \quad , \text{ and} \tag{11a & b}$$

α is defined by the following equation for the rotational acceleration in the
flow direction:

$$a_x = a_{x1} \left(\frac{x}{L} \right)^{\alpha} \tag{11c}$$

Equations (11) are applicable for a variety of curved conical surfaces provided the surface heat flux is known. When $\alpha = o$, the wall profile reduces to a conical surface of constant slope, as seen earlier in Figure 11(a).

In 1971, Ballback's analysis was extended by Newton [50], who included the effects of interfacial shear and vapor pressure drop while allowing the slope of the condensate film $d\delta/dx$ to be the same order as $\tan \phi$. His model and numerical solution have been described by Marto [32, 51]. The familiar assumptions of Nusselt's theory of laminar film condensation were made, including negligible subcooling, and negligible momentum changes and convection effects in the condensate film. In addition, it was assumed that centrifugal acceleration due to rotation was much larger than the normal acceleration of earth gravity, the thickness of the condensate film was much less than the radius of the condenser wall, the vapor density was much less than the liquid density, and the half-cone angle ϕ was small. It was reasoned that the last restriction should apply to most practical rotating heat pipe designs.

With these assumptions, a differential momentum analysis on the condensate film and the vapor core, and a differential energy balance from the vapor out to the ambient yielded two first order non-linear differential equations for the film thickness $\delta(x)$ and the vapor velocity $v(x)$. Newton [50] numerically integrated these equations for the case of a heat pipe whose inside diameter is stepped between the evaporator and condenser. This geometry is similar to that described by Figure 1(c). The integration was started by assuming initial conditions $\delta=\delta_i$ and $v=0$ at $x=0$ (the cold end of the heat pipe). After the integration was completed for $\delta(x)$ and $v(x)$, the energy equation was integrated over the length of the condenser to find the total heat transfer rate of the heat pipe. With the proper choice of δ_i, the calculated $\delta(x)$ reached a maximum along the length of the condenser, and then tapered off to a minimum value δ_{min} at the stepped overfall (see Figure 11(a)). It was observed that the heat transfer rate was insensitive to an exact value of δ_{min} which agreed with the results of Nimmo and Leppert [46] for laminar film condensation on a finite horizontal surface with drainage over the edges. Newton obtained results for half-cone angles of 0^0, 0.1^0, 0.2^0, and 0.3^0. As the half-cone angle increased, the solution of the differential equations became extremely sensitive to the initial values of δ_i. He also noted that, in general, $d\delta/dx$ can be neglected in comparison to $\tan \phi$.

This additional approximation, together with an order of magnitude analysis of the terms in the non-linear differential equations, yielded the following simplified equations:

$$\frac{dv}{dx} = \frac{2(R_o+x \sin \phi)\,(T_{sat}-T_\infty)}{\rho_v h_{fg} R(x)^2 (\delta/k_f+t/k_w+1/h)} - \frac{2v \sin \phi}{R(x)} \tag{12}$$

and,

$$\frac{\delta^3}{3} R(x) \sin \phi(\rho_f \omega^2 R(x) + \frac{2\rho_v v^2}{R(x)}) - \frac{\delta^2}{4}\rho_v fv^2 R(x) - \rho_v v \frac{\mu_f R(x)^2}{2\rho_f} = 0 \tag{13}$$

where R(x) is the local radius to the edge of the condensate film and f is the friction factor. Equation (12) includes the overall temperature difference between the vapor and the ambient ($T_{sat} - T_\infty$) and the overall thermal resistance ($\delta/k_f + t/k_w + 1/h$), where h is the external heat transfer coefficient between the rotating heat pipe and the ambient. Newton integrated eq'n (12) using eq'n (13), with initial conditions that $\delta=0$, $v=0$ at $x=0$. Results showed that the approximate solution (eq'ns (12) and (13)) was slightly conservative and became more accurate as the half-cone angle ϕ increased. At $\phi = 0.3°$, the approximate solution was in error by -1.1 percent. Since most practical designs will involve half-cone angles at least as large as 0.5° to 1.0°, it is felt that this approximate solution can be used with little error.

Daniels and Al-Jumaily [38, 52] independently extended Ballback's analysis. They assumed that $d\delta/dx$ was much smaller than $\tan \phi$, and included the effect of vapor drag while neglecting vapor pressure drop. Their analysis for a constant wall temperature yielded the following expression for the average Nusselt number:

$$Nu = \frac{h_m L}{k} = \frac{4}{3} Sh_L \left(\frac{\delta_L}{L}\right)^3 - \frac{1}{2} Dr_L \left(\frac{\delta_L}{L}\right)^2 - \frac{1}{2} Re_{vL}\left(\frac{\delta_L}{L}\right) \qquad (14)$$

where

$$Dr_L = \frac{\rho \tau_v h_{fg} x^2 \cos \phi}{\mu k (T_{sat} - T_w)} \qquad , \qquad (14a)$$

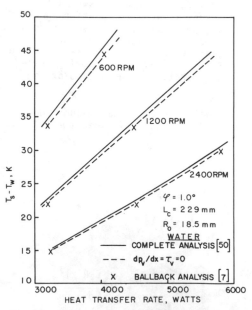

Fig.12. Influence of Vapor Pressure Gradient and Shear Stress on Theoretical Performance [32].

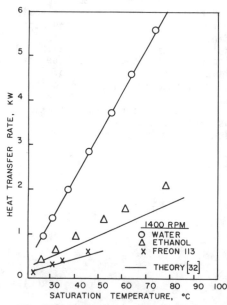

Fig.13. Comparison of Experimental Data with Theoretical Prediction [32].

$$Sh_L = \frac{\rho^2(\omega^2\bar{R}-g)h_{fg}x^3 \sin\phi}{4\mu k(T_{sat}-T_w)} \quad , \tag{14b}$$

and

$$Rev_L = \frac{\rho \, u_v \, x \cos\phi}{\mu} \quad . \tag{14c}$$

If the effect of vapor drag is neglected, the last two terms in eq'n (14) can be dropped, giving:

$$Nu \simeq \frac{4}{3} Sh_L \left(\frac{\delta_L}{L}\right)^3 \quad . \tag{15}$$

Upon substituting in the expressions for the Sherwood number Sh_L, eq'n (14b), and for $\frac{\delta_L}{L}$ from [38], eq'n (15) reduces to:

$$Nu \simeq 0.943 \left(\frac{\omega^2\bar{R}^2 L^2}{\nu^2} \cdot \frac{Pr}{C_p(T_{sat}-T_w)/h_{fg}}\right)^{\frac{1}{4}} \beta^{\frac{1}{4}} \tag{16}$$

for high rotational speeds.

It is interesting to note that, since the average condenser radius $\bar{R}$ is approximately equal to the minimum condenser radius R_0, eq'n (16) is identical to the solution of Ballback, eq'n (9) for small values of β.

The more complete analysis of Newton [50] is compared to the Ballback analysis in Figure 12 where the heat transfer rate is plotted versus the temperature difference between the fluid saturation temperature and the wall temperature for a given set of heat pipe operating parameters. The Ballback analysis predicts a slightly higher heat transfer rate since he neglected interfacial shear and vapor pressure drop effects. When these effects are neglected in the Newton analysis, the results agree very well with those of Ballback. Marto [32] compared his experimental data for water, ethanol and Freon 113 to the Newton analysis and this is shown in Figure 13. The agreement appears to be favorable although the experimental data for ethanol and Freon 113 were always higher than the theory. Similar results were observed by Daniels and Williams [53] for Freon 113.

Daniels and Williams [53, 54] studied the effect of noncondensable gas loading on the performance of rotating heat pipe condensers by making a series of measurements for various combinations of working fluids, noncondensable gases, and external cooling water conditions. Figures 14(a) and (b) show some of their results, where condenser wall temperature is plotted versus position. Figure 14(a) shows that for Freon 113 and nitrogen gas, the introduction of noncondensable gases raises the wall temperature over the first part of the condenser but lowers the temperature at the cold end. This indicates that the nitrogen gas (having a lighter molecular weight than Freon 113) was forming a pocket primarily at the cold end, shutting off this end of the heat pipe in a similar way that noncondensable gases influence stationary heat pipes. Figure 14(b) shows that for a mixture of acetone and CO_2 gas, where the molecular weights of the vapor and gas are closer to one another, the gas does not collect as neatly as before. In this situation, the centrifugal

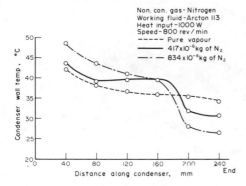

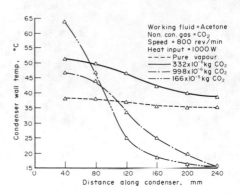

Fig.14(a). Condenser Wall Temperature
          Profiles for Freon 113-N$_2$
          Gas [53].

Fig.14(b). Condenser Wall Temperature
          Profiles for Acetone - CO$_2$
          Gas [53].

forces are acting on the gas and vapor molecules almost to the same extent, causing mixing over a larger percentage of the condenser. This mixing causes the wall temperature to have a more gradual variation rather than the abrupt change as shown in the previous case. Daniels and Al-Baharnah [55] have performed an analysis of noncondensable gas effects while including axial conduction effects in the condenser wall. Recent results [56] show that their model predicts heat pipe performance very well for Freon 113 (Figure 15(a)). Their work with water, however, has not been as successful (Figure 15(b)).

The general problem of condensation in a rotating two-phase radial thermosyphon was analyzed by Chato [57] in 1965. His results, which include the effects of vapor shear, may be applied to rotating heat pipes when heat is transferred radially. He pointed out that coriolis forces may have to be studied further. Maezawa, Susuki and Tsuchida [58] performed an analysis of disk-shaped rotating heat pipes. Their analysis is similar to that of Daniels and Al-Jumaily [38]. Chan, Kanai and Yang [59] performed an approximate analysis of conical rotating heat pipes. Condensation on the outside of rotating cylinders has been studied by Nicol and Gacesa [60] and Singer and Preckshot [61].

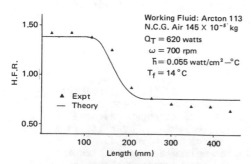

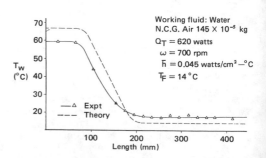

Fig.15(a). Ratio of Heat Flux with
          Gas to Heat Flux with Pure
          Vapor for Freon 113 and
          Air [56].

Fig.15(b). Condenser Wall Tempera-
          ture Distribution for
          Water and Air [56].

4.2. Enhanced Rotating Condensation

In general, heat transfer in the condenser section of a rotating heat pipe may be increased by increasing the half-cone angle ϕ, by operating at high rotational acceleration, by improving upon the external cooling mechanism, and by the proper choice of working fluid. Upon examination of eq'n (9), it is evident that the condenser heat transfer performance will depend upon the following group of fluid properties:

$$N = \frac{\rho^2 h_{fg} k^3}{\mu} \tag{17}$$

where N may be referred to as a Figure of Merit. This grouping of properties is plotted in Figure 16 [62] and it is clear that water is by far the best rotating heat pipe fluid.

Marto [32] showed that a marked improvement in rotating heat pipe condenser performance can occur if dropwise condensation conditions are promoted on the condenser surface. This was done for two experimental runs by wiping the surface with a very thin layer of vacuum grease. His results for water at 700 RPM are shown in Figure 17. Marto and Wagensil [63] obtained dropwise measurements for water at 700, 1400 and 2800 RPM. They used a promoting

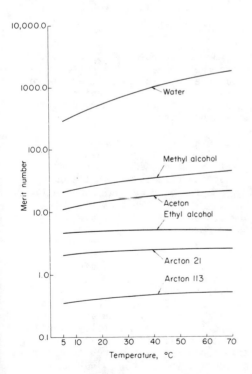

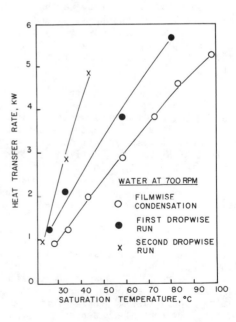

Fig.16. Comparison of Working
        Fluids [62].

Fig.17. The Effect of Dropwise
        Condensation on Condenser
        Heat Transfer [32].

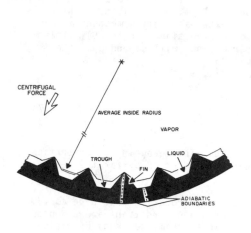

Fig.18(a). A Segment of a Finned
            Condenser [64].

Fig.18(b). Influence of Number of Fins
            Upon Condenser Heat Transfer
            [64].

solution of one percent by weight n-octadecyl mercaptan ($C_{18}H_{37}SH$) in octanoic
acid, and lightly swabbed the condenser before their experimental runs. Their
results showed that dropwise condensation made the most improvement at low rotational speeds and this was attributed to the fact that at high rotational
speeds the drops tend to flatten out more and coalesce, forming more complete
wetting of the surface as occurs in film-wise conditions.

    The use of straight, triangular fins within a conical condenser has been
recently analyzed by Salinas and Marto [64]. Figure 18(a) shows a schematic
representation of a cross-section of a segment of the condenser which they
analyzed. They included two-dimensional conduction effects within the condenser wall (axial conduction was neglected) and followed an analysis similar
to that of Newton [50] outlined earlier. Figure 18(b) shows that for a heat
pipe with a half-cone angle of 1 degree, the presence of the fins can dramatically increase the condenser heat transfer rate. The results shown are
for a copper wall, with water as the heat pipe fluid, and with an external
heat transfer coefficient assumed to be 28.4 $KW/m^2 °C$. For a heat pipe rotating at 3000 RPM, the addition of 50 fins improves the heat transfer rate by
160 percent. Of course, other results would be obtained as the heat pipe geometry, wall material, working fluid, external heat transfer coefficient, and
rotational speed are changed. The use of a helically finned cylindrical condenser was studied by Marto and Wagenseil [63] and later by Marto and Weigel
[65]. Figure 19(a) shows a photograph of a section of the internally finned
condenser they tested. It was purchased "off the shelf" from a heat transfer
tubing manufacturer. The performance of this condenser as compared to a
smooth, cylindrical condenser, is shown in Figure 19(b). It is clear that
by replacing a smooth hollow shaft with one containing helically wrapped

internal fins which act as pump impellers, the condenser performance improves
markedly.

5. VAPOR FLOW

The consideration of vapor flow in rotating heat pipes has not received
the same degree of attention that has been given to the heat transfer processes
within the evaporator and condenser sections.

The fundamental liquid and vapor flow processes which occur in conventional, stationary heat pipes have been reviewed by Tien [66]. He pointed out
that the treatment of vapor flow in the core region of a heat pipe will vary
depending upon whether viscous forces or inertia forces dominate. When viscous forces predominate, the flow can be analyzed as laminar, incompressible
flow in a pipe with either injection or suction. When the vapor velocity becomes very large, inertia forces predominate and the flow can be analyzed as
inviscid, compressible flow. In this later situation, the flow is limited by
the gas dynamic phenomenon of 'choking' where the vapor velocity equals the
local sonic velocity. Except during heat pipe start-up, it is expected that
before sonic conditions are reached, interfacial shear forces between the
counter-flowing liquid and vapor will become large enough such that entrainment of the condensate or 'flooding' will occur.

Sakhuja [67] studied the entrainment limit in gravity-fed, wickless heat
pipes (i.e., thermosyphons) with organic fluids. He postulated that for a
given liquid flow along the walls of the heat pipe, there is a fixed vapor
flow rate which leads to entrainment. He followed the analysis of Wallis
[68] and arrived at an empirical equation for the entrainment limit:

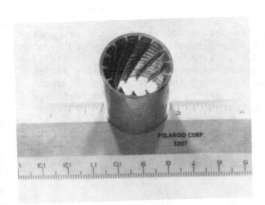

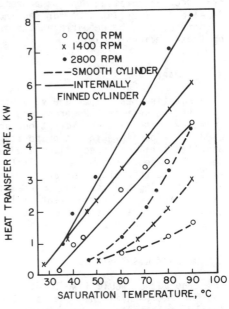

Fig.19(a). Photograph of Helically   Fig.19(b). Influence of Helical
           Finned Condenser [63].   Fins Upon Condenser
                                                       Performance [63].

$$q_{max} = C^2 h_{fg} \frac{\pi D_i^2}{4} \frac{\sqrt{g D_i \rho_v (\rho_f - \rho_v)}}{\left\{ 1 + (\rho_v / \rho_f)^{\frac{1}{4}} \right\}^2} \tag{18}$$

In the above expression, the value of C depends upon the edge condition between the evaporator and condenser. A smooth transition will give a C close
to unity, whereas a sharp edge will lower C to 0.725. Sakhuja carried out
several experiments with Dowtherm-A and 4-Methyl Pyridene using a thermosyphon
with a step change in diameter between the evaporator and condenser. He found
good agreement between eq'n (18) and his data for a value of C = 0.725.

An extensive series of measurements were made recently on an inclined
copper-water thermosyphon by Nguyen-Chi and Groll [69]. They compared all
their data with existing entrainment models and concluded that their data
could be well-correlated with eq'n (18) provided C was changed to 1.05. They
pointed out that further experimentation and additional analytical work are
needed to be able to determine the influence of heat pipe operating parameters
(such as liquid fill, internal surface roughness, working fluid, etc.) upon
the entrainment limit.

The influence of rotation upon vapor-liquid interaction and the stability
of surface waves has received little attention, and is an area worthy of future
research. Some preliminary measurements with air and water at low rotational
speeds showed qualitatively that rotation reduced the onset of flooding [70].
In addition to the effect of rotation, the entrainment limit for a rotating
heat pipe with a stepped evaporator (Figure l(c)) may be considerably different from that which would exist in the heat pipes shown in Figures l(a) and
l(b) which have a smooth transition between the evaporator and condenser. The
influence of rotating heat pipe geometry therefore upon the entrainment limit
is also an important area for research. A much better understanding of these
phenomena will be required as the demand for increased thermal performance
of these devices occurs.

6. SUMMARY

Rotating heat pipes are very effective devices to transfer thermal energy
in rotating systems. Numerous applications are possible depending upon system
geometry, materials considerations and cost. Very high heat transfer coefficients are possible in the evaporator and condenser sections of these devices
provided the usual precautions are taken regarding the elimination of noncondensable gases and the compatibility of the working fluid with the heat pipe
walls. The thermal performance of these devices can be adequately predicted
using existing analyses and correlations. More research is needed, however,
with regard to the entrainment limit in high speed, high performing devices
especially where heat pipe diameter undergoes an abrupt change between evaporator and condenser.

ACKNOWLEDGMENTS

The author would like to dedicate this paper to the late Vernon H. Gray
who correctly envisioned the usefulness of these devices. In addition, he is
deeply indebted to numerous individuals who provided valuable information for
this paper, particularly to: T. C. Daniels, P. Fries, M. Groll, L. Langston,
S. Maezawa, T. Ogushi, F. Polasek, W. Roetzel and L. L. Vasiliev.

NOMENCLATURE

a = rotational acceleration, m/s^2

a_x = rotational acceleration component along the flow direction, m/s^2

A = dimensionless grouping, defined by eq'n (11b)

C = coefficient in entrainment limit, eq'n (18)

C_p = specific heat, kJ/kg K

D = heat pipe diameter, m

Dr_L = friction number, defined by eq'n (14a)

f = friction factor

F = wetted fraction of passage circumference

g = acceleration of gravity, m/s^2

h = average heat transfer coefficient, W/m^2 K

h_{fg} = latent heat of vaporization, kJ/kg

H = thickness of porous insert, m

k = thermal conductivity, W/m K

K = permeability, m^2

L = length, m

N = figure of merit, defined by eq'n (17)

Nu = average Nusselt number

Pr = Prandtl number

q = heat transfer rate, W

q_{max} = maximum heat transfer rate, W

q'' = heat flux, W/m^2

q''_{max} = critical heat flux, W/m^2

Q_{in} = volumetric flow rate of water in radial pipe, m^3/s

r = radius of radial pipe, m

R = radial distance from axis of rotation, m

$R(x)$ = local radius to edge of condensate film, m

R_o = minimum inside radius of truncated cone condenser, m

$\bar{R}$ = average radius of condenser, m

Ra = Rayleigh number

Re = Film Reynolds number

Rev = two phase Reynolds number, defined by eq'n (14c)

Sh_L = Sherwood number, defined by eq'n (14b)

t = thickness of condenser wall, m

T = temperature, K

u_v = average vapor velocity at entrance to condenser, m/s

$v(x)$ = local vapor velocity, m/s

x = distance along flow direction, m

α = exponent defined in eq'n (11c)

β = dimensionless grouping, $\dfrac{L \sin \phi}{R_o}$

$\delta(x)$ = local thickness of condensate film, m

δ_i = initial value of condensate film thickness, m

δ_L = value of condensate film thickness at end of condenser, m

ρ = density, kg/m^3

ϕ = condenser half-cone angle

τ = interfacial shear stress, N/m^2

μ = dynamic viscosity, N s/m^2

ν = kinematic viscosity, m^2/s

ω = angular velocity, s^{-1}

Subscripts

b   = insert

c   = condenser

f   = liquid

i   = inside

L   = end of condenser

m   = mean

sat = saturation

t   = transport zone

v   = vapor

w   = wall

∞ = ambient

REFERENCES

1. Gray, V.H. 1969. The rotating heat pipe - a wickless, hollow shaft for transferring high heat fluxes. ASME Paper No. 69-HT-19.

2. Hoffmann, M. and Fries, P. 1969. Cooling the rotor of an electric motor. Patent application No. 1,900,411, Federal Republic of Germany.

3. Vasiliev, L.L. 1981. Heat exchangers based on heat pipes. Nauka i Tekhnika, Minsk, pp. 86-87. (in Russian)

4. Groll, M. Krähling, H. and Münzel, W.D. 1978. Heat pipes for cooling of an electric motor. Proceedings of the III International Heat Pipe Conference, Palo Alto, California, pp. 354-359.

5. Groll, M., Krähling, H. and Münzel, W.D. 1979. Heat pipes for cooling electric motors. Elektrizitätswertung, 2 January, pp. 10-15. (in German)

6. Japikse, D. 1973. Advances in thermosyphon technology. Advances in Heat Transfer, Vol. 9, Academic Press, pp. 77-91.

7. Ballback, L.J. 1969. The operation of a rotating, wickless heat pipe, M.S. Thesis, Naval Postgraduate School, Monterey, California.

8. Gray, V.H. 1974. Methods and apparatus for heat transfer in rotating bodies. U.S. Patent No. 3,842,596.

9. Vasiliev, L.L. 1981. Heat exchangers based on heat pipes. Nauka i Tekhnika, Minsk, pp. 129-136 (in Russian)

10. Cohen, H. and Bayley, F.J. 1954. Heat transfer problems of liquid-cooled gas-turbine blades. Institution of Mechanical Engineers, pp. 1063-1080.

11. Szogyen, J.R. 1979. Cooling of electric motors, Institution of Electrical Engineers, Journal on Electric Power Applications, Vol. 2(2), pp. 59-67.

12. Fries, P. 1970. Experimental results with a wickless centrifugal heat pipe. International Journal of Heat and Mass Transfer, Vol. 13, pp. 1503-1504. (in German)

13. Fries, P. and Schulze, P. 1973. Arrangement for cooling rotating bodies. German Patent No. 1,928,358. (in German)

14. Fries, P. 1974. Cooling arrangement for the rotor of an electric motor with a centrifugal heat pipe. German Patent No. 2,251,841. (in German)

15. Oslejsek, O. and Polasek, F. 1976. Cooling of electrical machines by heat pipes. Proceedings of the II International Heat Pipe Conference, Bologna, Italy, pp. 503-514.

16. Ogushi, T. 1982. DC spindle motors cooled by rotating heat pipes. Private communication, Mitsubishi Electric Corp., Japan.

17. DC servo motor series bulletin. 1982. Fujitsu Fanuc Ltd, Japan.

18. Maezawa, S., and Tsuchida, A. 1982. Application of disk-shaped rotating heat pipe to brake cooling. XIV ICHMT Symposium on Heat and Mass Transfer in Rotating Machinery, Dubrovnik, Yugoslavia.

19. Hessel, M.M. and Lucatorto, T.B. 1973. The rotating heat pipe oven: a universal device for the containment of atomic and molecular vapors. Review of Scientific Instruments, Vol. 44, pp. 561-563.

20. Merte, H. and Clark, J.A. 1961. Pool boiling in an accelerating system. Journal of Heat Transfer, Vol. 83, pp. 233-242.

21. Costello, C.P. and Tuthill, W.E. 1961. Effects of acceleration on nucleate pool boiling. Chemical Engr. Progress Symposium Series, Vol. 57, pp. 189-196.

22. Adelberg, M. and Schwartz, S.H. 1968. Nucleate pool boiling at high G levels. Chemical Engr. Progress Symposium Series, Vol. 64, pp. 3-11.

23. Gray, V.H., Marto, P.J. and Joslyn, A.W. 1968. Boiling heat transfer coeffi-
    cients, interface behavior, and vapor quality in rotating boiler operating
    to 475 G's. NASA TN D-4136.

24. Marto, P.J. and Gray, V.H. 1971. Effects of high accelerations and heat
    fluxes on nucleate boiling of water in an axisymmetric rotating boiler.
    NASA TN D-6307.

25. Körner, W. 1970. Influence of higher acceleration on heat flow in boiling.
    Chemie Ing. Techn. Vol. 42, No. 6 (in German).

26. Fishenden, M. and Saunders, O.A. 1950. An Introduction to Heat Transfer,
    Clarendon Press, Oxford.

27. Vasiliev, L.L. and Khrolenok, V.V. 1976. Centrifugal coaxial heat pipes.
    Proceedings of the II International Heat Pipe Conference, Bologna, Italy,
    pp. 293-302.

28. Costello, C.P. and Adams, J.M. 1961. Burnout heat fluxes in pool boiling at
    high accelerations. Paper 30, International Heat Transfer Conference, Den-
    ver, Colorado.

29. Lienhard, J.H. 1981. A Heat Transfer Textbook, Prentice-Hall, Inc., p. 406.

30. Usenko, V.I. and Fainzil'berg, S.N. 1974. Effect of acceleration on the
    critical heat load with the boiling of freons on elements having small
    transverse dimensions, High Temperature, Vol. 12, May/June.

31. Morozkin, V.I., Amenitskii, A.N. and Alad'ev, I.T. 1963, Experimental study
    of the effect of acceleration on the boiling crisis in liquids at the sat-
    uration temperature. Teplofizika Vysokikh Temperatur, Vol. 1, pp. 107-111.

32. Marto, P.J. 1976. Performance characteristics of rotating, wickless heat
    pipes. II International Heat Pipe Conference, Bologna, Italy, pp. 281-291.

33. Cichelli, M.T. and Bonilla, C.F. 1945. Transactions AIChE, Vol. 41, p. 755.

34. Roetzel, W. 1977. Laminar film surface evaporation with uniform heat flux
    in a fast rotating drum. International Journal of Heat and Mass Transfer,
    Vol. 20, pp. 549-553.

35. Roetzel, W. 1977. A new rotary plate evaporator. Wärme-und Stoffübertragung,
    Vol. 10, pp. 61-70.

36. Dakin, J.T. 1978. Vaporization of water films in rotating radial pipes.
    International Journal of Heat and Mass Transfer, Vol. 21, pp. 1325-1332.

37. Dukler. 1960. Fluid mechanics and heat transfer in vertical falling-film
    systems. Chemical Engineering Progress Symposium Series, Vol. 30.

38. Daniels, T.C. and Al-Jumaily, F.K. 1975. Investigations of the factors af-
    fecting the performance of a rotating heat pipe. International Journal of
    Heat and Mass Transfer, Vol. 18, pp. 961-973.

39. Khrolenok, V.V. 1981. Methods of designing a centrifugal heat pipe with a
    porous insert in the heated zone. Heat and Mass Transfer in Systems with
    Porous Components, L.L. Vasiliev (Editor), Liukov Institute for Heat and
    Mass Transfer, Minsk, pp. 29-34. (in Russian)

40. Sparrow, E.M. and Gregg, J.L. 1959. A theory of rotating condensation. Journal of Heat Transfer, Vol. 81, pp. 113-120.

41. Sparrow, E.M. and Hartnett, J.P. 1961. Condensation on a rotating cone. Journal of Heat Transfer, Vol. 83, pp. 101-102.

42. Nandapurkar, S.S. and Beatty, K.O. 1959. Condensation on a horizontal rotating disk. Chemical Engineering Progress Symposium Series, Heat Transfer - Storrs, pp. 129-137.

43. Sparrow, E.M. and Gregg, J.L. 1960. The effect of vapor drag on rotating condensation. Journal of Heat Transfer, Vol. 82, pp. 71-72.

44. Dhir, V. and Lienhard, J. 1971. Laminar film condensation on plane and axisymmetric bodies in nonuniform gravity. Journal of Heat Transfer, Vol. 93, pp. 97-100.

45. Leppert, G. and Nimmo, B. 1968. Laminar film condensation on surfaces normal to body and inertial forces. Journal of Heat Transfer, Vol. 90, pp. 178-179.

46. Nimmo, B. and Leppert, G. 1970. Laminar film condensation on a finite horizontal surface. Heat Transfer 1970, Vol. 6, Elsevier Pub. Co., Amsterdam.

47. Marto, P.J. 1973. Laminar film condensation on the inside of slender, rotating truncated cones. Journal of Heat Transfer, Vol. 95, pp. 270-272.

48. Roetzel, W. 1975. Improving heat transfer in steam-heated fast rotating paper drying drums. International Journal of Heat and Mass Transfer, Vol. 18, pp. 79-86.

49. Roetzel, W. and Newman, M. 1975. Uniform heat flux in a paper drying drum with a non-cylindrical condensation surface operating under rimming conditions. International Journal of Heat and Mass Transfer, Vol. 18, pp. 553-557.

50. Newton, W.H. 1971. Performance characteristics of rotating, non-capillary heat pipes. M.S. Thesis, Naval Postgraduate School, Monterey, California.

51. Marto, P.J. 1973. An analytical and experimental investigation of rotating, noncapillary heat pipes, NASA CR-130373, National Aeronautics and Space Administration, Washington, D.C.

52. Daniels, T.C. and Al-Jumaily, F.K. 1973. Theoretical and experimental analysis of a rotating wickless heat pipe. I International Heat Pipe Conference, Stuttgart, Germany.

53. Daniels, T.C. and Williams, R.J. 1978. Experimental temperature distribution and heat load characteristics of rotating heat pipes. International Journal of Heat and Mass Transfer, Vol. 21, pp. 193-201.

54. Daniels, T.C. and Williams, R.J. 1979. The effect of external boundary conditions on condensation heat transfer in rotating heat pipes. International Journal of Heat and Mass Transfer, Vol. 22, pp. 1237-1241.

55. Daniels, T.C. and Al-Baharnah, N.S. 1980. Temperature and heat load distribution in rotating heat pipes. AIAA Journal, Vol. 18, pp. 202-207.

56. Daniels, T.C. 1982. Private communication.

57. Chato, J.C. 1965. Condensation in a variable acceleration field and the condensing thermosyphon. Journal of Engineering for Power, Vol. 87, pp. 355-360.

58. Maezawa, S., Suzuki, Y. and Tsuchida, A. 1981. Heat transfer characteristics of disk-shaped, rotating wickless heat pipe. Advances in Heat Pipe Technology, D.A. Reay (Editor), Pergamon Press, Oxford, pp. 725-733.

59. Chan, S.H., Kanai, Z. and Yang, W.T. 1971. Theory of a rotating heat pipe. Journal of Nuclear Energy, Vol. 25, pp. 479-487.

60. Nicol, A.A. and Gacesa, M. 1970. Condensation of steam on a rotating, vertical cylinder. Journal of Heat Transfer, Vol. 92, pp. 144-152.

61. Singer, R. M. and Preckshot, G.W. 1961. The condensation of vapor on a horizontal rotating cylinder. Proceedings of the Heat and Mass Transfer Institute, Stanford, California, pp. 205-217.

62. Dunn, P. and Reay, D.A. 1976. Heat Pipes, Pergamon Press, Oxford, p. 176.

63. Marto, P.J. and Wagenseil, L.L. 1979. Augmenting the condenser heat transfer performance of rotating heat pipes. AIAA Journal, Vol. 17, pp. 647-652.

64. Salinas, D. and Marto, P.J. 1982. Finite element analysis of an internally-finned rotating heat pipe. Journal of Numerical Heat Transfer (submitted).

65. Marto, P.J. and Weigel, H. 1981. The development of economical rotating heat pipes. Advances in Heat Pipe Technology, D.A. Reay (Editor), Pergamon Press, Oxford, pp. 709-723.

66. Tien, C.L. 1975. Fluid mechanics of heat pipes. Annual Review of Fluid Mechanics, Vol. 7, Annual Reviews Inc., Palo Alto, Calif., pp. 167-185.

67. Sakhuja, R.K. 1973. Flooding constraints in wickless heat pipes. ASME Paper No. 73-WA/HT-7.

68. Wallis, G.B. 1969. One Dimensional Two-phase Flow, McGraw-Hill Publishing Co., New York, pp. 336-343.

69. Nguyen-Chi, H. and Groll, M. 1981. Entrainment or flooding limit in a closed two-phase thermosyphon. Advances in Heat Pipe Technology, D. A. Reay (Editor), Pergamon Press, Oxford, pp. 147-162.

70. Boss, R.A. 1970. A study of rotating two-phase annular countercurrent flow. M.S. Thesis, Naval Postgraduate School, Monterey, California.

Optimum Charge of Working Fluids in Horizontal Rotating Heat Pipes

W. NAKAYAMA and Y. OHTSUKA
Mechanical Engineering Research Laboratory
502 Kandatsu, Tsuchiura, Ibaraki, Japan

H. ITOH and T. YOSHIKAWA
Hitachi Works, Hitachi, Ltd.
3-1-1 Saiwai-cho, Hitachi, Ibaraki, Japan

ABSTRACT

The performance of wickless straight heat pipes rotating about their horizontal axes was investigated. The data reported herein were obtained with the copper pipes of 28 and 37mm ID, 480mm long with the evaporator and condenser sections each 170mm long, and distilled water as the working fluid. The transition of two-phase flow in the heat pipe from the stratified to the annular structure occurs at a certain rotational speed (Froude number), and this affects the heat transfer performance. The volumetric percentage of liquid phase in the heat pipe (volumetric charge) determines the transition Froude numbers. For a given Froude number and a heat load, a too lean volumetric charge invites dry-out of the evaporator wall. A too high volumetric charge reduces the area for thin film evaporation and condensation on the rotating wall which dips and leaves the liquid reservoir of the stratified fluid. In the range of Froude numbers less than 13 which include many cases of heat pipe applications to conventional rotating machines, the volumetric charge of 10-14 per cent minimizes the wall temperature difference between the evaporator and the condenser.

1.   INTRODUCTION

The rotating heat pipe studied in this paper is wickless, rotates about its own horizontal axis, and has the following distinct features compared to those rotating heat pipes studied previously.

(1)   The heat pipe has a uniform cross section throughout its length, i.e., it is made of a simple straight pipe.

(2)   The pipe has a relatively long evaporator section, its ratio to the inside diameter being over five.

(3)   The rotation rate of interest is relatively low. The range of rotation rates is 50 - 1600 rpm (revolution per minute) for the pipes of 28 and 37 mm ID.

The present work was motivated to provide an effective yet economical means of cooling for rotating machines. One instance is cooling of the bearings of the hydraulic turbine. Heat from the bearings is transferred through the heat pipe imbedded in the rotating shaft and disharged to water in the draft tube. Another instance is cooling of variable speed motors, where heat generated in the rotor is to be transferred through the heat pipe in the shaft. There are many other instances, where machine's structure, operating conditions, and more than anything else, economy, dictate the use of a wickless, slender heat pipe in a horizontal rotating shaft.

The present paper describes what physical parameters are involved and how they impose limitations on the performance of this simple scheme.

2. MODES OF FLOW AND HEAT TRANSFER IN A HORIZONTAL ROTATING HEAT PIPE

The item (3) of the features described above makes the flow of the working fluid inside the heat pipe gravity controlled; the Froude number defined in terms of the pipe's inner radius r (m), the rotation rate ω (rad/sec), and the earth's gravity acceleration g (9.807m/ϵ^2), r ω^2/g, ranges well below unity to over ten in this study.

Most published works including those reported recently by Marto and Wagenseil [1], and Daniels and Al-Baharnah [2], have dealt with the cases where the centrifugal force has a predominant influence on the flow in the heat pipe. The liquid phase undergoes a solid body rotation, adhering to the inner wall of the pipe with a uniform thickness around the circumference. The heat pipe has a tapered wall in the condenser section to drive the condensate toward the evaporator section under centrifugal accelerations. In [1], a straight pipe was also tested, revealing that the heat transfer rate was less than one half of those obtained with a tapered heat pipe under equal operating conditions.

The heat transfer performance of a heat pipe in the present scheme is affected by the change of flow structure from a stratified one like that shown in Fig. 1 to a solid body rotation at high Froude numbers. The transition of two-phase fluid structure is a complicated but interesting phenomena. Karwail and Corrsin [3] observed the development of cellular patterns as the liquid is scraped up by the rotating wall. Theoretical analysis to date has been made only on the perturbed state from a solid body rotation due to influence of the gravity [4]. There is no way to predict theoretically the transition of two-phase fluid structure from stratified to solid body rotation.

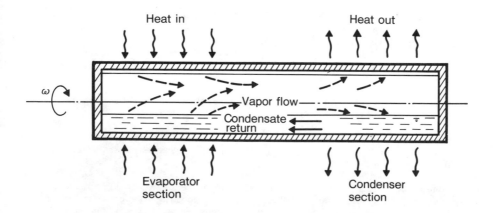

Fig. 1 Schematic diagram of horizontal rotating heat pipe

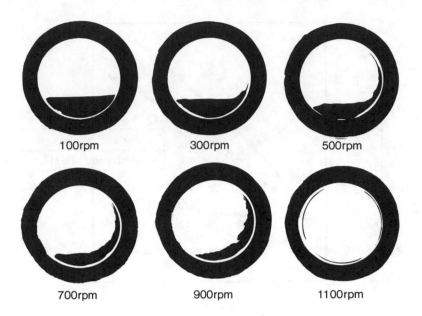

Fig. 2 Transition of two-phase fluid structure in a horizontal
             rotating pipe

Fig. 2 shows the results of visualization experiments conducted by the authors.
The liquid is colored R-113, and the view is from the end of a transparent
(acrylic) pipe (28mm dia. × 430mm long) rotating about its horizontal axis.
The liquid occupies 20 per cent of the inner space of the pipe. In this example,
the liquid comes to follow the solid body rotation at the rotational speed of
1100 rpm, which corresponds to a Froude number of 18. The transition Froude
number was found insensitive to whether the fluid was R-113 or water, material
of the pipe surface, or even the surface roughness. It is, however, dependent
on the quantity of liquid contained in the pipe. Moreover, there is a hysteresis
in transition in increasing and decreasing the rotational speed. Another complication was found; that is the non-uniformity of transition along the length
of the pipe during increasing the rotational speed. The transition commences
near one end of the pipe and propagates toward another end. Fig. 3 shows the
transition characteristics. With a given liquid content (expressed in terms of
the percentage of the internal volume),the transition proceeds in a following
manner. During increasing the rotational speed, at a Froude number marked by
an open circle, a   substantial volume of liquid near one end of the pipe begins
to make a full turn with the pipe's wall. At Froude numbers in the range between
a solid circle and an open circle, the region of annular structure extends toward
another end of the pipe as the rotational speed is increased. Cellular patterns
like those reported in [3] were observed, besides, there were numerous small-
scale irregular waves. At Froude numbers higher than the one marked by a solid
circle, the liquid undergoes a solid body rotation, and irregular waves virtually
disappears. Decreasing the rotational speed, one observes the maintenance of
the solid body state until the Froude number reaches the level marked by a
square symbol. Decrease of the rotational speed below this level prompts return
to the stratified structure over the entire length of the pipe. The liquid
content has large influence on the transition characteristics when increasing

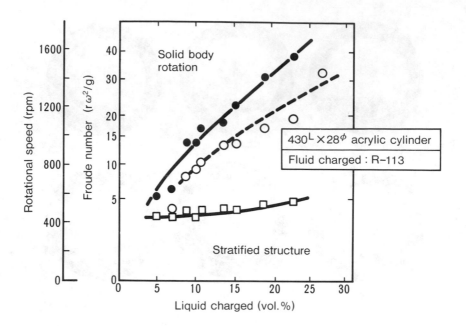

Fig. 3 Regime map of two-phase fluid structure in a horizontal
rotating pipe

● : Marking the completion of solid body rotation while increasing
the rotation rate, ○ : Marking the onset of solid body rotation on a
part of the cylinder while increasing the rotation rate, □ : Marking
the appearance of stratified structure while decreasing the rotation rate

the rotational speed. With larger liquid contents, higher Froude numbers are
required to achieve solid body rotations. Froude numbers marking the return to
the stratified structure are insensitive to the volumetric liquid content.

Primary modes of heat transfer in the different flow regimes are listed in
Table 1. The following are some supplementary notes.

In a heat pipe having a long evaporator section (the item (2) in INTRO-
DUCTION). dry-out of the evaporator wall may result when the rate of axial flow
of liquid is reduced seriously. It is very likely that the mobility of liquid
is reduced with the flow transition from the S-regime to the T- and A- regimes.
The depth of liquid in the direction of a primary force is smaller in the
annular structure than in the stratified structure. Hence, although the centrifugal force increases, the head available to spread the liquid film axially
decreases as the Froude number exceeds the transition Froude number. Moreover,
in the T-regime, the presence of secondary flows in the form of cellular patterns
and waves may pose a great resistance to the axial flow. In previous works, the
ratio of the evaporator length to the diameter is close to unity. Moreover, the
diameter of the evaporator section is made larger than that of the condenser
section, eliminating the possibility of liquid depletion in the evaporator.

The above description of flow and heat transfer points to the fact that the
quantity of working fluid charged in the heat pipe is an important factor for

Table 1 Primary modes of heat transfer in
different flow regimes

Designation	Two-phase flow structure	Primary mode of heat transfer	
		Evaporator *	Condenser
S	Stratified **	(1) Evaporation of a thin liquid film on a part of the wall outside the liquid reservoir	(1) Condensation on a part of the wall outside the liquid reservoir
		(2) Evaporation from the superheated liquid reservoir	(2) Condensation to the subcooled liquid reservoir
T	Partly annular or annular over the entire length, with cellular waves ***	Evaporation from the vapor-liquid interface of an adhered liquid film	Condensation on the vapor-liquid interface of an adhered liquid film
A	Annular without cellular waves (perfect solid body rotation)		

*   Nucleate boiling heat transfer becomes a primary mode where the wall
superheat is sufficiently high and the liquid layer has a thickness
greater than the diameter of bubbles.

**   The liquid reservoir occupies a lower part of the pipe's cross section,
with a recirculation under the effects of the gravity and the shear force
exerted by the moving wall. The pipe's wall repeatedly dips and leaves
the liquid reservoir, accompanying a thin liquid film when leaving the
reservoir. Due to recirculating motion in the liquid reservoir, the
vapor-liquid interface of the reservoir is likely to be superheated in the
evaporator section, and subcooled in the condenser section. Heat transfer
on the wall out of the reservoir is considered more effective than that
through convective motions in the reservoir, primarily because large
temperature differences are available there.

*** The description of heat transfer modes applies to the area covered by an
annular structure.

the heat transfer performance.

3.   TEST HEAT PIPE AND EXPERIMENTAL APPARATUS

     Several copper pipes of different dimensions were tested. Since two pipes
among them were tested in a systematic manner, the following reporting will
focus on their data. Only difference in the geometry of these two pipes is the
inside diameter, the one has 37mm, another 28mm. The pipe's total length is 480
mm, the evaporator section and the condenser section are equal in length, each
spans 170mm. An adiabatic section of 60mm long is present between the evaporator
and condenser sections. There are margins of 25mm and 55mm on the far ends of
the evaporator and condenser sections, respectively. There, the pipe wall is
heated or cooled only by axial thermal conduction through the wall.

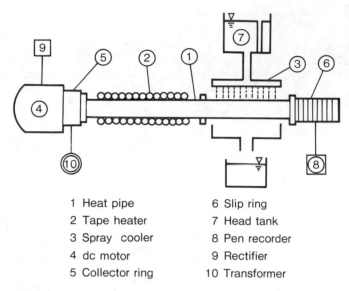

1 Heat pipe	6 Slip ring
2 Tape heater	7 Head tank
3 Spray cooler	8 Pen recorder
4 dc motor	9 Rectifier
5 Collector ring	10 Transformer

Fig. 4 Schematic diagram of experimental apparatus

Fig. 4 shows a schematic diagram of the experimental apparatus. The heat
pipe (1) is supported on both ends by the ball bearings, and rotated by the dc
motor (4) whose speed can be varied from 50 to 2000 rpm. Heat was supplied by
the electrical resistance heater (2) made of 0.2mm dia. constantan wires wound
around the evaporator section. All over the evaporator and adiabatic sections,
a cylindrical glass-wool cover was mounted to provide thermal insulation.
Electric current was supplied through the collector ring (5) , and the voltage
was controlled by the variable transformer (10) . Cooling of the condenser
section was provided by the pipe (3) having many holes through which water
from the tank (7) was sprayed onto the pipe.

The wall temperature was measured by using the pipe material (copper) and
constantan wires as thermocouples. The constantan wire has the diameter of 0.2mm,
its one end was imbedded in the pipe wall by the depth of 0.6mm and firmly held
by adhesives. Its another end was connected to the slip-ring (6) . Wall temperatures were measured at five locations along the length; three locations in
the evaporator section at distances from the end of the pipe 55, 100, 145mm;
two locations in the condenser section at distances from the end of the pipe
110, 160mm. Measurements of the temperature of the working fluid on the axis
of rotation were also attempted. Copper-constantan thermocouples of 0.2mm dia.
were inserted through the holes drilled on the pipe wall. After insertion, the
holes were sealed by adhesive bond. The locations of measurement will be clear
from Fig. 5.

The heat pipe has disc flanges on its ends to facilitate connection to
the motor and the slip-ring. At the center of one of the flanges, a check valve
coupling was mounted to allow degassing the interior and charging of the working
fluid. The interior surface was rinsed by the working fluid before charging
operation. Charging was performed first by degassing the interior by a vacuum
pump until the internal pressure reached 0.13Pa. The working fluid was
then let in by a predetermined volume. In another procedure, the working fluid
was charged first, then, the interior was degassed. This procedure was equally

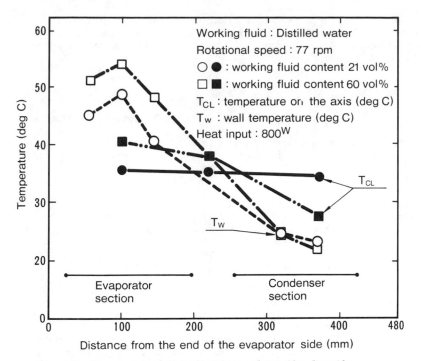

Fig. 5 Temperature distributions along the length
of a heat pipe

satisfactory in purging the air, and less time-consuming. This was adopted in
later operations. Finally, the check valve coupling was decoupled and its end
was sealed by adhesive bond. The valve was used for convenience of charging and
recharging the working fluid. Its reliability for keeping the internal vacuum
during a period of test run was confirmed by comparing the performance of a
heat pipe so prepared and the one prepared by using a capillary tube and crimp-
ing it with end sealing. After charging, the heat pipe was subjected to a static
heat transfer test, where the pipe was set vertical with the condenser section
above the evaporator section. From wall temperature distributions of the condenser section and the analysis of the thermosiphon performance, the presence of
non-condensable gas was checked. The static test was also conducted after the
rotation run. After the test run, the check valve was opened, and the quantity
of the working fluid was again measured. In reporting the data, the quantity
of the working fluid will be given in terms of the percentage of the volume
occupied by the liquid in the total internal volume of the heat pipe. This will
be called "volumetric charge" or simply "charge" hereafter.

The working fluid was distilled water or R-113 as already mentioned. The
effect of the volumetric charge on the heat transfer performance was found to be
similar for both fluids, hence, the data of water will be reported.

The accuracy of temperature measurements is within 0.5 deg C. A great
uncertainty was involved in the measurement of heat transfer rate. From
measurements of the flow rate of the cooling water and the temperature of water
at the inlet and outlet of the cooling chamber, the heat carried away by the
cooling water was estimated. It was found to be 60 - 80 per cent of the
electrical power input. In view of large uncertainty in measurements of heat

flow rate from the cooling water and estimation of heat loss, the electrical
power input will be referred in reporting the data. Heat transfer by axial
thermal conduction through the pipe wall was estimated to be at most 1 per cent
of the power input to the evaporator section.

4. RESULTS AND DISCUSSION

Figs. 5 - 7 show the data obtained with the pipe having 37mm ID.

Fig. 5 shows the temperature distributions of the pipe wall and the working fluid on the axis of rotation. The rotational speed was 77 rpm. (the
Froude number 0.123) and the flow was obviously in the S-regime. The
electrical power input was 800W.

The data for a 21 per cent volumetric charge of the working fluid show that
the vapor is at a uniform temperature of 35 deg C. The maximum wall temperature in the evaporator section is higher than the vapor temperature by 12 deg
C, and the minimum wall temperature in the condenser section is lower by an
almost equal degree. The wall temperature near the end of the evaporator side
of the pipe indicates that there is a heat loss to the flange and beyond.

In the case of a 60 per cent volumetric charge, the thermocouples on the
axis of rotation were apparently dipped in the liquid reservoir. The measured
temperatures indicate that the liquid was superheated in the evaporator section
and subcooled in the condenser section. Free convective circulation driven
by this temperature differential may contribute to heat transfer, however, the
decrease of areas for thin film evaporation and condensation (Table 1, the
regime S) causes a degradation of heat transfer performance. The wall temperature difference between the evaporator and condenser sections becomes larger
than the one for the 21 per cent volumetric charge. In the present experiment,
the heat transfer resistance on the spray-cooled exterior surface of the condenser section is negligibly small. The degradation of heat transfer performance was reflected in the rise of the evaporator temperature. In the following
report of the data, the wall temperature difference is defined as the difference
between the averaged temperature of the evaporator wall and that of the condenser wall.

Fig. 6 shows the wall temperature difference ΔT plotted against the rotational speed with the volumetric charge as a parameter. At low rotational
speeds, ΔT increases as the volumetric charge increases above 12 per cent. The
probable cause is the decrease of effective area for thin film evaporation and
condensation. A too lean charge of 7 per cent causes an increase of ΔT, probably due to dry area formation in the evaporator section. Above 700 rpm, the
wall temperature of the evaporator shoots up, indicating that a major part of the
evaporator wall is dried up. A steep rise of ΔT was also observed for the case
of 12 per cent charge at above 900 rpm. In the range of rotational speeds below
this critical value, the 12 per cent charge yields a minimum ΔT. For larger
volumetric charges of 40 and 61 per cents, ΔT shifts to higher levels at 1200
and 1400 rpm, respectively. The corresponding Froude numbers indicate the
transition of flow from the S-regime to the T-regime. Establishment of the
annular structure is particularly detrimental to heat transfer in the condensor
section, since heat has to be transferred through a thick liquid layer. It was
found that ΔT closely followed the increase of temperature difference between
the vapor and the condenser wall. Further increase of the rotational speed
up to 1950 rpm in the case of 40 per cent charge causes a steep rise of ΔT,
reflecting the rise of the wall temperature in the evaporator section. The

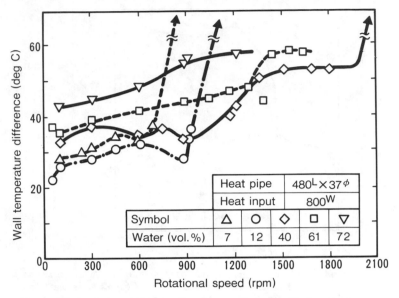

Fig. 6 Heat transfer performance of a smooth heat pipe
(distilled water)

corresponding Froude number seems to indicate the establishment of the A-regime. It may well be probable that, in the range of rotational speeds between 1200 and 1950 rpm, a part of the pipe near the evaporator end is occupied by the stratified structure and the rest by the annular, with the latter expanding as the speed is increased. Above 1950 rpm, the annular structure occupies a whole region, inviting dry-out of the evaporator wall. This is, however, a pure conjecture for lack of visualization information at high speeds.

The Δ T for an excessively high charge of 72 per cent rises gradually as the rotational speed is increased, reflecting the decrease of area for thin film evaporation and condensation as more liquid tends to be moved up by the rotating wall.

The data shown in Fig. 6 were obtained by increasing the rotational speed. The hysteresis of transition of flow regimes described in Section 2 was also observed in heat transfer experiments. Fig. 7 shows the data for the case of 12 per cent charge. The rotational speed was increased to 1350 rpm, where Δ T reached 65 deg C, then reduced. With the decrease of rotational speed, Δ T remained high until the speed reached about 400 rpm. This hysteresis phenomenon should be cautioned when applying the present scheme to cooling of a variable speed machine.

From data plots like that shown in Fig. 6 one can determine the optimum volumetric charge of the working fluid, which yields a minimum ΔT at a given heat load and a given rotational speed (a Froude number). Fig. 6 exhibits that the dry-out of the evaporator wall makes the 12 per cent charge no longer optimum at rotational speeds higher than 800 rpm. In this manner, the optimum charge increases as the Froude number increases beyond certain critical values. Fig. 8 shows the optimum volumetric charge determined from the heat transfer data of the two pipes, the one has 28 mm ID and another 37 mm ID. The data for the 28 mm ID. pipe are shown against the nominal heat flux.

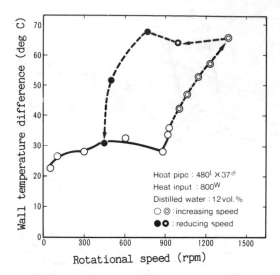

Fig. 7 Hysteresis of temperature **difference** in increasing and reducing the rotational speed (◎ ● Measured in short periods 4 ~ 5 min, the others in sufficiently long periods)

The nominal heat flux is defined as the electrical power input divided by the nominal evaporator area which is $\pi \times 0.028 \times 0.17 = 0.015m^2$ for the 28mm ID. pipe and $0.0198m^2$ for the 37mm ID pipe. For the Froude numbers of less than 12.7 and in the range of heat fluxes less than $3.3W/cm^2$, the optimum volumetric charge is insensitive to the change of heat flux. At higher heat fluxes, the optimum charge for the case of Fr < 5.6 increases with increasing heat flux. This can be explained such that more liquid is required to prevent the evaporator wall from drying while the wall is outside the liquid reservoir during rotation.

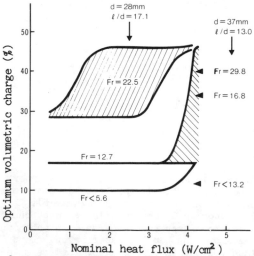

Fig. 8 Optimum volumetric charge of the working fluid as a function of nominal heat flux and Froude number

In the case of Fr = 12.7, ΔT was found to become less sensitive to the volumetric charge, therefore, the optimum charge is indicated by a shaded area. Intensification of nucleate boiling in the liquid reservoir is considered responsible for this behavoir.

At a relatively high Froude number of 22.5, the optimum charge was found to scatter in the shaded area. The variation of ΔT was found to be within 2 deg C when the volumetric charge was in the indicated range. It is conceivable that nucleate boiling is responsible for the spread of optimum charge. The wall superheat in those cases of high volumetric charge was found to be above 6 - 10 deg C, so that the onset of nucleate boiling was most likely. For higher heat fluxes, the lower boundary of the shaded area increases. This may be attributed to the nucleation of bubbles whose effect in obstructing the axial liquid flow becomes more pronounced for smaller volumetric charges and higher heat fluxes.

For three cases of different Froude numbers in Fig. 8, the approximate values of ΔT are recorded as follows; $\Delta T = 8 - 25$ deg C for Fr<5.6, $\Delta T = 17 - 23$ deg C for Fr = 12.7, $\Delta T = 14 - 26$ deg C for Fr = 22.5; the lower the heat flux, the smaller the ΔT.

The data for the 37mm ID pipe are shown only by triangular marks to save the diagram from becoming complicated. The triangular marks are indicating the level of optimum charge which is relatively insensitive to the change of heat flux. The dependence of optimum charge on the Froude number is somewhat different from that of the 28mm ID pipe. This may be due to the difference in the length - to - diameter ratio, ℓ/d, however, more data are needed to establish a correlation which incorporates the effect of ℓ/d.

Although more works are needed, a useful conclusion is reached by the present work such that the volumetric charge of 10 - 14 per cent is optimum for Froude numbers less than 13, which include many cases of heat pipe applications to conventional rotating machines.

5. MISCELLANEOUS NOTES

The use of R-113 in place of distilled water lessens the possibility of air leakage into the heat pipe because of high saturation pressures at temperatures of interest. However, the heat flow rate reduces by 50 - 70 per cent from that of a water heat pipe at equal ΔT.

Several enhancement measures were applied to the surfaces of evaporator and condenser. The porous surface structure described in [5] was manufactured on a copper strip, then the strip was rounded up and seam-welded to make a pipe having the internal porous surface. The heat transfer coefficient on the evaporator surface was found to double at the wall superheat of 5 deg C and rotational speeds of 150 - 900 rpm. An attempt was also made to improve heat transfer in the condenser section by means of spiral fins. The dimensions of fins were; height 0.8mm, pitch of spiral 0.7mm. Although the improvement was observed in a static test, little improvement was found in a rotating test. Works on other enhancement measures such as longitudinal grooves provided over the entire length of the pipe are currently in progress. The effect of non-condensable gases has not yet been investigated in a systematic manner. From a few data which was apparently affected by the presence of non-condensable gas, one finds that it accumulates near the end of the condenser section. This can be explained by the rotating fluid mechanics which assumes the predominance of Coriolis force in the gaseous phase.

6. CONCLUDING REMARKS

From the investigation into the performance of horizontal rotating heat pipes, the following conclusions are obtained.

(1) The two-phase flow in a horizontal rotating pipe makes transitions from the stratified structure to the annular structure accompanied by cellular patterns and irregular waves, and further to the solid body rotation without perturbations, as the rotational speed (the Froude number) is increased. The transition Froude numbers depend on the content of liquid phase in the pipe. Hysteresis of transition was observed when reducing the rotational speed.

(2) Systematic measurements were performed of the temperature difference between the evaporator wall and the condenser wall. It was found to reflect the transition of flow regimes as the rotational speed was varied. Predominant heat transfer mechanisms working in the heat pipe are discussed. Thin film evaporation and condensation on the rotating wall are important in the stratified regime.

(3) There exists an optimum volumetric charge of the working fluid which minimizes the wall temperature difference for a given rotational speed and a given heat load. For Froude numbers less than 13 which include many cases of heat pipe applications to conventional rotating machines, the optimum volumetric charge is 10 - 14 per cent of the total internal volume of the heat pipe.

ACKNOWLEDGEMENT

The authors wish to thank Mr. O. Sugimoto and Mr. S. Ishizaka, The Kansai Electric Co.. for their giving us the motivation to conduct the present fundamental work. Thanks are also due to Mr. T. Egashira of Hitachi Cable Industrial Products Ltd. for his manufacturing the test heat pipes.

NOMENCLATURE

d   inner diameter of the heat pipe, mm
F   rotational Froude number,
g   gravitational acceleration, 9.807 m/s^2
ℓ length of the heat pipe, mm
r   inner radius of the heat pipe, m
ΔT temperature difference between the evaporator wall and the condenser wall, deg
ω angular velocity, rad/sec

REFERENCES

1. Marto, P.J. and Wagenseil, L.L. 1979. Augmenting the Condenser Heat-Transfer Performance of Rotating Heat Pipes. AIAA Journal, Vol. 17, No.6, PP.647-652
2. Daniels, T.C. and Al-Baharnah, N.S. 1980. Temperature and Heat Load Distribution in Rotating Heat Pipes. AIAA Journal, Vol.18, No.2,pp 202-207
3. Karweit, M.J. and Corrsin, S. 1975. Observation of cellular patterns in a partly filled, horizontal, rotating cylinder. Phys. Fluids, Vol.18,No.1, pp. 111 - 112.
4. Gans, R. 1977. On steady Flow in a Partially Filled Rotating Cylinder. J. Fluid Mech., Vol. 82, pt. 3, pp. 415-427
5. Nakayama, W., Daikoku, T., Kuwahara, H., and Nakajima, T. 1980. Dynamic Model of Enhanced Boiling Heat Transfer on Porous Surfaces, Part 1 : Experimental Investigation. ASME Journal of Heat Transfer, Vol.102,No.3,pp.445-450

A Proposed Regenerative Thermosyphon Blade Cooling System for High Efficiency Gas Turbines

MAHER A. EL-MASRI
Department of Mechanical Engineering
Massachusetts Institute of Technology
Cambridge, Mass. 02139, USA

ABSTRACT

The increasing need for more efficient gas turbine engines has resulted in raising turbine inlet temperatures. This necessitates cooling the turbine hot components. Since the cooling system imposes thermodynamic losses on the cycle, it becomes imperative to devise schemes which adequately cool the blades while incurring a minimal thermodynamic penalty.

In this paper a novel cooling scheme is proposed, briefly analyzed and evaluated. A sealed fluid loop connects cooling channels in the turbine blades to the last row(s) of compressor blades. Due to the larger radius of the turbine it is shown that such a loop can be self-circulating. By using a coolant of suitable properties, such as a liquid metal, the circulation rates can be so large that the loop acts as a heat pipe of "infinite" conductance. General dimensionless relations are derived for the circulation rate and coolant temperature changes around the loop in terms of geometry and operating conditions. The turbine nozzles and compressor stators are to be connected by a forced-convection loop. Since this should not pose exceptional problems it is not discussed in detail.

The flow stability and distribution in the multiple parallel rotating loops is discussed and stable configurations identified.

A simple thermodynamic model for cycle cooling losses is developed and used to evaluate the potential efficiency advantage of using this scheme over the current technology of air-cooling. A gain of five to eight percentage points is shown to be possible. This may well justify the extensive research and development efforts required for practical implementation of the proposed scheme.

NOMENCLATURE

Latin

A	area
C	minor loss coefficient for loop referred to turbine channel diameter
C_f	friction factor
C_p	specific heat at constant pressure

Latin

d	diameter of turbine blade cooling channel
erf	denotes the error function
f	denotes functional dependence
G	mass flux
Gr	Grashoff Number

h	heat transfer coefficient, enthalpy	R	radius from axis of rotation
K	hydraulic resistance coefficient for loop	$\tilde{R}$	turbine radius ratio = R_1/R_2
k	thermal conductivity	$\tilde{R}_c$	compressor radius ratio = R_4/R_5
L_e	equivalent length of loop referred to turbine cooling channel diameter	$\tilde{R}_t$	tip radius ratio = R_5/R_2
		Re	Reynolds Number
ℓ	turbine blade perimeter	r	isentropic temperature ratio for gas turbine cycle = T_{g2}/T_{g1} = T_{g3}/T_{g4}
M	average Mach number of gas flow relative to blades		
		s	blade spacing
$\dot{m}$	mass flow rate of coolant per blade channel	St	Stanton Number
Nu	Nusselt Number	T	absolute temperature
Pr	Prandtl Number	V	coolant mean velocity
p	pressure		

Greek   Greek

α	adiabatic compressibility of coolant, defined in eq.2	θ	ratio of peak cycle temperature to ambient = T_{g3}/T_{g1}
β	expansion coefficient of coolant based on enthalpy	ϕ	heat flux
γ	specific heat ratio of gas	ξ	total wetted perimeter of cooling channels per blade
μ	viscosity	ω	radian rotational speed
ρ	density		

Subscripts   Subscripts

g	gas	w	coolant channel wall
m	mean coolant properties	unsubscripted properties refer to coolant	

Superscripts

$\sim$	dimensionless

1. INTRODUCTION

It is well known that the efficiency of the gas-turbine cycle increases at higher firing temperatures. Since the strength and longevity of hot-section components decrease at elevated temperature, cooling becomes necessary. This is especially critical for the highly-stressed rotating blades. Unfortunately, the cooling process, normally accomplished by air bleed from the compressor, imposes thermodynamic penalties upon the cycle tending to off-set the benefits of raising the combustion temperature. A point is eventually reached such that further increasing the cycle peak temperature becomes counter-productive. This temperature has been steadily rising over the last thirty years due to progress on two fronts: (i) metallurgical improvements such as new alloys, coatings, and manufacturing techniques, which allow an increase in the metal working temperature; and (ii) development of better cooling systems with more effective coolant utilization, thus minimizing compressor air-bleed and the associated thermodynamic penalty.

Armstrong [1], gave a concise review of those developments showing the historic trend of turbine inlet temperatures rising at a pace of about 20 K per year to the current level of 1550K - 1600K in high-performance aircraft engines.

It has long been realized that liquid-cooling systems, properly designed, can cope with the large thermal loads while imposing lower thermodynamic penalties than air-cooling. The first water-cooled turbine, intended for aircraft application, was designed and successfully tested by Schmidt in 1942/43 [2]. Since then, a large variety of liquid-cooling schemes have been proposed, some have been researched in laboratory experiments, and a few have actually been tested in experimental gas-turbines. Reviews of much of this work have been provided by Bayley and Martin [3], and by Van Fossen and Stepka [4]. Of particular interest to readers of this paper is the work reported by Cohen & Bayley [5], and by Le Grives and Genot [6], both proposing the use of closed thermosyphon inserts, full of liquid metal or a two-phase medium, to convey the blade heat load to an air-cooled cavity in the turbine disc. This cavity may be cooled by airflow through the engine due to motion of the aircraft. In both of those works, rotating closed thermosyphons were tested and found to have large thermal conductance. Tests on a small thermosyphon-cooled gas turbine for vehicular use were reported by Gabel [7]. In this engine a closed, steam-filled, thermosyphon loop conveyed the blade heat to a fuel-cooled heat exchanger, thus pre-heating the fuel before combustion.

2. PROPOSED COOLING SCHEME

The cooling system proposed in this paper is illustrated in figure 1. A sealed fluid-filled loop runs through the turbine blades, the rotating shaft or drum, and the last row(s) of compressor blades. Natural circulation in this loop is motivated by the large centrifugal acceleration. The fluid thus cools the turbine blades and conveys the heat to the compressor delivery air, pre-heating it before the combustor. The advantages of this system are:
(1) Due to the large coolant circulation rate and heat transfer coefficient, extremely high firing temperatures and pressure ratios may be realized.
(2) No loss of power due to bleeding of compressor air.
(3) Avoiding the aerodynamic losses associated with boundary layer separation due to coolant injection in film-cooled blades.
(4) Using the heat absorbed from the turbine blades to pre-heat the combustion air, thus further decreasing the cooling losses.
On the other hand, the chief drawbacks of the proposed system are:
(1) The complexity of manufacturing the leak-proof circulation loop required in this scheme. A large number of long axial holes are required through the shaft or drum. Also the connections through the turbine blade roots, which traditionally have some freedom of movement, can be quite difficult.
(2) Complicated design of the blades themselves, which would require complex channels lined with a durable material to be embedded in a high conductivity matrix covered by an outer blade skin. Blades of such construction are being developed for water-cooled industrial turbines [8], however in the latter application the channel configuration proposed is less complex and the blade size larger than for aircraft applications.
(3) Doubts about the structural integrity of the system due to the high coolant pressures developed and the consequences of a loss-of-coolant accident.

The above, as well as other questions, have to be resolved if this scheme is to be applied. The purpose of this paper is not to show the feasibility of every aspect of the proposed scheme, but rather to demonstrate its workability insofar as the fluid flow and heat transfer processes are concerned. This, together with the potential benefits, may encourage further study of this system. It is the author's view that the proposed arrangement may find long-term application in

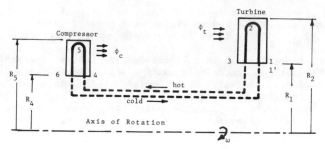

Figure 1. Schematic of Proposed System.

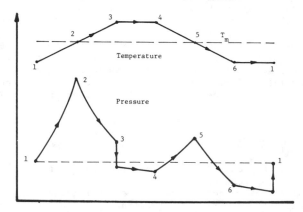

Figure 2. Temperature and Pressure Variations .

large, long-range, transport aircraft where efficiency is of particular importance.

It should be added that in what follows, a similar forced convection closed loop system is assumed to apply to turbine and compressor stators. This is not thought to pose difficult problems and is not discussed in detail.

3. SYSTEM MODEL

3.1. Description and Assumptions

Consider the schematic of the proposed loop shown in figure 1. If the fluid expands when heated, the turbine section 1-2-3, taken alone, is clearly unstable. A slight perturbation in either direction, taken to be 1-2-3, would generate circulation in that direction. This is because of the greater centrifugal force on the cooler, denser column 1-2 than on 2-3. For a given circulation rate, lower than that at which wall friction would counterbalance the driving effect mentioned above, the section 1-2-3 would act as a pump causing p_3 to be larger that p_1. Applying a similar argument to the cooled (compressor) section, 4-5-6, it is seen to be stable and will resist circulation, thus for a given flow rate, it requires $p_4 > p_6$ to drive the flow. Fortunately, due to the higher temperature, speed of sound, and lower density of the turbine gas relative to the compressor air, the turbine blades are usually at a larger radius than those of the compressor in most gas-turbine engines. Thus the positive pressure developed by the turbine section of the loop is greater than the resistance of the compressor section. The net difference between them overcomes friction and hydraulic resistance in the loop, and this balance determines the circulation rate. Figure 2 is a sketch of the temperature and pressure distributions around the loop.

For the purpose of analysis, the following assumptions are made, the non-obvious to be subsequently justified,

(1) Changes in fluid temperature as it circulates around the loop are much less than the difference between mean fluid and external gas temperatures.

(2) The internal conductance between the fluid and channel walls is much larger than that from the gas to the external blade surface.

(3) The heated and cooled sections of the loop may be treated as having uniform wall heat fluxes as a consequence of assumptions 1 and 2.

(4) The enthalpy change of the loop fluid due to heat flux is much larger than that due to rotation, i.e.,

$$\phi\pi d(R_2 - R_1) \gg \dot{m} \cdot \frac{1}{2} \omega^2 (R_2^2 - R_1^2)$$

(5) Changes in fluid density are much smaller than the mean density so that the functional dependance $\rho(p, h)$ may be linearized,

$$\rho - \rho_m = \alpha(p - p_m) - \beta(h - h_m) \tag{1}$$

where

$$\alpha \equiv \partial\rho/\partial p\big|_h , \qquad\qquad \beta \equiv -\partial\rho/\partial h\big|_p \tag{2}$$

are treated as positive constant coefficients.

(6) Changes in momentum flux, being of order $V^2\Delta\rho$, are negligible compared to the "centrifugal buoyancy" driving force, of order $\omega^2 R^2 \Delta\rho$.

(7) The mutual influence between centrifugal pressure distribution upon frictional pressure drop and vice-versa may be neglected. Thus we may imagine the loop to be frictionless with all the hydraulic resistance lumped at some point, calculate the pressure difference due to rotation at that point and equate that to the pressure drop due to the loop's hydraulic resistance.

(8) Heat transfer and wall friction factors in the loop may be calculated using forced-convection relations for a stationary channel. This neglects two effects. The first is the distortion of the velocity and the temperature profiles due to density variations over the channel cross section. The second is the influence of secondary flows. The effect of those assumptions will be discussed later.

3.2. Determination of Circulation Rate

In accordance with assumption 7, the circulation rate will be determined by assuming all hydraulic resistance of the loop to be concentrated at point 1. Following the loop in the direction of circulation 1-2-3-4-5-6-1', the centrifugal components of pressure difference can be summed up to yield a "driving pressure" $(p_1' - p_1) = \Delta pd$; equating this to the hydraulic resistance enables the circulation rate to be found. It may be added that this approach is more economical with algebra than a rigorous statement of pressure gradient integrals around the loop, and that both yield the same final result.

Starting at 1, we seek $(p_3 - p_1)$ due to rotation only, to this end, and in view of the above assumptions, the energy and momentum equations are

$$\frac{dh}{dR} = \pm \frac{\phi\pi d}{\dot{m}} \tag{3}$$

$$\frac{dp}{dR} = \pm \rho \omega^2 R \tag{4}$$

with the (+) applying for section 1-2 and (-) for section 2-3. Integrating equation 3 prescribes a linear enthalpy variation with distance, which when used in conjunction with eq. 1, allows the density in eq. 4 to be expressed as a function of pressure and radial location. Integrating eq. 4 around 1-2-3 then gives

$$P_3 - P_1 = \frac{\beta \phi \pi d \omega^2 R_2^3}{\dot{m}} f_1 (\tilde{R}, \tilde{\omega}) \tag{5}$$

where

$$f_1 (\tilde{R}, \tilde{\omega}) \equiv \frac{1 - \tilde{R}}{\tilde{\omega}^2} - \frac{\sqrt{\pi}}{2} \frac{e^{\tilde{\omega}^2 \tilde{R}^2}}{\tilde{\omega}^3} [\text{erf } \tilde{\omega} - \text{erf } \tilde{\omega} \tilde{R}] \tag{6}$$

$$\tilde{\omega} \equiv \sqrt{\frac{\alpha}{2}} \omega R_2 \tag{7}$$

the centrifugal contribution between 3 and 4 is

$$P_4 - P_3 = \rho_3 \frac{\omega^2}{2} (R_4^2 - R_3^2) \tag{8}$$

and for the compressor loop 4-5-6, as for the turbine, we find

$$P_6 - P_4 = - \frac{\beta \phi_c \pi d_c \omega^2 R_5^3}{\dot{m}} f_1 (\tilde{R}_c, \tilde{\omega} \tilde{R}_t) \tag{9}$$

we complete the loop by noting

$$P_1' - P_6 = \rho_1 \frac{\omega^2}{2} (R_1 - R_6) \tag{10}$$

adding equations 5, 8, 9, and 10, and noting the $R_4 = R_6$, and that eq. 1 and energy balances imply

$$\rho_1 - \rho_3 = \frac{\beta \phi \pi d \, 2(R_2 - R_1)}{\dot{m}} \quad \text{and} \quad \phi \, d \, (R_2 - R_1) = \phi_c \, d_c \, (R_5 - R_4),$$

gives after some manipulation

$$\Delta P_d = P_1' - P_1 = \frac{\beta \phi \pi d \omega^2 R_2^3}{\dot{m}} f_2 (\tilde{R}, \tilde{R}_c, \tilde{R}_t, \tilde{\omega}) \tag{11}$$

where

$$f_2 (\tilde{R}, \tilde{R}_c, \tilde{R}_t, \tilde{\omega}) \equiv (1 - \tilde{R})(\tilde{R}_c^2 - \tilde{R}_c^2 \tilde{R}_t^2) + f_1(\tilde{R},\tilde{\omega}) - \frac{1 - \tilde{R}}{1 - \tilde{R}_c} f_1(\tilde{R}_c, \tilde{\omega} \tilde{R}_t) \tag{12}$$

The function f_2 is shown for typical values of its arguments in figure 3. In the limit when the engine tip speed is very much less than the adiabatic speed of sound in the coolant, $\tilde{\omega} \to 0$, f_2 depends on geometry only,

$$f_2 \to (1 - \tilde{R}) \left\{ (\tilde{R}^2 - \tilde{R}_c^2 \tilde{R}_t^2) + \frac{(\tilde{R} - \tilde{R}_c)}{3} [1 - 2(\tilde{R} + \tilde{R}_c)] \right\} \quad (\tilde{\omega} \to 0) \tag{13}$$

We can express the hydraulic resistance of the loop as $K \cdot 1/2 \, \rho_m v^2$, thus

$$\Delta P_d = K \cdot 1/2 \, \rho_m v^2 = (C + 4C_f \frac{Le}{d}) \cdot 1/2 \, \rho_m v^2 \tag{14}$$

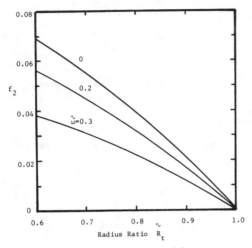

Figure 3. The Function f_2 for $\tilde{R}=\tilde{R}_c=0.85$

C being the sum of minor loss coefficients, and L_e and equivalent length for the loop, both being referred to the turbine channel diameter d and velocity V. Substituting for Δp_d from eq. 11 and re-arranging we may relate the flow rate expressed as the Reynolds Number to the heat flux and operating conditions,

$$Re = \left[\frac{8}{K} \left(\frac{d}{R_2}\right)^2 \tilde{\phi} f_2 \right]^{1/3} \tag{15}$$

where

$$\tilde{\phi} \equiv \frac{\beta \phi (\omega^2 R_2) \rho_m R_2^4}{\mu^3} \tag{16}$$

is a dimensionless turbine channel heat flux. It should be noted that the dependance of the factor K on Re through the friction factor C_f of eq. 14 may be neglected, since at the expected Reynolds Numbers, a value of C_f = 0.005 can be used with little error. The order of magnitudes likely for L_e/d and C are 400 and 8 respectively, the latter being equivalent to 8 right angle bends at turbine channel diameter. Thus the order of magnitude for the factor K is expected to be around 16. This can only be determined precisely for a detailed loop design.

Equation 15 gives the circulation rate in terms of the heat flux . The latter may be related to the gas-to-blade temperature difference by the heat balance

$$\phi \xi = \phi_g \ell = St_g G_g Cp_g (T_g - T_b) \ell \tag{17}$$

combining together with definition (16) gives

$$\tilde{\phi} = \left[\frac{\beta Cp(\omega^2 R) \rho_m R_2^3 (T_g - T_b)}{\mu^2} \right] St_g \left(\frac{\mu_g}{\mu}\right) \left(\frac{Cp_g}{Cp}\right) \left(\frac{R_2}{\xi}\right) Re_{g\ell} \tag{18}$$

the term in brackets is a Grashoff Number based on coolant properties, tip radius, tip acceleration, and gas-to-blade temperature difference.

3.3. Coolant Temperature Rise

The temperature rise for the coolant in the loop can now be found from

$$(T_3 - T_1) = (T_4 - T_6) = \frac{2 \phi \pi d (R_2 - R_1)}{\dot{m} Cp} \tag{19}$$

this can be compared to the gas-to-blade difference to yield

$$\frac{T_3 - T_1}{T_g - T_b} = 8(1 - \tilde{R}) \frac{Re_{g\ell}}{R_e} \frac{R_2}{\xi} \frac{Cp_g}{Cp} \frac{\mu_g}{\mu} St_g \tag{20}$$

for typical conditions, such as those cited in Appendix I, this is a small number. Assumption (1) is valid for such cases.

3.4. Internal Heat Transfer

For turbulent flow of liquid metals, the Nusselt number may be given by the correlation [9],

$$Nu = 7 + 0.025 \ (RePr)^{0.8} \tag{21}$$

the ratio of internal (channel to fluid) to external (gas to blade) thermal conductance is

$$\frac{hA}{h_g A_b} = \frac{T_g - T_b}{T_w - T} = \frac{Nu}{Nu_{g\ell}} \ \frac{k}{k_g} \ \frac{\xi}{d} \tag{22}$$

which is typically a large number, as shown in Appendix I, thus justifying assumption 2.

To determine the conditions of validity of assumption 7, we form the ratio Gr_w/R_e^2 where Gr_w is based upon the wall-to-bulk fluid temperature difference

$$(T_w - T) = \frac{\phi}{Nu} \ \frac{d}{k} \tag{23}$$

and the channel diameter, d,

$$Gr_w = \frac{(\omega^2 R) \ \beta \ Cp \ \rho_m \ d^3 (T_w - T)}{\mu^2} \tag{24}$$

dividing by Re^2 and using eq. 15 gives

$$\frac{Gr_w}{R_e^2} = \frac{\phi^{1/3} \ Pr \ (d/R_2)^{8/3}}{Nu \ (8 \ f_2/k)^{2/3}} \tag{25}$$

Studies of turbulent mixed-convection are reported in references 10 - 13. Bara [14] summarized their results and concluded that the influence of buoyancy was small (less than 15% effect on wall shear and heat transfer) if $Gr_w/Re^2 \lesssim 0.05$. The example in Appendix I approaches this constraint. In any event, the influence of secondary flow, which is neglected, can be quite considerable in mixing the fluid over the cross section and minimizing the buoyancy effect.

Thus we may conclude that the thermosyphon loop will act as a highly effective heat pipe between the turbine and compressor sections. The principal thermal resistances are the external heat transfer and blade conduction. Indeed it is the latter that has to be carefully designed to maintain the required blade surface temperature for a given gas temperature. It may be added that the loop mean temperature will be approximately the average between turbine gas and compressor air temperatures. This will be the case if the Stanton Numbers, the ratio of gas flow cross-sectional area to blade surfaces area, and the blade metal conductance were the same at both sides.

3.5. Channel Configurations and Stability

The schematic of figure 1 shows a single closed loop. If a practical system is designed, it will necessarily include a number of parallel channels, both within each blade as well as between the rows of blades. Construction is

simplified by having common
plenums or headers feeding such
channels, the former being connected by the least possible
number of conduits.

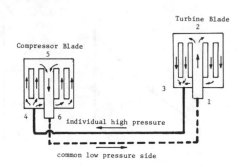

Figure 4. Suggested Channel Configuration for Stability

For systems involving heated
channels in parallel between common plenums, problems of flow ex-
cursion, maldistribution and oscillation may occur. In single-
phase systems, the principal
source of such phenomena is a neg-
ative-sloping pressure-drop vs.
flow rate characteristic. Thus
the system will be stable if
$\partial \Delta p / \partial \dot{m} > 0$ for any channel of a group between common phenums.

Following the arguments presented in references 15 and 16, we examine section 1-2-3. If common connections between a number of parallel channels were
made at 1 and 3, the bank of channels would be stable, since the buoyancy term
alone, represented by eq. 5, gives $\partial(p_1 - p_3)/\partial \dot{m} > 0$, adding the distributed
frictional term only contributes to stability. By a similar argument, common
connections across a bank of compressor loops, at points 4 and 6, will cause
unstable behaviour since by equation 9 $\partial(p_4 - p_6)/\partial \dot{m} < 0$, and the wall friction
in this section of loop may be shown to be inadequate to stabilize it at the
steady-state operating point. If the common connections were at 1 and 6 the
bank of loops is seen to be stable. Thus it is permissible to have a common
return conduit, or group of conduits, connecting headers at 1 and 6. The connection 3 - 4, however, will have to be made individually for each blade.

Turning attention to the channels within each blade it can be concluded
that the arrangement depicted in figure 4 possesses considerable merit. It
allows stable flow within each blade, allows the channels to be distributed around
the blade periphery, and, in addition, makes for an even greater circulation rate
in the loop than the previous calculations show. This is because the central
blade hole, radially outflowing in the turbine and inflowing in the compressor,
is subject to a minimal heat flux. Thus the function f_2 for this configuration
will be greater than its value as calculated with uniform heat flux throughout
each loop.

ESTIMATE OF CYCLE EFFICIENCY WITH PROPOSED SCHEME

The efficiency of a heat engine may be defined as

$$\eta = \frac{W}{Q} \tag{26}$$

For the ideal Brayton Cycle 1-2-3-4 of figure 5 the efficiency is

$$\eta_{ideal} = 1 - \frac{1}{r} \tag{27}$$

If compressor and turbine isentropic efficiencies were η_c and η_T respectively,
this may be shown to be

$$\eta = \frac{\frac{1}{\eta_c}(1 - r) + \theta \eta_T \left(\frac{r - 1}{r}\right)}{\theta - \frac{1}{\eta_c}(r - 1) - 1} \tag{28}$$

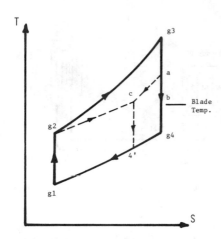

Figure 5. Idealized Cycle for Cooling Loss Model

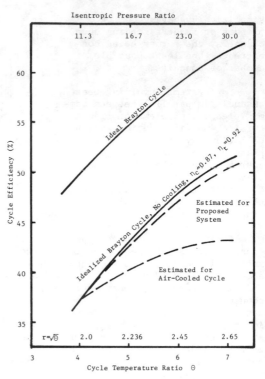

Figure 6. Estimated Efficiency Differential for the Case $r=\sqrt{\theta}$, $T_b=4T_{g1}$.

both equation 27 and 28 assume an ideal gas working fluid. They are plotted in figure 6 for the special case $\eta_c = 0.87$, $\eta_T = 0.92$ and $r = \sqrt{\theta}$, which corresponds to the maximum specific work in an ideal cycle. Advanced engines operate close to these conditions.

For a cycle employing the cooling scheme proposed, the efficiency would be affected due to changes in both work output and heat addition. Taking the logarithmic derivative of 26 we get

$$\frac{\delta\eta}{\eta} = \frac{\delta W}{W} - \frac{\delta Q}{Q} \tag{29}$$

this may be used to evaluate the relative change in efficiency due to cooling. This relative change, calculated for an ideal cycle, is still a very good representation for a real cycle, since the impact of departure from ideality upon $\delta\eta/\eta$ is of second order in smallness. All we have to do is use η for the real cycle to calculate the real $\delta\eta$.

A heat balance applied to N rows of blades shows that the drop in stagnation temperature of the gas due to heat transfer through the blade surfaces is

$$\Delta T_g = N \, St_g \, (T_g - T_b) \, (\tfrac{\ell}{s}) \tag{30}$$

Referring to figure 5, we note that the mean value of gas temperature over the cooled section of the expansion path 3-b is Ta. Defining a fraction x as

$$x = \frac{T_{g3} - T_a}{T_{g3} - T_{g4}} = \frac{1}{2} \frac{T_{g3} - T_b}{T_{g3} - T_{g4}} \tag{31}$$

enables the relative loss of net work, reduction of heat added, and of cycle efficiency to be expressed as

$$\frac{\delta W}{W} = - N St_g \left(\frac{\ell}{s}\right) \frac{x \theta}{\theta - r} \tag{32}$$

$$\frac{\delta Q}{Q} = - N St_g \left(\frac{\ell}{s}\right) \frac{x \theta (r - 1)}{r(\theta - r)}, \tag{33}$$

and

$$\frac{\delta \eta}{\eta} = - N St_g \left(\frac{\ell}{s}\right) \frac{x \theta}{r(\theta - r)} \tag{34}$$

to estimate this quantity, we assume that $T_b = 4T_{g1}$, corresponding to 1200K blade surface if ambient is 300K, which gives, in view of 31,

$$x = \frac{r(\theta - 4)}{2\theta (r - 1)} \tag{35}$$

using $r = \sqrt{\theta}$, $N = 4$ cooled blade rows, $\ell/s = 3$, and $St_g = 0.005$ [17], provides the estimate

$$\frac{\delta \eta}{\eta} \simeq - 0.06 \frac{r^2 - 4}{2r (r - 1)^2} \tag{36}$$

which is used with the result of eq. 28 to produce the curve on figure 6.

Turning attention to an air-cooled cycle, the losses are modelled as due to mixing a fraction m of the flow at state 2 into the expansion path at point a, producing a stagnation pressure drop Δp to accelerate the coolant, and cooling the gas from a to c. The work losses due to the mass deficit m for the section 3-a, the stagnation temperature reduction a - c, and the stagnation pressure drop a-c, may be shown to be respectively

$$\frac{\delta W_1}{W} = - m x \frac{\theta}{\theta - r} \tag{37}$$

$$\frac{\delta W_2}{W} = - m \frac{[r(\theta-r) - x \theta(r-1)](1-x)}{(\theta-r) [r - x (r-1)]}, \tag{38}$$

$$\frac{\delta W_3}{W} = - m (\gamma-1) M^2 \frac{\theta [r- x(r-1)]}{(r-1) (\theta-r)} \tag{39}$$

whereas the reduction in heat addition due to the mass deficit m in the combustor is

$$\frac{\delta Q}{Q} = - m \tag{40}$$

assuming $(\gamma-1) M^2 \stackrel{\simeq}{=} 0.2$, $r = \sqrt{\theta}$, and calculating x from equation 35 as above, enables the relative efficiency penalty $\frac{\delta \eta}{m\eta}$

to be determined as a sole function of r. This, together with a breakdown of its various components, is shown in figure 7.

In order to calculate the impact on cycle efficiency, the fraction m must be determined. Since the art of film-cooling is

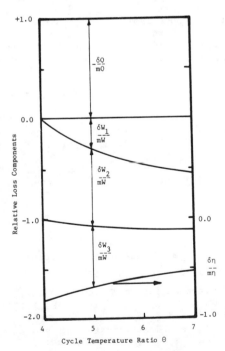

Figure 7. Breakdown of Cooling Losses for Air-Cooled Cycle Model.

highly empirical, as well as proprietary, it is quite hard to find a closed-form, simple expression. Based upon the author's judgement, and for the purpose of estimates, values of m of 0, 0.10, 0.20, and 0.30 were used for θ = 4, 5, 6 and 7 respectively to generate the curve on figure 6.* The peak of this curve is close to state-of-the-art operating conditions for jet engines.

Figure 6 shows that an efficiency improvement of about 5-8 percentage points may result from adopting the proposed cooling scheme. Even though the figures for cycle efficiency are approximate, it is thought that the results insofar as the differences in efficiency are quite representative.

5. CONCLUSIONS

1. A cooling scheme consisting of a self-pumping thermosyphon loop for the rotors, and a closed forced-convection loop for stators is proposed.
2. With a suitable coolant, such as Lithium, the rotating thermosyphon loop is self-pumping. The circulation rates can be sufficiently large that the loop acts as a heat pipe of large conductance between turbine and compressor.
3. Stable configurations are believed possible for the loop channels.
4. An efficiency improvement of the order of five to eight percentage points is possible with the proposed scheme.

REFERENCES

1. Armstrong, F. W., "Gas Turbine Evolution", Inst. Mech. Engrs. Symposium on "Gas Turbines - Status and Progress", paper C1/76, 1976.

2. Schmidt, E., "Heat Transmission by Natural Convection at High Centrifugal Acceleration in Water-Cooled Gas Turbine Blades", Proc. Gen. Disc. on Heat Transfer, IV, 361-363, Instn. Mech. Engrs., London 1951.

3. Bayley, F. J. and B. W. Martin, "A Review of Liquid Cooling of High Temperature Gas-Turbine Rotor Blades", Proc. Instn. Mech. Engrs., 185, 18/71, 219-227 (1970/71).

4. Van Fossen, G. J. and F. S. Stepka, "Liquid Cooling Technology for Gas Turbines, Review and Status", NASA TM-78906, 1978.

5. Cohen, H. and F. J. Bayley, "Heat Transfer Problems of Liquid-Cooled Gas Turbine Blades", Proc. Instn. Mech. Engrs. 169, 53, p. 1063, 1955.

6. LeGrives, E. and J. Genot, "Refroidissement des Aubes de Turbines par les Metaux Liquides", O.N.E.R.A., T. P. 872, 1970, also AGARD Conference on High Temperature Turbines, Florence, Italy, 1970.

7. Gabel, R. M., "Feasibility Demonstration of a Small Fluid-Cooled Turbine at 2300°F", SAE Paper No. 690034, 1969.

8. Caruvana, A., et al., "Design and Test of a 73-MW Water-Cooled Gas Turbine", ASME 80-GT-112.

9. Kays, W. M. and H. C. Perkins, "Forced Convection Internal Flow in Ducts," Section 7 of Handbook of Heat Transfer, W. M. Rohsenow Ed., McGraw-Hill 1973.

10. Easby, J. P., "The Effect of Buoyancy on Flow and Heat Transfer for a Gas Passing Down a Vertical Pipe at Low Turbulent Reynolds Numbers", Int. J. of Heat and Mass Transfer 21, 791-801, 1978.

*Those values are on the basis that θ defines the stagnation temperature prior to the first stage.

11. Axcell, B. P. and W. B. Hall, "Mixed Convection to Air in a Vertical Pipe", Int. Heat Transfer Conference, Toronto 1978, MC-7.

12. Carr, A. D., M. A. Connor and H. O. Buhr, "Velocity, Temperature and Turbulence Measurements in Air for Pipe Flow with Combined Free and Forced Convection", J. of Heat Transfer, Nov. 1973, pp. 152-158.

13. Buhr, H. O., E. A. Horsten and A. D. Carr, "The Distortion of Turbulent Velocity and Temperature Profiles on Heating Mercury in a Vertical Pipe", J. of Heat Transfer, May 1973, pp. 152-158.

14. Bara, R. J., "Mixed Convection in a Vertical Thermosyphon Loop", S. M. Thesis, Dept. of Mech. Eng., M.I.T., September 1981.

15. El-Masri, M. A., "Fluid Mechanics and Heat Transfer in the Blade Channels of a Water-Cooled Gas Turbine", Ph.D. Thesis, Dept. of Aero. and Astro., M.I.T., September 1978.

16. El-Masri, M. A. and J. F. Louis, "Design Considerations for the Closed-Loop Water Cooled Turbine", ASME 79-GT-71.

17. Demuren, H. O., "Aerodynamic Performance and Heat Transfer Characteristics of High Pressure Ratio Transonic Turbines", Ph.D. Thesis, Dept. of Aero. and Astro., M.I.T., Feb. 1976.

18. Gierszewski, P., B. Mikic and N. Todreas, "Property Correlations for Lithium, Sodium, Helium, Flibe and Water in Fusion Reactor Applications", Plasma Fusion Center Report PFC-RR-80-12, M.I.T., 1980.

APPENDIX I

NUMERICAL EXAMPLE

Engine Dimensions:

R_1 = 0.47 m R_5 = 0.4 m

R_2 = 0.55 m $\tilde{R}$ = $\tilde{R}_c$ = 0.85

R_4 = 0.34 m $\tilde{R}_t$ = 0.73

Turbine Blade Chord = 0.04 m

Turbine Blade Perimeter = 0.10 m

Speed = 8,000 RPM

Turbine Tip Speed = 458 m/s

Turbine Tip Acceleration = 3.82×10^5 m/s^2

Average gas speed relative to blade = 600 m/s

Turbine blade surface temperature = 1200K

Gas Properties and Conditions:

T_{gas} = 1550 K (adiabatic wall temperature)

T_{air} = 700 K (compressor discharge)

P_{gas} = 20 bar

ρ_g = 4.36 kg/m^3

μ_g = 5.61 x 10^{-5} N-s/m^2

Cp_g = 1.18 kJ/kg

$Re_{g\ell}$ = 4.65 x 10^6

Coolant Channel Assumptions

d = 0.003 m R_2/d = 183.3

coolant Lithium, mean temperature is average between turbine gas and compressor air, T_m = 1125K. Properties from reference 18 are:

ρ = 429 kg/m^3

μ = 2 x 10^{-4} N-s/m^2

Cp = 4140 J/kg K

k = 55 W/mK

β = 2.44 x 10^{-5} kg-s^2/m^5

Operating Conditions

Assuming $\tilde{\omega} \simeq 0$ f_2 = 0.0506

from eq. 16, $\tilde{\phi}$ = 1.5 x 10^{20}

assuming K = 16, eq. 15 gives Re = 48,150

eq. 21 gives Nu = 11.82

from eq. 25 Gr/Re^2 = 0.068 which indicates some buoyancy effects, but in view of secondary flow mixing those are not expected to influence the assumptions considerably.

from eq. 20 the coolant maximum-to-minimum temperature difference is 52K, maximum-to-mean temperature difference is 26K.

Application of Disk-Shaped Rotating Heat Pipe to Brake Cooling

S. MAEZAWA, M. TAKUMA, and A. TSUCHIDA
Department of Mechanical Engineering
Seikei University
Musashino, Tokyo, Japan

ABSTRACT

The rise in temperature resulted from friction heat is a cause in deterioration of the performance, and relate to the thermal failure which is caused of destruction of brake comporrents.

In this work, the rotating heat pipe is applied to heat dissipation from caliper disk brake.

As the results of experimental investigation tests, the temperature at the source of heat generation decreased, and the uniform temperature distributions of the wall of the disk were obtained.

1. INTRODUCTION

The present brake, particularly those on larger cars, are operating at or near maximum capacity, and that improvement in braking capacity is an early requirement. Especially, this need is caused to factors such as higher speeds resulting from increased engine horse powers and better roads.

Until recently, brake size was influenced mainly by its holding ability or torque capacity and not so much by a requirement for heat dissipation capacity. Fortunately, many are working on this problem, as evidenced by the innovation of highly efficient disk brakes, impeller hubcap air cooling etc.

The disk brakes have excellent characteristic performance on safety, a sence of stability and cooling. The ability of the brake to dissipate heat is improved by increasing the disk surface area in contact with the surrounding air, and by causing more air to circulate over the surfaces. Specifically the disk brake has an advantage that the heat dissipation capacity can be increased by exposing the larger braking surface to air flow. But as a result of the larger exposed braking surface, the area of pad must be subject to restriction, and therefor the pressure per unit area and the source of heat generation are locally concentrated.

In this work, the rotating heat pipe developed from earlier work/I/is applied to heat dissipation from caliper disk brake. In essence, this disk is a sealed hollow, and containing a fixed amount of working fluid. Heat is transferred from friction heat by which means braking through the wall of evaporator to the working fluid causing evaporation. The vapor flows radially toward the condenser section where it condenses and ultimately transfers its latent heat through the condenser layer and wall to an external air flow. Rotation about the center axis of the disk generetes a centrifugal force field. This force pump the condensate from the central region of a disk brake to the peripheral region to complete the cycle.

2. EXPERIMENTAL APPARATUS

A sketch of the general arrangement of the experimental apparatus is shown in Fig. 1 . In essence, it consisted of the caliper disk brake system, the slip ring assembly for the thermocouple pick up, the motor drive arrangement, the oil pump, the pressure gauge, safety vacuum valve, and the rotating speed and temperature measuring instruments.

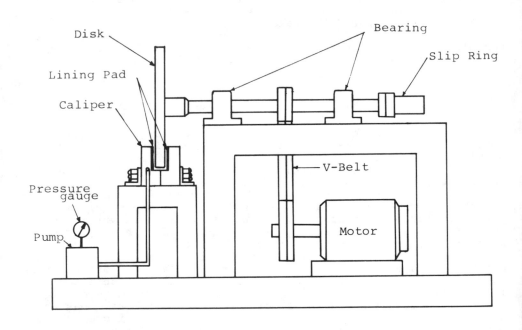

Fig. 1 Sketch of Experimental Apparatus

2.1 Caliper

A drawing of the caliper brake (Airflex 225DP100 ; Japan Fawick CO, LTD) is shown in fig. 2, and its specification is shown in Table 1 . This caliper disk brake is used for industrial in high inertia stopping and tensioning application.

<u>Table 1</u>

Items	225DP (2 Piston)
Caliper Weight	7.5 Kg
Pad Area(Friction Area)	40.2 cm^2 x 2
Lining thickness	16.0 mm
Disk Diameter	380 mm
Disk thickness	25.4 mm
Disk Material	S45C or FCD-45
Max. Safe Temperature	200 °C

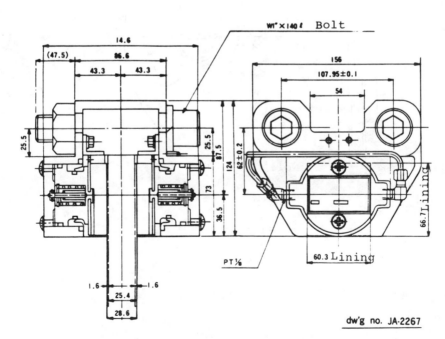

dw'g no. JA-2267

Fig. 2 Drawing of Caliper

2.2 Disk

A schematic view of the rotating heat pipe disk is shown in
Fig. 3 . The disk is of fixed geometry machined out and welded of
S45C metal. The overall diameter was 380mm and the width was 25.4
mm. On the inside sealed hollow of the disk,the diameter of the
evaporator end was 360mm , whilst the diameter of condenser end
was 110mm and wall thickness was 5.0mm. A number of 18 fins were
set up at the wall surface of the central cooling region. The
dimension of the fin was 15mm height x 4mm thickness. Copper-Constantan thermocuples located at appropriate points in the wall placed
at positions No.1 - No.9 were set up along radius.

No.1, No.2,···No.9 ; thermocouples

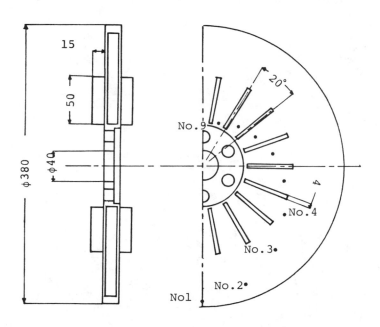

Fig. 3   Heat Pipe Disk

3. EXPERIMENTAL PROCEDURES

3.1 Filling procedure

The filling and deaerating procedure for the heat pipe was
as follows. The vacuum pressure gauge and safety valve were opened
and the heat pipe disk was evacuated. A known quantity of working

fluid,about 350cc(plus a predetermined extra quantity to allow for deaeration) was drawn into the heat pipe disk. The exclusion of air from the heat pipe disk was then ensured by partial evaporation of the charge and bleeding off some of the gas before the heat pipe disk was sealed and tested for leaks. By considering the working fluid's compatibility with S45C metal so that corrosion and non-condesable gas generation does not occour, Methyl Alcohol was used for working fluid.

3.2 Running Procedure

Throughout this investigation, the following running pro-cedure was used. A series of wall temperature distributions along radius was measured at the constant pressure 2kg/cm^2 on the friction material and the constant rotational speed. The operating ro--tational speed was selected to set at each 200rpm , 300rpm and 400rpm.

3.3 Data Reduction

The outputs of all thermocouples passing along leads through the hollow centered shaft to a slip ring, and were recorded. During unsteady state operation, the temperatures were read off in °C directly from "thermodac" instrument which had a resolution of 0.1°C.

4. RESULTS AND DISCUSSION

A series of temperature meassurement experiments was begun with a study of the standard disk. The results obtained were found to be consistent with published data and thus provided a confident base from which to enter the heat pipe disk studies.

The distributions of the thermocouples was as follows ; No.1, No.2 , No.3 , No.4 , No.5 are at the source of heat generation (Friction surface), particularly No.3 is placed at the center of operating brake pressure. The temperature distributions along the cooling region was given by No.6 , No.7 , No.8 , No.9 thermo-couples.

The effect of the heat pipe disk is shown in plots of temper-ture versus the lapse of time for different constant rotational speeds 200, 300, 400 rev/min at constant brake pressure 2kg/cm^2 in Fig.5, from which it is apparent that the heat pipe disk increases the cooling effectiveness. In this figure , the temperature change of the standard disk is shown as solid lines, and the result for the heat pipe disk is shown as dotted lines. All of temperatures were measured by No.3 thermocouple. As expected, the rise in tempe-rature of the heat pipe disk decreases as compared with that of the standard disk. Especially , the rate of temperature decreasing is very high at much higher speed of rotation. An explanation for this is assumed to be that increasing rotational speed produces increases in the centrifugal flow rates resulting in larger heat transfer rate inside sealed hollow. And the effectiveness of the

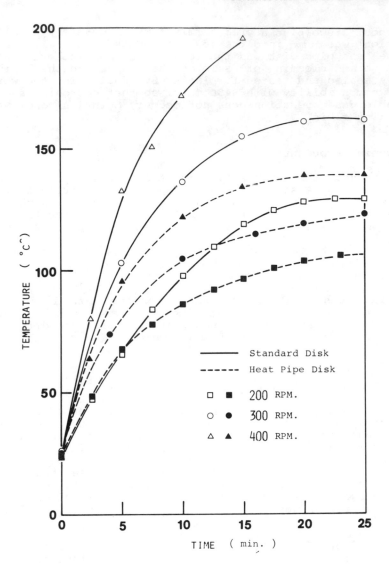

Fig. 4   Temperature rising at center of friction region

heat pipe is smaller at lower temperature owing to be that
vaporization of Methyl Alcohol causes more than 70°c.

Fig.5, Fig.6, Fig.7 support this contention by showing the
development of radial temperature distributions.   The temperature
distributions along radius at different typical time are shown for
each constant rotational speed. Temperatures increase with rota-

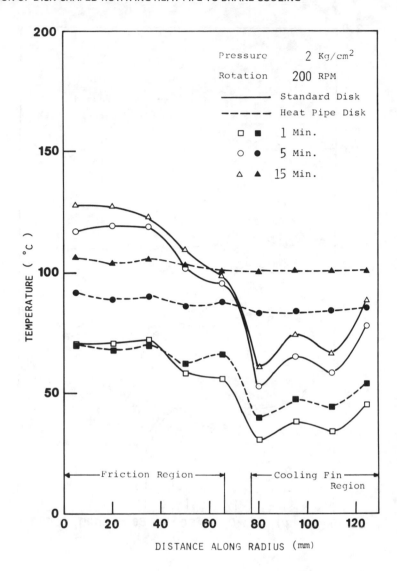

Fig. 5   Temperature Distribution along Radius

tional speed and time. Specifically, on the standard disk, temperature rises nearly at 200°c, and there is a substantial temperature
gradient in the radial direction between the friction region and
the cooling region. It is apparent that generated heat at the
friction region is transferred to the cooling region by conductive
heat transfer. As is well known, when their temperature exceeds a
certain amount, they tend to develop heat checks, causing in turn
brake shudder that can assume alarming proportions.   On the other
hand, the temperature distributions on the heat pipe disk are uni-

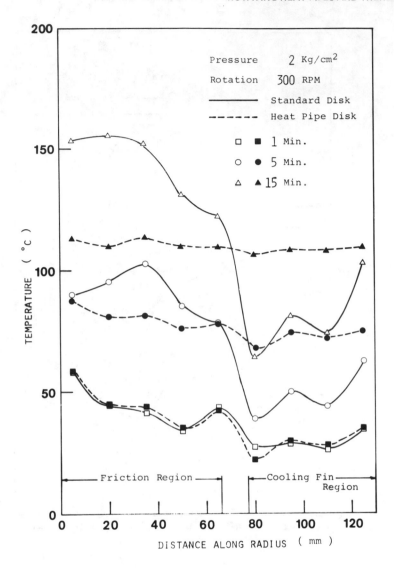

Fig. 6 Temperature Distribution along Radius

formly in the radial direction between evaporator(friction region)
and condenser(cooling region). At the friction region , temperature
is lower as compared with that of the standard disk,and also a
existence of maximum temperature is not recognized. Those uniform
temperature distributions show that this disk operates effectively
as the heat pipe. Namely, the disk is heated by generated friction
heat at the peripheral friction surface causing the internal working fluid to evaporate and the vapour condenses at the central
cooling region giving up its latent heat. In the braking the

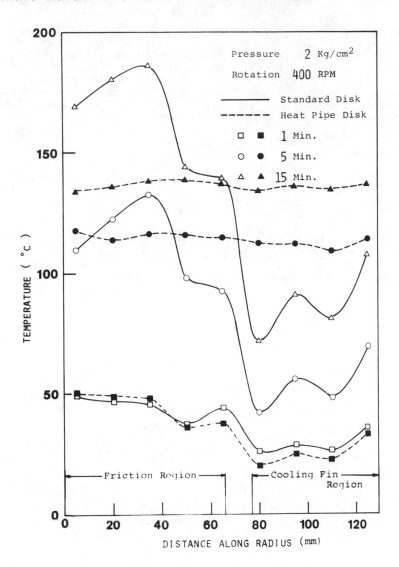

Fig. 7   Temperature Distribution along Radius

heat pipe disk there is continuous flow of vapour from the evaporator to the condenser and working fluid is continuously re-cycled
by centrifugal force in the opposite direction back to the evaporator. At the start of braking, temperatures of both the heat pipe
disk and the standard disk show the same distribution and existence
of only a small gradient.   This may well be due to lower temperature for heat pipe operation.

The purpose of this paper was to present an initial study of a heat pipe system for application to a disk brake in which thermal characteristics have been explored. The excellent cooling effectiveness recorded for the heat pipe disk offer great promise for the application cited earlier and may generate new applications. However, it is very apparent that a rotating heat pipe behaviour and intensity of a disk for both high speed of rotation and high braking pressure are not fully understood.

5.   REFERENCES

1. S. Maezawa, Y. Suzuki and A. Tsuchida (1981). Heat Transfer Characteristics of Disk-Shaped Rotating, Wickless Heat Pipe Proceeding 4th Int.Heat pipe Conf.,London , England, 725-733
2. George T. Ladd and Sidney B. Dew (1953) . Safe Brakes for Passenger Cars SAE Transactions Volume 61 623-639
3. G.A.G.Fazekas (1953) Temperature Gradients and Heat Stresses in Brake Drums SAE Transactions Volume 61 279-308
4. T. Irie (1965). Development and Problem of Disk-Brake Journal of the J.S.M.E. Volume 68, No. 555 32-39

Heat Transfer in Rotating Co-Axial and Parallel Heat Pipes and Their Application in Machinery

JIŘI SCHNELLER, BOHUMIL POKORNÝ,
and FRANTIŠEK POLÁŠEK
National Research Institute for Machine Design
Prague, Běchovice, Czechoslovakia

ABSTRACT

Results of theoretical and experimental research of heat transfer in co-axial and parallel rotating heat pipes with water filling in horizontal and vertical position of the pipe are presented.

Examples of cooling of electric motors with performance in the range from 4,5 to 80 kW are presented in the case of horizontal rotating heat pipes; in the case of vertical heat pipes unconventional solution of cooling of high temperature ventilator for lift-bell furnaces is described.

1. INTRODUCTION

Rotating heat pipes represent closed heat exchanging systems with two-phase circulation of heat transfer medium, in which return condensate flow from the condenser to the evaporator part of a heat pipe is ensured by centrifugal forces. According to the relative position of the heat pipe axis and axis of rotation rotating heat pipes can be devided into three basic groops (Fig. 1):

a) co-axial   - heat pipe axis and axis of rotation are identical (R 1-8, 10, 11, 12, 14),

b) parallel   - heat pipe axis and axis of rotation are parallel (R 6, 20),

c) non-parallel - the angle θ between heat pipe axis and axis of rotation is from 0 to 90 degrees.

As there is a substantial difference in rotating heat pipe performance under different inclination, mainly horizontal heat pipes and vertical heat pipes with condensate transport against gravity were studied. In industry rotating heat pipes have found useful application in intensification of rotating machinery cooling, e.g. of electric motors (R 1, 2, 3, 4,7,23-29) and rotating heat exchangers (20) etc. In National Research Institute for Machine Design in Prague - Běchovice theoretical and experimental research of chosen rotating heat pipe types is being carried out with respect to rotating machines cooling. The aim of this paper

Fig. 1   Schematic drawing of rotating heat pipes

a)   co-axial-axis of rotation and heat pipe axis are identical

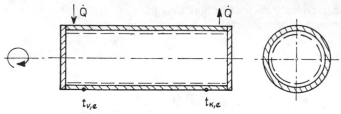

b)   parallel - axis of rotation and heat pipe axis are parallel

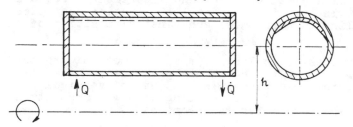

c)   non-parallel - heat pipe axis is tilted at an angle Θ against
     axis of rotation

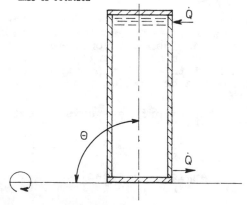

is to present some results of heat transfer in co-axial and parallel rotating heat pipes and their application in industry.

2. HORIZONTAL ROTATING HEAT PIPES

2.1 Horizontal Co-axial Rotating Heat Pipe

With regard to production simplicity the investigation was directed mainly towards cylindrical inner space heat pipes in which as a result of phase changes in the evaporator and condenser parts the diameter of condensing fluid layer changes along the tube.

Out of experimental works of National Research Institute for Machine Design, partial results of which have already been published (R 4, 6, 8, 18), following conclusions can be drawn:

a) owing to centrifugal forces liquid film forms at the inner wall of the tube (the assumption that centrifugal forces are much greater than gravity forces is in most technical applications fully justifiable). The film surface is smooth and evaporation in it is similar to that in the film flowing down the wall. Only critical heat fluxes reach higher values and are increasing with the fourth root of centrifugal acceleration.

To calculate heat transfer coefficients known equations for evaporation in thin film flowing down the vertical plane (R 8) and modified Nusselt's equation for vapour condensation in condenser part of the tube can be used (R 21, 22);

b) under the assumption of smooth liquid film surface and thickness of the liquid film at the end of the evaporator part of the tube (X = 0) equal to the so called critical value

$$\delta_L = \delta_{L,Kr} = \sqrt{\frac{\dot{v}^2}{4 \cdot \pi^2 \cdot r^3 \cdot w^2}} \tag{1}$$

the shape of the curve of free film surface can be expressed in therms of equation

$$\frac{d\delta_L}{dx} = \frac{\frac{1}{32} \cdot \xi \cdot \dot{v}^2}{w^2 \cdot r^3 \cdot \delta_L^3 \cdot \pi^2 - \dot{v}} \cdot \tag{2}$$

For laminar flow and technical applications of rotating heat pipes the following equation is valid

$$\dot{v} \ll w^2 \cdot r^3 \cdot \delta_L^3 \cdot \pi^2 \tag{3}$$

and equation (2) results in

$$\frac{d\delta_L}{dx} = \frac{3 \cdot \dot{v} \cdot \nu}{8 \cdot \pi \cdot w^2 \cdot r^2 \cdot \delta_L^3} \tag{4}$$

Out of the course of the surface curve calculated above not
only the optimal amount of working fluid for given geometrical and thermal parameters of the heat pipe but also the
thickness of the liquid film in the evaporator and condenser part can be determined. They are necessary for the determination of thermal resistences.

For the liquid phase temperature 100 $^{\circ}$C and 1800 and 3000
revolutions per minute dependences of the amount of working
fluid V_L on the heat performance $\dot{Q}$ are presented in
Fig. 2. It is obvious that the amount of working fluid depends of flow conditions in the layer, thermophysical properties of working fluid, working temperature, geometrical
parameters and heat performance. For this reason it is not
possible to produce universal co-axial heat pipes, but it
is necessary to design them for maximum heat and temperature load under which they will be working. Under different
working conditions the heat pipe will than be working less
efficiently.

The comparison of theoretical and experimental dependances of
heat performance on the outside surface temperature difference
between evaporator and condenser parts of a heat pipe is presented in Fig. 3 for n = 1800 rpm.

The analysis of thermal resistances of the co-axial rotating
heat pipe with cylindrical inner space shows that thermal resistance is higher in condenser that in evaporator part in most
technical cases (see example in Fig. 4).

The most simple way of heat transfer intensification is to
intensify return condensate flow, e.g. by cone condenser part
(R 14, 25, 29) or grooves (R 24, 29).

2.2 Horizontal Parallel Rotating Heat Pipe

Further works of National Research Institute for Machine Design were directed towards parallel heat pipes with axis of rotation and heat pipe axis parallel and located in the distance
of eccentricity h . Eccentricity changed from the value approximately equal to the liquid film thickness ($h \approx \delta_L < 1$ mm) to
the value several times higher than tube radius ($h > 10$ mm)
(Fig. 5). The case with small eccentricity may happen e.g. when
production is not precise enough, the case with great eccentricity e.g. when heat pipes are used for cooling of winding of an
electric motor rotor. For eccentricity increasing from zero value the ring of liquid which was at the beginning co-axial
starts to move eccentrically and acquires changing thickness
(Fig. 5a). With increasing eccentricity the layer thickness decreases in the place nearest to the axis of rotation up to the
point when dryout of the inner wall occurs. If the eccentricity
increases more only part of the tube inner surface is wetted
(Fig. 5b).

For thermal calculations similar assumptions can be used to
that for co-axial rotating heat pipe. The mean values of heat
transfer coefficient relative to the unit of inner surface were

Fig. 2 Optimal fluid charge versus heat performance for co-axial rotating heat pipe

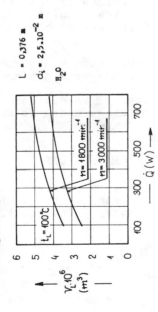

$L = 0,376$ m

$d_i = 2,5.10^{-2}$ m

H_2O

$t_L = 100°C$

$n = 1800$ min^{-1}

$n = 3000$ min^{-1}

$V_L.10^6$ (m^3)

$\dot{Q}$ (W)

Fig. 3 Comparison between theoretical and experimental values of heat performance

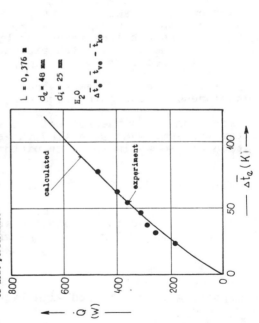

$L = 0,376$ m

$d_a = 48$ mm

$d_i = 25$ mm

H_2O

$\Delta \bar{t}_e = \bar{t}_{ve} - \bar{t}_{ke}$

calculated

experiment

$\dot{Q}$ (W)

$-\Delta \bar{t}_e$ (K)

Fig. 4 Partial temperature differences in co-axial rotating heat pipe

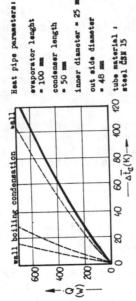

Heat pipe parameters:

evaporator lenght = 100 mm

condenser length = 50 mm

inner diameter = 25 mm

out side diameter = 48 mm

tube material : steel ČSN 15

$\Delta \bar{t}_e = \bar{t}_{ve} - \bar{t}_{ke}$

a) $N = 1800$ rpm

wall boiling condensation wall

$\dot{Q}$ (W)

$-\Delta \bar{t}_e$ (K)

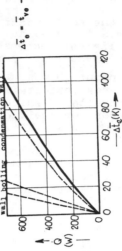

b) $N = 3000$ rpm

wall boiling condensation wall

$\dot{Q}$ (W)

$-\Delta \bar{t}_e$ (K)

determined by integration of local values of this coefficient.

Results of approximate calculation of local liquid film thermal resistances and corresponding heat transfer coefficients show that in the eccentricity range $0 < h < \delta_L$, (where δ_L is the layer thickness for co-axial rotation) the mean value of heat transfer coefficient increases considerably in the condenser as well as in the evaporator part.

For the point $h = \delta_L$, α reaches infinitely great value to which corresponds infinitely great heat performance. For the calculation results of which are presented in Fig. 6 local value of heat transfer coefficient was limited to the technically real value 15 000 $W.m^{-2}.K^{-1}$.

For $h > \delta_L$ rapid decrease in heat transfer coefficient value occurs in the evaporator part as a result of heat exchanging surface area decrease; on the other hand in condenser part thermal conditions are more advantageous because there is rapid return flow of condensate out of the condenser surface.

It can be seen from the results presented in Fig. 6 that a small eccentricity which may originate in production (several tenths of mm) won't have a negative effect on the magnitude of heat performance transfered by the heat pipe.

Fig. 7 shows heat pipe heat performance as a function of the temperature difference between the evaporator and condenser wall calculated by the above mentioned procedure for 550 rpm and values determined experimentally. It can be seen that there is a good agreement between theory and experiment up to the heat performance of approximately 220 W. For higher values of heat performance the assumptions taken for hydrodynamics, meniscus and transfer phenomena inside the pipe don't fully correspond with the thermal process.

Similar agreement was reached for 100 and 1060 rpm. Experimental results for both cases are shown in the same picture.

For comparison theoretical and experimental values determined for the same heat pipe (and the same filling) under co-axial rotation are shown in the same picture.

3. VERTICAL ROTATING HEAT PIPES WORKING AGAINST GRAVITY

The surface of liquid forms an envelope of a rotary paraboloid (Fig. 8) due to the simultaneous influence of gravitaional and centrifugal forces. Its height is given by the equation

$$H = \frac{w^2}{2g} \cdot r_i^2 \tag{5}$$

Corresponding volume filled with liquid is

$$V_L = \frac{\pi^3 \cdot w^2}{4g} \cdot r_i^4 \tag{6}$$

So the height of the paraboloid in a heat pipe depends mainly on

Fig. 5   Schematic illustration of parallel rotation

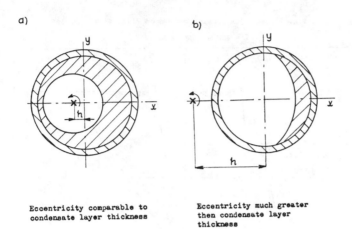

a)

b)

Eccentricity comparable to
condensate layer thickness

Eccentricity much greater
then condensate layer
thickness

Fig. 6   Mean value of heat transfer coefficient in the
evaporator part of heat pipe versus eccentricity

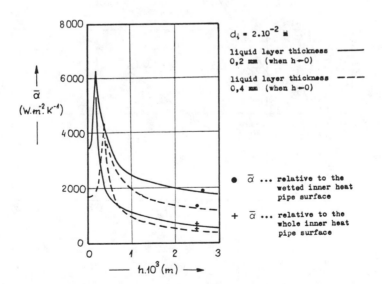

$d_i = 2.10^{-2}$ m

liquid layer thickness ——
0,2 mm (when h→0)

liquid layer thickness ----
0,4 mm (when h→0)

● $\overline{\alpha}$ ... relative to the
wetted inner heat
pipe surface

+ $\overline{\alpha}$ ... relative to the
whole inner heat
pipe surface

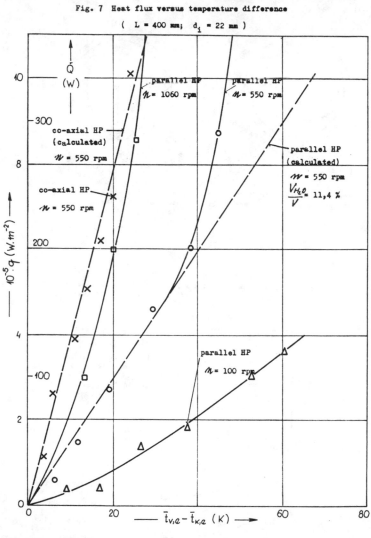

Fig. 7 Heat flux versus temperature difference

(L = 400 mm; d_i = 22 mm)

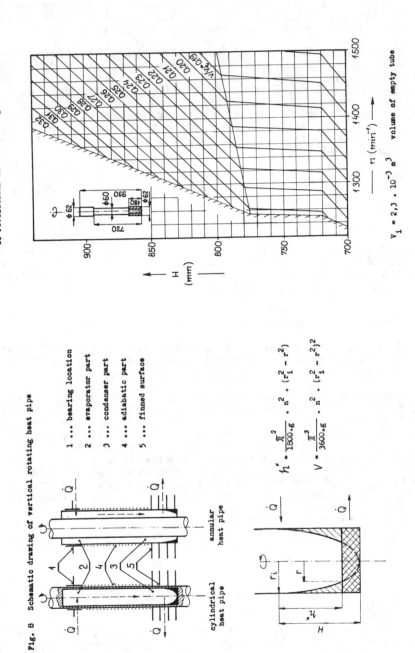

Fig. 9 Lifting height of rotating heat pipe versus number of revolutions and fluid charge

$V_1 = 2,3 \cdot 10^{-3} \, \text{m}^3$ volume of empty tube

Fig. 8 Schematic drawing of vertical rotating heat pipe

1 ... bearing location
2 ... evaporator part
3 ... condenser part
4 ... adiabatic part
5 ... finned surface

annular heat pipe

cylindrical heat pipe

$$h'' = \frac{\pi^2}{1800 \cdot g} \cdot n^2 \cdot (r_1^2 - r^2)$$

$$V = \frac{\pi^3}{3600 \cdot g} \cdot n^2 \cdot (r_1^2 - r^2)^2$$

the number of revolutions and the heat pipe diameter. The liquid
film thickenss in the evaporator and in the condenser part will
be considerably different. The lifting height of a rotating heat
pipe as a function of rpm's and fill charge is given in Fig. 9.

When common liquids with small thermal conductivity are used
liquid films represent the main thermal resistance first of all
in the condenser part. To decrease this resitance a conical insert with high thermal conductivity (Cu, Al) was pressed into
the condenser part. It intensifies heat transfer from the condenser surface inside the heat pipe to the cooling fins outside
(Fig. 10). To increase the circumferential evennes and eventually
wetting in the upper paraboloid part a screen capillary structure has been used (Fig. 10).
    In Fig. 11 outside surface temperatures of a co-axial rota-
ting heat pipe at constant heat performance 800 W and 1400 rpm
are shown.

4.   APPLICATION OF ROTATING HEAT PIPES IN INDUSTRY

    Horizontal co-axial and parallel rotating heat pipes were
used for cooling of electric motors of different types, vertical
co-axial rotating heat pipes were used for cooling of bearings
and seals of high temperature fans.

4.1 Cooling of electric motors by rotating heat pipes

    Rotating heat pipes were designed for cooling of three ty-
pes of asynchronous electric motors with electric performance of
10,4,5 and 80 kW (R1, 2, 3, 4, 5, 7 , 23-29).

    For illustration temperatures of the rotor and stator win-
ding as a function of performance of an electric motor with a
rotating heat pipe in the axis of 10 kW motor shaft are presented in Fig. 12 (schematic drawing is in Fig. 13). After rotating
heat pipes were built into electric motors smaller values of
stator winding temperature increase were reached in all cases so
that it was possible to increase performance of the electric motors by up to 25 per cent. Rotating heat pipes find special application in the electric motor with massive rotor which it is
not possible to run in the whole range of working parameters
when cooled by conventional air cooling (R 3). If the rotor is
produced as a rotating heat pipe (Fig. 14) favourable electric
parameters of the electric motor are achieved as a results of
lower rotor surface temperature.

4.2 Cooling of Bearings and Seals of High Temperature Fans
     (R 15, 16, 17, 30)

    Some types of industrial devices use draft or recirculating
fans, propellers of which work at high temperature, e.g. fans
ensuring circulation of inert atmoshpere in lift bell furnaces
work at temperature over 700 °C (Fig. 15). Operating reliability
of fans depends on propeller design as well as on cooling of
shaft bearings and seals. As the fan shaft operates at relative-

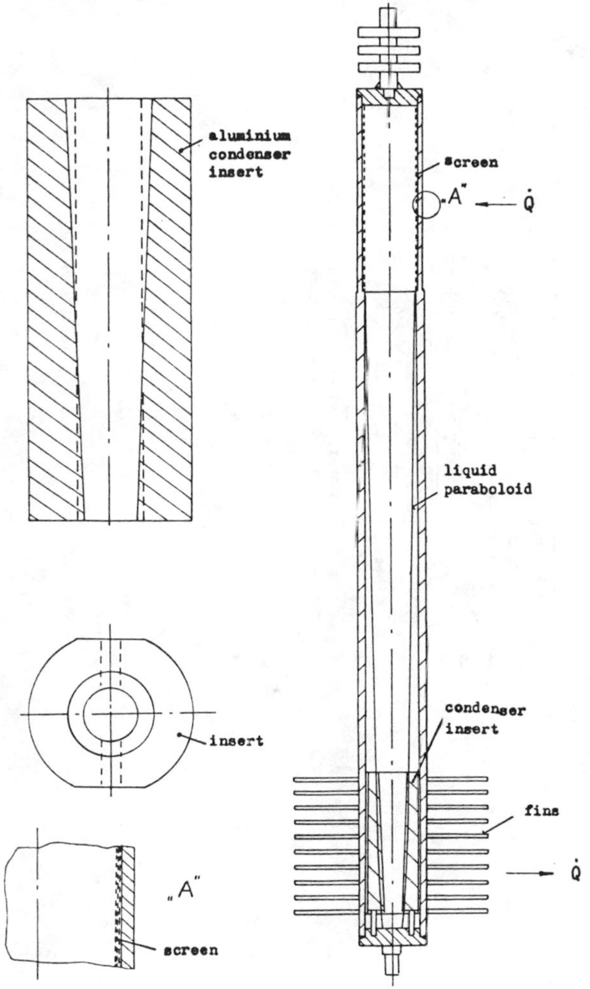

Fig. 11 Temperature distribution along vertical rotating heat pipe

Fig. 10 Schematic drawing of co-axial vertical rotating heat pipe

679

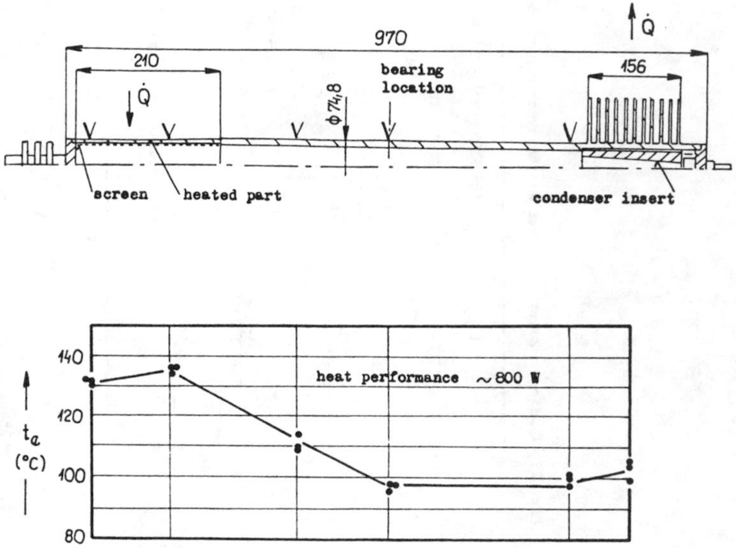

Fig. 12 Winding temperature versus electric motor performance

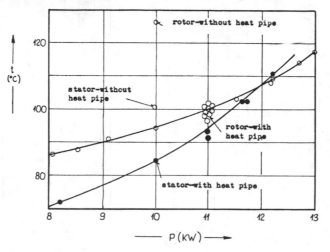

Fig. 13 Schematic drawing of the 10 kW electric motor with a
rotating heat pipe

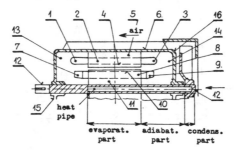

Stator		Rotor	
1, 3	face of winding	8	bar
2	channels of winding	10	tooth
4	tooth	11	yoke
5	yoke	12	shaft
6	carcass	13, 14	inner air
7, 9	ring	15	bearing
		16	ventilator

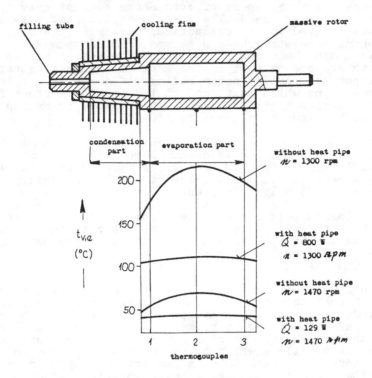

Fig. 14 Electric motor with a massive rotor in the shape
of the rotating heat pipe

filling tube

cooling fins

massive rotor

condensation part

evaporation part

without heat pipe
n = 1300 rpm

with heat pipe
Q = 800 W
n = 1300 Rpm

without heat pipe
n = 1470 rpm

with heat pipe
Q = 129 W
n = 1470 rpm

$t_{v,e}$
(°C)

200

150

100

50

1   2   3

thermocouples

Electric motor output P = 4,5 kW

Fif. 15 Schematic drawing of lift bell furnace

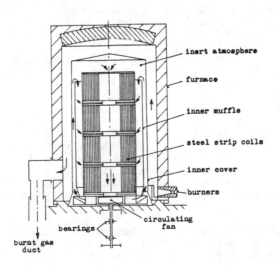

inert atmosphere

furnace

inner muffle

steel strip coils

inner cover

burners

circulating fan

bearings

burnt gas duct

ly high revolutions (n = 1400 rpm) and vertical position, vertical co-axial cylindrical and annular (Fig. 8) rotating heat pipes with transport of condensate against gravity by
centrifugal forces were used. The co-axial cylidrical rotating
heat pipe was created in the cylindrical cavity in the shaft
axis. The annular rotating heat pipe was created by a special
casing between the shaft and its bearing. The heat pipes were
designed with evaporator part in the area of thermally strained
bearing. Evaporated heat transfer medium condenses in the lower
part of a heat pipe which is finned at the outside surface for
better heat transfer to the ambient air. Maximum operating bearing and seal temperature allowed is 140 ^{o}C.

Fig. 16 shows calculated outside surface temperature of an
uncooled shaft and a shaft with co-axial rotating heat pipe. The
heat pipe influence on the temperature profile is obvious. To
keep temperature below required value of 140 ^{o}C the heat pipe
heat performance 155 W is sufficient.

In experiments heat performances from 400 to 1200 W were
achieved. Those are values several times higher than minimum
required value of 155 W. It means that the suggested way of cooling enables to decrease the operating bearing temperature to as
much as 80 ^{o}C at $\dot{Q}$ = 1000 W.

When an annular rotating heat pipe is used (Fig. 17a) a
certain additional thermal resistance between shaft and bearing
is created. Contrary to the case with co-axial rotating heat pipe no
changes in shaft shape have to be made. It is important first of
all as far as strenght of material is concerned. Part of the pipe is formed by a bellows which compensates dilatations in the
axial directions. The inner tube is strengthened by pair of six
supports. It was desired to make it as simple to produce as possible; unlike in the former alternative (cylidrical heat pipe)
the screen capillary structure and pressed insert in the condenser part were left out here.

For better understanding of heat transfer phenomena in the
heat pipe a resistance model has been devised (Fig. 17a). It is
based on heat transfer and electrical current analogy.

The comparison of resistance model results with experiments
is presented in Fig. 17b. Points determined experimentally are
connected by a dashed line, theoretical values by a full one. It
can be concluded from the picture that the heat pipe cooling ability is to a considerable extent influenced by the value of heat
transfer coefficient from the finned surface to the ambient air; the
bearing temperature decrease of 30 K corresponds to the increase
of α_2 from 7 to 45 $W.m^{-2}.K^{-1}$. Functions assuming boiling (α_1 =
= 6400 $W.m^{-2}.K^{-1}$), and evaporation (λ/s = 1315 $W.m^{-2}.K^{-1}$ - limit value) are shown in the same figure. It can be seen that the
changes in inner transfer characteristics won't have as big consequences as in heat transfer from the heat pipe to the ambient
air.

Fig. 18 shows the bearing temperature, the temperatures of
both seals and the heat performance transferred by the condenser
part calculated as a function of the shaft temperature. In agree-

Fig. 16 Temperature distribution along the ventilator shaft

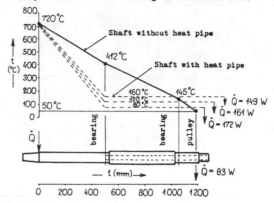

Fig. 17 Annular vertical heat pipe

a) resistance model

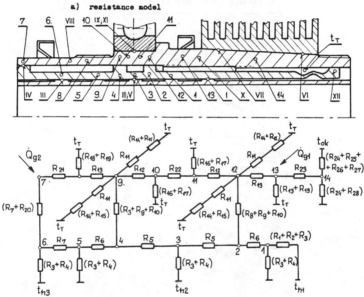

b) comparison between results of resistance model
and experiment

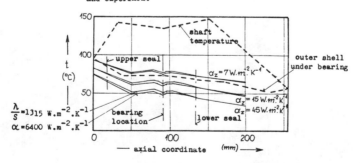

Fig. 18 Temperature conditions and heat performance of annular heat pipe

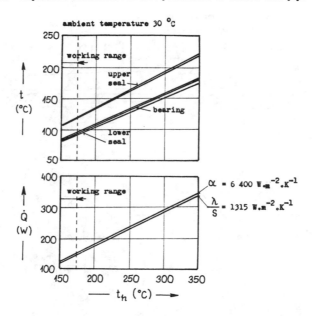

ment with the experiment the most favourable conditions appear
to be in the area of the upper bearing.

5.   CONCLUSION

Experimental study verifying different embodiments of rotating heat pipes prooved that co-axial as well as parallel heat
pipes achieve high heat fluxes and thus enable to intensify cooling in machinery,e.g. of electric motors, bearings, seals etc.

Out of works carried out further design and technology
knowledge was gained which will help to solve a broader range of
problems in cooling of other thermally exposed machine parts.

6.   REFERENCES

1. Bubeníček M. and Polášek F. 1973. Cooling AC-motor by means
    of heat pipes. Proceedings of the First International Heat
    Pipe Conference, Stuttgart.

2. Ošlejšek O., Bubeníček M. and Polášek F. 1974. Cooling of
    an electric motor rotor by a rotating heat pipe. Elektro-
    technický obzor, Vol. 63, No. 1, pp. 40.

3. Polášek F. and Dušek V. 1974. Cooling of a regulation in-
    duction motor with massive rotor by a rotating heat pipe.
    Proceedings of the Symposium on Cooling of Electric Machi-
    nes, Brno.

4. Bubeníček, M., Polášek F. and Štulc P. 1975. Perspective of
    new directions in cooling of electrical machines and equip-
    ment. Proceedings of the International Conference on the
    Future Progress of Electrical Engineering, Prague.

5. Schneller, J. and Polášek F. 1975. Heat pipes - new heat
    exchanging elements. Strojírenství, Vol. 25, No 8, pp. 488.

6. Bubeníček, M., Polášek F. and Štulc, P. 1976. Heat and mass
    transfer in a rotating heat pipe with the axis of rotation
    parallel to the axis of the tube. Proceedings of the 5th
    Conference on Heat and Mass Transfer, Minsk.

7. Ošlejšek, O. and Polášek F. 1976. Cooling of electric ma-
    chines by heat pipes. Proceedings of the II. International
    Heat Pipe Conference, Bologna.

8. Bubeníček, M. 1977. Calculation and intensification possi-
    bilities of horizontal rotating heat pipes. 24th CHISA Con-
    ference, Bratislava.

9. Hwangbo, H. and Eby, R.J. 1976. Performance of heat pipe
    rotating about the vertical axis and its applications to a
    spinning spacecraft. II. International Heat Pipe Conference,
    Bologna.

10. Chrolenok, V.V. 1981. A method of calculation of rotating

heat pipes with porous structure in the heated zone. Heat and mass transfer in systems with porous elements, Minsk, pp. 29-34.

11. Krivošejev, B.N. and Kucharskij, M.P. 1979. Investigation of heat transfer in evaporator part of rotating heat pipes at low number of revolutions. Inž.-fiz. ž., Vol. 37, No 1.

12. Kucharskij, M.P. and Krivošejev, B.N. 1976. A calculation method of condensate distribution along a cylindrical rotating heat pipe. Aerodynamics and heat transfer in electrical machines, Vol. 6, pp. 64-68.

13. Maezawa, S., Suzuki, Y. and Tsuchida, A. 1981. Heat transfer characteristics of disk-shaped rotating wickless heat pipe. Proceedings of the 4th International Heat Pipe Conference, London, pp. 725-733.

14. Marto, P.J. and Wagenseil, L.L. 1978. Augmenting the condenser heat transfer performance of rotating heat pipes. Proceedings of the 3rd International Heat Pipe Conference, Palo Alto.

15. Berka, A. and Schneller, J. 1978. High temperature vertical fan with a device for cooling of bearings and seals. Czechoslovak patent No 188.816

16. Schneller, J. 1975. Heat exchanging casing for air cooling of bearings and seals. Czechoslovak patent application No PV 7024-75.

17. Pokorný, B. and Janata, J. 1977. Cooling of bearings and seals of high temperature fans by rotating heat pipe. The 24th CHISA Conference, Bratislava.

18. Bubeníček, M. 1977. Calculation and intenzification possibilities of rotating horizontal heat pipes. National Research Institute for Machine Design report.

19. Marto, P.J. and Weigel, H. 1981. The development of economical rotating heat pipes. Proceedings of the 4th International Heat Pipe Conference, London, pp. 709-724.

20. Niekawa, J., Matsumoto, K., Koizumi, T., Hasegawa, K., Kaneko, M. and Mizoguchi, Y. 1981. Performance of revolving heat pipes and application to a rotary heat exchanger. Proceedings of the 4th International Heat Pipe Conference, London, pp. 225-234.

21. Leppert, G. and Nimmo, B. 1968. Laminar film condensation on surfaces normal to body and internal forces. Trans. ASME, J. Heat Transfer, Vol. 90, pp. 178-179.

22. Nimmo, B. and Leppert, G. 1970. Laminar film condensation on a finite horizontal surface. Elsevier Pub. Co., Amsterdam.

23. Bubeníček, M. , Londin, J., Ošlejšek, O., Polášek, F. and
     Schneller, J. 1975. Cooling device of stators of rotating
     electric machines. Czechoslovak patent No 159.178.

24. Bubeníček, M., Londin, J., Ošlejšek, O., Polášek, F. and
     Schneller, J. 1975. Cooling device of rotors of electric
     machines containing a heat pipe. Czechoslovak patent No
     161.179.

25. Bubeníček, M., Londin, J., Ošlejšek, O., Polášek, F. and
     Schneller, J. 1975. Cooling device of rotors of electric
     machines containing heat pipes fixed in axial position.
     Czechoslovak patent No 161.181.

26. Bubeníček, M., Londin, J., Ošlejšek, O., Polášek, F. and
     Schneller, J. 1975. Cooling device of rotors of electric
     machines equipped with heat pipes. Czechoslovak patent No
     161.182.

27. Bubeníček, M., Londin, J., Ošlejšek, O., Polášek, F. and
     Schneller, J. 1975. Cooling device of rotors of electric
     machines containing a heat pipe. Czechoslovak patent No
     161.180.

28. Bubeníček, M., Londin, J., Ošlejšek, O., Polášek, F. and
     Schneller, J. 1975. Cooling device of rotors of electric
     machines containing axial heat pipes placed in the rotor
     circumference. Czechoslovak patent No 161.576.

29. Heřman, J., Hurych, A., Dušek, V., Bubeníček, M., Polášek, F.
     and Štulc, P. 1976. Rotor of an asynchronous motor cooled
     by a rotating heat pipe. Czechoslovak patent No 165.259.

30. Schneller, J. 1978. Device for air cooling of bearings and
     seals. Czechoslovak patent No 185.909.

7.0 NOMENCLATURE

d_i inner diameter (m)

g acceleration of gravity ($m.s^{-2}$)

H max. lifting height (m)

h eccentricity (m)

h^x lifting height (m)

L length (m)

n revolutions per minute (min^{-1})

P output (kW)

$\dot{Q}$ heat performance (W)

q heat flux ($W.m^{-2}$)

r radius (m)

r_i inner radius (m)

t temperature ($^{\circ}C$)

t_h temperature of shaft ($^{\circ}C$)

t_{ke} outer temperature of the cond. part ($^{\circ}C$)

t_{ve} outer temperature of the evap. part ($^{\circ}C$)

$\dot{V}$ volume flow rate ($m^3.s^{-1}$)

V_i volume of empty tube (m^3)

V_L volume of working fluid (m^3)

x, y coordinates (m)

α heat transfer coefficient ($W.m^{-2}.s^{-1}$)

α_z heat transfer coefficient of fins ($W.m^{-2}.s^{-1}$)

δ_L thickness of liquid layer (m)

θ angle of inclination ($^{\circ}$)

ξ friction coefficient (-)

ν kinematic viscosity ($m^2.s^{-1}$)

w angular velocity (s^{-1})

t_L liquid temperature ($^{\circ}C$)

Heat Pipe Cooled Twin Airfoil Blade as an Element for Higher Efficiency of Longlife Gas Turbine

MARKO MAJCEN and NENAD ŠARUNAC
Power Plant Department
Faculty of Mechanical Engineering and Naval Architecture
University of Zagreb
Salajeva 5, 41000 Zagreb, Yugoslavia

ABSTRACT

The present state of the art in gas turbine engines is closely tied to improvements in design techniques that have resulted, over the years, in a steady increase in operating temperatures. Higher firing temperatures are essential for development of smaller, lighter, more efficient engines. One possible way to meet aforesaid trend, a double gas turbine cycle based on heat pipe cooled twin airfoil blade is described in this paper. The basic and improved flow diagrams of the double gas turbine cycle, its performances, heat transfer analysis on, across and from twin airfoil blade and some calculated examples are presented.

1. INTRODUCTION

As a result of the trend to higher gas turbine cycle pressure ratios and turbine inlet temperature as well as to longer service lives and intervals between engine overhaul, it has been necessary to resort to cooling of the turbine blades and vanes. Its purpose is not simply to decrease the metal temperature but to ensure the smooth temperature distribution within the blade. Increase of firing temperature and cycle pressure ratio intensifies heat transfer on gas turbine components.

During the cooling system design it becomes necessary to solve complex problems of thermodynamics of the cycle, construction, technology, strength, cooling losses and others. The most important is, of course, the heat transfer estimation on and within the blade.

2. DOUBLE GAS TURBINE CYCLE AND ITS VARIATIONS

More than 25 years ago a double gas turbine cycle was proposed working with two media, combustion gases an air. This cycle is founded on the twin airfoil blade cooled with built-in heat pipe, subjected to the strong gravitational field of the rotating turbine.

All existing low and medium-temperature gas turbine cycles are very unfavourable from the thermodynamic point of view. Instead of compressing quantity of air needed for stoichiometric combustion, a greater quantity of compressed air enters combustor in order to ensure, by mixing, a suitable temperature level and profile at the turbine inlet.

Various turbine manufacturers have introduced different types of design to increase the firing temperature and ensure successful long-life operation of the gas turbine parts exposed to high temperatures , as for example, by introducing of long-shank blade, Ref. 10, where long slender shank provides higly effective thermal insulation between the hot gas path and much cooler wheel rim.

The proposed design is a twin airfoil design (Fig. 4), consisting of the upper and lower airfoil, the former beeing in the hot flow and latter in the cold flow. The significant difference between the proposed twin air-foil blade design, and long-shank blade design is that the upper airfoil in the former design is separated from the wheel rim not by the shank but by the airfoil around which the cold air is expanding. Twin airfoil blade enables two gas turbine cycles to be carried out simultaneously, the hot and the cold one. The hot cycle operates on high firing temperature and is, therefore thermodynamically efficient. The purpose of the cold cycle is to supply the quantity of air for cooling of the blades. The cold cycle maximum temperature is equal to the compressor discharge tempera-ture. Despite of the fact that the cold cycle is righthandwise, it can also be a work-consumer because of the relative low maximum temperature existing there.

The advantage of the proposed double cycle and twin blade is that the blade temperature is expected to be near the average of the hot flow and cold flow temperatures. In case of efficient liquid metal heat pipe cooling, extracted heat due to cooling of the hot path of the blades can be partially converted to mechanical work during expansion in the cold flow. This reduces cooling losses.

2.1. "A" Double Gas Turbine Cycle

The basic flow diagram of the double gas turbine cycle, named A_1 cycle is shown in Fig.1.Air as a working fluid in both flows enters the compressor (C). Discharged air is divided in two flows. The first flow passes through high pressure heat exchanger (HPHE) where it is heated to state point 3. In the combustor (CC) the cycle maximum temperature is reached, state point 4. Combustion gases are expanding in the hot flow of the high pressure (HPT) and low pressure turbine (LPT), to state point 7. By passing high pressure and low pressure heat exchanger (LPHE) gases are cooled to state point 8,9 respectively and exhaused into the atmosphere. The second flow is expanded in the cold flow of the high pressure turbine to state point 10. Then it is heated to state point 11 with a part of the heat extracted from the hot flow exhaust gases, in the low pressure heat exchanger. Later on the cold flow is expanded in the low pressure turbine cold flow and exhausted into the atmosphere. Obviously, a large quantity of cooling air can be obtained by a small amount of work done. Heat transfer between the hot and cold flow takes place only in the low pressure heat exchanger and twin airfoil blade, neglecting cooling losses in vanes and undesirable mixing of the hot and cold flow.

Some results of cycle calculations, cycle thermal efficiency and specific work are plotted vs. cycle pressure ratio and firing temperature and shown in Figs. 2 and 3.

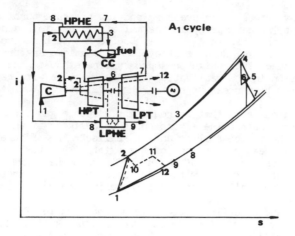

Fig. 1 A_1 Double gas turbine cycle

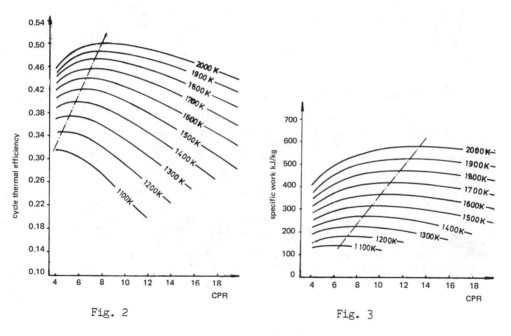

Fig. 2   Fig. 3

The suppositions done in cycle calculations are as follows: the hot and
cold flow turbine efficiency is the same o.84 ; compressor efficiency is
correlated to cycle pressure ratio and expressed as: η_c = o.9o5 - o.oo8 ·
CPR; cycle pressure drops are: in the intake duct 2%, in the exhaust
duct 2%, in the combustor 2%, in the heat exchanger 2%. Thermodynamic
cooling loss is equal 6%. Mass rates in the hot and cold flow are
equal.

Reheat in hot flow between high pressure and low pressure turbine increases the work output. The maximum cycle thermal efficiency is reached at
higher cycle pressure ratio than in A_1 cycle. Double gas turbine cycle
with reheat is named A_2 cycle. Results of cycle calculations are presented
in Ref. 4. The temperature of the high pressure turbine exhaust gases
represents the limitation in the realisation of the A_2 cycle with regard
to the material used for high pressure heat exchanger.

All these calculations show that by realising a double gas turbine
cycle a high efficiency gas turbine engine could be designed if an
efficient liquid metal heat pipe cooling can be realised. The efficiency
of such turbine will be greater than that of the existing gas turbines
and even of diesel engines. In spite of the high firing temperature, the
peak metal temperature of the hot flow turbine parts is expected to be
at the level permitted for available temperature-resistance materials.

The mixing of hot and cold flow could have an undesirable effect upon
the application of the twin airfoil blade because of the negative mixing
effect in the heat transfer process, in the high pressure and low pressure
heat exchanger. Consequentlly, the efficiency of the cycle could be lowered, in the case.

Therefore it becomes necessary to examine further possibilities in
double cycle simplification and improvement in order to facilitate its
realisation. The supposition done about efficient operation of the twin
airfoil blade has to be preliminary carefully tested.

3. BLADE TEMPERATURE ESTIMATION

The actual three-dimensional temperature field existing within the
blade is replaced with quasi two-dimensional one. The average temperatures
within the blade cross-sectional area are estimated in several sections,
by supposition of non-existence of the temperature gradient across the
blade.

The conditions for heat transfer occuring over the blade feature a
complicated distribution along the blade surface and are dependent upon
the condition of the boundary layer on this surface. The position of
the transition area from laminar to turbulent boundary layer affects,
to a great extent, the distribution of the local values of the heat
transfer coefficient, while the position of the zone of minimum pressure
determines the coordinates of the transition area within the boundary
layer. The heat transfer coefficients have a neglible effect upon the
temperature distribution within the blade, under steady conditions, Ref.
14, but affects greatly stress conditions.

Because the aim of this paper is not to evaluate the strength of
the twin airfoil blade, but to testify the possibility of its successful
operation, the average blade temperature is estimated by using the average heat transfer coefficient.

Because of the high temperatures occuring in the hot flow, the most efficient blade cooling system, i.e. liquid metal heat pipe cooling (thermosyphon cooling) has been selected. Metals as K, Na, Li and Na-K alloys could be selected as coolants in high-temperature gas turbine. Twin airfoil blade and cooling channels, performing cavities of the closed thermosyphon, are shown in Fig. 4.

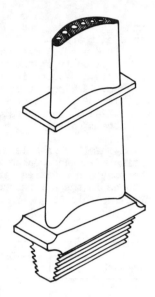

Fig. 4   Twin Airfoil Blade

Heat extracted from hot flow is transported by blade built-in heat pipe-thermosyphon to the cold flow working fluid (air). Due to heat exchange in the turbine stage, entalpy drop, velocity and temperature of the working fluid at the stage exit are changed when compared to non-cooled stage, Ref. 1, Ref. 7.

The closed thermosyphon, according to Ref. 11, 12, 13 could be treated as two open thermosyphons with open ends joined together. Heat transfer between the surface of the thermosyphon cavity and coolant is evaluated in accordance with the correlation:

$$Nu_1 = 3 \cdot (Gr \cdot Pr)^{0.25} \, \bar{I}^{-0.7}$$

where Gr is Grasshoff number and $\bar{I} = 1/d$ relative length of the thermo-syphon, Ref. 8.

To obtain the temperatures of the blade and thermosyphon cavity surface in hot and cold flow and also the temperature of the coolant, it is necessary to solve the system of equations describing the heat transfer on, across and from the blade.

3.1. Blade Temperature in A_1 Double Gas Turbine Cycle

A_1 double gas turbine cycle with firing temperature of 14oo K and cycle pressure ratio lo is analysed in detail. On the basis of the results obtained, power turbine is designed as a two-stage one. In order to determine the gas temperature upstream the rotor blade row accuratelly, the vane air convective cooling is considered.

As a result of vane cooling (heat exchanged on one vane is equal 2232 W), gas temperature and velocity at the vane row exit are lowered. Temperature and velocity drops due to cooling are equal 14 K and 1.3 m/s respectively.(In the case of non-cooled vanes, temperature and velocity at the vane row exit will be 1321 K and 492 m/s). Vane cooling air is exhausted in the gas stream through the openings at the vane trailing edge. As a result of the air and gas stream mixing, the temperature and velocity of the gas stream are lowered. After mixing, gas stream temperature reaches 1292 K and its velocity 481 m/s. (Cooling air rate is equal 2.5% of the gas stream rate).

As a coolant in heat pipe a 25% K + 75% Na alloy is selected. Thermophysical proporties of the coolant are estimated in accordance with Ref. 5. Within the blade six cooling channels (thermosyphon cavities) are designed, Fig. 5. Average blade and coolant temperatures in the hot and cold flow of the power turbine first stage are presented in Table 1.

TABLE 1

Average temperatures in the hot and cold flow of the power turbine first stage at the mean diameter	Temperature K
Blade surface in the hot flow	1084
Thermosyphon cavity surface in the hot flow	984
Coolant	973
Thermosyphon cavity surface in the cold flow	961
Blade surface in the cold flow	867

Average temperatures presented in Table 1, are calculated for turbine speed of 6ooo RPM, for hot and cold flow mean diameter of 7oo mm, 595 mm respectively and for blade length in hot and cold flow 58 mm and 55 mm, respectively. Gravitational field on mean diameter due to turbine rotation in hot flow reaches 14o85·g and in cold flow 11972·g.

As a result of the blade cooling (heat exchanged on one blade is equal 2482 W), temperature and velocity of the working fluid at the blade row exit in the hot flow are decreased, while in the cold flow are increased. In the hot flow the temperature drop and drop of the relative velocity, due to cooling, are equal 22 K and 4o.2 m/s, respectively. (In the case of non-cooled blade, temperature and relative velocity at its exit will be 1283 K and 416 m/s). Temperature of the air, in the cold flow, at the blade cascade exit due to heating is increased from 535 K to 54o.7 K. Relative velocity is increased from 14o m/s to 141,5 m/s. Total heat, exchanged in the power turbine first stage reaches 438.2 KW, in other words, thermodynamic cooling loss in the hot flow is equal 5.976%.

Three-dimensional flow estimation in the stage is done by using free vortex method. As a result, the form of the blade channel is changing along the blade height. Radial variations in fluid conditions and channel form affect radial variation of the heat transfer coefficient along the blade. Some results of the three-dimensional flow calculations are shown in Table 2.

TABLE 2

Blade Hot flow, cold flow	Temperature K		
	diameter		
	root	mean	tip
Blade surface in the hot flow	1o92	1o84	1o82
Thermosyphon cavity surface in the hot flow	971	984	977
Thermosyphon cavity surface in the cold flow	942	961	958
Blade surface in the cold flow	829	867	878

Temperatures, presented in Table 2, should be treated as approximate because of neglecting heat conduction along the blade and heat transfer intensification at the blade channel edges due to secondary flow. Blade tip temperature is expected to be higher than calculated. Due to heat conduction into the root, the root area will be at lower temperature than estimated.

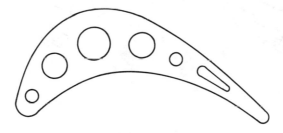

Fig. 5   Cross-sectional Area of the Blade in the Hot Flow

3.2. Other Possibilities in Thermosyphon Cavity Design

The blade with cooling channels, Fig. 4, is advantageus, because the cross-sectional area of the channels is small and, therefore, the additional force and additional stress, due to coolant (liquid metal) are small too. For the blade profile, shown in Fig. 5, the cross-sectional area of the channels equals $4.867 \cdot 10^{-5}$ m^2 . Hydraulic pressure of the coolant, due to turbine rotation, reaches $100 \cdot 10^5$ N/m^2 , hence the additional force on the blade is equal 486 N and additional stress $\sigma_{add} = 1.08$ MN/m^2 .

There is no doubt that manufacturing technology for such blade will be complicated and expensive. The blade could be simplified for manufacture by making its airfoil hollow. Here, the whole cavity within the blade performs the cooling channel, Fig. 6. Calculations show that such, single-channel cooling will be more efficient than multi-channel cooling. This is shown in Table 3.

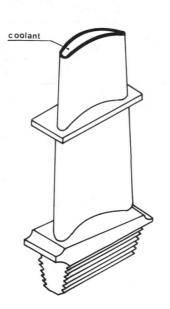

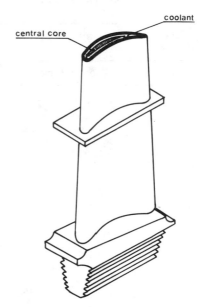

Fig. 6 Hollow, single-channel,   Fig. 7 Hollow twin airfoil blade
        twin airfoil blade   with central core

One can see that the average blade temperature in the hot flow is decreased, while in the cold flow is increased. This reduces radial temperature gradient within the blade. The additional stress due to coolant, is, in such blade, considerably higher, $\sigma_{add} = 17.77$ MN/m^2 , compared to a multi-channel blade. This additional stress can be lowered by inserting a central core within the thermosyphon cavity, Fig. 7. The average blade temperatures in such case remain nearly the same as in hollow blade, but additional stress is significantly lowered ($\sigma_{add} = 7.7$ MN/m^2).

TABLE 3

	Temperature K	
	hollow, single-channel blade	multi-channel blade
Blade surface in the hot flow	1o48	1o84
Thermosyphon cavity surface in the hot flow	989	984
Thermosyphon cavity surface in the cold flow	971	961
Blade surface in the cold flow	911	867

4. "B" DOUBLE GAS TURBINE CYCLE

Efforts in double gas turbine cycle improvement resulted in double cycle named B cycle, where twin airfoil blade is applied in the high pressure turbine only. This reduces the negative effect of the hot and cold flow mixing. Low pressure turbine can be cooled by some conventional cooling method.

This is the principal advantage of the B cycle, compared to A cycle. Moreover, the maximum cycle efficiency is reached at higher cycle pressure ratio than in A cycle, which results in the reduction of the engine size.

The basic B cycle is named B_1 cycle. The cycle, modified by addition of intercooling is named B_2 cycle. Thermal efficiency of the B_2 cycle is better than of the B_1 cycle. Furthermore, the great change in B_2 cycle pressure ratio affects a small change in its thermal efficiency. More details about B_2 and B_1 cycles can be found in Ref. 4. The most attractive B - cycle, semi-closed B_3 cycle is displayed in Fig. 8.

After compression, discharged air flow enters the heat exchanger, where is heated to state point 5. Latter on it is divided in two flows: one flow enters the combustor and another one is piped to the cold flow of the high pressure turbine, where is expanded to the state point 11.Heat exchanger surface required for transferring the same heat rate is reduced by entering both air flows in the heat exchanger, Fig. 8. By supposition: $\dot{m}_{AIR} = \dot{m}_{GAS}$, $\dot{Q}_{GAS}/\dot{Q}_{AIR}$ ratio becomes equal o.5. Heat exchanger efficiency is increased from η_{ex} = o.8; o.6; o.4 (for $\dot{Q}_{GAS}/\dot{Q}_{AIR}$ = 1) to η_{ex} = o.94; o.66; o.45 (for $\dot{Q}_{GAS}/\dot{Q}_{AIR}$ = o.5) respectively. Heat exchanger surface required for transferring the same heat rate is reduced by factor o.55; o.7; o.81 respectively. Thermal efficiency and specific work of the B_3 double cycle are plotted vs. cycle pressure ratio, firing temperature and heat exchanger efficiency and shown in Figs. 9 and 1o.

The favourable characteristic of the B_3 cycle is that the great change in CPR affects a small change in cycle thermal efficiency and specific work.

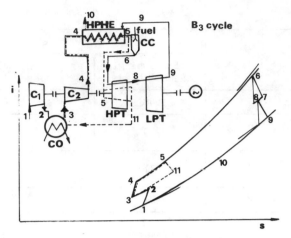

Fig. 8 B₃ Double Gas Turbine Cycle

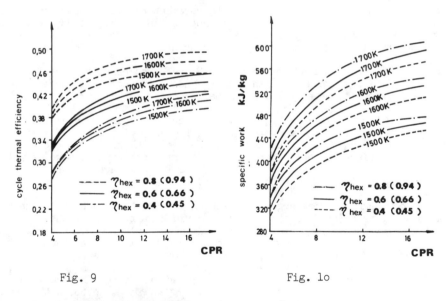

Fig. 9   Fig. 1o

4.1. Blade Temperature in B₃ Double Gas Turbine Cycle

B₃ double gas turbine cycle with firing temperature of 15oo K and 16oo K and cycle pressure ratio 1o is analysed. Calculated cycle thermal efficiency reaches o.456 and o.478 respectively. Average blade and coolant temperatures in the hot and cold flow are presented in Table 4.

Average temperatures, presented in Table 4 are calculated for turbine speed of 1o162 RPM.

TABLE 4

| B_3 - cycle | Temperature K | |
cycle pressure ratio = 1o	1500 K	1600 K
Blade surface in the hot flow	1165	1236
Thermosyphon cavity surface in the hot flow	1115	1193
Coolant	1111	1190
Thermosyphon cavity surface in the cold flow	11o7	1187
Blade surface in the cold flow	1o38	1125

B_3 double gas turbine cycle with firing temperature of 16oo K could be realised by using avaliable high temperature resistance materials. Cross-section of the high pressure turbine is shown in Fig. 11. (T_{firing} = 15oo K, CPR = 1o).

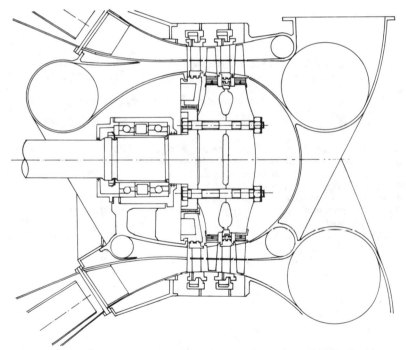

Fig. 11   Cross-section of the High Pressure Gas Turbine

5. CONCLUSION

Results presented indicate some new possibilities in long-life high temperature gas turbine engine design. Twin airfoil blade and B$_3$ cycle enable high firing temperatures, and, therefore, high cycle thermal efficiency which is reached at high cycle pressure ratio. Despite of the high firing temperature, the peak metal temperature in the hot flow will be at the level permitted for avaliable high temperature resistance materials.

In the continuation of this work the effort is done to estimate the temperature field within be blade with greater accuracy . Finite element approach is applied. The further step will be the model research of the heat transfer process on twin airfoil blade cooled by the liquid metal heat pipe.

Successful realisation of the twin airfoil blade would be the main, and the only new element in the production of the new type of the gas turbine engine.

The double cycle described has many attractive features, could be **economically attractive in the near future, and B double gas turbine** cycle engine would be especially suitable for marine propulsion.

REFERENCES

1. Žirickij, G.S. and Lokaj, V.I. and Maksutova, M.N. and Strunkin, V.A. Gazovie turbinie dvigateli letateljnih aparatov. Mašinostroenie, Moskva 1971.
2. Švec, I.T. and Diban, E.P. Vozdušnoe ohlaždenie detalej gazovih turbin. Naukova dumka, Kiev 1974.
3. Šnez, J.I. and Kapinos, V.M. and Kotljar, I.V. Gazovie turbini 1. Viša škola, Kiev 1976.
4. Majcen, M. Neke daljnje mogućnosti razvoja plinskih turbina uz upotrebu intenzivnog hlađenja lopatica rotora. Disertacija. Zagreb 1979.
5. Čečetkin, A.V. Visokotemperaturnie teplonositeli. Energija, Moskva 1971.
6. Muller, K.J. Thermische Strömungsmachinen, Auslegung und Berechnung. Springer-Verlag, Wien/New York 1978.
7. Traupel, V. Thermische Turbomachinen I. Springer-Verlag, Berlin/ Göttingen/Heidelberg 1958.
8. Zisina, L.M. and Zisin, L.V. and Poljak, M.P. Teploobmen v turbomašinah. Mašinostroenie, Lenjingrad 1974.
9. Šarunac, N. Projekt dvostrukog kružnog, plinsko-turbinskog procesa. Diplomski rad. Zagreb 1977.
10. Starkey, N.E. Long-life Base Load Service at 16oo F Turbine Inlet Temperature. Journal of Engineering for Power, January 1967.
11. Lighthill, M.J. Theoretical considerations on free convection in tubes. Q.Jl Mech.appl.Math. 1953 6,398.
12. Bayley, F.J. and Lock, G.S.H. Heat transfer caracteristics of the closed thermosyphon. Am.Inst.Chem.Engrs.Am.Soc.Mech.Engrs, Paper No.64-HT-6 1964.
13. Romanov, A.G. O teploobmene v gluhom kanale s estestvenoj konvekcii. Izd. AN SSSR-OTN 1956, No 6.
14. Molchanov, E.I. Calculation of thermal stresses in Turbine Components. Proceeding of the International Conference: Thermal stresses and Thermal Fatigue, Berkley Castle UK, September 1969.

Index